Springer Tracts in Modern Physics
Volume 155

Managing Editor: G. Höhler, Karlsruhe

Editors: J. Kühn, Karlsruhe
Th. Müller, Karlsruhe
R. D. Peccei, Los Angeles
F. Steiner, Ulm
J. Trümper, Garching
P. Wölfle, Karlsruhe

Honorary Editor: E. A. Niekisch, Jülich

Springer
Berlin
Heidelberg
New York
Barcelona
Hong Kong
London
Milan
Paris
Singapore
Tokyo

Springer Tracts in Modern Physics

Springer Tracts in Modern Physics provides comprehensive and critical reviews of topics of current interest in physics. The following fields are emphasized: elementary particle physics, solid-state physics, complex systems, and fundamental astrophysics.
Suitable reviews of other fields can also be accepted. The editors encourage prospective authors to correspond with them in advance of submitting an article. For reviews of topics belonging to the above mentioned fields, they should address the responsible editor, otherwise the managing editor. See also http://www.springer.de/phys/books/stmp.html

Managing Editor
Gerhard Höhler
Institut für Theoretische Teilchenphysik
Universität Karlsruhe
Postfach 69 80
D-76128 Karlsruhe, Germany
Phone: +49 (7 21) 6 08 33 75
Fax: +49 (7 21) 37 07 26
Email: gerhard.hoehler@physik.uni-karlsruhe.de
http://www-ttp.physik.uni-karlsruhe.de/

Elementary Particle Physics, Editors
Johann H. Kühn
Institut für Theoretische Teilchenphysik
Universität Karlsruhe
Postfach 69 80
D-76128 Karlsruhe, Germany
Phone: +49 (7 21) 6 08 33 72
Fax: +49 (7 21) 37 07 26
Email: johann.kuehn@physik.uni-karlsruhe.de
http://www-ttp.physik.uni-karlsruhe.de/~jk

Thomas Müller
Institut für Experimentelle Kernphysik
Fakultät für Physik
Universität Karlsruhe
Postfach 69 80
D-76128 Karlsruhe, Germany
Phone: +49 (7 21) 6 08 35 24
Fax: +49 (7 21) 6 07 26 21
Email: thomas.muller@physik.uni-karlsruhe.de
http://www-ekp.physik.uni-karlsruhe.de

Roberto Peccei
Department of Physics
University of California, Los Angeles
405 Hilgard Avenue
Los Angeles, CA 90024-1547, USA
Phone: +1 310 825 1042
Fax: +1 310 825 9368
Email: peccei@physics.ucla.edu
http://www.physics.ucla.edu/faculty/ladder/peccei.html

Solid-State Physics, Editor
Peter Wölfle
Institut für Theorie der Kondensierten Materie
Universität Karlsruhe
Postfach 69 80
D-76128 Karlsruhe, Germany
Phone: +49 (7 21) 6 08 35 90
Fax: +49 (7 21) 69 81 50
Email: woelfle@tkm.physik.uni-karlsruhe.de
http://www-tkm.physik.uni-karlsruhe.de

Complex Systems, Editor
Frank Steiner
Abteilung Theoretische Physik
Universität Ulm
Albert-Einstein-Allee 11
D-89069 Ulm, Germany
Phone: +49 (7 31) 5 02 29 10
Fax: +49 (7 31) 5 02 29 24
Email: steiner@physik.uni-ulm.de
http://www.physik.uni-ulm.de/theo/theophys.html

Fundamental Astrophysics, Editor
Joachim Trümper
Max-Planck-Institut für Extraterrestrische Physik
Postfach 16 03
D-85740 Garching, Germany
Phone: +49 (89) 32 99 35 59
Fax: +49 (89) 32 99 35 69
Email: jtrumper@mpe-garching.mpg.de
http://www.mpe-garching.mpg.de/index.html

Matthias Hein

High-Temperature-Superconductor Thin Films at Microwave Frequencies

With 134 Figures

 Springer

Dr. Matthias Hein
University of Wuppertal
Department of Physics
Gauss-Strasse 20
D-42097 Wuppertal, Germany
Email: mhein@wpos4.physik.uni-wuppertal.de

Physics and Astronomy Classification Scheme (PACS):
06.30.Lz, 41.20.Jb, 44.50.+f, 72.30.+q, 73.50.Mx, 74.25.Nf, 74.60.Ec, 74.70.Ad, 74.72.Bk, 74.72.Fq, 74.76.Bz, 75.30.Hx, 77.22.Gm, 81.40.Rs, 84.30.Ey, 84.30.Vn, 85.25.-j

ISSN 0081-3869
ISBN 3-540-65646-4 Springer-Verlag Berlin Heidelberg New York

Library of Congress Cataloging-in-Publication Data.

Hein, Matthias, 1961-, High-temperature superconductor thin films at microwave frequencies/Matthias Hein. p. cm. – (Springer tracts in modern physics, v. 155). Includes bibliographical references and index. ISBN 3-540-65646-4 (hc.: alk. paper). 1. High temperature superconductors. 2. Thin films. 3. Microwaves. I. Title. II. Series: Springer tracts in modern physics; 155. QC1.S797 vol. 155 [QC611.98.H54] 539 s–dc21 [537.6'23] 99-21052

This work is subject to copyright. All rights are reserved, whether the whole or part of the material is concerned, specifically the rights of translation, reprinting, reuse of illustrations, recitation, broadcasting, reproduction on microfilm or in any other way, and storage in data banks. Duplication of this publication or parts thereof is permitted only under the provisions of the German Copyright Law of September 9, 1965, in its current version, and permission for use must always be obtained from Springer-Verlag. Violations are liable for prosecution under the German Copyright Law.

© Springer-Verlag Berlin Heidelberg 1999
Printed in Germany

The use of general descriptive names, registered names, trademarks, etc. in this publication does not imply, even in the absence of a specific statement, that such names are exempt from the relevant protective laws and regulations and therefore free for general use.

Typesetting: Data conversion by Satztechnik Katharina Steingraeber, Heidelberg
Cover design: *design& production* GmbH, Heidelberg
Computer-to-plate and printing: Mercedesdruck, Berlin
Binding: Universitätsdruckerei H. Stürtz AG, Würzburg

SPIN: 10709224   56/3144/tr - 5 4 3 2 1 0 – Printed on acid-free paper

To Sabine, Anneli, and Michael

Preface

About the Content

The discovery of the oxide high-temperature superconductors (HTS) by Bednorz and Müller [1] in 1986 added one further Nobel prize to the continuing history of superconductivity, which was initiated by the fundamental work of Kammerling Onnes in 1908 and 1911 [2, 3]. The high transition temperature and the related short coherence length, the quasi-two-dimensional electronic structure and the related anisotropy as well as unconventional features in the electronic density of states belong to the specific properties of HTS. These have had multiple impact on interdisciplinary sciences, and are still attracting increasing attention from scientists and economists. Chemists, physicists and engineers merged their intellects and interests in order to develop a comprehensive understanding and a consequent exploitation of the wide-spread potential that these novel compounds promise. Since their discovery, enormous technological improvements in the preparation and physical characterization of high-quality thin HTS films and Josephson junctions have been achieved. Extended treatments of the phenomenology, possible application and functional demonstration of HTS films and devices at high frequencies accompanied this development [4–11]. Theoretical and experimental physicists are still puzzling over spectroscopic data and measurement techniques in order to resolve the nature of superconductivity in HTS in detail [12]. While this debate has markedly enriched the academic view of superconductivity, potential customers have become aware of promising market opportunities, and efforts are currently being undertaken to commercialize HTS in communication and remote sensing systems [13–15].

This book summarizes those physical and technological aspects of HTS films at microwave frequencies which are considered basic for the present state of the art. The parallel discussion of the various specific theoretical and empirical issues aims at emphasizing recent progress in this field. However, the analysis is supplemented by a review of the corresponding frameworks, in order to make this exciting field accessible not only to specialists, but also to advanced students and interested scientists from other disciplines. Regarding the potential for microwave device applications of HTS, the discussion is focused on $YBa_2Cu_3O_{7-x}$ and $Tl_2Ba_2CaCu_2O_y$.

In line with the general and the specific features of superconducting films, the microwave properties of the oxide superconductors are compared to those of the "conventional high-temperature superconductor" Nb_3Sn. The compounds with cubic A15 lattice type displayed the highest transition temperatures known before the discovery of the superconducting oxides [16]. Due to the much simpler structure and the well-verified theoretical description, Nb_3Sn presents a valuable reference material for HTS. Such a comparison becomes even more exciting when considering the strong pair coupling and the short coherence length, which are common to both materials.

About the Strategy Behind this Book

The microwave responses of superconducting films at low excitation levels (linear regime, Chaps. 1 and 2) and at high levels (nonlinear regime, Chaps. 3 and 4) are treated separately. Both topics are analyzed in terms of well-approved models and theoretical or numerical approaches as well as in terms of recent experimental results. The theoretical treatment (Chaps. 1 and 3) forms a bridge from famous summaries of the fundamental aspects of superconductivity (like Tinkham's book [17] or the book edited by Parks [18]) to the most specific properties at high frequencies. The experimental data discussed (Chaps. 2 and 4) refer to the worldwide state of the art, as reported by many international groups. Of special importance is the development of a consistent and comprehensive understanding of the microwave response of the two materials considered, which was not available in the literature before.

Chapter 1 starts with the electrodynamic and the microscopic descriptions of the surface impedance Z_s of superconductors, including the impact of finite film thickness. Some emphasis is put on the phenomenological treatment of Z_s in the framework of the two-fluid model. While this approach developed from the initial understanding of microwave superconductivity by F. London and H. London [19], recent extensions account for granularity and layered structures. In Chap. 2, the various techniques and procedures to deduce the surface impedance of superconducting films from measurement are summarized from a systematic point of view. The related physical aspects bear an exciting symmetry between the real and the imaginary parts of the surface impedance, which is explained and illustrated in terms of comparative analyses and analogies.

The nonlinear microwave response of superconductors (Chaps. 3 and 4) deserves special attention since it enables the identification of various intrinsic and extrinsic effects. Furthermore, it currently limits many promising applications of HTS devices in communication systems. Chapter 3 starts with the theoretical description of the known mechanisms which introduce microwave field and power dependences to the surface impedance. Adiabatic, i.e. quasi-stationary, types of nonlinear response are distinguished from dynamic effects, which are related to the intrinsic time scales of the superconductors. The sources and consequences of microwave heating which dominate the nonlinear surface impedance of nonideal films over an extended temperature range are also analyzed. Experimental results on the DC and microwave magnetic field dependent surface impedance at field levels around the lower critical field are analyzed in detail in Chap. 4. The knowledge gained from the DC measurements serves as a valuable basis for identifying magnetic field limitations at high frequencies and distinguishing them from thermal effects.

The book continues with a brief description of the preparation, handling and mounting techniques appropriate for both superconductors (Chap. 5). While the A15 compound has been known for about 30 years, large-grained phase-pure Nb_3Sn films on dielectric substrates became available only recently. Chapter 5 also sketches the preparation of HTS films, and remarks on the fabrication of engineered grain-boundary Josephson junctions. The latter part is kept brief, since excellent treatments on the details are available elsewhere [20]. Chapter 5 finally contains a summary of the present status of refrigeration technology, which acknowledges its relevance to the integration of functional superconducting microwave devices.

The last chapter is devoted to a general discussion of possible applications of high-temperature superconductors in passive microwave devices. The first part constitutes a bridge from single resonators to filters, which represent the most prominent class of near-term applications of HTS. Selected examples of resonant devices are then discussed schematically, in order to illustrate the potential of HTS at frequencies between 10 MHz and 10 GHz. Passive microwave components based on the Josephson effect are the subject of the last part of Chap. 6. Though presently of less economic relevance than passive film devices, this field combines basic aspects of both microwave and Josephson technology, and thus bears the potential for yielding a large variety of novel integrable microwave devices.

About the Support for this Work

Many aspects described in this book reflect the advanced state of research and development at the Departments of Physics and Electrical Engineering at the University of Wuppertal. The expertise of my colleagues therefore contributed to the physical, social and financial basis of this work. I am especially grateful to Dr. B. Aminov, Dr. B. Aschermann, C. Bauer, A. Baumfalk, Prof. H. Chaloupka, W. Diete, M. Getta, S. Hensen, F. Hill, M. Jeck, Dr. T. Kaiser, J. Kallscheuer, Dr. S. Kolesov, Dr. M. Lenkens, Prof. G. Müller, Dr. S. Orbach-Werbig, M. Perpeet, Prof. H. Piel, J. Pouryamout, M. Reppel, S. Schmöe, P. Seidel, and H. Schlick.

Furthermore, I appreciate very much the kind support of, fruitful discussions with, and partially unpublished information from many colleagues: Prof. S. M. Anlage (University of Maryland), Dr. S. Beuven (ISI, FZ Jülich), Prof. A. Braginski (ISI, FZ Jülich), Dr. E. H. Brandt (MPI Stuttgart), Dr.

A. Cassinese (University of Naples), Dr. T. Dahm (MPI Dresden), Dr. M. Golosovsky (Hebrew University), Prof. R. Gross (Universität Köln), Dr. H.-U. Häfner (Leybold), Dr. J. Halbritter (IMF, FZ Karlsruhe), Dr. R. Heidinger (IMF, FZ Karlsruhe), Dr. E. Ilíchev (IPHT Jena), Dr. Chr. Jooss (MPI Stuttgart), Dr. M. Klauda (Bosch), Dr. N. Klein (IFF, FZ Jülich), Dr. V. Z. Kresin (Lawrence Berkeley Laboratory), Prof. M. J. Lancaster (University of Birmingham), Dr. M. Manzel (IPHT Jena), Dr. H.-G. Meyer (IPHT Jena), Dr. M. Nisenoff (Naval Research Laboratory), Dr. D. E. Oates (Lincoln Laboratory), Prof. Ya. G. Ponomarev (Moscow State University), Dr. A. Porch (University of Birmingham), Prof. A. M. Portis (University of Berkeley), Prof. K. Scharnberg (Universität Hamburg), Dr. E. Sodtke (ISI, FZ Jülich), Dr. G. Thummes (Universität Gießen), Prof. R. Vaglio (University of Naples), Prof. I. B. Vendik, Prof. O. G. Vendik (Technical University St. Petersburg), Dr. B. Willemsen (STI), Dr. R. Withers (Bruker), Dr. R. Wördenweber (ISI, FZ Jülich), Dr. J. Wosik (University Houston), and Dr. C. Zuccaro (ISI, FZ Jülich).

This review of supporters would be incomplete by considering merely the scientific aspects. I am very much indebted to my wife Sabine, my daughter Anneli and my son Michael as well as to my mother, who often encouraged me during the intense time of writing and reviewing before publication. They have been very patient and knowledgeable companions during numerous "Highs" and "Lows".

Solingen, April 1999   *Matthias Hein*

Contents

1. **Temperature and Frequency Dependent Surface Impedance**   1
   1.1 Physical Framework 1
       1.1.1 Field Equations and Characteristic Lengths 1
       1.1.2 Surface Resistance and Surface Reactance 5
       1.1.3 Effect of Finite Film Thickness
             and Metallic Cap Layers 8
   1.2 Extended Two-Fluid Model 10
       1.2.1 Equivalent Circuits and Interpretation
             for a Homogeneous Superconductor 11
       1.2.2 Interpretation for a Granular Superconductor 12
       1.2.3 Scaling Laws 16
   1.3 Microscopic Picture 19
       1.3.1 BCS Theory:
             Cooper Pairs, Energy Gap and Density of States 20
       1.3.2 BCS Theory: Complex Conductivity
             and Surface Impedance 23
       1.3.3 Non-BCS Densities of States
             and Consequences for the Surface Impedance 27
       1.3.4 Tunneling Spectroscopy of the Density of States
             in $Y(Yb)Ba_2Cu_3O_{7-x}$ 41

2. **Measurements of the Surface Impedance at Linear Response**   43
   2.1 Measurement Methods 43
       2.1.1 Characteristic Parameters
             of Superconducting Resonators 44
       2.1.2 Modes of Operation 50
       2.1.3 Selected Examples of Integral Measurement Systems .. 55
       2.1.4 Survey of Local Measurement Systems 63
   2.2 Surface Impedance of Nb_3Sn Films on Sapphire 68
       2.2.1 Properties of Nb_3Sn Thin Films
             on Dielectric Substrates 69
       2.2.2 Superconductive Properties
             of the Nb_3Sn Films Prepared by Tin Diffusion 71

		2.2.3	Effect of Film Thickness and Grain Size on the Surface Impedance 78

    2.3 Surface Impedance of Oxide Superconductor Films 82
          2.3.1 Granular $YBa_2Cu_3O_{7-x}$ Films 83
          2.3.2 Epitaxial $YBa_2Cu_3O_{7-x}$ and Tl-Ba-Ca-Cu-O Films... 85
          2.3.3 Effects of Composition 94
          2.3.4 Theoretical Description 96

3. Field and Power-Dependent Surface Impedance 103
    3.1 Critical Field Levels and Corresponding Surface Impedance .. 104
          3.1.1 Overview 105
          3.1.2 Thermodynamic and Upper Critical Fields 110
          3.1.3 Adiabatic Superheating 114
          3.1.4 Lower Critical Fields 117
          3.1.5 Implications of Film Thickness, Anisotropy and Unconventional Pairing 126
    3.2 Microwave Heating 130
          3.2.1 Description of the Problem 132
          3.2.2 Analytical Approach 136
          3.2.3 Numerical Simulations 139
    3.3 Dynamic Nonlinear Surface Impedance 146
          3.3.1 Dynamic (Nonequilibrium) Microwave Response 146
          3.3.2 Two-Tone Frequency Intermodulation in the Adiabatic Limit 150
          3.3.3 Two-Tone Frequency Intermodulation in Dynamic Systems 155
    3.4 Summary: Consequences for Nb_3Sn and $YBa_2Cu_3O_{7-x}$ 159
          3.4.1 Anomalous Field-Effect in Superconductors: Coupled-Grain Models 159
          3.4.2 Anomalous Field-Effect in Superconductors: Pair Breaking Effects 163
          3.4.3 Typical Field Ranges for Nb_3Sn and YBCO 166

4. Measurements of the Surface Impedance at Nonlinear Response 171
    4.1 Microwave Field Dependence of the Surface Impedance 171
          4.1.1 Phenomenological Picture 172
          4.1.2 Critical Field Levels and Power-Law Behavior 176
          4.1.3 Local Measurements of the Nonlinear Surface Impedance 185
          4.1.4 Survey of Data on Nb_3Sn Films 194
    4.2 DC Magnetic Field Dependence of the Surface Impedance ... 199
          4.2.1 Phenomenological Description of $Z_s(B_{DC})$ 199
          4.2.2 Flux Penetration and Surface Impedance in Perpendicular Fields 204

		4.2.3	Surface Impedance in Parallel Fields 213

    4.2.3 Surface Impedance in Parallel Fields 213
4.3 Identification of Magnetic
and Thermal Field Limitations in YBCO Films 218
    4.3.1 Identification of Magnetically Induced Nonlinearities .. 218
    4.3.2 Anomalous Field Effect in $YBa_2Cu_3O_{7-x}$ Films 225
    4.3.3 Identification of Thermally Induced Nonlinearities 231
    4.3.4 Summary 236

5. Technology of High-Temperature Superconducting Films and Devices 239

5.1 Thin-Film Deposition Techniques 239
    5.1.1 Nb_3Sn Films on Sapphire Substrates 240
    5.1.2 Preparation of $YBa_2Cu_3O_{7-x}$ Films 247
    5.1.3 Preparation of $YBa_2Cu_3O_{7-x}$ Grain Boundary
          Junctions 260
5.2 Integration of Films into Devices 264
    5.2.1 Choice of Substrate Materials:
          Structural Compatibility 264
    5.2.2 Choice of Substrate Materials: Complex Permittivity... 267
    5.2.3 Microwave Design 271
5.3 Refrigeration 273
    5.3.1 General Considerations 273
    5.3.2 Liquid Coolants 278
    5.3.3 Closed-Cycle Cryocoolers 279

6. Passive High-Temperature Superconducting Microwave Devices 283

6.1 Basic Features of Filters 283
    6.1.1 From Resonators to Filters 284
    6.1.2 Different Types of Filter Characteristics 288
    6.1.3 Relations Between the Properties of Filters
          and the Constituting Resonators 296
6.2 Passive Devices Related to Resonant Elements 299
    6.2.1 Survey of Passive Microwave Device Applications 300
    6.2.2 Remaining Challenges 307
6.3 Passive Devices Based on Josephson Junctions 318
    6.3.1 Unique Features of Josephson Junctions and SQUIDs . 318
    6.3.2 Survey of Josephson Junction Microwave Devices 325
    6.3.3 Single-Junction SQUIDs Coupled
          to Microwave Devices 332

Appendix 345

A.1 Fundamental Constants 345
A.2 Frequently used Abbreviations,
    Symbols and Definitions 346

A.3　Physical Quantities 347
　　A.4　HTS Compounds, Substrates
　　　　 and Deposition Techniques 348

References 349

Subject Index 391

1. Temperature and Frequency Dependent Surface Impedance

> *Only fractions of our world behave linearly, but it is amazing how far linear modeling has proceeded.*

1.1 Physical Framework

This section introduces the electrodynamics needed to understand the basic concept of surface impedance of superconductors. Starting with Maxwell's equations and the kinetics of the charges which carry the microwave currents, the treatment is kept on a phenomenological level. Three characteristic lengths are identified to distinguish between different limiting regimes of the electrodynamics of superconductors. The physical concept of the surface resistance and the surface reactance (or, equivalently, the penetration depth) are subsequently illustrated in terms of the energy dissipated and stored, respectively, in the superconductor. Finally, the effect of finite film thickness on the concept of the surface impedance is discussed.

1.1.1 Field Equations and Characteristic Lengths

Penetration of Microwave Fields into Superconductors

The interaction between a plane electromagnetic wave at circular frequency $\omega = 2\pi f$ and a metal of conductivity σ is described by Maxwell's equations. These present expressions for the microscopic electric field vector $\boldsymbol{E}$ and the magnetic induction $\boldsymbol{B}$, and the macroscopic displacement $\boldsymbol{D}$ and the magnetic field $\boldsymbol{H}$. A detailed treatment of the relations between $\boldsymbol{E}$ and $\boldsymbol{D}$ and between $\boldsymbol{B}$ and $\boldsymbol{H}$ can be found, e.g., in [1]. Taking into account the absence of magnetic monopoles (i.e., $\operatorname{div} \boldsymbol{B} = 0$), and the absence of free or polarized charges in metals (i.e., $\operatorname{div} \boldsymbol{D} = 0$), the remaining Maxwell's equations are

$$\operatorname{curl} \boldsymbol{E} = -\frac{\partial \boldsymbol{B}}{\partial t} \tag{1.1a}$$

and

$$\operatorname{curl} \boldsymbol{H} = \boldsymbol{J} + \frac{\partial \boldsymbol{D}}{\partial t} . \tag{1.1b}$$

The magnetic induction $\boldsymbol{B}$ can be expressed as the rotation of the magnetic vector potential $\boldsymbol{A}$: $\boldsymbol{B} = \operatorname{curl} \boldsymbol{A}$. In the absence of magnetizing effects and in nonmagnetic materials, $\boldsymbol{B}$ is proportional to the magnetic field $\boldsymbol{H}$, $\boldsymbol{B} = \mu_0 \cdot \boldsymbol{H}$ with $\mu_0 = 4\pi \times 10^{-7}$ Vs/Am being the magnetic permeability of vacuum. In order to find solutions for $\boldsymbol{E}$ and $\boldsymbol{B}$, (1.1) needs to be supplemented by a current–field relation like, for instance,

$$\boldsymbol{J} = \sigma \cdot \boldsymbol{E} . \tag{1.2}$$

With σ real, (1.2) describes Ohm's law for isotropic metals. However, as will be shown below, it also remains valid for superconductors, the conductivity of which is complex:

$$\sigma = \sigma_1 - \mathrm{i}\sigma_2 = \sigma_1(1 - \mathrm{i}\tan\theta) . \tag{1.3}$$

Let us consider a plane electromagnetic wave with harmonic time dependence, with the electric and the magnetic fields spanning the x-y plane and propagating into a metal along the z direction: $E(z,t) = E_0 \exp[\mathrm{i}(\omega t - kz)]$. Combination of (1.1) and (1.2) then yields the wave equation for the electric field

$$\frac{\partial^2 E}{\partial z^2} = \mu_0 \sigma \frac{\partial E}{\partial t} + \varepsilon\varepsilon_0\mu_0 \frac{\partial^2 E}{\partial t^2} , \tag{1.4}$$

with ε and ε_0 the permittivity of the metal and of vacuum ($\varepsilon_0\mu_0 = c^{-2}$ with c the velocity of light). The first term on the right-hand side of (1.4) results from the microwave transport current J, while the second term is due to the displacement current $\partial D/\partial t$. The ratio of the magnitudes of these two contributions is weighted by $\sigma/\omega\varepsilon\varepsilon_0$. Since this parameter is large compared to unity for any metal at microwave frequencies (ω of order 10^{10} Hz), the displacement currents can usually be neglected. The remainder of (1.4) represents the equation of motion of a wave decaying in the z direction:

$$E(z,t) = E_0 \exp\{\mathrm{i}(\omega t - kz)\} \exp(-z/\lambda) , \tag{1.5a}$$

with the supplementing condition

$$\left(\mathrm{i}k + \frac{1}{\lambda}\right)^2 = \mathrm{i}\omega\mu_0\sigma . \tag{1.5b}$$

The characteristic length λ is the microwave penetration depth. It describes the surface region of a superconductor within which the external field has decayed to $1/e$ of its value at the surface. Equation (1.5b) leads to the general expression

$$\lambda^2 = \frac{2}{\mu_0\omega\sigma_1} \cdot \frac{\sqrt{1+y^2}-1}{y} = \frac{2}{\mu_0\omega\sigma_1}\frac{1-\sin\theta}{\cos\theta} , \tag{1.6}$$

where the substitution $y = \sigma_1/\sigma_2 = \cot\theta$ was used to indicate the "degree of normal conductivity" (see Sect. 1.2). For normal conductors, $\sigma_2 = 0$ and thus $y \to \infty$ or $\theta = 0$. In this limit, (1.6) reproduces the skin depth $\lambda = \delta = (2/\omega\mu_0\sigma_1)^{1/2}$, with the wave number $k = 1/\delta$ following from (1.5b). In the opposite case $\sigma_1 = 0$ ($y = 0$ or $\theta = \pi/2$), which corresponds to a perfect superconductor at zero temperature, $\lambda = (1/\omega\mu_0\sigma_2)^{1/2}$ and $k = 0$ are obtained.

Complex Conductivity

The concept of complex conductivity was initially introduced by Gorter and Casimir in the framework of the two-fluid model (TFM) [2], and applied by F. London and H. London to describe the finite conductivity of superconductors at microwave frequencies [3]. However, as discussed in several books [4,5], the general idea can be applied to various physical situations (Sect. 1.2). It shall suffice here to reconstruct the qualitative features of the original TFM to develop a consistent picture of the electrodynamics of superconductors.

At finite temperatures below the transition temperature T_c, a superconductor contains normal electrons and Cooper pairs. The equation of motion of normal electrons (index "n", mass m and charge $-e$) takes into account damping, which results from scattering at time intervals τ. The scattering time is assumed to result solely from collisions between electrons and phonons, as expected for a free electron gas, and to be independent of frequency. This leads to the Drude model of conductivity of metals:

$$m\frac{d\boldsymbol{v}_n}{dt} + m\frac{\boldsymbol{v}_n}{\tau} = -e \cdot \boldsymbol{E} \ . \tag{1.7}$$

The range that the electrons can travel between subsequent scattering events is described by the mean free path $\ell = v_F \cdot \tau$, with the velocity evaluated at the Fermi level. In contrast, Cooper pairs behave, due to their quantum-statistical bosonic nature, like free charges of mass $2m$ and charge $-2e$ (index "s"). The corresponding equation of motion is

$$2m\frac{d\boldsymbol{v}_s}{dt} = -2e \cdot \boldsymbol{E} \ . \tag{1.8}$$

In an electric field, both kinetic contributions add to the total velocity $\boldsymbol{v} = \boldsymbol{v}_n + \boldsymbol{v}_s$. Furthermore, the flux of charges causes a current $\boldsymbol{J} = -n \cdot e \cdot \boldsymbol{v}$ where n is the number density of charge carriers. In the London gauge (div $\boldsymbol{A} = 0$ and $\boldsymbol{E} = -\partial \boldsymbol{A}/\partial t$), (1.8) thus presents a simplified version of the London equation, which relates the local supercurrent $\boldsymbol{J}_s$ to the vector potential $\boldsymbol{A}$ at the same position [6]:

$$\boldsymbol{J}_s = -\frac{n_s e^2}{m}\boldsymbol{A} \ . \tag{1.9}$$

Combining (1.7) and (1.8), considering the harmonic time dependence of the fields, and converting velocities into currents, leads to the current–field relation of superconductors:

$$\boldsymbol{J} = \sigma_0 \left(x_n \frac{\omega\tau - \mathrm{i}(\omega\tau)^2}{1+(\omega\tau)^2} - \mathrm{i}x_s \right) \cdot \boldsymbol{E} \ . \tag{1.10}$$

Here, $\sigma_0 = n_e e^2/m\omega = (1/\omega\tau)\sigma_n$, and $x_n = n_n/n_e$ ($x_s = 2n_s/n_e$) denotes the fractional number densities of normal (paired) electrons, with n_e and σ_n the number density of electrons at the Fermi level and the conductivity in the normal state, respectively. Comparing (1.10) with (1.2) illustrates the physical meaning of the complex conductivity $\sigma = \sigma_1 - i\sigma_2$. In addition, the penetration depth λ of a perfect superconductor at $T = 0\,\mathrm{K}$ equals the London penetration depth $\lambda_\mathrm{L} = m/\mu_0 e^2 n_\mathrm{s}$.

In the framework of the two-fluid model, Gorter and Casimir assumed complete pairing, $x_\mathrm{n} + x_\mathrm{s} = 1$, with the temperature dependence of x_n described by $(T/T_\mathrm{c})^4$. We will see below that the conditions for complete pairing can be violated, although the merit of the formalism developed in this section remains unaffected.

Characteristic Lengths

The current–field relations in (1.2), (1.9) and (1.10) imply local conditions, i.e., the charges carry current at a position $\boldsymbol{r}$ which follows the electric field at the same position $\boldsymbol{r}$. Such a local relationship (called "normal skin effect") is fulfilled in normal metals if the mean free path ℓ is small compared to the skin depth δ which describes the spatial variation of $\boldsymbol{E}(\boldsymbol{r})$ (see (1.5a)). Since δ decreases with increasing frequency and ℓ increases with decreasing temperature, there are ranges of parameters where the application of local electrodynamics is no longer justified. Rather, the current originates from the average electric field sensed by the electrons along their mean free path [6]:

$$\boldsymbol{J}(\boldsymbol{r}) = \int K(\boldsymbol{r} - \boldsymbol{r}') E(\boldsymbol{r}') \mathrm{d}^3 \boldsymbol{r}' \,. \tag{1.11}$$

The kernel $K(\boldsymbol{r} - \boldsymbol{r}')$ hence has a range of about ℓ. This limit, which is called the "anomalous skin effect", was theoretically analyzed by Reuter and Sondheimer [7]. Further details of this case and corresponding results for the surface impedance can be found, e.g., in [8, 9].

Similar to the normalconducting case, the local current–field relation fails if the electrodynamic "coherence" range of Cooper pairs, ξ_P, exceeds the superconducting penetration depth λ. This was first realized, and argued in analogy to the anomalous skin effect, by Pippard [8]. He proposed to substitute London's equation (1.9) by the nonlocal relation relation

$$\boldsymbol{J}(\boldsymbol{r}) = -\frac{3ne^2}{4\pi m \xi_0} \int \frac{\boldsymbol{R}[\boldsymbol{R} \cdot \boldsymbol{A}(\boldsymbol{r}')]}{R^4} \exp(-R/\xi_\mathrm{P}) \mathrm{d}^3 \boldsymbol{r}' \,, \tag{1.12}$$

where $\boldsymbol{R} = \boldsymbol{r} - \boldsymbol{r}'$ and $R = |\boldsymbol{R}|$. The "intrinsic coherence length" ξ_0 is a constant of the pure metal, determined by v_F and the thermodynamic time constant $\hbar/k_\mathrm{B} T_\mathrm{c}$ characterizing the superconducting state ($k_\mathrm{B} = 1.38 \times 10^{-23}$ J/K is the Boltzmann constant) [6, 10]:

$$\xi_0 = \frac{\gamma}{\pi^2} \frac{\hbar v_F}{k_\mathrm{B} T_\mathrm{c}} \tag{1.13}$$

where $\gamma \approx 1.781$ is Euler's constant (see Sect. 1.3.1). The value of ξ_P varies between ξ_0 for clean superconductors ($\ell \gg \xi_0$) and ℓ for dirty superconductors ($\ell \ll \xi_0$):

$$\frac{1}{\xi_P} = \frac{1}{\xi_0} + \frac{1}{\ell} \,. \tag{1.14}$$

In summarizing the above results, the appropriate electrodynamic description of superconductors depends on the three basic lengths λ, ℓ and ξ_0. The penetration depth λ itself scales with ℓ in the local limit, but still for sufficiently large mean free paths, like [11, 12]:

$$\lambda(\ell) \approx \lambda_L \sqrt{1 + \frac{\xi_0}{\ell}} \,. \tag{1.15}$$

Consequently, four limits can be defined by the dimensionless ratios λ/ξ_0 and ℓ/ξ_0 as indicated in Table 1. Nonlocal effects need to be considered especially in superconductors of the first kind. Mean free path effects become increasingly important in alloyed superconductors. In the case of superconducting films, the film thickness d_F places additional constraints on the locality conditions. We assume here, for simplicity, that d_F is sufficiently larger than ℓ and ξ_0. It shall further be noted that the electrodynamic ("Pippard") coherence length is, in general, to be distinguished from the "Ginzburg–Landau" coherence length ξ_{GL} which describes the spatial variation of the superconducting order parameter in the presence of magnetic fields. As given in [10], the microscopic theory yields expressions for ξ_{GL} and the Ginzburg–Landau parameter $\kappa = \lambda_{GL}/\xi_{GL}$ (see Chap. 3) as functions of ξ_0 and ℓ, with the limiting cases reproduced in Table 1.1.

Table 1.1. The four limiting cases describing the electrodynamics of superconductors, and corresponding equations relating electrodynamic and magnetic parameters [10, 11, 13]

	Electrodynamics		Ginzburg–Landau
	Nonlocal, $\lambda/\xi_0 \ll 1$ Eq. (1.12)	Local, $\lambda/\xi_0 \gg 1$ Eqs. (1.9), (1.10)	(coefficients evaluated at $T \approx T_c$).
Clean limit, $\ell/\xi_0 \gg 1$	$\xi_P = \xi_0$ $\lambda \approx 0.852(\lambda_L^2 \xi_0)^{1/3}$	$\xi_P = \xi_0$ $\lambda = \lambda_L$	$\xi_{GL} = 0.739\xi_0$ $\kappa = 0.957\lambda_L/\xi_0$
Dirty limit, $\ell/\xi_0 \ll 1$	$\xi_P = \ell$ $\lambda \approx \lambda_L(\xi_0/\ell)^{1/2}$	$\xi_P = \ell$ $\lambda = 1.25\lambda_L(\xi_0/\ell)^{1/2}$	$\xi_{GL} = 0.852(\xi_0 \ell)^{1/2}$ $\kappa = 0.720\lambda_L/\ell$

1.1.2 Surface Resistance and Surface Reactance

Concept of the Surface Impedance

The concept of the surface impedance facilitates the description of the electrodynamics at the interface between vacuum and the surface of a (super)conductor. The superconductor is assumed here to extend from $z = 0$

to infinity. The effect of finite thickness $d_\mathrm{F} \leq \lambda$ of a superconducting sheet is discussed in Sect. 1.1.3. In accordance with the usual concept of the complex-valued AC impedance, the "bulk" surface impedance Z_s is defined by the ratio of the electric field at the surface to the total current flowing across a unit line in the surface:

$$Z_\mathrm{s} = \frac{E_\mathrm{s}}{\int_0^\infty J(z)\mathrm{d}z} = \frac{E_\mathrm{s}}{H_\mathrm{s}} \;. \tag{1.16}$$

The integration can be performed by using (1.1b), leaving Z_s as the ratio of the tangential electric field to the tangential magnetic field at the surface of the superconductor. Alternatively, by application of (1.2) and (1.5) to (1.16), Z_s can be expressed in the local limit in terms of the complex conductivity σ [9, 14]:

$$Z_\mathrm{s} = \sqrt{\frac{\mathrm{i}\omega\mu_0}{\sigma}} = R_\mathrm{s} + \mathrm{i}X_\mathrm{s} \;. \tag{1.17}$$

The real part of Z_s is called the surface resistance R_s, the imaginary part the surface reactance X_s. The expression for $Z_\mathrm{s}(\sigma)$ for arbitrary values of the conductivity ratio y is found to be

$$Z_\mathrm{s} = \sqrt{\frac{\omega\mu_0}{2\sigma_1}}(\varphi_- + \mathrm{i}\varphi_+) \tag{1.18a}$$

with

$$\varphi_\pm^2 = \frac{y}{1+y^2}\left(\sqrt{1+y^2} \pm 1\right) = \cos\theta(1 \pm \sin\theta) \;. \tag{1.18b}$$

The limiting case for a normal conductor is

$$R_\mathrm{s} = X_\mathrm{s} = R_\mathrm{n} = (\omega\mu_0/2\sigma_1)^{1/2} \;, \tag{1.18c}$$

while for a superconductor in the limit $y \ll 1$

$$R_\mathrm{s} = 1/2\mu_0^2\omega^2\sigma_1\lambda^3 \quad \text{and} \quad X_\mathrm{s} = (\omega\mu_0/\sigma_2)^{1/2} = \omega\mu_0\lambda \;. \tag{1.18d}$$

According to [15], the two-fluid model can be modified in order to derive semi-analytical formulations of Z_s for arbitrary values of ℓ/ξ_0 in relation to the microscopic understanding of σ. Further details of this approach are postponed until Sect. 1.3.

The implication of nonlocal behavior to the surface impedance of superconductors was analyzed by Mattis and Bardeen [16] and discussed in more detail in [17, 18]. As before, Z_s can be obtained from the frequency dependent complex conductivity, which now depends additionally on the wave vector of the charge trajectories. For diffuse scattering of electrons at the surface of the metal, Z_s is given by

$$Z_\mathrm{s} = \mathrm{i}\pi\omega\mu_0\left[\int_0^\infty \ln\left(1 + \frac{\mu_0\omega\sigma(,k)}{k^2}\right)\mathrm{d}k\right]^{-1} \;. \tag{1.19a}$$

This equation reduces in the extreme anomalous limit to the relation

$$\frac{Z_\mathrm{s}}{Z_\mathrm{n}} = \left(\frac{\sigma_\mathrm{n}}{\sigma_1 - \mathrm{i}\sigma_2}\right)^{1/3}, \tag{1.19b}$$

which then replaces the first part of (1.17). Equation (1.19a) transforms into

$$Z_\mathrm{s} = \frac{2}{\pi}\int_0^\infty \left\{\sigma(\omega,k) - \mathrm{i}\frac{k^2}{\mu_0\omega}\right\}^{-1} \mathrm{d}k \tag{1.19c}$$

for specular reflection of the electrons.

The local relations between Z_s and the conductivity σ (1.17) and (1.18) remain valid in the dirty limit. However, the condition $x_\mathrm{n} + x_\mathrm{s} = 1$ fails, which describes the perfect pairing of the charge carriers in clean superconductors. In order to evaluate the surface impedance, tabulated values for $\sigma_1/\sigma_\mathrm{n}$ and $\sigma_2/\sigma_\mathrm{n}$, which are independent of ℓ, must be used [16].

Referring to the data on Z_s of oxide high-temperature superconductors (presented in Sect. 2.3), it is noted that the analysis within the clean, local limit is justified over a wide range of temperatures since $\xi_0 (\approx 2\text{ nm}) \ll \ell\, (\geq 10\text{ nm}) \ll \lambda\, (\geq 150\text{ nm})$ [12, 18, 19]. The situation is less clear for Nb$_3$Sn, where $\xi_0 (\approx 7\text{ nm}) \approx \ell\,(3\text{--}12\,\text{nm}) \ll \lambda\,(80\text{--}100\,\text{nm})$ (see [10, 20, 21] and Sect. 2.2). While the electrodynamic response can still be expected to be local, it seems to be close to the transition between clean and dirty superconductors.

Physical Meaning of Surface Resistance and Surface Reactance

The propagation of an electromagnetic wave in a lossy medium provokes an electric field $\boldsymbol{E}$ along the path of the current density $\boldsymbol{J}$:

$$P_\mathrm{diss} = \frac{1}{2}\mathrm{Re}\left\{\int_\mathrm{volume} \boldsymbol{J}\cdot\boldsymbol{E}\,\mathrm{d}^3\boldsymbol{r}\right\}. \tag{1.20}$$

Applying the local current–field relation (1.2), this expression reduces to the volume integral of $\sigma_1\cdot|\boldsymbol{E}|^2$. Since the electric field decays exponentially into the superconductor (see (1.5a)), the integral can be split into a surface integral and an integration along the propagation in the z direction. Evaluating the latter integral, and using (1.16), we obtain

$$P_\mathrm{diss} = \frac{1}{4}\int_\mathrm{surface}(\sigma_1\lambda|Z_\mathrm{s}|^2)\cdot|\boldsymbol{H}(\boldsymbol{r})|^2\,\mathrm{d}\boldsymbol{r}\,. \tag{1.21}$$

The factor in brackets can be calculated on the basis of (1.6) and (1.18) to equal $2R_\mathrm{s}$. Accordingly, the average surface resistance quantifies the power dissipated by the electromagnetic wave in the superconductor:

$$P_\mathrm{diss} = \frac{1}{2}\overline{R}_\mathrm{s}\int_\mathrm{surface}|\boldsymbol{H}(\boldsymbol{r})|^2\mathrm{d}^2\boldsymbol{r}\,. \tag{1.22}$$

The physical meaning of the surface reactance can be illustrated similarly by evaluating the electromagnetic energy W_S stored in the superconductor. Two mechanisms contribute to W_S: the magnetic field inside the superconductor, and the kinetic energy associated with the frictionless motion of the charge carriers:

$$W_S = \frac{1}{2} \int_{\text{volume}} [\mu_0 |\boldsymbol{H}(\boldsymbol{r})|^2 + \text{Im}\{\boldsymbol{J} \cdot \boldsymbol{E}\}] d^3 \boldsymbol{r} . \tag{1.23}$$

By treating (1.23) analogously to (1.20), JE can be identified with $|\boldsymbol{E}|^2/\omega\sigma_2$. After splitting the integral as before, we arrive at

$$W_S = \frac{1}{4\omega} \int_{\text{surface}} \mu_0 \omega \lambda \left(1 + \frac{\sigma_2 |Z_s|^2}{\mu_0 \omega}\right) \cdot |\boldsymbol{H}(\boldsymbol{r})|^2 d^2 \boldsymbol{r} , \tag{1.24}$$

which can be simplified in the framework of the two-fluid model to the result

$$\omega W_S = \frac{1}{2} \overline{X}_s \int_{\text{surface}} |\boldsymbol{H}(\boldsymbol{r})|^2 d^2 \boldsymbol{r} . \tag{1.25}$$

We conclude that the average surface reactance quantifies the energy stored in the superconductor during one microwave cycle $1/\omega$.

1.1.3 Effect of Finite Film Thickness and Metallic Cap Layers

Finite Film Thickness

Until now it was assumed that the superconductor forms an infinite half-plane along the z direction. In reality, superconductor films on dielectric substrates are usually used to measure the surface impedance, and to constitute microwave devices. As the film thickness d_F decreases from large values, $d_F/\lambda \gg 1$, to the order of unity or less, corrections have to be applied to Z_s to account for the modified field penetration. The physical situation is depicted in Fig. 1.1: An electromagnetic wave propagating in vacuum (region I) experiences the free-space wave impedance $Z_0 = (\mu_0/\varepsilon_0)^{1/2} = 120\pi [\Omega]$.

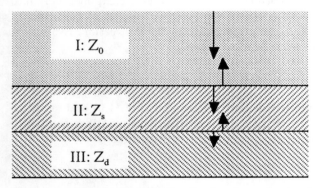

Fig. 1.1. Schematic field distribution in a stack of three layers. I, vacuum; impedance Z_0. II, superconductor Z_s; III, dielectric substrate Z_d. The arrows indicate the propagation of an electromagnetic wave.

1.1 Physical Framework

At the surface of the superconductor (surface impedance Z_s, region II), part of the wave is reflected while another part propagates through the film. At the boundary between the film and the dielectric (impedance Z_d, region III), the wave is again reflected and can, to a minor part, be transmitted to the substrate, which is assumed to be thick enough to neglect further reflections (e.g., at the backside of the substrate). While the effects at the boundary I–II were inherently taken into account in the concept of the surface impedance of bulk superconductors, the boundary II–III modifies these previous results.

The analytical problem can be solved in a straightforward way by setting up the field solutions, corresponding to (1.5), for each of the three regions, and evaluating the continuity conditions at the boundaries. An alternative approach is based on the transformation rules of impedances along a stripline [22] (k is the magnitude of the wave vector):

$$Z_s(d_F) = Z_{s,\infty} \frac{Z_{s,\infty} \tanh(kd_F) + Z_d}{Z_{s,\infty} + Z_d \tanh(kd_F)} . \quad (1.26)$$

Equation (1.26) simplifies the mathematical procedure, though at the expense of intuition. In both cases, realistic assumptions are required to find analytic expressions for the surface impedance in terms of the intrinsic surface impedance (i.e., that of infinitely thick films, $Z_{s,\infty}$), and in terms of the reduced film thickness $d^* = d_F/\lambda$ [23]:

$$\begin{aligned} R_s(d^*) = & R_{s,\infty} \left(\coth(d^*) + \frac{d^*}{\sinh^2(d^*)} \right) \\ & + \sqrt{\varepsilon_r} \frac{X_{s,\infty}^2}{Z_0} \frac{1}{\sinh^2(d^*)} \end{aligned} \quad (1.27a)$$

$$\begin{aligned} X_s(d^*) = & X_{s,\infty} \coth(d^*) \\ & - \sqrt{\varepsilon_r} \frac{2R_{s,\infty} X_{s,\infty}}{Z_0} \left(\coth^2(d^*) + \frac{d^* \coth(d^*)}{\sinh^2(d^*)} - 1 \right) . \end{aligned} \quad (1.27b)$$

Equation (1.27) was deduced in the limit $y \ll 1$, where $R_s \ll X_s = \omega\mu_0\lambda$. It was further assumed that the surface impedance of the dielectric substrate

$$Z_d = \frac{Z_0}{\sqrt{\varepsilon_r}} \sqrt{\frac{1}{1 - i\tan\delta}} \approx \frac{Z_0}{\sqrt{\varepsilon_r}} (1 + i/2 \tan\delta) \quad (1.28)$$

was real, i.e., had a loss tangent $\tan\delta < 10^{-4}$. Finally, the inequality $\tanh^2(d_F/\lambda) \gg \omega\varepsilon_0\varepsilon_r/\sigma_2$ also entered (1.27), which is in general fulfilled for superconductors and for typical dielectrics. While the first term in the sum of (1.27a) and in the difference of (1.27b), respectively, originates from the modified field distribution in the superconductor, the second term (having $\varepsilon_r^{1/2}$ as a multiplier) indicates the effect of transmission into the substrate. This term can generally be neglected if high-permittivity substrates (like SrTiO$_3$) or standing waves in the dielectric are avoided [23]. In fact, as further analyzed in [4, 24], the thin-film approximation of (1.26) and (1.27),

$$Z_s(d_F) \approx Z_{s,\infty} \coth(d_F/\lambda) \,, \tag{1.29}$$

was found to give correct results in the range $0.01 \leq d_F/\lambda \leq 1$. The thin-film limit approached $Z_{s,\infty}$ within 10.5%, 3.7%, 1.4%, 0.5% at $d_F/\lambda = 1.5$, 2, 2.5, 3, respectively. Since the penetration depth varies with temperature, especially at temperatures close to T_c, the deduction of the surface impedance from measurements usually requires numerical methods [23, 25].

Metallic Cap Layers

The implementation of superconducting films in microwave devices often requires the deposition of cap layers, e.g., to provide protection, or well-defined electrical contact to the normal-conducting environment [26]. In this context it is interesting to analyze the effect of a metallic cap layer on the measurable surface impedance of the superconducting film. This problem can be solved with (1.26) by substituting Z_s with the surface impedance Z_c of the cap layer, and Z_d with that of the superconductor $Z_{s,\infty}$. We assume, as a representative example, the validity of the normal skin effect $Z_c = R_c(1+\mathrm{i})$, and that the superconductor can be regarded to be thick. We further simulate realistic conditions by focusing on cap layers with a thickness d_c much smaller than the skin depth δ_c of the metal. Equation (1.26) then yields for the effective surface impedance Z_{eff}

$$Z_{\mathrm{eff}} = \frac{R_c(1+\mathrm{i})(R_s+\mathrm{i}X_s) + (2\mathrm{i}d_c/\delta_c)R_c^2}{R_c(1+\mathrm{i}) + d_c/\delta_c(R_s+\mathrm{i}X_s)} \,. \tag{1.30}$$

For metallic layers, $R_c \gg R_s$ is obviously valid. Furthermore, $R_c \gg X_s$ can safely be assumed since this condition is equivalent to $\delta_c \gg 2\lambda$. Equation (1.30) can now be solved and expanded in terms of the small parameters d_c/δ_c, R_s/R_c and X_s/R_c, leading to the approximative result

$$Z_{\mathrm{eff}} \approx Z_s + \frac{d_c}{\delta_c} R_c \,, \tag{1.31}$$

where quadratic terms were neglected. With $R_c = \omega\mu_0\delta_c/2$, the additional surface impedance reduces to $\omega\mu_0 d_c/2$, which depends only on the thickness of the cap layer. Furthermore, the surface resistance of the metallized superconductor, R_{eff}/R_s, is seen to increase much more strongly with d_c than the surface reactance X_{eff}/X_s.

1.2 Extended Two-Fluid Model

The concept of complex conductivity was discussed in the preceding section in relation to the microwave surface impedance of clean homogeneous superconductors in the local limit. Since this formalism presents a powerful tool to also describe various linear and nonlinear properties of superconductors in general [5], it is applied in this section to problems related to granularity. Referring to the "two-fluid model" should thus not be confused with the specific ideas introduced by Gorter and Casimir [2]. It is used in a more general sense, to account for the two transport channels giving rise to a normal and an imaginary part of the conductivitiy. The section concludes with general scaling laws of the TFM surface impedance.

1.2.1 Equivalent Circuits and Interpretation for a Homogeneous Superconductor

The complex conductivity of a superconductor can be represented by an equivalent circuit, constructed such that it yields the required current–field relation. Correspondingly, the lumped-element representation of a superconductor consists of a resistivity, $\rho = 1/\sigma_1$, shunted by the kinetic inductivity $l_{\mathrm{kin}} = 1/\omega\sigma_2$, as sketched in Fig. 1.2 [5]. The dimensionless conductivity ratio $y = \sigma_1/\sigma_2$ is to be identified with the impedance ratio $y = \omega l_{\mathrm{kin}}/\rho$. It is related to microscopic quantities in the weak-scattering limit by $y = \omega\tau x_{\mathrm{n}}/x_{\mathrm{s}} = \omega\tau x_{\mathrm{n}}/(1 - x_{\mathrm{n}})$. The second equality implies the validity of the clean limit and weak pair coupling. An increasing fraction x_{n} of quasi-particles enhances y, which therefore provides a sensitive measure of pair breaking effects. In addition, nonvanishing conductivity ratios also have to be expected at zero temperature in the case of incomplete quasi-particle pairing [27].

In the limit of large y, the microwave current experiences mainly the resistive (i.e., dissipative) nature of the superconductor. In the opposite extreme $y \ll 1$, its inductive nature $l_{\mathrm{kin}} = \mu_0 \lambda^2$ prevails. It is worth mentioning that the penetration depth of HTS is anisotropic in all three directions of the crystal structure. For $YBa_2Cu_3O_{7-x}$ (YBCO) $\lambda_{\mathrm{La}} = 160\,\mathrm{nm}$, $\lambda_{\mathrm{Lb}} = 103\,\mathrm{nm}$ and $\lambda_{\mathrm{Lc}} = 1100$ nm were reported [28] (b denotes the direction of the Cu–O chains in YBCO, and c the direction perpendicular to the CuO_2 planes). At $T = 0$ and weak scattering $\omega\tau \ll 1$ (1.10), $l_{\mathrm{kin}} = 1.8 \times 10^{-13}$ Hm is obtained for the average value $\lambda_{\mathrm{av}} = 140\,\mathrm{nm}$. The kinetic inductivity of Nb_3Sn is about

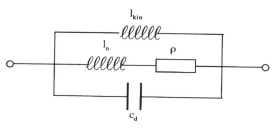

Fig. 1.2. Equivalent circuit representatiy of a homogeneous superconductor, according to the conductivity (1.7)–(1.10). The capacitive branch c_{d} is related to the displacement current [see (1.1b)] and can usually be neglected. In the limit $\omega\tau \ll 1$, the inductive contribution of the normal carriers, l_{n}, also becomes negligible.

a factor of 2–4 lower than that of YBCO because of the smaller penetration depth.

With $\omega\tau$ increasing towards unity, quasi-particle retardation leads to additional shielding, and therefore to an improved inductive channel (represented by l_n in Fig. 1.2). Tunneling, infrared and microwave spectroscopy on YBCO yielded highly temperature dependent quasi-particle scattering rates between T_c and 4.2 K between 0.1 to 10 THz ([19, 29–31], Sect. 2.3.2). Since $\tau^{-1} = 1\,\mathrm{THz}$ corresponds to a mean free path of $\ell = 100\,\mathrm{nm}$, which is of the order of λ_L, the validity of the local limit even becomes questionable at low temperatures. Finally, as ω approaches the gap frequency $\Delta/\hbar \approx 35$–40 THz for HTS and $\Delta/\hbar \approx 5\,\mathrm{THz}$ for $\mathrm{Nb_3Sn}$ (Sect. 1.3.4), photon-assisted pair breaking forces y to grow to infinity.

1.2.2 Interpretation for a Granular Superconductor

The extremely short coherence length in HTS, which is comparable to the unit cell dimensions [32], causes these materials to be highly sensitive to defects. If such defects extend over a range larger than ξ_0, they are expected to be normalconducting or insulating, depending on their number density of free charge carriers or localized electronic states. In contrast, small defects like grain boundaries are found to act as Josephson junctions in HTS [33], and suspected to do so in $\mathrm{Nb_3Sn}$ [34], i.e., to allow the transmission of Cooper pairs across a barrier of thickness comparable to ξ_0 [35, 36]. These results initiated intense studies of the effect of Josephson junctions (JJ) on the microwave response of HTS films [37]. Detailed discussions of this topic are left for Chaps. 3 and 6 because of the specific nonlinearities. However, since the linear microwave response of JJs can be represented in a simplified picture by a complex conductivity closely similar to homogeneous superconductors, the basic formalism is introduced here.

Microwave Response of Josephson Junctions

The transport properties of Josephson junctions are characterized by the characteristic voltage $V_\mathrm{c} = I_\mathrm{c} R_\mathrm{J}$, where I_c is the critical current of the junction and R_J its resistance in the normal state. This parameter defines the characteristic response time $\tau_\mathrm{c} = \Phi_0/V_\mathrm{c}$, where $\Phi_0 = h/2e = 2.067 \times 10^{-15}\,\mathrm{Vs}$ is the magnetic flux quantum. For ideal tunnel junctions, where the superconducting electrodes are separated by an insulating barrier (S-I-S junction), absolute value and temperature dependence of V_c below $T_\mathrm{c}/2$ are related to the energy gap [38]:

$$V_\mathrm{c}(T) = \frac{\pi}{2e}\Delta(T)\tanh\left(\frac{\Delta}{2k_\mathrm{B}T}\right). \tag{1.32}$$

The situation is different in HTS junctions, the barrier of which seems to be intrinsically shunted by charge transport via localized states [36]. In many cases, R_J is almost constant, and the characteristic voltage is strongly suppressed, $eV_c \ll \Delta$. The temperature dependence of I_c can be described by a power-law behavior

$$I_c(T) = I_c(0) \cdot \left(1 - \frac{T}{T_c}\right)^p, \tag{1.33}$$

with p ranging between 1 and 2.

At frequencies $f \ll 1/\tau_c$ and $f \ll \Delta/h$, the microwave response of JJs can be adequately described by the "resistively shunted junction" (RSJ) model [39]. Neglecting capacitive coupling across the junction, the relation between the time dependent current density $J(t)$ through the junction and the voltage drop $V(t) = E(t)d_J$ across it are given by

$$J(t) = J_c \cdot \sin\{\Delta\varphi(t)\} + \frac{E(t)}{\rho_J}, \tag{1.34a}$$

where the phase difference $\Delta\varphi$ of the Cooper pair functions in the two superconducting electrodes is related to the electric field gradient according to the second Josephson equation

$$E(t) = \frac{\Phi_0}{2\pi d_J} \cdot \frac{\partial(\Delta\varphi)}{\partial t}. \tag{1.34b}$$

The parameter $d_J = d_0 + 2\lambda$ denotes the total thickness of the junction, consisting of the physical dimension d_0 and the magnetic extension 2λ of the junction into the two superconducting electrodes (which are assumed to have identical penetration depths). Since d_0 is comparable to ξ_0, $d_J \approx 2\lambda$ for extreme type-II superconductors like HTS and Nb$_3$Sn. Combining both parts of (1.34) yields the current–field relation of JJs (index "J")

$$J = \left(\frac{1}{\rho_J} + \frac{1}{i\omega\ell_J}\right) \cdot E, \tag{1.35a}$$

with the Josephson inductivity

$$\ell_J = \frac{\Phi_0}{2\pi d_J} \frac{1}{J_c \cos(\Delta\varphi)}. \tag{1.35b}$$

The inductive term is related to the Josephson penetration depth λ_J, which marks the length scale of the variation of the magnetic field in Josephson junctions:

$$\lambda_J = \sqrt{\frac{\Phi_0}{2\pi\mu_0 d_J J_c}}. \tag{1.36}$$

The electrodynamic properties of JJs having a physical width w larger than λ_J are rather complicated [35]. The following analysis is therefore restricted to "small" junctions, $w/\lambda_J < 1$. Furthermore, only the small-signal limit ($J \ll J_c$ or $\Delta\varphi \approx 0$) is considered here.

The current–field relation of Josephson junctions (1.35a) has the same form as that of a homogeneous superconductor, (1.10). The appropriate equivalent circuit is therefore identical to the one shown in Fig. 1.2. However, the physical properties are distinctly different for the two cases. The resulting conductivity ratio y_J of JJs is

$$y_J = \frac{\omega \Phi_0}{2\pi d_J \rho_J J_c} = \frac{\omega \Phi_0}{2\pi I_c R_J} = \frac{\omega \tau_c}{2\pi} \, . \tag{1.37}$$

While y measured the Cooper pair density and the reduced scattering time $\omega\tau$ in the homogeneous situation, y_J senses the maximum Josephson coupling energy $E_J = \Phi_0/2\pi I_c$, or the corresponding time scale $\omega\tau_c$.

Application to Granular Superconductors

Grain boundaries in superconducting films can be adequately modeled in many cases by a parallel circuit of resistivity ρ_J and inductivity $\mu_0 \lambda_J^2$. In high-quality epitaxial HTS or large-grained phase-pure Nb_3Sn films, grain-boundary junctions form, beside the superconducting grains themselves, only part of the complete system. The question arises, therefore, how should grains and boundaries be coupled, and what should the resulting equivalent circuit look like. The initial approach in [37] considered grains and grain boundaries to constitute a series-connected effective medium (index "em"), giving rise to an average penetration depth

$$\lambda_{em} = \sqrt{\lambda_g^2 + \lambda_{J,\,eff}^2} \, , \tag{1.38}$$

with the index "g" denoting values associated with the grains (as treated in Sect. 1.1). The second term under the root represents the modified Josephson penetration depth, which takes into account the areal contribution of JJs weighted by the grain size a: $(\lambda_{J,\,eff}/\lambda_J)^2 = 2\lambda_g/a$ [40]. To obtain the effective surface impedance in this framework, one needs to calculate the complex conductivity of the equivalent circuit depicted in Fig. 1.3. Finally, the surface impedance $Z_{s,\,em}$ can be calculated from (1.17), yielding for the limits $y_g \ll 1$ and $y_J \ll 1$:

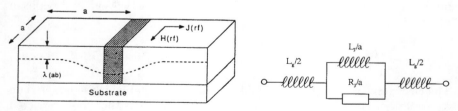

Fig. 1.3. Equivalent circuit of the effective-medium approach to describe granular superconductors. Grains and Josephson-type grain boundaries are connected in series [37].

$$R_{s,\,em} = \frac{2\lambda_g/a}{\sqrt{2\lambda_g/a + 2\lambda_g^3/a\lambda_J^2}} \omega\mu_0\lambda_J \frac{\omega\Phi_0}{4\pi V_c} \qquad (1.39a)$$

and

$$X_{s,\,em} = \omega\mu_0\lambda_g\sqrt{1 + 2\lambda_J^2/a\lambda_g}\,. \qquad (1.39b)$$

Obviously, small grains and weakly coupled junctions lead to an enhanced surface impedance.

This "two-fluid model" for granular superconductors has been adopted and analyzed for its temperature and magnetic field dependences by many groups (see [41] and Sect. 3.4). It described the microwave properties of granular HTS films reasonably well. However, for strongly coupled grains in high-quality films, the effective-medium approach seems no longer justified.

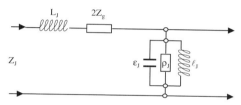

Fig. 1.4. Equivalent circuit of the stripline model for granular superconductors [42]. The inductance L_J represents flux within the junction. Z_g denotes the surface impedance of the grains. The specific impedance across the junction is represented by permittivity ε_J, resistivity ρ_J and inductivity ℓ_J.

An alternative model was proposed in [42], allowing for heterogeneous penetration into the grains and the junctions, respectively. It considers the grain boundary junction acting as a stripline between the two superconducting grain electrodes as shown in Fig. 1.4. The surface impedance $Z_{s,\,sl}$ (index "sl" for stripline) of this composite is the sum of that of the grains, Z_g, and that of the Josephson stripline, Z_J, weighted by $2\lambda_g/a$:

$$Z_{s,\,sl} = Z_{s,\,g} + \frac{2\lambda_g}{a}\sqrt{\frac{i\omega\mu_0}{\sigma_J}}\,, \qquad (1.40)$$

where (1.17) was applied to the current–field relation (1.35) of Josephson junctions, and the junction permittivity was neglected. Evaluating (1.40) in the limit of small y_J leads to the surface resistance $R_{s,\,sl}$ and the surface reactance $X_{s,\,sl}$:

$$R_{s,\,sl} = R_{s,\,g} + \frac{2\lambda_g}{a}\omega\mu_0\lambda_J\frac{\omega\Phi_0}{4\pi V_c} \qquad (1.41a)$$

and

$$X_{s,\,sl} = \omega\mu_0\lambda_g\left(1 + \frac{2\lambda_J}{a}\right), \qquad (1.41b)$$

with obvious differences in the weighting factors compared to (1.39). Only in the limit of weak coupling, the effective-medium approach and the stripline model yield identical results.

Residual Resistance Resulting from Normal-Conducting Inclusions

The above discussion covered elementary examples of equivalent circuits with complex conductivity in order to evaluate the resulting surface impedance. Yet, there are many more combinations and refinements reported in the literature. For instance, the analysis of the field dependent surface impedance of granular films revealed the necessity of considering an inductive shunt across the grain-boundary junctions [43, 44]. The surface impedance of bicrystal YBCO films was modeled in terms of the stripline model, but additionally taking into account the finite film thickness [45].

As a concluding example, the two-fluid model was also applied to calculate the enhanced residual surface resistance originating from normal-conducting inclusions [46, 47]. The size of and the spacing between the metallic spheres were assumed small compared to the penetration depth. The relative volume of the superconducting matrix occupied by the inclusions was η, their conductivity σ_{incl}, and the corresponding skin depth δ_{incl}. Assuming a complete condensation of the quasi-particles of the matrix at zero temperature yielded the residual surface resistance

$$R_{\text{res}} = \frac{9}{4}\eta \frac{\omega^2 \mu_0^2 \lambda^3 \sigma_{\text{incl}}}{2} \frac{1}{1+(\lambda/\delta_{\text{incl}})^2} \ . \tag{1.42}$$

The prefactor reflects the geometrical arrangement of the normal inclusions, the middle factor gives the TFM expression of R_s of a material with quasi-particle conductivity σ_{incl}, (1.18d), and the last factor is close to unity since $\lambda \ll \delta_{\text{incl}}$. It is worth noting that a finite conductivity of non-pairing charge carriers $\sigma_{1,\text{res}} = \omega \tau x_{n,\text{res}}/(1 - x_{n,\text{res}})$ at low temperatures is phenomenologically indistinguishable from the result (1.42), when equating $\sigma_{1,\text{res}} = 9\eta \sigma_{\text{incl}}/4$.

1.2.3 Scaling Laws

The preceding sections illustrated the applicability of the complex conductivity to various problems. All examples had in common the use of an impedance ratio y, resulting from parallel circuits of resistivities and inductivities. This concept can easily be extended to account also for permittivities (e.g. [5]). In summary, (1.18) provides a comprehensive phenomenological description of the surface impedance. It appears, therefore, adequate to draw some conclusions about the scaling of Z_s with y. If the meaning of y is clear for a certain problem, the variations of Z_s with the relevant physical input parameters can then be estimated. This procedure is especially helpful if the variation of y is mainly given by that of σ_2, with σ_1 approximately constant like in the case of Josephson junctions. More precise conclusions can be deduced for homogeneous superconductors in the weak-scattering limit ($\omega\tau \ll 1$) if the normal carrier fraction x_n, instead of y, is considered. In the most general case, σ_1 and σ_2 should be deduced from the surface impedance, rather than trying to induce Z_s from the conductivity:

$$\sigma = \omega\mu_0 \left(\frac{2R_\mathrm{s}X_\mathrm{s}}{(X_\mathrm{s}^2 + R_\mathrm{s}^2)^2} - \mathrm{i}\frac{X_\mathrm{s}^2 - R_\mathrm{s}^2}{(X_\mathrm{s}^2 + R_\mathrm{s}^2)^2} \right). \tag{1.43}$$

Scaling of the Surface Impedance with the Conductivity Ratio y

Figure 1.5 displays the functional dependence of R_s and X_s on y, as given by the normalized functions φ_- and φ_+ in (1.18). The commonly applied assumption $R_\mathrm{s} \ll X_\mathrm{s}$ is obviously fulfilled for $y \leq 1$. For larger conductivity ratios, R_s and X_s approach unity, as expected for normal conductors. A specific feature in Fig. 1.5 is the occurrence of a slight maximum of X_s at $y = \sqrt{3}$, which results from a redistribution of the microwave current through the resistor and the inductor. A consequence of this feature is a negative slope $\partial\varphi_-/\partial y < 0$, which was indeed observed in various experiments on granular HTS films [43,45] (Sect. 3.4).

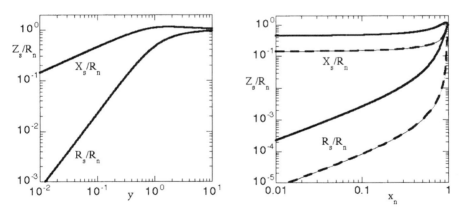

Fig. 1.5. Dependence of the normalized surface impedance on the conductivity ratio y (*left*), and on the normal fraction x_n (*right*), corresponding to (1.18) and (1.10). The solid and dashed lines in the right-hand diagram represent the cases $\omega\tau = 10^{-1}$ and 10^{-2}, respectively.

The variation of $\varphi_\pm(y) \propto y^a$ with y can be deduced from (1.18) on the basis of its logarithmic derivative, $\partial(\log\varphi_\pm)/\partial(\log y)$. The two-fluid formalism yields in the regime $y \ll 1$ an exponent $a = 3/2$ for $\varphi_-(y)$ and $a = 1/2$ for $\varphi_+(y)$. This result corresponds to the fact that R_s scales like λ^3 while X_s is linear in λ.

Scaling of the Surface Impedance with the Normal Fraction x_n

While the scaling of Z_s with y has proven valuable to describe the microwave response of (small) HTS grain boundary junctions, homogeneous superconductors should be analyzed, in the weak-scattering limit $\omega\tau \ll 1$, in terms of the normal fraction x_n, which controls at the same time both σ_1 and σ_2. With (1.10), equation (1.18) can be completely expressed in terms of x_n:

$$Z_s = R_n[\tilde{\varphi}_-(x_n) + i\tilde{\varphi}_+(x_n)] \tag{1.44a}$$

with the modified scaling functions

$$\tilde{\varphi}_\pm^2(x_n) = \omega\tau \frac{\sqrt{(1-x_n)^2 + (\omega\tau x_n)^2} \pm (1-x_n)}{(1-x_n)^2 + (\omega\tau x_n)^2}, \tag{1.44b}$$

as displayed on the right-hand side of Fig. 1.5. $R_n = (\omega\mu_0/2\sigma_n)^{1/2}$ is the surface resistance in the normal state ($x_n = 1$). The basic features of $\varphi_\pm(y)$ (left-hand side of Fig. 1.5) are reproduced, including the maximum in X_s/R_n which shifts closer to $x_n = 1$ as $\omega\tau$ decreases from 10^{-1} to 10^{-2}. The scaling exponents of $Z_s(x_n)$ can be found from the logarithmic derivative of (1.44b), yielding $1 + 3x_n/2$ for R_s and $x_n/2$ for X_s. This result corresponds at low temperatures, where $x_n \ll 1$, to the well-known relations $R_s \propto x_n$ and $X_s \approx$ constant.

Frequency Dependence of the TFM Surface Impedance

Another important feature of the TFM surface impedance is its frequency dependence. This can be obtained again from (1.18) (which is mathematically more handsome than (1.44)) since σ_1 is frequency independent for $\omega\tau \ll 1$. Assuming the power-law behavior $R_s \propto f^b$, and taking into account the explicit frequency dependence in (1.18a), the exponent b of the surface impedance is found to be

$$b_\pm = \frac{1}{2} + \frac{\sqrt{1+y^2} \pm (1-y^2)}{2(1+y^2)\left(\sqrt{1+y^2} \pm 1\right)}. \tag{1.45}$$

Here, the upper and lower signs refer to the frequency exponent of the surface reactance and resistance, respectively. The functional variation $b_\pm(y)$, which is displayed in Fig. 1.6, shows that the TFM surface resistance should not generally be scaled quadratic in frequency, even in the superconducting regime $y \leq 1$. Deviations from $b_- = 2$ result from a finite value of $y \propto \omega/\rho$, and thus depend on the absolute values of the measurement frequency and the normal-state resistivity $\rho = 1/\sigma_n$. At $y \gg 1$, (1.45) yields $b_- = 1/2$ in accordance with the normal skin effect. In the opposite limit $y \ll 1$, b_- approaches 2 like $b_- = 2 - 5y^2/4$ (dashed line in Fig. 1.6).

Regarding the applications of superconductors to dispersive microwave devices, the frequency dependence of the surface reactance also has to be considered, especially if scattering effects cannot be excluded or if operating temperatures close to T_c are envisaged. Equation (1.45) yields at small y: $b_+ = 1$ or $X_s \propto \omega$, which corresponds to a constant effective penetration depth $\lambda_{\text{eff}} = X_s/\omega\mu_0$. The frequency exponent decreases to $X_s \propto \omega^{1/2}$ at large y ($\lambda_{\text{eff}} \propto \omega^{-1/2}$) with an initial decrease of $b_+ = 1 - 3y^2/4$. The extremal behavior of b_+ reflects that of $X_s(y)$ at $y \geq \sqrt{3}$ (Fig. 1.5).

For homogeneous superconductors, the temperature dependence of b enters (1.45) through that of $x_n(T)$ and $\tau(T)$. Since scattering effects depend sensitively on the crystalline quality of the superconductor films, no general prediction can be deduced for $b(t)$. Experimental results on epitaxial HTS films at 87 GHz indicate only shallow changes of $|b_-(0.8\,T_c) - b_-(0)| \leq 0.3$ [23]. Finally, a deviation from $b_+ = 1$ has to be expected at temperatures $t \leq 0.5$ in a frequency range between 10 and 100 GHz, or at frequencies above 300 GHz, as indicated by data on $\tau(t)$ and $x_n(\omega, t)$ of YBCO samples (Sect. 1.1.1).

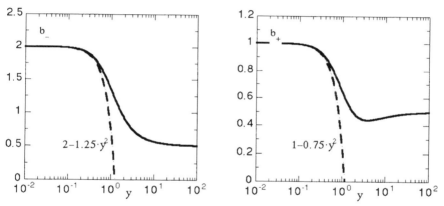

Fig. 1.6. Dependence of the frequency exponents $b_\pm$ of the two-fluid surface impedance on the conductivity ratio y, corresponding to (1.45). The dashed lines indicate the asymptotic behavior at small y values.

1.3 Microscopic Picture

The preceding sections dealt with the calculation of the surface impedance Z_s of superconductors from their complex conductivity σ on a purely phenomenological level. In order to relate both σ and Z_s to the nature of the charge carriers, to the interaction between them and to the density of energy states (DOS) they can occupy, microscopic theories of superconductivity are needed. Among them, the theory of Bardeen, Cooper and Schrieffer (BCS) [48] plays the major role in comprehensive understanding and has an extremely wide range of validity [17]. Therefore, the basic features of BCS are sketched in the first section. These are followed in Sect. 1.3.2 by the microscopic formulation of the current–field relation (Sect. 1.1.1). Based on this expression, analytical results for σ and Z_s are derived in the low-frequency limit $\hbar\omega \ll \Delta$, where the photonic energy is much less than the energy gap. The last section covers different cases including strong pair coupling, magnetic impurities, and an order parameter with d-wave symmetry, in which the DOS deviates from the BCS shape with its characteristic singularity at $E = \Delta$. Since absolute level and temperature dependence of Z_s are determined by the DOS, these cases are of special importance in the measurable microwave response of high-temperature superconducting films [49] (Sect. 2.3).

1.3.1 BCS Theory: Cooper Pairs, Energy Gap and Density of States

The BCS Assumptions

A detailed treatment of the BCS theory can be found in various prominent textbooks such as [11, 50]. In order to stay comprehensive, only a few milestones of the theory are repeated here. We start by recalling the basic assumptions underlying the original approach of BCS. They assumed a quantum-mechanical system which contains independent quasi-particles with infinite lifetime. Quasi-particles are elementary excitations from the coherent ground state, and electrons and holes appear equivalent in this respect. Coulomb repulsion between the charge carriers was not taken into consideration, as was inelastic scattering which limits the lifetime of the quasi-particles. Furthermore, no collective excitations were considered, and the superconducting metal was assumed isotropic, i.e., to have a sperically shaped Fermi surface. It is noted that all these additional aspects have since then been considered in subsequent papers, and some of them will be mentioned in Sect. 1.3.3.

The BCS Hamiltonian

In the BCS theory the Hamilton operator H_p sums the contribution H_0 of the free quasi-particle states at momentum $\hbar\boldsymbol{k}$ and spin σ (the spin variable σ in this section should not be confused with the electric conductivity), and the interaction H_int between pairs (the Cooper pairs) of momenta $\boldsymbol{k}$ and $\boldsymbol{l}$ and opposite spin directions:

$$H_\mathrm{p} = H_0 + H_\mathrm{int} = \sum_{\boldsymbol{k},\sigma} \varepsilon_{\boldsymbol{k}} c^*_{\boldsymbol{k},\sigma} c_{\boldsymbol{k},\sigma} + \sum_{\boldsymbol{k},\boldsymbol{l}} V_{\boldsymbol{k},\boldsymbol{l}} c^*_{\boldsymbol{k},\uparrow} c^*_{-\boldsymbol{k},\downarrow} c_{-\boldsymbol{l},\downarrow} c_{\boldsymbol{l},\uparrow} \ . \qquad (1.46)$$

The operators c^* and c create and annihilate the states $\{k,\sigma\}$ defined by the corresponding indices. In evaluating this Hamiltonian, the large total number of charge carriers forming the ground state, and the interaction between single pairs leads to the notation of an average expectation value

$$b_{\bm{k}}^0 = \langle c_{-\bm{k},\downarrow} c_{\bm{k},\uparrow} \rangle_{\text{av}} \,. \tag{1.47}$$

The interaction term in (1.46) can then be rewritten in terms of the order parameter $\Delta_{\bm{k}}$

$$\Delta_{\bm{k}} = -\sum_l V_{\bm{k},l} b_l^0 \,. \tag{1.48}$$

The pairing Hamiltonian can be diagonalized on the basis of a linear canonical transformation of the creation and annihilation operators. This transformation is named after Bogoliubov [51] and Valatin [52] who independently introduced this formalism:

$$c_{\bm{k},\uparrow} = u_{\bm{k}} \gamma_{\bm{k}0} + v_{\bm{k}}^* \gamma_{\bm{k}1}^* \tag{1.49a}$$

and

$$c_{-\bm{k}\downarrow}^* = -v_{\bm{k}} \gamma_{\bm{k}0} + u_{\bm{k}}^* \gamma_{\bm{k}1}^* \,. \tag{1.49b}$$

To obtain the diagonal form of H_{p}, the parameters $u_{\bm{k}}$, $v_{\bm{k}}$, and $\Delta_{\bm{k}}$ are found to be related by

$$|u_{\bm{k}}|^2 = \frac{1}{2}\left(1 + \frac{\varepsilon_{\bm{k}}}{E_{\bm{k}}}\right), \tag{1.50a}$$

and

$$|v_{\bm{k}}|^2 = \frac{1}{2}\left(1 - \frac{\varepsilon_{\bm{k}}}{E_{\bm{k}}}\right), \tag{1.50b}$$

where

$$E_{\bm{k}} = \sqrt{\varepsilon_{\bm{k}}^2 + |\Delta_{\bm{k}}|^2} \,. \tag{1.50c}$$

The quantities $|u_{\bm{k}}|^2$ and $|v_{\bm{k}}|^2$ denote the probability that the pair $(\bm{k},\uparrow)$, $(-\bm{k},\downarrow)$ is unoccupied and occupied, respectively, and $\varepsilon_{\bm{k}}$ measures the energy of the quasi-particle state at wave vector $\bm{k}$ relative to the Fermi energy E_{F} (i.e., $\varepsilon_{\bm{k}} = 0$ at E_{F}). Equation (1.50c) describes the energy of quasi-particle excitations, showing that the pair state is energetically separated from excitations by an energy gap $\Delta_{\bm{k}}$. In further evaluating the gap equation (1.48), and considering the quantum-statistical nature of the quasi-particles in terms of the Fermi–Dirac distribution function

$$f(E) = \frac{1}{1 + \exp(E/k_{\text{B}}T)}, \tag{1.51}$$

one arrives at the following expression which presents the eigenvalue equation for T_{c} (since $\Delta_{\bm{k}}(T = T_{\text{c}}) = 0$):

$$\Delta_{\bm{k}} = \sum_l V_{\bm{k},l} \Delta_l \frac{1 - 2f(E_{\bm{k}})}{2 E_{\bm{k}}} \,. \tag{1.52a}$$

The BCS Energy Gap and the Density of States

The nonlinear equation (1.52a) can be simplified by following Cooper's suggestion about the pairing potential V. He assumed V to be constant for quasi-particle energies below the cutoff energy $\hbar\Omega_D$, given by the Debye frequency Ω_D of the superconductor, and to be zero otherwise. Equation (1.52a) yields, under this assumption, an energy-independent gap value, and transforms to the integral equation

$$1 = N(0)V \int_0^{\hbar\Omega_D} d\varepsilon \frac{1 - 2f(\varepsilon)}{\varepsilon} , \tag{1.52b}$$

where $N(0)$ is the density of quasi-particle states at the Fermi level. Equation (1.52b) is strictly valid only for $N(0)V \ll 1$, i.e., weak pair coupling. Its solution leads to the well-known expression for the critical temperature T_c, and for the BCS reduced energy gap $\Delta(0)/kT_c$:

$$k_B T_c = \frac{2\gamma}{\pi} \hbar\Omega_D \exp{-\frac{1}{N(0)V}} \tag{1.53a}$$

and

$$\frac{\Delta(0)}{k_B T_c} = \frac{\pi}{\gamma} \approx 1.764 , \tag{1.53b}$$

where $\gamma \approx 1.7811$ is Euler's constant:

$$\ln\left(\frac{2\gamma}{\pi}\right) = -\frac{1}{2} \int_0^\infty dx \frac{\ln x}{4\cosh^2(x/2)} \approx 0.1256 . \tag{1.53c}$$

The superconducting state is characterized by two energies: the gap energy and the Fermi level, which is inversely proportional to $N(0)$. It shall be noted that the ratio Δ/E_F is of order 10^{-4} for the metallic superconductors while it is much larger, namely of order of 10^{-1}, for the oxide superconductors. Specific consequences of this drastic difference are discussed, e.g., in [53].

The temperature dependence of Δ has to be computed, in general, from the gap equation. However, the following expression proved to yield an excellent approximation [54]:

$$\frac{\Delta(T)}{\Delta(0)} = \left[\cos\frac{\pi}{2}\left(\frac{T}{T_c}\right)^2\right]^{1/2} . \tag{1.54}$$

The exponent under the root exceeds 2 for strongly coupled superconductors (see below).

The occurrence of an energy gap in the spectrum of the quasi-particle energies causes the density of states $N_s(E)$ in a BCS superconductor to vanish at $E < \Delta$, and to rise above the gap like

$$N_{\rm s}(E) = N(0) \frac{E}{\sqrt{E^2 - \Delta^2}} \ . \tag{1.55}$$

This expression reveals a singularity at $E = \Delta$ which governs the temperature dependence of various physical properties like the electronic specific heat or, as will be discussed in the next section, the complex conductivity and the surface impedance.

1.3.2 BCS Theory: Complex Conductivity and Surface Impedance

General Approach

On the basis of the BCS theory, it is now straightforward to obtain expressions for the interaction of a superconductor with a transverse electromagnetic field that is represented by the vector potential $\boldsymbol{A}$ (Sect. 1.1.1). As usual, the kinetic particle momentum $\boldsymbol{p}$ is supplemented by $\boldsymbol{p} = m\boldsymbol{v} + e\boldsymbol{A}$. It is helpful to evaluate $\boldsymbol{A}(\boldsymbol{r})$ in terms of its spatial Fourier components $\boldsymbol{a}(\boldsymbol{q})$ with $\boldsymbol{q}$ being the wave vector. Since we are interested in weak fields, i.e., in linear response, it is sufficient to expand the operator for the kinetic energy, $(\boldsymbol{p} - e\boldsymbol{A})^2$, to first order in field. This yields the interaction Hamiltonian (index "em" for electromagnetic):

$$H_{\rm em} = \frac{e\hbar}{m} \sum_{\boldsymbol{k},\boldsymbol{q}} \boldsymbol{k} \cdot \boldsymbol{a}(\boldsymbol{q}) c^*_{\boldsymbol{k}+\boldsymbol{q},\sigma} c_{\boldsymbol{k},\sigma} \tag{1.56}$$

and the corresponding operator for the current density

$$\boldsymbol{J}_{\rm op}(\boldsymbol{q}) = -\frac{ne^2}{m} \boldsymbol{a}(\boldsymbol{q}) - \frac{e\hbar}{m} \sum_{\boldsymbol{k}} \boldsymbol{k} c^*_{\boldsymbol{k}-\boldsymbol{q}} c_{\boldsymbol{k}} \ . \tag{1.57}$$

Equation (1.57) can be evaluated in terms of $\boldsymbol{a}(\boldsymbol{q})$ to yield a generalized current–field relation

$$\boldsymbol{J}(\boldsymbol{q},\omega,T) = \frac{-1}{\mu_0} K(\boldsymbol{q},\omega,T) \cdot \boldsymbol{a}(\boldsymbol{q},\omega) \ . \tag{1.58}$$

Since the electric field $\boldsymbol{E}$ is given by $\boldsymbol{E} = -\mathrm{i}\omega \boldsymbol{A}$, both the conductivity σ and the surface impedance $Z_{\rm s}$ can be expressed in terms of the Kernel K:

$$\sigma = \sigma_1 - \mathrm{i}\sigma_2 = \frac{-\mathrm{i}}{\omega\mu_0} K(\boldsymbol{q},\omega,T) \tag{1.59a}$$

and

$$Z_{\rm s} = R_{\rm s} + \mathrm{i}X_{\rm s} = \mathrm{i}\omega\mu_0 \left[K(\boldsymbol{q},\omega,T)\right]^{-1/2} \ . \tag{1.59b}$$

The procedure indicated by (1.58) and (1.59) covers the calculation of σ from the local to the extreme anomalous limit (the exponent $-1/2$ in (1.59b) must be modified for the latter, see Sect. 1.1.1). It is furthermore not only applicable to the evaluation of the BCS particle operators in (1.57), but can be extended to strong-coupling superconductors or such compounds which display a d-wave symmetry of the order parameter (see Sect. 1.3.3).

Temperature Dependent Penetration Depth

The London penetration depth can be evaluated from the response kernel $K(q \to 0, \omega = 0, T)$ for fields varying slowly in space (i.e., in the limit $q \to 0$) [11]. This can be verified, e.g., from (1.57) and (1.58) for $T = 0\,\text{K}$: The first term in (1.57) reduces then to the London equation (1.9) with the notation $K_\text{L}(0) = 1/\lambda_\text{L}^2$, according to our earlier results $\sigma_2 = 1/\omega\mu_0\lambda_\text{L}^2$ and $X_\text{s} = \omega\mu_0\lambda_\text{L}$. The temperature dependence of the penetration depth (and equivalently that of σ_2) is therefore determined by $K(T) = 1/\lambda_\text{L}^2(T)$. Following the analysis given in [11], we obtain the local-limit result

$$\frac{\lambda_\text{L}^2(0)}{\lambda_\text{L}^2(T)} = 1 - 2\int_\Delta^\infty dE \frac{N_\text{s}(E)}{N(0)}\left(-\frac{\partial f(E)}{\partial E}\right). \tag{1.60a}$$

At temperatures $T < T_\text{c}/2$, where the gap $\Delta(T) = \Delta$ can be considered constant, the integral equation (1.60a) can be simplified, giving for the reduced change in penetration depth [55, 56]

$$\frac{\lambda(T) - \lambda(0)}{\lambda(0)} = \sqrt{\frac{\pi\Delta}{2k_\text{B}T}}\exp\left(-\frac{\Delta}{k_\text{B}T}\right). \tag{1.60b}$$

The exponential temperature dependence uniquely reflects the singularity in the BCS density of states, which is convoluted in the integral (1.60a) with the bell-shaped weight function $\partial f/\partial E$.

Complex Conductivity

Mattis and Bardeen, and independently Abrikosov, Gorkov and Khalatnikov, derived analytical expressions for the complex conductivity within the framework of the BCS theory [16, 57] on the basis of (1.57)–(1.59) in the low-frequency limit, $\hbar\omega \ll \Delta$:

$$\frac{\sigma_1}{\sigma_\text{n}} = \frac{2}{\hbar\omega}\int_\Delta^\infty dE \frac{[E(E+\hbar\omega)+\Delta^2]|f(E)-f(E+\hbar\omega)|}{\sqrt{E^2-\Delta^2}\sqrt{(E+\hbar\omega)^2-\Delta^2}} \tag{1.61a}$$

and

$$\frac{\sigma_2}{\sigma_\text{n}} = \frac{1}{\hbar\omega}\cdot\int_{\Delta-\hbar\omega}^\Delta dE\frac{[E(E+\hbar\omega)][1-2f(E+\hbar\omega)]}{\sqrt{\Delta^2-E^2}\sqrt{(E+\hbar\omega)^2-\Delta^2}}. \tag{1.61b}$$

Equation (1.61) is strictly valid only in the dirty ($\ell/\xi_0 \ll 1$) or the extreme anomalous ($\lambda_\text{L}/\xi_0 \ll 1$) limits [11], but the expressions seem to contain qualitatively the main features characteristic of all BCS superconductors.

The term $E(E+\hbar\omega) + \Delta^2 = E(E+\hbar\omega)F(E,\omega,\Delta)$ in the numerator of (1.61a) deserves special attention. The factor F is known as the type-II coherence factor, which describes the interaction of quasi-particles with

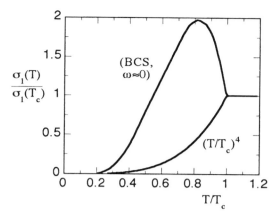

Fig. 1.7. Temperature dependent quasi-particle conductivity $\sigma_1(T)/\sigma_1(T_c)$ evaluated from BCS theory, in comparison with the expectation from the two-fluid model. The form of the coherence peak corresponds to a broadening of the BCS DOS of about 0.02Δ [11].

an electromagnetic field [11]. It enhances the effect of the DOS singularity, since it contributes most to the integration for energies E, $E + \hbar\omega$ close to the gap value. As a result, $\sigma_1(T)$ exceeds the normal-state value $\sigma_1(T_c)$ at temperatures close below T_c. This behavior, which is usually referred to as the coherence peak (or Hebel–Slichter peak after the experimentalists who first verified this unique BCS feature experimentally [58]), is demonstrated in Fig. 1.7 in comparison with the classical prediction of the two-fluid model $\sigma_1(T)/\sigma_1(T_c) = (T/T_c)^4$. The enhancement factor $\max\{\sigma_1(T < T_c)\}/\sigma_1(T_c)$, which would grow to infinity in the case of an ideal DOS singularity and at low frequency, is usually limited to finite values and broadened in width due to impurity scattering (i.e., in dirty superconductors) or finite lifetimes of the quasi-particles (e.g., in strong-coupling superconductors, Sect. 1.3.3) [11]. Physically, the conductivity peak results from the phase-coherent superposition of the occupied quasi-particle states due to the pair interaction: The scattering, creation or annihilation of quasi-particles reflects the coherence in the scattering, creation or annihilation of the pair constituents. Such a coherence is not contained in the two-fluid model, resulting in the obvious differences displayed in Fig. 1.7. As the temperature decreases further below T_c, more and more quasi-particles join the superconductive condensate. The conductivity $\sigma_1(T)$ vanishes, as discussed in [55,56] for $T < T_c/2$ on the basis of an expression similar to (1.60a), like:

$$\sigma_1 \propto \frac{1}{k_B T} \ln\left(\frac{4k_B T}{\hbar\omega}\right) \exp\left(-\frac{\Delta}{k_B T}\right), \tag{1.62}$$

where the exponential decrease of $\sigma_1(T)$ is again reminiscent of the DOS singularity at $E = \Delta$. In comparing (1.62) with the two-fluid formulation

(1.10), the exponential term can be regarded as a Boltzman factor, which describes the vanishing number x_n of unpaired charge carriers as the thermal energy drops below the gap energy. However, the phase coherence of the quasi-particles violates the two-fluid assumption $x_\mathrm{n} + x_\mathrm{s} = 1$ at temperatures close to T_c.

The low-temperature behavior of σ_2 can be concluded from (1.60) for the local limit. Expressions for the dirty limit are found in [11].

Surface Impedance

Expressions like (1.61) for the complex conductivity of BCS superconductors cannot, in general, be derived analytically but require numerical computation. Computer codes were developed by Halbritter [59] and Turneaure [60], and analyzed in terms of the dependences of the surface impedance on the characteristic lengths ξ_0, λ, and ℓ.

Since the above equations were evaluated for low temperatures $T < T_\mathrm{c}/2 \ll \Delta/k_\mathrm{B}$ and at low frequencies $\omega \ll \Delta/\hbar$, it is common to argue that the surface impedance can be approximated by $R_\mathrm{s} \propto \omega^2 \lambda_\mathrm{L}^3 \sigma_1$ and $X_\mathrm{s} \propto \omega \lambda_\mathrm{L}$, since $\sigma_1 \ll \sigma_2$ in these limits (Sect. 1.2.1 and (1.18)). Consequently, R_s should exponentially vanish at low temperatures while X_s approaches a constant. Vice versa, measurements of the temperature dependent surface resistance provide essential information about the existence and the magnitude of the reduced energy gap $\delta \equiv \Delta/k_\mathrm{B} T_\mathrm{c}$. As the temperature approaches T_c, $R_\mathrm{s}(T)$ becomes steeper than exponential due to the rapidly decreasing conductivity $\sigma_2(T)$.

The TFM frequency exponent b (in $R_\mathrm{s} \propto f^b$; see (1.45)) can be compared with the microscopic theory in a similar argumentation. The "two-fluid" factor ω^2 yields $b = 2$, which is reduced by the logarithmic frequency term in (1.62). The reduction below the value 2 depends on temperature, with the scaling $2k_\mathrm{B}/h = 41.4\,\mathrm{GHz/K}$, and on frequency. These features agree qualitatively with the results of Sect. 1.2.3, although the deduction of absolute b values requires some care. Furthermore, the scaling behavior of the BCS surface impedance depends on the characteristic lengths via the nonlocality parameter λ_L/ξ_0, and via the scattering of the quasi-particles described by ℓ/ξ_0 [55, 56]. Figure 1.8 summarizes representative results on $R_\mathrm{s}(f)$ for different LTS and HTS materials [56].

A semi-analytical approach to obtain the BCS surface impedance was presented by Hinken [61] and later by Linden et al. [15]. They put the complex conductivity in a form which accounted for the BCS density of states. In this way, it was possible to parametrize the surface impedance in the local limit on the basis of (1.17) or (1.59b). Further details were analyzed, e.g., in [4].

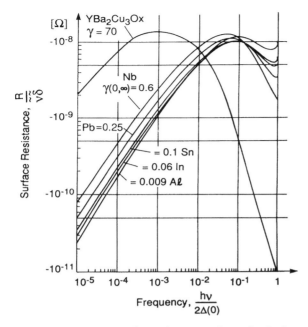

Fig. 1.8. Frequency dependence, evaluated relative to the gap frequency, of the BCS surface resistance, normalized to the frequency ($\nu = 1\tilde{\nu}$ GHz) and to the clean-limit penetration depth $\lambda_\mathrm{L}(0) = 10\tilde{\delta}$ nm for different materials with $\gamma = 2\lambda_\mathrm{L}(0)/\pi\xi_0$ [56].

1.3.3 Non-BCS Densities of States and Consequences for the Surface Impedance

The preceding paragraph illustrated the consequences of the BCS singularity of the density of states for the temperature and frequency dependent surface impedance. The main features are the coherence peak below T_c, the exponential temperature variation $\exp(-\Delta/k_\mathrm{B}T_\mathrm{c})$ at $T \ll T_\mathrm{c}$, and the logarithmic frequency contribution $\ln(4k_\mathrm{B}T/\hbar\omega)$ (see (1.62)). In turn, the absence of the DOS singularity at the gap energy leads to a "smearing" of the coherence peak and to a modified scaling of the surface impedance with temperature and frequency. As will be shown below, the singularity is suppressed in real superconductors. The physical reasons for such discrepancies are mainly

1. an anisotropic Fermi surface,
2. finite quasi-particle lifetimes (e.g., due to inelastic scattering of quasi-particles at thermally excited phonons in strongly coupled superconductors), or
3. mechanisms breaking the time-reversal symmetry which is characteristic of a Cooper pair (e.g., due to the retarding electron–phonon interaction in

strongly coupled superconductors or due to spin-flip scattering at magnetic impurities).

The following paragraphs summarize the basic features of these mechanisms and their consequences for the surface impedance.

Superconductors Containing Magnetic Impurities

The effect of paramagnetic impurities on the properties of superconductors was originally investigated by Abrikosov and Gor'kov [62]. They found a strong decrease of the critical temperature with increasing impurity concentration. Furthermore, in a certain range of concentrations, the energy gap Δ_G dropped to zero, while significant pair correlation (nonzero number density of Cooper pairs or nonzero order parameter Δ_OP) was maintained. This region, which is called gapless superconductivity, is supplementary to the BCS theory, where both parameters merge into a single one, $\Delta_\mathrm{G} = \Delta_\mathrm{OP}$, which is described by (1.52).

The electromagnetic response of BCS-like superconductors containing paramagnetic impurities was analyzed in detail in [63], with some results reproduced in the following. The interaction Hamiltonian evoked by the impurities is given by

$$H_\mathrm{imp}(\boldsymbol{r}) = \sum_i [V_1(\boldsymbol{r} - \boldsymbol{R}_i) + V_2(\boldsymbol{r} - \boldsymbol{R}_i)\boldsymbol{S}_i\boldsymbol{\sigma}] \,, \tag{1.63}$$

where $\boldsymbol{R}_i$ denotes the position of the impurity i with the free spin $\boldsymbol{S}_i$, and $\boldsymbol{\sigma}$ is the spin operator of the quasi-particles. The positions of the impurities are assumed to be random, and the impurity spins uncorrelated. The first term in (1.63) represents collisions (at a scattering rate Γ_1) which do not involve the quasi-particle spin. This would be the only perturbation present if the impurities were not magnetic. Furthermore, it affects neither T_c, nor the energy gap, nor the DOS. The second term describes the spin interaction specific to magnetic scatterers, and permits the reversal of the quasi-particle spin in a collision (with the spin-flip scattering rate Γ_2). Based on (1.63), and similar to the procedure sketched in Sect. 1.3.1, the critical temperature is determined from

$$1 = N(0)V \int\limits_{-\hbar\Omega'_\mathrm{D}}^{\hbar\Omega'_\mathrm{D}} \mathrm{d}\varepsilon \frac{\varepsilon}{\varepsilon^2 + \hbar^2\Gamma^2} \tanh\left(\frac{\varepsilon}{2k_\mathrm{B}T}\right) \,, \tag{1.64}$$

where V denotes the BCS pairing potential. Equation (1.64) reduces to (1.52b) for $\Gamma = 0$, where $\Gamma \equiv \Gamma_1 - \Gamma_2$. Spin-flip events obviously lead to the substitution of the singularity $1/\varepsilon$ by the broadened function $\varepsilon/(\varepsilon^2 + \hbar^2\Gamma^2)$. As noted in [62], the limited lifetime of Cooper pairs is algebraically equivalent to a complex-valued order parameter (with its imaginary part being proportional to Γ) and complex-valued quasi-particle energies.

From the comparison of (1.52b) and (1.64) it follows that the transition temperature $T_c(\Gamma)$ in the presence of paramagnetic impurities is reduced below the BCS value $T_c(0)$. It vanishes at the critical scattering rate $\Gamma_{cr} = \Delta_{OP}(0)/2\hbar$, where $\Delta_{OP}(0)$ corresponds to the BCS energy gap introduced in Sect. 1.3.1. The left-hand part of Fig. 1.9 displays $T_c(\Gamma)$ [63], which can be approximated by $T_c(\Gamma) - T_c(0) \approx -\pi\hbar\Gamma_2/4k_B$ at small scattering rates [64]. The diagram also shows the results for the order parameter $\Delta_{OP}(\Gamma, T=0)$ [denoted by $\Delta(0)$] and of the energy gap $\Delta_G(\Gamma, T=0)$ [denoted by $\Omega_G(0)$, each being normalized to the BCS value ($\Gamma_1 = \Gamma_2 = 0$) denoted by $\Delta^P(0)$]. The order parameter vanishes at the critical value Γ_{cr}, while the energy gap drops to zero already at the lower value $2\exp(-\pi/4)\Gamma_{cr} \approx 0.912\Gamma_{cr}$. The superconductor is gapless within the interval $0.912 \leq \Gamma/\Gamma_{cr} \leq 1$. Both Δ parameters are related by the equation

$$\Delta_G(\Gamma, T) = \Delta_{OP}(\Gamma, T)\left[1 - \left(\frac{\hbar\Gamma}{\Delta_{OP}(\Gamma, T)}\right)^{2/3}\right]^{3/2}. \quad (1.65)$$

The resulting density of states is displayed on the right-hand side of Fig. 1.9 for different values of the reduced scattering rate $\hbar\Gamma/\Delta_{OP}(\Gamma, T)$. The smearing of the BCS singularity is clearly revealed, as is the vanishing of the energy gap at the critical scattering rate. Furthermore, the reduced order parameter $\Delta_{OP}(\Gamma, T=0)/k_B T_c(\Gamma)$ was found to be no longer constant but to increase with Γ, approaching a maximum of 7.837, which is $2^{1/2}\pi$ times the BCS value (see (1.53b)).

The surface impedance of superconductors containing paramagnetic impurities can be calculated from the response kernel $K(\mathbf{q}, \omega)$ (Sect. 1.3.2). For instance, the penetration depth is given by $K(0,0)^{-1/2}$, where

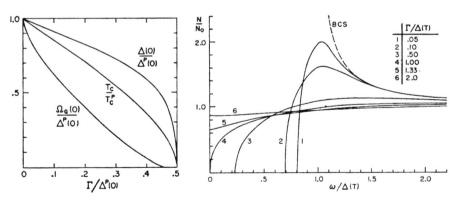

Fig. 1.9. *Left*: Critical temperature, order parameter $\Delta(0)$ and energy gap $\Omega_G(0)$ at $T = 0$ K, plotted as a function of the scattering rate Γ. The superscript "P" refers to the value when $\Gamma = 0$ [63]. *Right*: DOS as a function of the reduced energy $\omega/\Delta(\Gamma, T)$ for several reduced scattering rates $\Gamma/\Delta(\Gamma, T)$ (with $\hbar = 1$, after [63]). (© 1999 by the American Physical Society.)

$$K(0,0) = \frac{1}{\lambda_L^2} \Delta^2 \int_0^\infty d\varepsilon \frac{1}{(\varepsilon^2 + \Delta^2)[(\varepsilon^2 + \Delta^2)^{1/2} + \hbar\Gamma_2]}$$
$$\times \left(1 - \frac{\hbar\Gamma\Delta^2}{(\varepsilon^2 + \Delta^2)^{3/2}}\right), \tag{1.66}$$

and the index "OP" at the order parameter Δ has been omitted. Equation (1.66) yields an increase of the penetration depth with Γ, the slope of which is determined by the ratio Γ_2/Γ. As Γ approaches the critical value, λ/λ_L diverges.

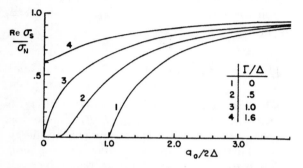

Fig. 1.10. Real part of the complex conductivity for $q = 0$ and $T = 0\,\text{K}$, plotted as a function of the reduced frequency $q_0/2\Delta$ for various values of $\Gamma/\Delta(\Gamma,0)$ [63]. (© 1999 by the American Physical Society.)

The real part of the complex conductivity is given by the imaginary part of the response kernel, evaluated at the frequency ω (see (1.59a)). The corresponding integral equation requires numerical computation, with some results shown in Fig. 1.10. The effect of magnetic pairbreaking on the microwave absorption is clearly seen. Curves 3 and 4 represent results for the gapless regime. Analytical expressions for σ_1/σ_n in the extreme gapless regime were deduced for the dirty limit in [64].

Anisotropic Superconductors

Early experimental [65], theoretical [66] and numerical work [67] showed that the BCS singularity in the DOS is also smeared if the energy gap is anisotropic, i.e., takes different values along different directions in k space. As one consequence, the frequency dependence of R_s comes closer to quadratic, and thus deviates from the isotropic BCS result [67, 68].

In relation to the layered high-temperature superconducting compounds, the temperature and frequency dependence of the surface impedance of a two-dimensional anisotropic superconductor was theoretically investigated in [69]. The angular dependence of the order parameter was modeled by

$$\Delta(\boldsymbol{k}) = \Delta_0 + \Delta_1 \cos(4\varphi) , \tag{1.67}$$

where $\Delta_1 < \Delta_0$ and φ is the azimuthal angle of the $\boldsymbol{k}$ vector in the CuO_2 planes of the HTS material. The response kernel K was investigated in the local limit and in linear response theory. The absolute level and the temperature dependence of the surface resistance are both affected by the gap anisotropy, as depicted in Fig. 1.11 for different mean free paths at a frequency (10 GHz) much below the assumed gap frequency. For the anisotropic order parameter, the exponential temperature variation of R_s is shifted to lower temperatures compared to the isotropic case. Moreover, $R_s(T)$ became shallower at intermediate temperatures, dependent on the degree of anisotropy. In contrast, the surface reactance was not sensitive to the ratio Δ_1/Δ_0. The numerical results were found to agree with experimental data.

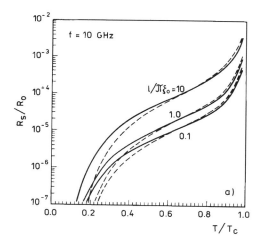

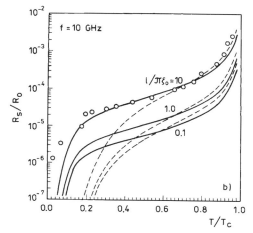

Fig. 1.11. Surface resistance R_s, normalized to $R_0 = 2\mu_0 \lambda_L k_B T_c/\hbar$, versus reduced temperature at $f = 10$ GHz for three values of the mean free path. The dashed curves refer to an isotropic energy gap, and the solid lines to a gap anisotropy of $\Delta_1/\Delta_0 = 0.4$ (*top*) and 0.8 (*bottom*). The open circles represent experimental data on YBCO films [70], from [69]. (© 1999 by the American Physical Society.)

Strong-Coupling Superconductors

The validity of the BCS theory is restricted to superconductors with weak pair interaction V such that $\lambda_{\text{coupl}} \equiv N(0)V \ll 1$ (Sect. 1.3.1). Consequently, its application is appropriate for superconductors with low values of $k_B T_c/\hbar\Omega_D$ and with reduced energy gaps 1.764. Neither the metallic A15 nor the oxide high-temperature superconductors belong to this class strictly, thus asking for an extended theoretical frame. Strong quasi-particle–phonon interaction is incompatible with the BCS assumptions for two reasons:

1. The scattering of the quasi-particles at thermally excited phonons, especially at energies corresponding to the Debye frequency, limits their lifetime so strongly that the particle concept loses its merit.
2. The phonon-mediated interaction causes retardation, thus violating the correlated occupancy of time-reversed quasi-particle states, and evoking a frequency (or energy) dependent order parameter [71].

As further consequences of the strong interaction with phonons, the effective mass of the quasi-particles is enhanced (or: renormalized) by the factor $(1 + \lambda_{\text{coupl}})$, and the original two-fluid condition $x_n + x_s = 1$ is no longer justified.

The theory of strong pair coupling in superconductors was originally treated by Eliashberg [72a]. It involves the solutions of a complicated, strongly interacting, many-body system, based on the evaluation of the self-energies of the two types of charge carriers in the superconductor. The key result can be presented in terms of two coupled (the so-called Eliashberg) equations for the energy renormalization function $Z(\omega)$ and for the order parameter $\Delta(\omega)$. In the dirty limit and at zero temperature the two expressions become (with $\hbar = 1$) (see also, e.g., [72b]):

$$[1 - Z(\omega)]\omega = \int_{\Delta_0}^{\infty} d\omega' \operatorname{Re}\left\{\omega'/[\omega'^2 - \Delta^2(\omega')]^{1/2}\right\}$$

$$\times \int_0^{\infty} d\Omega\, \alpha^2(\Omega) F(\Omega) G_-(\omega, \omega', \Omega) , \qquad (1.68\text{a})$$

$$\Delta(\omega) = \frac{1}{Z(\omega)} \int_{\Delta_0}^{\infty} d\omega' \operatorname{Re}\left\{\Delta(\omega')/[\omega'^2 - \Delta^2(\omega')]^{1/2}\right\}$$

$$\times \int_0^{\infty} d\Omega\, \alpha^2(\Omega) F(\Omega) G_+(\omega, \omega', \Omega)$$

$$- \frac{\mu^*}{Z(\omega)} \int_0^{\omega_c} d\omega' \operatorname{Re}\left\{\Delta(\omega')/[\omega'^2 - \Delta^2(\omega')]^{1/2}\right\} . \qquad (1.68\text{b})$$

Here, $\Delta_0 = \text{Re}\{\Delta(\hbar\omega = \Delta_0)\}$ is the edge value of the frequency dependent energy gap, and $\mu^* = N(0)U$ is a Coulomb pseudo-potential taking into account the screened Coulomb interaction between the charge carriers. The interaction between quasi-particles and phonons is represented by the product $\alpha^2(\Omega)F(\Omega)$ of the phonon density of states $F(\Omega)$ and an effective quasi-particle–phonon coupling $\alpha^2(\Omega)$. Finally, ω_c is a cutoff frequency, and the functions $G_\pm$ are given by

$$G_\pm(\omega, \omega', \Omega) = \frac{1}{\omega' + \omega + \Omega - i\vartheta} \pm \frac{1}{\omega' - \omega + \Omega - i\vartheta}, \quad (1.68c)$$

where the small parameter ϑ represents energy-conserving interactions. Due to the complex nature and the energy dependence of Δ, the DOS does not exhibit a singularity, but is smeared with quasi-particle excitations extending to low energies. The temperature dependence of the gap edge, $\Delta_0(T) = \text{Re}\{\Delta(\hbar\omega = \Delta_0(T), T)\}$, was found to resemble closely that of the BCS energy gap (Sect. 1.3.1). The critical temperature of strongly coupled superconductors was evaluated in terms of the coupling constant λ_{coupl} for intermediate coupling strengths $\lambda_{\text{coupl}} \leq 1$ in [73, 74]. A general formulation of T_c for arbitrary values of λ_{coupl} was given in [75]:

$$k_B T_c = \frac{\hbar \tilde{\Omega}}{4} \frac{1}{\sqrt{\exp(2/\lambda_{\text{coupl}}) - 1}}, \quad (1.69)$$

with $\tilde{\Omega}$ a characteristic phonon frequency of the superconductor which replaces the phonon spectrum by a Dirac delta function: $\alpha^2(\Omega)F(\Omega) \propto \delta(\Omega - \tilde{\Omega})$ [76]. The coupling constant is related to the phonon spectrum by

$$\lambda_{\text{coupl}} = 2 \int_0^\infty d\omega \frac{\alpha^2(\omega) F(\omega)}{\omega}. \quad (1.70)$$

The Coulomb pseudo-potential was assumed to be zero in (1.69). However, finite values of μ^* affect the T_c equation only in the form of a suitably modified coupling constant $\lambda_{\text{coupl}}^{\text{eff}}$ [75]. The strong electron–phonon interaction leads not only to T_c values exceeding the BCS value but also to enhanced values of the reduced energy gap δ. The dependence of δ on λ_{coupl} was found to saturate at $\delta = 3.35$ for $\lambda_{\text{coupl}} \gg 1$ [75].

The surface impedance was calculated within a general theoretical approach including strong pair coupling, magnetic scattering and anisotropy by Nam [77]. The scaling of the penetration depth and the electronic coherence length with the strength of the pair coupling was investigated by Marsiglio and coworkers [72b, 78]. As one specific result of strong coupling, the slope $\partial \lambda_L^{-2}(T)/\partial T$ at $T \leq T_c$ increases over that of a BCS superconductor. Numerical computations of the complicated expressions were performed by Blaschke and Blocksdorf [79]. Phenomenologically, an exponential temperature dependence of the surface impedance is found in an intermediate temperature range $T \leq T_c/2$, but with δ exceeding the BCS value (Sect. 2.2). As could be expected from the smeared DOS of strongly coupled superconductors, the temperature dependence of the penetration depth [80] and of the surface resistance [81] at very low temperatures were found to be no longer exponential, but to follow power laws. Furthermore, the frequency dependence of the surface resistance exceeds the BCS value, approaching square-law behavior [68]. At $T = 0$, the surface resistance vanishes at frequencies below $2\Delta_0/\hbar$, in accordance with the BCS results obtained for a constant energy gap [16].

It is finally worth mentioning that the definitions of the microscopic BCS parameters can be scaled to strongly coupled superconductors ("scaled BCS theory", see [10] and references [11–20] therein). This scaling transformation comprises two factors, according to

$$Z(T) = \eta_Z(T) \cdot Z^{\mathrm{BCS}}(X^*) \,. \tag{1.71}$$

Here, $Z(T)$ denotes any particular physical quantity (e.g., the penetration depth) and Z^{BCS} the BCS expression for this quantity. X^* stands for the normal-state parameters determining Z, and the asterisk indicates that in evaluating Z^{BCS} the mass renormalization $(1 + \lambda_{\mathrm{coupl}})$ of these parameters has to be applied. The prefactor $\eta_Z(T)$ represents the correction due to the results of the strong-coupling theory.

Two-Band Model of High-Temperature Superconductivity

The preceding paragraphs described specific features of superconductors, each resulting in deviations from the BCS density of states. The discovery of the high-temperature superconductors revealed a group of materials that display several of these features simultaneously:

1. Due to the layered lattice structure (Chap. 5), the electronic properties are highly anisotropic along directions within (a,b) and perpendicular (c direction) to the CuO_2-planes [82]. Furthermore, in contrast to most metallic superconductors, the extremely short coherence length ξ_0/ℓ prevents averaging effects across the Fermi surface.
2. Especially in the $Y(Yb)Ba_2Cu_3O_{7-x}$ compounds, the CuO chains (extending along the b direction), were suspected to contain unsaturated magnetic moments at the position of the Cu atoms at finite oxygen deficiency $x > 0$ [83].
3. The high values deduced for the energy gap from tunneling spectroscopy (Sect. 1.3.4) lead to a normalized energy gap of $\delta \approx 3$–4, thus indicating strong pair coupling.

Based on this situation, in 1987 Kresin and Wolf introduced a two-band model accounting for the existence of two electronic subsystems [84], S^{p} and S^{c}, associated with the CuO_2 planes (system "p") and the CuO chains (system "c"). The total Hamiltonian is given by

$$H_{2\,\text{band}} = H_0^{\text{p}} + H_0^{\text{c}} + \sum_{\chi,\kappa,q} g_{\chi\kappa q}^{\text{pc}} c_\chi^{\text{p}*} c_\kappa^{\text{c}} d_q + \sum_{\chi,\kappa} T_{\chi\kappa}^{\text{pc}} c_\chi^{\text{p}*} c_\kappa^{\text{c}} + \text{c.c.} \qquad (1.72)$$

H_0^{p} and H_0^{c} describe the individual subsystems, with H_0^{p} containing the interaction $\Sigma g_{\chi,\chi'}^{\text{p}} c_\chi^{\text{p}*} c_{\chi'}^{\text{p}} b_q$ which leads to pairing in "p". The quantum numbers describing the electronic and phononic states are denoted by χ, κ and q; $g_{\chi,\kappa,q}^{\text{pc}}$ is the electron–phonon matrix element for the $p \to c$ transition, and $T_{\chi,\kappa}^{\text{pc}}$ is the tunneling matrix element. The last two terms in (1.72) describe two charge transfer mechanisms between the two subsystems. The third term represents transitions mediated by phonons (as indicated by the operator d_q), and the fourth term is a tunneling Hamiltonian which reflects an intrinsic proximity effect. The planes are considered intrinsically superconducting with a large coupling constant $\lambda_{\text{coupl}}^{\text{p}} \approx 3$, while the chains are assumed to be normalconducting. By means of the intrinsic proximity effect and the inelastic interband scattering (represented by small scattering rates Γ^{pc}, $\Gamma^{\text{cp}} \leq k_\text{B} T_\text{c}/\hbar$), superconductivity can be induced in the chains at a strength $\lambda_{\text{coupl}}^{\text{pc}} \approx \lambda_{\text{coupl}}^{\text{cp}} \approx 0.2$ [85, 86]. The resulting set of coupled Eliashberg equations is formally similar to the case of a one-band superconductor with strong gap anisotropy.

A specific consequence of the coupling between the two subsystems is the occurrence of two order parameters (and energy gaps Δ^{p} and Δ^{c}), but a single transition temperature with no universal relation between T_c and the two gaps. It shall be noted explicitly that the set of coupling strengths mentioned above is consistent with a high transition temperature of 90 K. This value is the result of two competitive interactions [84]: The proximity channel tends to reduce T_c while it is enhanced by the phonon-mediated charge transfer. In contrast to the competing influence on T_c, both channels interfere constructively in the formation of the induced energy gap.

Since the two bands are separated in $\boldsymbol{k}$ space rather than in real space, the presence of magnetic impurities (represented by the scattering rate Γ_m^c) in the chains affects both subsystems simultaneously. It may even lead to gapless superconductivity in S^p as well as in S^c. As calculated in [84b], the critical impurity concentration corresponds to a mean free path of the order of the b axis coherence length (about 2.5 nm in $YBa_2Cu_3O_{7-x}$ [87]). Small oxygen deficiencies $x \geq 0.1$ are therefore sufficient to induce the gapless state. However, in marked contrast to the situation in single-band superconductors containing magnetic impurities, the gapless state in the two-band system is achieved without a significant suppression of the critical temperature. The fractional T_c reduction corresponding to the critical impurity concentration was estimated in [84c] to be only of the order of percent. An important result of this is that the layered structure of the oxide superconductors leads to a noticeable extension of the gapless regime.

The surface resistance was analyzed separately for the two subsystems in [88] for the case of YBCO. For fully oxygenated chains, R_s^c was in accordance with the BCS results presented before, thus reflecting the assumed weak pair interaction. The occurrence of gapless superconductivity naturally resulted in enhanced residual losses. The surface resistance of the "p" system corresponded to that of a strongly coupled superconductor ($R_s \propto \omega^2$) dominated by an exponential temperature dependence according to the large gap value.

More detailed calculations of the surface impedance and comparison with experimental data on YBCO films (Sect. 2.3) were based on the density of states [89, 90]. Nonmagnetic scattering in both subsystems was considered with the corresponding rates $\Gamma^p = \Gamma^c$ adjusted to the normal-state conductivity at 100 K [85, 86]. The DOS of the two-band system reveals two gap-like structures [84], as displayed in Fig. 1.12. The higher energy peak, corresponding to the plane-system "p", resembles the behavior of a one-band superconductor with strong quasi-particle–phonon interaction and a reduced energy gap $\delta^p \approx 3$. The lower energy peak represents the DOS in the chains. It displays unusual properties which deviate from the BCS behavior although the pair coupling is weak. The position of the peak is almost independent of temperature (left diagram), in contrast to its width which strongly increases with increasing T. This behavior reflects the strong temperature dependence of the quasi-particle scattering rate, caused by the interband coupling and the strong pair coupling in system "p" [89].

At low temperatures and low concentration of magnetic impurities (right diagram), there exists a small energy gap $\delta^c \leq 0.8$, i.e., an energy interval where the DOS vanishes. If present, it is this small gap which governs the temperature dependence of the low-frequency penetration depth [85, 86] and of the total surface impedance ([89], Sect. 2.3). With the oxygen deficiency exceeding its critical value, the zero-DOS region vanishes. The low-temperature scaling of the surface impedance is then described by power-law behavior, turning from linear to quadratic with further decreasing oxygen content. Figure 1.13 displays representative results for the temperature dependence of λ [85].

The two-band model bears further implications for the interpretation of the surface impedance. The CuO chain fragments are considered not the only source of magnetic scatterers. As argued in [91], the out-of-plane oxygen in various HTS compounds adds contributions to the magnetic pair breaking, even if the chains are completely filled ($x = 0$). That is, superconductivity in most HTS compounds is inherently suppressed by their electronic structure, leading to the concept of an "intrinsic" $T_{c,0}$ [92]. Accordingly, the measurable critical temperature is lower than it would be in the absence of any free magnetic moment. Referring to (1.53b), the intrinsic T_c value is given at the critical scattering rate $\Gamma_{cr} = \Delta_{OP}(0)/2\hbar$ by

$$T_{c,0} = \frac{2\gamma}{\pi} \frac{\hbar}{k_B} \Gamma_{cr} . \tag{1.73}$$

The value of Γ_{cr} was deduced by analyzing the temperature dependence of the upper critical field $H_{c2}(T)$ at low temperatures [91], leading to $T_{c,0} \approx 160\,\mathrm{K}$

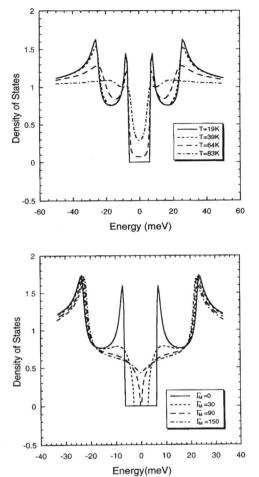

Fig. 1.12. Two-band-model DOS, normalized to the normal-state value, versus energy at $\Gamma_{\mathrm{m}}^{\mathrm{c}} = 0$ and various temperatures (*top*) and $T = 0$ and various scattering rates $\Gamma_{\mathrm{m}}^{\mathrm{c}}$ (in K, *bottom*) [90]. (© 1999 by the American Physical Society.)

for the HTS compounds. According to the theory of superconductors containing paramagnetic impurities (see above), the T_{c} reduction is accompanied by a smearing of the density of states and, in turn, by an increase of the surface resistance at finite temperatures, or at frequencies above the (reduced) gap value. If this model applies to the nature of HTS, a considerably high and weakly temperature dependent surface resistance should be expected in contrast to the behavior of one-band quasi-isotropic superconductors.

Surface Impedance of d-Wave Superconductors

In contrast to the two-band model, pairing mechanisms involving antiferromagnetic (AF) spin fluctuations have been proposed (e.g. [49, 93] and re-

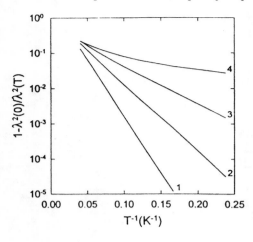

Fig. 1.13. Temperature dependence of the penetration depth on a logarithmic scale to emphasize the exponential nature for small values of the magnetic scattering rate, $\hbar\Gamma_m^c/k_B$ (K) = 0 (curve 1), 30 (2), 60 (3) and 90 (4). From [85]. (© 1999 by the American Physical Society.)

view [94]), mainly motivated by the complicated phase diagram of the HTS superconductors, which varies with doping from an antiferromagnetic insulator to a superconductor (Chap. 5). A simplified version of the relevant interaction Hamiltonian leads to the one-band Hubbard model with

$$H_{AF} = - \sum_{(ij),\sigma} t(c_{i,\sigma}^* c_{j,\sigma} + c_{j,\sigma}^* c_{i,\sigma}) + \sum_{i} U n_{i\uparrow} n_{i\downarrow} \,. \tag{1.74}$$

Here, $c_{i\sigma}^*$ creates a quasi-particle of spin σ on site i and n_i is is the number operator. The physical meaning of (1.74) is an effective one-electron transfer t between nominal "Cu sites" in the CuO_2 planes, experiencing an on-site Coulomb repulsion U that prevents two electrons from occupying the same position. Theoretical and numerical investigations of this and extended Hamiltonians always revealed, in the weak-coupling limit, a d-wave symmetry of the pair state, which apparently fits the structural symmetry of HTS. As pointed out in [94], the basic feature of the d-wave orbital symmetry of the pair function is the existence of nodes of the gap along lines on the (cylindrical) Fermi surface of the form

$$\Delta_{d_{x^2-y^2}}(\mathbf{k}) = \Delta_0(T) \cos(2\varphi) \,, \tag{1.75}$$

where φ specifies the orientation of the two dimensional momentum of the Cooper pairs. In contrast to the anisotropic order parameter described by (1.67), equation (1.75) yields a fourfold sign change of $\Delta(\varphi)$ along a contour around the Fermi surface, as illustrated in Fig. 1.14. Implications of these π jumps of the phase of the pair function were reviewed in [94] and are mentioned briefly here. Another important consequence of these nodes is a density of states which, for clean d-wave superconductors, varies at $E \ll \Delta_0$ proportional to the energy [95]

$$N_s(E) \approx N(0) \frac{E}{\Delta_0} \,. \tag{1.76a}$$

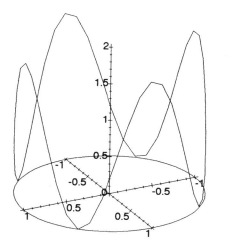 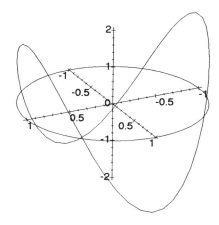

Fig. 1.14. Relative variation of the order parameter on a contour (*circle*) along the Fermi surface for an anisotropic s-wave (*left*, from (1.67) with $\Delta_0 = \Delta_1$), and for a d-wave superconductor (*right*, from (1.75) with $\Delta_0 = 2$).

Equation (1.76a) is in sharp contrast with the weak-coupling isotropic BCS result (1.55). Referring to the expression for the penetration depth ((1.60a) in Sect. 1.3.2), the d-wave DOS leads to a temperature dependence of $\lambda(T)$ corresponding to

$$\frac{\lambda(T) - \lambda(0)}{\lambda(0)} \approx \ln 2 \frac{k_B T}{\Delta_0} \, . \tag{1.77a}$$

A more detailed theoretical treatment of this problem was given, e.g., in [95–97]. In order to describe various experimental data showing $\lambda(T) \propto T^2$, the role of impurity scattering was analyzed in [98]. It was found that for strong scattering at a rate Γ and with a phase shift of $\pi/2$ (unitary limit), the DOS of a $d_{x^2-y^2}$ superconductor changes below a crossover temperature $T^* \propto (\Gamma \Delta_0)^{1/2}$ from the form (1.76a) to

$$N_s(E) \approx N(0) \frac{k_B T^*}{\Delta_0} + \mathrm{const}\, E^2 \tag{1.76b}$$

with the resulting temperature dependence of λ given by

$$\frac{\lambda(T) - \lambda(0)}{\lambda(0)} = \frac{1}{2} \sqrt{\frac{\hbar \Gamma}{\Delta_0}} + \mathrm{const} \frac{T^2}{T + T^*} \, . \tag{1.77b}$$

The last expression yields qualitatively the desired transition from T^2 behavior at low temperatures ($T \ll T^*$) to a linear T dependence above the crossover value T^*. Unfortunately, the T_c value of d-wave superconductors is suppressed by strong nonmagnetic scattering at a rate $\Delta T_c/T_c(0) \propto \hbar\Gamma/\Delta_0$ [95, 98], similarly to the T_c depression due to magnetic scatterers in s-wave superconductors. Such a strong reduction of T_c appears in conflict with experimental data.

The existence of quasi-particle states at all energies leads to a lower limit of the quasi-particle conductivity $\sigma_{1,\mathrm{min}}$, as found for the low-frequency limit at $T = 0\,\mathrm{K}$ [97,99]:

$$\sigma_{1,\mathrm{min}} = \sigma_\mathrm{n}\frac{\hbar\Gamma(T_\mathrm{c})}{\pi\Delta_0}, \tag{1.78}$$

and thus to a minimum achievable surface resistance $R_{\mathrm{s,min}} = \mu_0^2\omega^2\lambda_\mathrm{L}^3\sigma_{1,\mathrm{min}}/2$ (see (1.18d)). As argued in [100], this limit is of little practical consequence for experiments performed at finite frequencies unless the unitary limit applies to the quasi-particle scattering.

Exact calculations of the surface impedance of d-wave superconductors were presented in [97]. An analysis of the clean local limit was studied in [100] in relation to experimental data on YBCO films for the temperature range from 4.2 K to T_c (see Sect. 2.3). The calculations were performed for weak coupling with T_c as a parameter. It was found that the consideration of strong elastic scattering with intermediate phase shifts between the Born ($\Delta\varphi = 0$) and the unitary limit ($\pi/2$) was essential to describe the data, and to reproduce the residual value of the surface resistance at $T = 0\,\mathrm{K}$. At intermediate temperatures, $Z_\mathrm{s}(T)$ revealed a strongly temperature dependent scattering rate, which was approximated by a three-parameter test function. In addition, the $\Gamma(T)$ behavior calculated within the framework of the nested Fermi liquid model (NFL) of strongly correlated systems [101] yielded comparable results. The amplitude of the order parameter (see (1.75)) corresponded to a reduced gap value $\Delta_0/k_\mathrm{B}T_\mathrm{c} \approx 3$–$3.5$, but the electromagnetic response was not sensitive to the sign change of the order parameter. Further consequences of d-wave superconductivity for applications at high frequencies were considered, e.g., in [31,102,103].

Remarks on the Superconductivity in HTS Films

The aforementioned models still leave some questions open. In the d-wave approach, the pairing mechanism based on antiferromagnetic spin fluctuations cannot explain the observed high critical temperatures or coupling strengths [49,104,105]. In addition, the nature and the amount of impurities needed to explain the observed power-law dependences of Z_s on temperature imply a strong T_c suppression [103]. These features also seem to be in contradiction to the magnetic scattering in HTS. Furthermore, the orthorhombic symmetry of the HTS unit cells might favor a mixed symmetry of the order parameter like $s + \alpha d$ [106,107]. The situation becomes even more complex when regarding the variation of the mixing ratio α with temperature, doping level and distance to adjacent interfaces.

On the other hand, the existence of a finite, reproducibly high residual resistance and the results for Z_s of doped YBCO crystals [31] require quantitative explanations within the two-band approach. Also, the thoroughly tested data from the phase-sensitive Josephson junction and SQUID experiments indicate a reproducible sign change of the order parameter within the CuO_2 planes (e.g., [106,108]). This appears to support the d-wave symmetry of the order parameter, although alternative explanations based on magnetic scattering are available [109,110].

In order to develop a consistent and comprehensive understanding of the nature of the superconductivity in HTS materials, the analysis of the surface impedance plays an important role. The strong electronic correlation of the quasi-particles, the high sensitivity to defects and the important role of oxygen vacancies lead to significant, and strongly temperature dependent scattering. This feature must be considered in any theory for Z_s, and might be a clue for experimental verification. Finally, as always, the investigation of high-frequency properties needs to be supplemented by an analysis of other thermodynamic and transport properties (like specific heat, Knight shift, nuclear spin relaxation rate etc.), to provide an unambiguous characterization of the superconducting state.

1.3.4 Tunneling Spectroscopy of the Density of States in $Y(Yb)Ba_2Cu_3O_{7-x}$

Tunneling spectroscopy is best suited to imaging the energy distribution of quasi-particle excitations in superconductors, and thus to determining the presence or the absence of an energy gap, i.e., a region where the DOS vanishes.

The existence of a well-defined temperature dependent energy gap has been confirmed for all prominent metallic superconductors, like Nb or Nb_3Sn [111], from tunneling spectroscopy. The situation is more complicated for HTS due to the specific features described before. In order to verify interpretations of the surface impedance, experimental data on the bulk DOS are required. There is strong evidence from electron tunneling and photon emission spectroscopy for the existence of a large energy gap, $\Delta = 25$–$30\,\mathrm{meV}$, and significant quasi-particle scattering, $\hbar\Gamma/\Delta = 0.1$, especially for the $Bi_2Sr_{n+1}Ca_nCu_2O_y$ (Bi-2212 and 2223) and the yttrium (or ytterbium) 123 compounds [30, 112–115]. Figure 1.15 shows tunneling results on the density of states, which strikingly resembles that of conventional superconductors [116]. Taking into account the corresponding critical temperatures, the reported gap energies correspond to reduced values $\delta \approx 2.5$–4, indicating strong pair coupling. These large values imply a great potential benefit of HTS compared to LTS in terms of reduced thermal sensitivity ($\Delta/k \approx 10^2\,\mathrm{K}$), very short dynamic response times ($h/\Delta \approx 0.1\,\mathrm{ps}$, Sect. 3.1) and high operation frequencies ($\Delta/h \approx 10\,\mathrm{THz}$).

However, there are various indications for the existence of quasi-particle states at energies below Δ (for a recent overview, see [117]). This is not surprising since the tunneling density of states (TDOS) could easily deviate

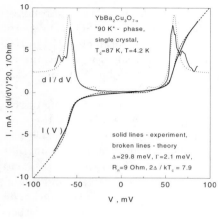
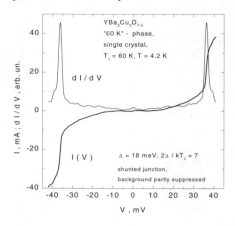
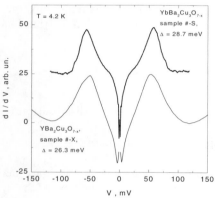

Fig. 1.15. Current–voltage characteristics and differential conductivities of four different Y-123 and Yb-123 break junctions at $T = 4.2\,\text{K}$ displaying different energy gaps and very low sub-gap currents. In the lower left-hand diagram, the coexistence of two gap features becomes visible (cf., Fig. 1.12) [116].

from the bulk DOS due to interface effects, especially when considering the coherence length being comparable to unit-cell dimensions.

Besides the absolute value of Δ, its isotropy and multiplicity also affect the high-frequency transport properties. Angle-resolved photo emission spectroscopy gives clear evidence of a pronounced gap anisotropy within the CuO_2 planes of the HTS compounds [113, 114]. However, the directions of maximum and minimum gap values, e.g., with respect to the CuO chains of $YBa_2Cu_3O_{7-x}$, are less clear [113, 114, 118, 119]. Although not yet experimentally quantified, anisotropic gap features should also be expected from the layered structure of HTS between directions within and normal to the CuO_2 planes. Data on tunneling and microwave spectroscopy have revealed the coexistence of two distinct gap features (Fig. 1.15 and [19, 23, 112, 116, 120]) with $\Delta_1 = 25\text{--}30\,\text{meV}$ and $\Delta_2 \leq 10\,\text{meV}$, supporting the two-band scenario described above (c.f., calculated DOS in Fig. 1.12).

2. Measurements of the Surface Impedance at Linear Response

> *In the physical world only the measurable is real, in fact, only that which is measurable twice, at least.*

2.1 Measurement Methods

There are many different methods to deduce the surface impedance in terms of its real and imaginary parts R_s and X_s (see (1.17) and (1.27)) from experiment. The discovery of the high-temperature superconductors further initiated the development of a great variety of systems, which were already thoroughly reviewed [1–5]. Instead of trying to review the appreciable amount of information, the following treatment focuses on the specific advantages of resonant measurement systems. Among these, special emphasis is put on resonators devoted to the characterization of unpatterned films, i.e., for fundamental investigations. The application of superconducting resonators to microwave devices is treated in Chap. 6.

Nonresonant superconducting waveguides or planar transmission lines display, by definition, a large bandwidth over which the characteristic impedance and the propagation constant are frequency independent. This feature makes them useful in the investigation of pulse shape deformation, intermodulation and harmonic generation in relation to the investigation of nonlinear effects (Chap. 3). In contrast, superconducting resonators bear the advantage of very low power dissipation (or high quality factors), thus making them highly sensitive for the investigation of superconducting samples with low surface impedance.

After introducing in Sect. 2.1.1 the basic quantities measurable with superconducting resonators, and their relationship to the physical quantities discussed in Chap. 1, Sect. 2.1.2 summarizes briefly the different modes of operation and their implications for analyzing data. The remaining sections contain brief reviews of high-resolution resonators devoted to the integral and local investigation of the surface impedance of superconducting films, respectively.

2.1.1 Characteristic Parameters of Superconducting Resonators

Quality Factor

The characteristic parameters of superconducting resonators can be obtained by starting with the differential equation of a free-decaying damped oscillator:

$$\ddot{x}(t) + \eta \dot{x}(t) + \omega_0^2 x(t) = 0 \,. \tag{2.1}$$

The time-dependent variable $x(t)$ can be considered, for instance, to be the angular amplitude of a pendulum, the "charge amplitude" of an electric R-L-C circuit, or the field components of an electromagnetic wave. The damping is represented by the coefficient η which has the dimension of a frequency. The resonant frequency of the frictionless oscillator is denoted by ω_0. The solution of (2.1) is well known from standard text books:

$$x(t) = \exp(-\eta t/2) \exp[i(\omega_1 t + \varphi)] \tag{2.2a}$$

with the resonant frequency ω_1, which is reduced by the damping according to

$$\omega_1^2 = \omega_0^2 - \frac{1}{4}\eta^2 \,. \tag{2.2b}$$

The total energy W stored in the oscillator decays from its initial value W_0 like

$$W(t) = W_0 \exp(-\eta t) \,. \tag{2.2c}$$

If the resonator is fed by a high-frequency generator at frequency ω, the average power dissipated during one cycle assumes the form

$$P(\omega) = P_0 \cdot \frac{\eta^2 \omega^2}{(\omega_0^2 - \omega^2)^2 + \eta^2 \omega^2} \,, \tag{2.3}$$

where $P_0 = P(\omega = \omega_0)$. It is convenient to characterize the response of the resonator to the excitation by the dimensionless quality factor Q

$$Q = 2\pi \frac{\text{stored energy}}{\text{average energy lost per cycle}} = \frac{\omega W}{P_{\text{diss}}} \,. \tag{2.4a}$$

Further analysis of the problem reveals the important relationship

$$Q = \frac{\omega}{\eta} \,, \tag{2.4b}$$

which identifies the quality factor to be the product of the circular frequency (at resonance) and the decay time $1/\eta$ of the stored energy. Equation (2.4b) allows us to rewrite (2.3) in terms of the reduced frequency $x \equiv \omega/\omega_0$:

$$P_{\text{diss}}(x) = P_0 \frac{1}{1 + Q^2(x - 1/x)^2} \,. \tag{2.5}$$

Inspection of the functional dependence $P_{\text{diss}}(x)$ shows that its full width at half maximum (FWHM), which spans the frequency interval $x_+ - x_-$ with $P_{\text{diss}}(x_\pm) = P_0/2$, is related to Q by

$$Q = \frac{1}{x_+ - x_-} = \frac{\omega_0}{\omega_+ - \omega_-}. \tag{2.6}$$

This equation means, in addition to the result (2.4b) for the time domain, that the quality factor of the resonant system determines the width of the resonance curve in the frequency domain.

The relation between the dissipated power and the surface resistance was defined in (1.22). Consequently, the definition in (2.4a) leads to the basic relation between the Q factor and the average R_s of the superconductor enclosing the resonant volume:

$$Q = \frac{\omega \mu_0}{\overline{R_s}} \frac{\int_{\text{resonator}} |\boldsymbol{H}(\boldsymbol{r})|^2 \mathrm{d}^3 \boldsymbol{r}}{\int_{\text{conductor}} |\boldsymbol{H}(\boldsymbol{r})|^2 \mathrm{d}^2 \boldsymbol{r}}, \tag{2.7}$$

where the surface integral extends over the superconducting surface A. For the following discussion it appears convenient to express (2.7) in the modified form

$$Q = \frac{G}{R_s} = \frac{\omega L_{\text{eff}}}{R_s}, \tag{2.8}$$

where the bar over R_s has been omitted for simplicity. The factor $G = \omega L_{\text{eff}}$ can be considered as the characteristic impedance of the resonator in the absence of losses. It is often referred to as the "geometry factor", since it depends at a given frequency only on the dimensions of the resonator and on the distribution of the electromagnetic fields in it (referred to as "mode"). In the case of rotational symmetric homogeneous resonators (e.g., cylindrical cavity or dielectric resonators), G can be calculated analytically (e.g., [1, 6, 7]). Numerical computations of G can be performed for arbitrary geometry and inhomogeneous filling by means of commercial computer codes (e.g., [8]). The effective inductance L_{eff} is given by

$$L_{\text{eff}} = \mu_0 \frac{V \overline{H}_V^2}{A \overline{H}_A^2}. \tag{2.9}$$

In analogy to the dissipated power, which is described by $R_s/2$ times the surface integral of H^2, $\omega L_{\text{eff}}/2$ times this integral characterizes the microwave power transported by an electromagnetic wave traveling along a medium of wave impedance $Z_L = \omega L_{\text{eff}}$. This power is referred to as the circulating (or oscillating) power P_{circ} (see also [6]). It can be derived from combining (2.4a) and (2.8):

$$P_{\text{circ}} = \omega W. \tag{2.10}$$

46   2. Measurements of the Surface Impedance at Linear Response

The quantity P_{circ} denotes the maximum power handled by the resonator. It is therefore an important parameter for high-power microwave devices (Sect. 6.3).

Finally, for a given mode, the achievement of high quality factors and high circulating power necessarily means to store an appreciable amount of energy in an extended volume.

Resonant Frequency

The resonant frequency f_0 of a resonator is determined by its geometrical dimensions. Deviations from the ideal geometry, e.g., due to mechanical tolerances or to the presence of coupling elements also affect the resonant frequency. The size of the resonator is furthermore subject to changes which may result from temperature variations. When the temperature changes, the length of the mechanical support, the permittivity of a dielectric within which field energy is stored, and the penetration depth of the superconducting material also change. Consequently, all these variations result in a shift of f_0 which is given by Slater's theorem [9]:

$$\frac{\Delta f_0}{f_0} = \frac{\int_{\Delta V} \{\mu |\boldsymbol{H}|^2 - \varepsilon |\boldsymbol{E}|^2\} \mathrm{d}^3 \boldsymbol{r}}{\int_{\text{Volume}} \{\mu |\boldsymbol{H}|^2 - \varepsilon |\boldsymbol{E}|^2\} \mathrm{d}^3 \boldsymbol{r}}. \qquad (2.11a)$$

ΔV stands for the varying part of the total volume, and $\mu = \mu_0 \mu_{\text{r}}$ and $\varepsilon = \varepsilon_0 \varepsilon_r$ are the absolute permeability and permittivity. Equation (2.11a) has two important consequences. First, the resonant frequency cannot be measured absolutely since this would require knowledge of the geometrical dimensions of the resonator to fractions of the sub-µm penetration depth λ. Second, if the frequency shift is related solely to changes of λ, i.e., to changes of the magnetic energy stored in the resonator, (2.11a) can be simplified, yielding a basic relation between Δf_0 and changes of the surface reactance $\Delta X_{\text{s}} = \omega \mu_0 \Delta \lambda$:

$$\frac{\Delta f_0}{f_0} = -\frac{\Delta X_{\text{s}}}{2\omega \mu_0} \frac{\int_{\text{surface}} |\boldsymbol{H}|^2 \mathrm{d}^2 \boldsymbol{r}}{\int_{\text{volume}} |\boldsymbol{H}|^2 \mathrm{d}^3 \boldsymbol{r}} . \qquad (2.11b)$$

Using the notation for G in (2.8), we arrive at the following equation which relates changes of the complex surface impedance to the two measurable quantities Q and Δf_0:

$$\Delta R_{\text{s}} + \mathrm{i} \Delta X_{\text{s}} = G \left(\Delta \left(\frac{1}{Q} \right) - \mathrm{i} \frac{2 \Delta f_0}{f_0} \right). \qquad (2.12)$$

As will be shown in later sections (see also Sect. 1.1.2 and [2, 10]), the mutual variation $r \equiv \partial X_{\text{s}} / \partial R_{\text{s}}$ contains important information about the microscopic nature of the surface impedance. The differential loss tangent r analysis is especially appealing in cases where the geometry factor is not (or not accurately enough) known. According to (1.18) and (2.12), the r value directly relates the conductivity ratio y to measurable quantities:

$$r = \frac{\partial \varphi_+(y)}{\partial \varphi_-(y)} = \frac{2\partial \ln f_0}{\partial Q} . \tag{2.13}$$

Field-to-Power Conversion Coefficient

A very important issue is the investigation of the performance of superconducting films at elevated power levels (and correspondingly microwave field levels, Chaps. 3 and 4). The maximum surface current $H_{s,\max}$ in a resonator is related to the stored energy through the field-to-power conversion coefficient γ_H:

$$H_{s,\max} = \gamma_H \sqrt{\omega W} = \gamma_H \sqrt{P_{\text{circ}}} , \tag{2.14a}$$

which can be written in a notation analogous to (2.9):

$$\gamma_H^2 = \frac{2}{\omega \mu_0 V} \cdot \frac{H_{s,\max}^2}{\overline{H}_V^2} = \frac{2}{\omega L_{\text{eff}} A} \frac{H_{s,\max}^2}{\overline{H}_A^2} . \tag{2.14b}$$

It can be concluded from (2.14) that large values of $H_{s,\max}$ on the HTS surface occur in high-Q (i.e., large-W) and/or in low-dimensional (i.e., small-V and/or small-A) devices. Microwave components designed for applications at large power levels can accordingly be fabricated advantageously with three dimensional cavity or dielectric-loaded resonators (Sect. 2.1.3) rather than with lower-dimensional devices (see below).

Low-Dimensional Resonators

A few examples illustrate the above formalism. Starting with a simple R-L-C circuit, the Q factor, the resonant frequency and the current-to-power conversion factor are found immediately from the corresponding definitions: $Q = \omega L/R$, $\omega_0 = (LC)^{-1/2}$, and $\gamma_I^2 = 2/\omega L$. This case corresponds to a "zero-dimensional" structure, where the dimensions along and transversal to the direction of current flow are much smaller than the guided wavelength. Furthermore, a homogenous current distribution across the width of the conducting elements is assumed, i.e., fringe fields at the contours of the circuit elements are neglected. It is then sufficient to consider currents and voltages, rather than electromagnetic fields. These are localized at the positions of inductor and capacitor, respectively. The transformation between both points of view can be found from Maxwell's equations (1.1) under suitable boundary conditions (e.g., [11]).

If the length ℓ of the resonator reaches an appreciable fraction of the guided wavelength, like a half-wavelength resonator, the reactive and dissipative contributions can no longer be localized, but are distributed along

48   2. Measurements of the Surface Impedance at Linear Response

the structure. According to standard transmission line theory [11], inductor, capacitor, and resistor have to be replaced by inductance, capacitance and resistance per unit length, L', C', and R'. This is correspondingly reflected in the characteristic parameters Q, f_0 and γ_I. From the general definition (2.4a), the quality factor is found to be $Q = \omega C' |Z_L|^2/R'$, where the wave impedance Z_L of the transmission line is given by $Z_L \approx (L'/C')^{1/2} \cdot (1 - iR'/2\omega L')$. In the case of weak dissipation $R' \ll \omega L'$, the result $Q = \omega L'/R' = \omega L/R$ is recovered. The resonant frequency of the fundamental mode of a half-wavelength resonator is given by $\omega_0 = (\ell^2 L'C')^{-1/2}$, and the current-to-power conversion factor by $\gamma_I^2 \approx 2/\omega L' \ell$.

Displayed in Fig. 2.1 are three different geometries which are commonly applied to transmission lines. Each of these displays specific advantages and disadvantages which were reviewed in detail elsewhere [1, 6, 12] and will not be considered here. However, the three geometries have different electromagnetic boundary conditions, which are basically contained in the aspect ratio w/h where w and h are the width of the central conductor and the height of the dielectric, respectively. As a result, different expressions are found to relate the quantities L', C' and R' with the surface impedance of the superconducting material. As one example, an analytic approximation for the inductance per unit length of a superconducting microstrip line was deduced in [13] on the basis of Maxwell's equations (1.1) and the London equation (1.9):

$$L' \approx \mu_0 \frac{h}{w} \frac{1}{K(w,h,d_1)} \left[1 + \frac{\lambda_1}{h}\coth\left(\frac{d_1}{\lambda_1}\right) + \frac{\lambda_2}{h}\coth\left(\frac{d_2}{\lambda_2}\right)\right]. \quad (2.15)$$

The thicknesses and penetration depths of the center line and the ground plane are denoted by the indices 1 and 2. $K(w,h,d_1)$ is a correction taking into account the effect of fringe fields. For $w \gg h$, K approaches unity, while it diverges as w drops below h. In terms of (2.8), equation (2.15) describes a "geometry factor" which depends on the absolute value of the penetration depth and thus on temperature.

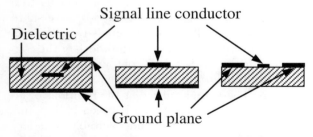

STRIPLINE   MICROSTRIP   COPLANAR

Fig. 2.1. Commonly used transmission line geometries. Stripline is the easiest to analyze, while microstrip and coplanar are important for applications.

The fringe fields at the contours of the center conductors give rise to a peaking of the current density. This current density distribution is not analytically accessible but requires numerical computation. Relevant analytical [14] and numerical work was performed for stripline [15] and coplanar geometries [16]. It is worth mentioning that the geometrical arrangement of the coplanar line allows for the deduction of absolute λ values from measuring the temperature dependent resonant frequencies of two samples with different ratios w/h [1, 16]. (In a coplanar line, h is the spacing between the central strip and the ground plane). In order to assure reasonable power handling (low γ_{I} values) of the transmission lines, thin dielectrics and wide center lines have to be used.

Total Power Balance: Sum Rule for Inverse Q Values

The only dissipation mechanism considered so far was power dissipation in the superconductor (see (2.7)). In a resonator there are usually further sources for power loss which can be distinguished as internal and external loss mechanisms. "Internal" means dissipation related to the resonator itself, while "external" comprises losses due to coupling of the resonator to an environment (shielding, external circuit, etc.). All these contributions can be assigned individual Q factors in accordance with the general definition (2.4a). The conservation of energy translates into a sum rule for inverse Q values:

$$\frac{1}{Q_{\mathrm{L}}} = \frac{1}{Q_{\mathrm{int}}} + \frac{1}{Q_{\mathrm{ext}}}. \tag{2.16a}$$

Q_{L} is the loaded Q value which appears in the measurement. It was involved in (2.4), (2.5) and (2.6). In contrast, Q_{int} is often referred to as the unloaded quality factor Q_0. It is determined by the surface resistance of the conductors and the loss tangent of the dielectrics:

$$\frac{1}{Q_{\mathrm{int}}} = \frac{1}{Q_{\mathrm{cond}}} + \frac{1}{Q_{\mathrm{diel}}} = \gamma_{\mathrm{cond}} \frac{R_{\mathrm{s}}}{Z_0} + \gamma_{\mathrm{diel}} \tan \delta, \tag{2.16b}$$

with γ_i weighting the contribution of each loss mechanism to the total dissipation, and $Z_0 = 120\pi$ (in units of Ω) the free-space wave impedance. Q_{cond} is described by (2.7)–(2.8) with the replacement $G \equiv Z_0/\gamma_{\mathrm{cond}}$. The dielectric filling factor is given by the electric field distribution and the volume within which the electric field energy is stored:

$$\gamma_{\mathrm{diel}} = \frac{1}{Z_0^2} \frac{\int_{V\mathrm{diel}} \varepsilon_r |\boldsymbol{E}|^2 \mathrm{d}^3 \boldsymbol{r}}{\int_{V\mathrm{res}} \mu_r |\boldsymbol{H}|^2 \mathrm{d}^3 \boldsymbol{r}}. \tag{2.16c}$$

If the resonant volume is entirely filled with the dielectric, $\gamma_{\mathrm{diel}} = 1$. The external quality factor consists of two further contributions, namely radiation losses due to coupling of the resonator to a source (Q_1) and a detector (Q_2) in a two-port arrangement (Sect. 2.1.2), and parasitic losses:

$$\frac{1}{Q_{\text{ext}}} = \frac{1}{Q_1} + \frac{1}{Q_2} + \frac{1}{Q_{\text{paras}}}. \tag{2.16d}$$

In a reflection measurement (one-port setup), $Q_2^{-1} = 0$. Q_{paras} comprises all additional losses which might result from fringe fields at the coupling elements, from power dissipation in a conducting shield (housing), or from any other previously unknown amount of power evanescent from the resonator.

In optimized measurement systems, Q_{paras} can be adjusted to high values, well above $Q_{1,2}$. Its inverse can thus be neglected in (2.16d). It is then convenient to rewrite (2.16a) in terms of the coupling coefficients β_1 and β_2:

$$Q_{\text{L}} = \frac{Q_0}{1 + \beta_1 + \beta_2}, \tag{2.16e}$$

where Q_0 was written instead of Q_{int}.

2.1.2 Modes of Operation

According to the physical meaning of the Q factor described in the preceding section, there are basically two ways to determine Q values and resonant frequency, namely from frequency-domain and from time-domain measurements. While the former method is facilitated by the availability of network analyzers, the latter has to be performed on the basis of microwave diodes and fast analog or digital oscilloscopes. However, it bears the great advantage of pulsed operation, e.g., to minimize microwave heating at elevated power levels.

Frequency Sweep Across the Resonant Frequency

The microwave response of a linear two-port is described in the frequency domain by the scattering parameters $S_{ij} = |S_{ij}| \exp\{-i\Phi_{ij}\}$ with $i, j = 1, 2$. For example, S_{11} describes the amplitude of the voltage reflected at port 1, normalized to the magnitude of the incident voltage. In case of a reciprocal device, the coefficients of voltage transmission from port 2 to 1 and vice versa, S_{12} and S_{21}, are identical. Any such device can be characterized by the frequency dependence of the dissipative response (insertion loss $a_{\text{IL}} = -20 \log |S_{21}|$) and by that of the dispersive response (group delay $\tau_{\text{gr}} = d\Phi_{21}/d\omega$). If the device can be assumed to be lossless, energy conservation requires that $|S_{11}|^2 = |S_{22}|^2 = 1 - |S_{21}|^2$ and $2\Phi_{21} = \Phi_{11} + \Phi_{22} \pm \pi$. The first identity provides a quick method for checking for dissipation losses in the resonator. The scattering parameters at the resonant frequency and the phase response can be related to the loaded quality factor Q_{L}:

$$|S_{ii}(f_0)| = 1 - \frac{2Q_{\text{L}}}{Q_i}, \tag{2.17a}$$

$$|S_{ij}(f_0)| = \frac{2Q_{\text{L}}}{\sqrt{Q_i Q_j}}, \tag{2.17b}$$

$$\tan(\Phi_{ij}) = \frac{1}{Q_{\mathrm{L}}} \frac{\omega^2}{\omega_0^2 - \omega^2}, \tag{2.17c}$$

and

$$\tau_{\mathrm{gr}} = \frac{2Q_{\mathrm{L}}}{\omega_0}. \tag{2.17d}$$

According to (2.16e), (2.17a) provides relationships between the scattering parameters at resonant frequency and the coupling coefficients:

$$|S_{11}(f_0)| = \frac{1 + \beta_2 - \beta_1}{1 + \beta_1 + \beta_2}. \tag{2.18a}$$

$|S_{22}(f_0)|$ results from (2.18a) by exchanging β_2 and β_1. The transmission is described by

$$|S_{12}(f_0)| = \frac{2\sqrt{\beta_1 \beta_2}}{1 + \beta_1 + \beta_2}. \tag{2.18b}$$

From the definition of the scattering parameters it is obvious that $|S_{12}(f)|^2 = P_2(f)/P_{\mathrm{inc}}$, where P_{inc} (index "inc" from "incident") is the power delivered from the high-frequency source to the resonator. Since $P_2 Q_2 = \omega W$, $|S_{12}(f)|^2$ displays the same frequency dependence as the stored energy, $W(f)$. The latter can be derived for a damped oscillator in relation to the dissipated power $P_{\mathrm{diss}}(f)$ in (2.5): $W(f)/W(f_0) = P_{\mathrm{diss}}(f)/P_{\mathrm{diss}}(f_0) \times (f^2 + f_0^2)/2f^2$. Because of the high Q values of superconducting resonators, the transmitted power is significant only in a frequency region close to f_0 with $|f - f_0|/f_0$ being of order Q^{-1}. As a result, the frequency dependence of $|S_{12}(f)|$ can be approximated by

$$|S_{12}(f)| = \frac{|S_{12}(f_0)|}{\sqrt{1 + Q_{\mathrm{L}}^2 (f/f_0 - f_0/f)^2}}. \tag{2.19}$$

The resonant frequency can be deduced from (2.19) as that value at which $|S_{12}|$ becomes maximum. Furthermore, the frequency response of the transmission coefficient is symmetric with respect to $f = f_0$ at sufficiently low power levels. This consistency can be easily proven in experiment by verifying the identity $f_0 = (f_+ + f_-)/2$. The loaded Q factor can be deduced from the FWHM of $|S_{12}|$, as discussed in Sect. 2.1.1.

It is advantageous to analyze Q_0 on the basis of the modified coupling coefficients $\beta_1' = \beta_1/(1 + \beta_2)$ (β_2' is analogously defined with β_1 replaced by β_2), which can be found from the reflection coefficients $|S_{11}(f_0)|$ and $|S_{22}(f_0)|$, respectively:

$$\beta_i' = \frac{1 \pm |S_{ii}|}{1 \mp |S_{ii}|}, \tag{2.20}$$

where $i = 1$ or 2. The solution for β_i' solely from (2.20) is ambiguous: it is either $\beta_i' < 1$, corresponding to undercritical coupling (lower sign), or $\beta_i' > 1$

at overcritical coupling (upper sign). This ambiguity can be removed by, e.g., inspecting the frequency gradient of the phase:

$$\left.\frac{d\Phi_{ii}}{df}\right|_{f=f_0} = \frac{2Q_L}{f_0}\frac{2\beta_i'}{1-\beta_i'}, \qquad (2.21)$$

which is either positive or negative in the two cases. If β_1' and β_2' are known, Q_1 and Q_2 follow from the identities $Q_i = Q_L(1+\beta_i')/\beta_i'$. Q_0 is then deduced according to (2.16a). Finally, the maximum excited field amplitude in the resonator can be found from the stored energy as described by (2.14a).

It might be convenient in some cases to couple the resonator only at one port, and to measure its response in the reflection mode. This case follows naturally from the previous analysis by setting $\beta_2 = 0$. The determination of the resonant frequency and of the loaded Q factor are based in this case on the frequency dependence of $|S_{11}(f)|^2 = 1 - P_{\text{diss}}(f)/P_0 - W(f)/W(f_0)$:

$$|S_{11}(f)| = \sqrt{\frac{|S_{11}(f_0)|^2 + Q_L^2(f/f_0 - f_0/f)^2}{1 + Q_L^2(f/f_0 - f_0/f)^2}}. \qquad (2.22)$$

Constant-Frequency Mode with Variable Pulse Duration

In the previously described mode of operation the source power is kept constant, while the frequency is swept across the resonant value. In contrast, the frequency can be kept adjusted at f_0 while modulating the source power. We might assume for simplicity the application of square pulses of duration t_0, starting at t_{on} modulo $\{n(t_0 + \Delta t), n = 1, 2, \ldots\}$ and being repeated after a time interval Δt. Regarding the instrumentation required for this type of measurement, it is more natural to talk about different powers rather than scattering parameters. In case of two-port coupling we need to distinguish between the incident power $P_{\text{inc}}(t)$, the power P_r reflected from the resonator under steady-state conditions, and the powers $P_1(t_{\text{off}})$ and $P_2(t_{\text{off}})$ radiated through ports 1 and 2 when the input pulse is turned off. The ratios P_r/P_{inc} and P_2/P_{inc} can be identified with $|S_{11}(f_0)|^2$ (see (2.18a)) and $|S_{12}(f_0)|^2$ (see (2.18b)), respectively. In addition, P_1/P_{inc} is expressed in terms of the effective input coupling coefficient β_1' [17]:

$$\frac{P_1}{P_{\text{inc}}} = \frac{4\beta_1'^2}{(1+\beta_1')^2}. \qquad (2.23)$$

The power balance, which has to be satisfied at each instant t, can be verified by monitoring either one of the combinations $\{P_{\text{inc}}, P_r \text{ and } P_2\}$, $\{P_{\text{inc}}, P_r \text{ and } P_1\}$ or $\{P_r, P_1 \text{ and } P_2\}$. The redundancy contained in the three parameter sets provides valuable opportunities to check consistency in the determination of the coupling Qs (which follows the same strategy as in the frequency-domain method). Furthermore, the value of β_1' (and similarly of β_2' if the output port is used as input) can be deduced without the need for phase measurements: $\beta_1' < 1$ if $P_1(t_{\text{off}}) < P_{\text{inc}}$, and $\beta_1' > 1$ if $P_1(t_{\text{off}}) > P_{\text{inc}}$ (see Fig. 2.2 below).

The resonant frequency can be determined from varying the source frequency slightly, such that the maximum of the transmitted power P_2 is adjusted. Since the resonance curve (2.19) is flat at its maximum, this procedure is much less sensitive to changes of f_0 than the frequency sweep, with the result that power dependent changes of the penetration depth are more complicated to resolve (e.g., through supplementary monitoring of the frequency response of the phases). The determination of the loaded Q factor is based on the deduction of the relaxation time of the resonator during energy loading at t_{on} or decaying at t_{off}, by either observing $P_1(t)$ or $P_2(t)$ (see (2.4b)). Since P_1 is subject to interference with standing waves in the input line, higher accuracy is usually achieved by monitoring the transmitted power. Figure 2.2 illustrates some pulse shapes for P_{inc}, P_1, P_{r} and P_2 for $\beta'_1 = 0.5$, taking into account the exponential load and decay curves of the resonator. Table 2.1 summarizes representative results of the power ratios $P_{\text{r}}/P_{\text{inc}}$ and P_1/P_{inc} for selected values of β'_1.

Table 2.1. Summary of representative power ratios for different input couplings

β'_1	$P_{\text{r}}/P_{\text{inc}}$	P_1/P_{inc}
1/3	1/4	1/4
1/2	1/9	4/9
1	0	1
∞	1	4

With Q_{L}, Q_1 and Q_2 known, Q_0 can be determined from (2.16a). Once the coupling Qs are known, Q_{L} can also be concluded from (2.18b), which reads in this context:

$$\frac{\overline{P}_2}{\overline{P}_{\text{inc}}} = \frac{4Q_{\text{L}}^2}{Q_1 Q_2} . \tag{2.24}$$

The bars above the Ps denote the steady state values, i.e, at times $\eta t \gg 1$ after the pulse was turned on. The field level excited in the resonator is determined, like in the previous case, by the value of P_2, given that Q_2 is known.

A special case of pulsed measurements is the limit of infinitely long pulse duration Δt. This mode is known as CW ("continuous wave") operation. The analysis of the measured data relies in this case basically on the steady-state values of the different powers involved, unless microwave heating causes instabilities (Sect. 3.2).

Transient Behavior of the Stored Energy

Until now it was assumed that the Q factor of the resonator could be considered constant, i.e., independent of source power (or field amplitude). If P_{inc} is

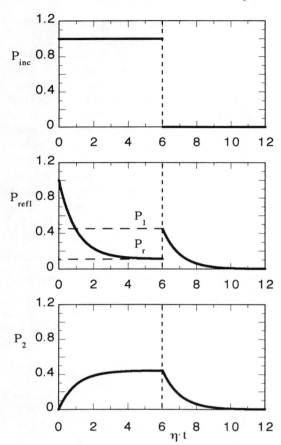

Fig. 2.2. Representative pulse shapes for the incident power (*top*, square pulse of duration $6/\eta$, amplitude normalized to 1), the power reflected from port 1 (*middle*, from (2.18a) and (2.23)) and the power radiated through port 2 (*bottom*, from (2.18b) and (2.24)) at linear response for an effective input coupling coefficient $\beta'_1 = 0.5$ and symmetric coupling ($\beta_1 = \beta_2 = 1$).

consecutively increased, this assumption becomes eventually invalid. In this case the shapes of the response pulses (Fig. 2.2) change in a characteristic way that reveals information on the transient $Q(P_2)$ dependence [18,19]. The basic features of this analysis can be illustrated in the following way. (We will not distinguish between Q_L and Q_0, which is experimentally justified in the case of weak coupling, $\beta_1, \beta_2 \ll 1$). According to (2.2c) and (2.4b), the power radiated off the resonator decays upon turning off the source by

$$P_2(t) = P_2(t=0) \exp\left[-\omega t/Q\left(P_2(t)\right)\right] \text{ at } t \geq t_{\text{off}}. \tag{2.25a}$$

It is important to note that, as the power (and therefore the field amplitude in the resonator) change with time, Q becomes implicitly time dependent, too. Differentiation of (2.25a) with respect to time yields

$$\frac{\partial \ln\left[P_2(t)/P_2(0)\right]}{\partial(\omega t)} = -\frac{1}{Q}\left(1 - \frac{\partial \ln Q}{\partial \ln(\omega t)}\right) \text{ at } t \geq t_{\text{off}}. \tag{2.25b}$$

Since $P_2(t)$, $P_2(0)$ and t can be recorded, e.g., with a digital oscilloscope, Q can be reconstructed for every instant t, and consequently for the field levels $\propto P_2(t)^{1/2}$. Figure 2.3 displays a representative example obtained in this way with a Nb accelerating cavity.

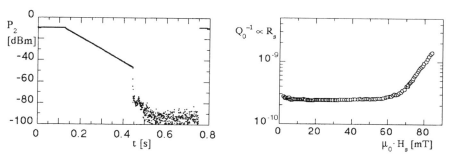

Fig. 2.3. Transient-time analysis of the transmitted power (*left*, logarithmic scale), resulting in the magnetic-field dependence $Q_0^{-1}(\mu_0 H_s)$ of a Nb accelerating cavity at 3 GHz (*right*) [19].

The transient-time analysis is not restricted to the free-decaying resonator as assumed in (2.25). It can similarly be applied to the "loading" curve of the resonator which varies with time as $P_2(t) = \overline{P_2}\left[1 - \exp\left(-\omega t/Q(t)\right)\right]$. Analysis of this behavior shows that nonlinear loss mechanisms, which are slow compared to the time constant $1/\eta$ of the resonator, are strongly damped and lead mainly to a reduced level of $\overline{P_2}$. In contrast, fast nonlinearities might cause a pronounced extremal behavior of $P_2(t)$. It was found phenomenologically that different loss mechanisms could be distinguished from their characteristic pulse shapes [20].

The transient-time analysis is very fast and allows quick "snap-shots" of the field dependence of the Q factor. Moreover, this method helps to identify the dynamic nature of the nonlinear effects, which might be overseen in slow measurement modes (like "sweep" or "cw"). Further aspects of this issue are covered in Sect. 3.3.1.

2.1.3 Selected Examples of Integral Measurement Systems

General Framework

The three resonators described in this section are selected examples among a large variety of similar devices built in many different groups (for references see Sect. 2.1.1). The main categories of different possible arrangements are sketched in Fig. 2.4. Cavity resonators are referred to as "type 1". The superconductor film under test (FUT) usually forms one endplate of a cylindrical cavity, either in a closed ("1a") or in an open ("1b") arrangement. In the

56   2. Measurements of the Surface Impedance at Linear Response

latter case, a gap between the host cavity and the FUT allows for an adjustment of the sample temperature independent of that of the host. Special care has to be taken in this geometry to prevent the formation of unwanted modes, and to keep radiation losses at a tolerable level (e.g., with a so-called choke-flange [21]). Instead of the cavities 1a and 1b, the resonators 2a and 2b employ a dielectric cylinder to store the electromagnetic field energy. It is placed inside a normal or a superconducting housing which aims to minimize radiation losses. In the symmetric and asymmetric dielectric resonators (3a) and (3b), two films rather than one are simultaneously tested. This arrangement maximizes the sensitivity to the surface impedance of the FUTs, but at the expense of averaging over the properties of two individual samples. Type 4 is a miniature form of 3a, where the dielectric serves at the same time as the substrate, and the two films to be studied are deposited on either side of it.

The unloaded Q factors of the resonators sketched in Fig. 2.4 can be rewritten from (2.8) and (2.16) in a more detailed form:

$$\frac{1}{Q_0(T,f)} = \frac{R_{s,a}(T,f)}{G_a} + \frac{R_{s,b}(T,f)}{G_b} + \gamma_{\text{diel}} \tan \delta(T,f) + \frac{1}{Q_{\text{paras}}}, \quad (2.26)$$

where the indices "a" and "b" refer to the FUT and the host resonator, respectively, with G_a and G_b denoting the corresponding geometry factors. According to the definition (2.7), G_a and G_b are related to the total geometry factor G_{tot} by $G_a^{-1} + G_b^{-1} = G_{\text{tot}}^{-1}$. The dielectric filling factor γ_{diel} is zero for type 1 and, within a few percent, close to unity for types 2, 3 and 4. The last term in (2.26) summarizes unwanted losses related to the mounting of the sample. These can be, e.g., losses at the interfaces between samples and host in the closed geometries 1a, 2a and 3, or radiation losses in the open systems 1b, 2b and 3b. The dependences on temperature and frequency were introduced explicitly in this equation since these allow one to distinguish between

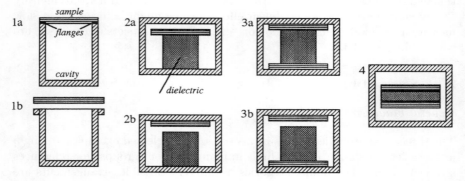

Fig. 2.4. Sketches of the most prominent arrangements of resonators for measurements of the surface impedance of unpatterned superconducting films.

the different contributions to Q_0. Dielectric losses are discussed further in Chap. 5 in relation to the properties of HTS compatible substrate materials.

An expression analogous to (2.26) can be formulated for the temperature dependence of the resonant frequency $f_0(T)$ on the basis of (2.11):

$$-\frac{1}{f_0}\frac{\partial f_0}{\partial T} = \frac{1}{2G_a}\frac{\partial X_{s,a}}{\partial T} + \frac{1}{2G_b}\frac{\partial X_{s,b}}{\partial T} + \frac{\gamma_{diel}}{2\varepsilon_r}\frac{\partial \varepsilon_r}{\partial T} + \frac{1}{2L}\frac{\partial L}{\partial T}. \qquad (2.27)$$

The first and second terms describe the temperature variation of the penetration depth of the FUT and the host, respectively. The third term arises from the temperature dependent permittivity of the dielectric (see also Sect. 6.2), and the last term summarizes the effect of thermal expansion on the total length L of the resonator.

One can conclude from (2.26) and (2.27) that a high sensitivity to $Z_{s,a}$ requires a small ratio G_a/G_b. This condition bears basic limitations for the choice of the resonator geometry (diameter D and aspect ratio D/L), and for the choice of the mode. High-sensitivity systems also favor a large ratio $R_{s,a}/R_{s,b}$. This requirement is usually achieved by choosing high-conductivity copper or superconducting Nb (or Nb_3Sn) as the host material. Further optimization of this ratio can be achieved with a normal-conducting housing by using a high resonant frequency, because $R_{s,a}$ increases with frequency more strongly than $R_{s,b}$ (Chap. 1). There are additional restrictions on the resonant frequency that result from the size of the samples to be measured, and from the instrumentation available for swept, pulsed and/or high-power measurements. Finally, the dielectric and parasitic losses should be kept as low as possible. This condition can be met by using high-purity sapphire, and by choosing modes that prevent the flow of microwave currents parallel to the contour of the samples (Sect. 2.1.1). In some cases, the resonator geometry can be chosen such that several high-Q_0 modes can be excited at sufficient frequency separation in the accessible band. This allows an in-situ investigation of the frequency dependence of R_s (over a narrow frequency interval) and, more importantly, an analysis of the homogeneity of the surface impedance of the FUT (Sect. 2.1.4).

The examples described below belong to resonator types 1a, 2b and 4. This selection illustrates some of the basic procedures needed to design the "optimum" resonator for a desired task. The described systems display specific advantages for measurement sensitivity and the achievement of high magnetic field levels, e.g., for the investigation of nonlinear phenomena.

Normal-conducting Cavity Resonator at 87 GHz

A Cu cavity resonator operating at 87 GHz ("type 1a") was developed for fundamental investigations on 1 cm × 1 cm HTS films as described in [22, 23]. The resonant frequency of cylindrical cavities is given by [7, 9b]

$$f_{mnp} = \frac{c}{\pi D} \cdot \sqrt{x_{mn}^2 + \left(\frac{\pi p}{2}\right)^2 \cdot \left(\frac{D}{L}\right)^2}, \qquad (2.28)$$

where x_{mn} denotes the nth zero of the mth Bessel function $J_m(x)$ for transverse-magnetic (TM) modes, and of the first derivative of $J_m(x)$ for transverse-electric (TE) modes. The field indices $m \geq 0$, $n \geq 1$ and p ($p \geq 0$ for TM and $p \geq 1$ for TE) describe the number of wavelengths along the azimuth, the number of maxima of the magnetic field along the radial direction, and the number of half-wavelengths along the axial direction, respectively. In order to minimize parasitic losses, rotational-symmetric ($p = 0$) TE_{0np} modes are best suited. The filling factor $\gamma_{\text{cond}} = Z_0/G$ (see (2.16b)) of a single endplate can be calculated analytically for this group of modes:

$$\gamma_{\text{cond}}(TE_{0np}) = \frac{1}{2} \frac{c^3 p^2}{2\pi L^3 f_{0np}^3}. \tag{2.29}$$

A large value of the axial field index, $p = 3$, provides a high sensitivity of f_0 to changes of the penetration depth of the FUT. Figure 2.5 illustrates the distribution of the magnetic field components along the axial direction (z, left), and along the radial direction (r, right) at the surface of the sample.

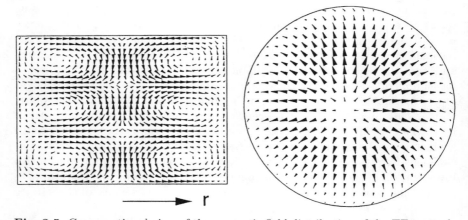

Fig. 2.5. Cross-sectional view of the magnetic field distribution of the TE_{013} mode along the axial direction (*left*) and along the radial direction at the surface of the sample (*right*) [23]. The direction of the fields is indicated by the *arrows*. The field amplitudes are reflected by the size of the arrows.

The geometric dimensions of the 87-GHz resonator were adjusted to $D = 8.11$ mm and $D/L = 1.34$, which is appropriate for the standardized sample size 1 cm × 1 cm. Within the accessible frequency band, the TE_{013} mode (86.6–87.0 GHz) as well as the TE_{021} mode (86.0–86.3 GHz) could be excited. The natural degeneracy of the TE_{0np} modes with the TM_{1np} modes was lifted due to field distortions at the coupling holes, located at two opposite sides of the cylinder wall, and due to a step-like mode trap at the bottom of the Cu cylinder. The analytically calculated ratio G_a/G_b amounted to 39.2% for the TE_{013} mode, which is close to the maximum value of 50% for a single film and at $D/L \to \infty$. The corresponding value was 5.4% for the TE_{021} mode. The first mode was applied for the determination of Z_s at low and moderate temperatures $T < T_c$ (where $R_s(\text{FUT}) < 0.3\,\Omega$) while the second mode allowed Z_s measurements at and above T_c, at a decisively reduced sensitivity. The resonator received power from a synthesized sweeper and a frequency multiplier ($f \times 6$). Due to the high measurement frequency, the source power was limited below 1 mW, corresponding to a maximum field level of $H_{s,\text{max}} = 20$ A/m.

Temperature dependent measurements were performed during warm-up of the evacuated cavity, after the system was cooled to 4.2 K and the coupling Qs were determined. Data on the surface resistance and the change of penetration depth of the films under test were extracted from (2.26) and (2.27). The contributions of the host cavity were determined from reference measurements, where the superconductor was replaced with a sample of known surface resistance (e.g., Cu or Nb plates). Weak, symmetric coupling ($Q_0 \approx Q_L$, and $\beta_1 = \beta_2 \ll 1$) improved the accuracy of the analysis further. Data were acquired with a computer from curve-fitting of the transmitted power, according to (2.19). A high number of measurement points allowed for a resolution of the FWHM to better than 0.2%. Corresponding relative and absolute accuracies of 0.3 and 1 mΩ were achieved for R_s, and 1–10 nm for $\Delta\lambda$. The R_s resolution was limited at low temperatures by parasitic losses at the interface between the FUT and the Cu cavity. The use of copper as the host material made it possible to study the dependence of $Z_{s,a}$ on a DC magnetic field (Sect. 4.2).

Nb-Shielded Sapphire Resonator at 19 GHz

The replacement of a cavity (type 1) by a dielectric (types 2 and 3) allows one to miniaturize a resonator, i.e., to achieve a desired or a lower resonant frequency at a smaller or similar sample size, respectively. The resonant condition results from total internal reflection at the dielectric–vacuum boundary rather than at the vacuum–metal interface. The electromagnetic energy is stored almost completely within a low-loss dielectric with $\gamma_{\text{diel}} \approx 1$, therefore yielding high quality factors, and allowing operation at high microwave power levels. However, the need to account for a finite loss tangent and for losses in the metallic housing are potential disadvantages of this approach.

The concept of dielectric resonators (DR) is well known [24], as well as its application to the investigation of HTS films [1, 6], which mainly relies on the specific advantages of the TE_{011} mode. The resonant frequency of this mode can be approximated in close similarity to (2.28) in terms of the dimensions D and L, and of the permittivity ε_r of the dielectric [24]:

$$f_0(TE_{011}) \approx \frac{c}{\pi D_{\text{diel}} \sqrt{\varepsilon_r}} \sqrt{x_{01}^2 + \frac{\pi^2}{4}\left(\frac{D_{\text{diel}}}{L_{\text{diel}}}\right)^2} \qquad (2.30)$$

with $x_{01} \approx 3.8317$ the first zero of the first Bessel function. Equation (2.30) describes the scaling of f_0 with D_{diel} and ε_r correctly but overestimates the exact values by about 10%. More accurate calculations of the resonant frequencies in DRs can be found in [25]. In order to reduce field leakage from the dielectric, large $D_{\text{diel}}/L_{\text{diel}}$ are desirable, with $D_{\text{diel}}/L_{\text{diel}} \geq 2$ typical. This leads, at a given frequency, to a trade-off between field confinement and resonator volume, which might be evaluated differently for fundamental investigations and for applications. The dissipative filling factor γ_{cond} of a single face of the dielectric cylinder, and the dielectric filling factor γ_{diel} are for the TE$_{011}$ mode [6]:

$$\gamma_{\text{cond}} = \frac{1}{2\varepsilon_r} \frac{c^3}{2\pi L^3 f_{011}^3} \frac{1+\varepsilon_r R}{1+R} \qquad (2.31\text{a})$$

and

$$\gamma_{\text{diel}} = \frac{1}{1+R} \, . \qquad (2.31\text{b})$$

R denotes the ratio of electric field energy stored outside the dielectric to the amount stored inside of it. Equation (2.31a) reduces to (2.29) for $R = 0$ and $\varepsilon_r = 1$ as expected. The field-to-power conversion factor γ_H can be deduced for the TE$_{011}$ mode from the formula given in [6]:

$$\gamma_H(\text{DR}) = \frac{1}{\sqrt{\varepsilon_r(1+R)}} \cdot \gamma_H(\text{cavity}), \qquad (2.32)$$

where the corresponding coefficient of a cavity resonator is described by (2.14).

A high-purity sapphire rod ($\varepsilon_r \approx 10$) with $D = 7$ mm and $D/L = 2$ formed the central part of a dielectric resonator of type 2b, operated around 19 GHz in the TE$_{011}$ mode [26]. The housing, to which the sapphire was glued, was made of superconducting Nb. Housing and dielectric were permanently kept at or below 4.2 K during the measurements. The FUT (diameters ranging between 1″ and 2″) was thermally and electrically isolated from the sapphire, allowing for a separate adjustment of its temperature. The inner diameter of the housing exceeded that of the sapphire rod by about a factor of 3. The degeneracy of the TE$_{011}$ mode with the TM$_{111}$ mode was lifted due to field distortions at the loop antennas reaching into the volume between the dielectric and the Nb housing, and due to the gap between sample and rod. The numerically calculated geometry factor of the FUT remained identical to that of the opposite face of the sapphire ($G = 454\,\Omega$) up to a gap width of 100 μm, and increased to 512 Ω at 300 μm. The dielectric filling factor was larger than 0.99 indicating that less than 1% of the total energy was stored outside the sapphire. This result is illustrated in Fig. 2.6 on the basis of computer calculations of the electric and magnetic fields. The field-to-power conversion factor was 11.6 A/m/W$^{1/2}$ or 14.6 μT/W$^{1/2}$ (1 A/m corresponds to $4\pi/10$ μT) at a gap width of 100 μm, and dropped

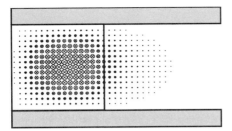 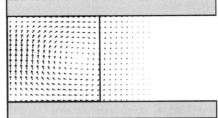

Fig. 2.6. Cross-section of the electric (*left*) and the magnetic (*right*) field distribution of the TE$_{011}$ mode of the Nb-shielded sapphire resonator along one radius [26]. The dielectric is indicated by the *bold rectangle*, the endplates are *grey*. The left vertical axis is the symmetry axis of the setup. The direction of the fields is indicated by the *arrows* and encircled *crosses*, respectively. Large symbols denote high field amplitudes.

to 10.6 A/m/W$^{1/2}$ at 300 µm. The resonator was excited by the integrated source of a network analyzer and a 25-W traveling-wave tube amplifier for high-power measurements.

Temperature dependent measurements were performed computer-controlled in a procedure similar to that described for the 87 GHz system. The sample was cooled to 4.2 K using He exchange gas. The resonator was evacuated for measurements at elevated temperatures in order to assure stable operation. The position of the coupling antennas could be adjusted in-situ for optimum matching of the source impedance to that of the resonator. Unloaded Q values above 3×10^7 were obtained at 1.8 K with a Nb reference film, indicating $\tan \delta < 10^{-8}$ at this temperature. The absolute reproducibility in the determination of R_s corresponded to 15–20 µΩ. Due to the presence of the gap between film and sapphire, and to the thermal expansion of the latter, the measurement system was not suited for accurate measurements of $\Delta\lambda(T)$. However, it provided the important advantage of applying magnetic microwave field amplitudes up to 50 mT (Sects. 4.1 and 4.3). The absolute accuracy in determining $H_{s,\max}$ was limited by the knowledge of the coupling Qs to about 30%. Relative variations of $H_{s,\max}$ could be monitored with $< 5\%$ accuracy. The field dependent R_s measurements were performed with the network analyzer in the sweep mode at low power levels. As nonlinear behavior of the FUT showed up, pulsed operation was applied using microwave diodes and a fast analog oscilloscope. The duty cycle (ratio of pulse duration to time interval between successive pulses) was chosen sufficiently above the time constants of the resonator and the detector, but still small enough to avoid excessive warming of the sample. Typical values were $t_0 = 200$ µs and $\Delta t = 1$ s. Indications for continuous microwave heating could be deduced from a temperature sensor located in a Cu block upon which the FUT was mounted. Investigation of the pulse shapes revealed additional information in some cases (Sect. 4.3).

Microstrip Disk Resonator

Significant miniaturization of dielectric resonators can be achieved in quasi-two dimensional geometries. A special example is the parallel-plate resonator ("type 4", Fig. 2.4) with circular cross section (therefore called "disk resonator"). The distance between the superconducting films (0.5–1 mm) in such a device is much smaller than the diameter (≥ 20 mm) and the guided wavelength (typically 15 mm at 4 GHz and $\varepsilon_r = 25$). As a result, the electromagnetic field enclosed in the dielectric becomes independent of the axial extension L (the thickness of the dielectric). In contrast to the two resonators described before, the construction of a disk resonator requires double-sided films. Furthermore, the microwave response results from the film sides facing the substrate, rather than from the film surfaces. Finally, patterning is usually required to obtain the desired dimensions.

The characteristic parameters f_{mn}, γ_{cond} of both film sides, and γ_{H}, of the mode (m,n) can be estimated analytically in a magnetic-wall model which assumes the azimuthal magnetic field component to vanish at the diameter D. The resulting expression for f_0 agrees with the previously obtained results by setting $p = 0$ (the modes are therefore identified as TM$_{mn0}$ modes):

$$f_{mn} = \frac{c}{\pi D}\frac{x_{mn}}{\sqrt{\varepsilon_r}}, \tag{2.33a}$$

$$\gamma_{\mathrm{cond}} = \frac{Z_0}{\pi\mu_0 f_{mn} L}, \tag{2.33b}$$

$$\gamma_{\mathrm{H}} = y_{mn}\sqrt{\frac{\varepsilon_0\varepsilon_r f_{mn}}{L}}. \tag{2.33c}$$

In (2.33) x_{mn} and y_{mn} denote the nth zero of the first derivative of $J_m(x)$ and the maximum of $\{x_{mn}J_m(x); mJ'_m(x)\}$, respectively. The finite value of D/L induces stray fields which affect the resonant frequency in the form of an enhanced effective diameter [27]. Another consequence of the stray fields is that γ_{diel} drops below 1 as D/L increases.

As discussed in relation to low-dimensional resonators (Sect. 2.1.1), inhomogeneous current distributions can be deleterious to the attainable quality factor and power handling capability. The optimum approach with respect to miniaturization and, at the same time, good power handling, is therefore to use planar structures with homogeneously distributed microwave fields over as large a conductor surface as possible [12c, 28, 29] (see (2.8)). This concept has been pursued and successfully verified with the TM$_{010}$ mode disc resonator [30–32].

The magnetic fields of the fundamental TM$_{110}$ mode and the edge-current-free TM$_{010}$ mode are compared in Fig. 2.7. The azimuthal and the radial field components of the former exhibit sources ("+") and sinks ("-"), inducing localized polarization charges at the center and the terminations of the main diagonal. In contrast, $H_\varphi \equiv 0$ in the TM$_{010}$ mode, and H_r is rotationally symmetric with an induced concentric current pattern. Due to the absence of polarization charges and of microwave currents parallel to the contour of the superconducting disks, operation in this mode has the potential of improving the attainable Q factors because of reduced radiation losses. Furthermore, the power handling might be strongly improved compared to linear stripline devices, and thus approach typical features of three-dimensional resonators. The TM$_{010}$ disk resonator concept has therefore gained, besides its interesting features for fundamental investigations, great interest for high-power microwave devices (Sect. 6.2).

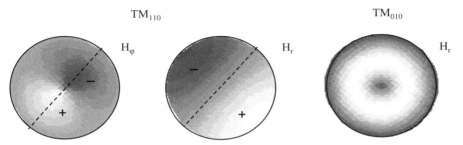

Fig. 2.7. Top-view contour plots of the magnetic field components of the fundamental mode (TM$_{110}$, H_ϕ, *left*; H_r, *middle*), and of the edge-current-free TM$_{010}$ mode (H_r, *right*). The field amplitudes are encoded by the grayscale.

2.1.4 Survey of Local Measurement Systems

Description of the Problem

The measurement of the surface impedance of superconducting films yields an integral value over the area over which the electromagnetic fields are distributed. It gives access, via the dissipated power, to the average surface resistance (see (1.22)) or, via the circulating power (2.10) or the resonant frequency (2.27), to the average surface reactance (see (1.25)) and hence to the average penetration depth. Microwave investigations of any kind of superconductor have revealed the decisive role of local defects or inhomogeneities on the surface impedance [33], which cannot be evaluated appropriately with the previously described methods. This is especially true for high-T_c superconductors like Nb$_3$Sn and YBCO which display short coherence lengths and therefore a high sensitivity to microscopic defects. The locality problem is further illustrated by rewriting (1.22) and (1.25):

$$P_\mathrm{diss} + \mathrm{i}\omega W_\mathrm{S} = \frac{1}{2} \int\limits_\mathrm{surface} [R_\mathrm{s}(\boldsymbol{r}) + \mathrm{i}X_\mathrm{s}(\boldsymbol{r})] \, |H(r)|^2 \mathrm{d}^2\boldsymbol{r} \,, \qquad (2.34)$$

where $R_s(\mathbf{r})$ and $X_s(\mathbf{r})$ depend in general on the position across the investigated surface. The challenge is to develop measurement methods which allow, at the same time, high accuracy in the determination of local Z_s values ($\leq 10^{-4}\overline{Z}_s$), high spatial resolution (μm to mm range), large field of view ($\leq \varnothing 4''$) and reasonable measurement times ($\leq$ hours). The signal-to-noise ratio decreases generally with decreasing filling factors, thus necessitating the application of frequency-modulation techniques in order to enhance the measurement resolution.

There are presently three known different approaches to attack this problem, especially for HTS films. These are listed in Table 2.2 and discussed in some detail below. Passive imaging methods employ the scanning of a detector probe across a superconducting sample that forms part of a separately excited microwave circuit. In contrast, active imaging involves probes which excite the sample locally, and serve at the same time as detector. In the third category, a local perturbation (mostly thermal in nature) is scanned across the sample with the resulting changes of the integral circuit parameters being detected. In principle, low-frequency transport or microstructural properties could also be monitored instead of microwave parameters, in order to judge the homogeneity of a superconducting film, given that the detected quantity is correlated with the surface impedance. Promising candidates are magneto-optic imaging [50–52], Hall-probe scanning [53, 54], eddy-current testing [55–57], scanning Raman microscopy [58, 59], and the usual types of scanning probe microscopy.

Table 2.2. Methods capable of local surface impedance measurements.

Category	Physical parameters	Type of detection/ perturbation	Ref.
Passive imaging	Q_0 and f_0 of resonators reflected microwave voltage dissipated microwave power microwave flux magnitude	multimode analysis near-field antenna thermometry SQUID monitor	34, 52 35, 36, 37 38, 52 39
Active imaging	Q_0 and f_0 of resonators	open-resonator endplate scanning	40, 41, 42
	Q_0 and f_0 of resonators microwave voltage pick-up kinetic inductance	dielectric resonator near-field antenna Josephson eddy currents	26, 43, 44 35, 36, 37 39
Perturbation scanning	Q_0 and f_0 of resonators Q_0 and f_0 of resonators	scanned electron beam scanned laser beam	45, 46 47, 48, 49

Passive Imaging

The simplest way to obtain information about the homogeneity of Z_s of a superconducting film is to compare unloaded Q values and resonant frequencies of different modes [34, 52]. This method is especially appealing for "linear" resonators which display a well-defined mode spectrum over many harmonics. However, the different frequencies might complicate a conclusive analysis. Furthermore, the spatial resolution is given by the considered modes and is of the order of millimeters to centimeters.

Another well-known approach is the imaging of the complex conductivity of a metallic, semiconducting or dielectric sample by means of a coaxial or a small-aperture near-field antenna [60, 61], which allows one to circumvent the diffraction-limited resolution of far-field methods. The coaxial approach was recently applied to the investigation of HTS films [35, 37]. The spatial resolution was found to be limited by the diameter of the inner conductor, but could be as low as 2 µm, compared to a guided wavelength of $\lambda_g = 30$ mm ($f \approx 10\,\mathrm{GHz}$). Imaging contrast could be obtained by the magnitude of the reflected voltage and the observable shift of the resonant frequency and the Q value of a transmission line, which was decisively impedance-mismatched with the surrounding microwave network. One of the major drawbacks of this approach is the necessity of following the surface contour of the sample at a constant separation from the probe that should be comparable to the diameter of the inner conductor. Furthermore, the electromagnetic resolution is limited to resistivity values which are typical for normalconductors. Forming the center line of a coaxial resonator as the tip of a scanning microscope [37] improved the spatial resolution to the sub-µm regime. Surface resistance could be monitored with a sensitivity of about 50 µΩ at 12 GHz. A slot-antenna resonant at 80 GHz was employed in [36] to investigate the local properties of HTS films with a spatial resolution of $\lambda_g/80$. In contrast to the coaxial near-field antennas, the sensitivity of the resonant slit was anisotropic and required deconvolution of scans along perpendicular directions. The high operating frequency prevented this method also from being applicable to high-power measurements. The resistivity resolution was comparable to the coaxial approaches.

A completely different approach to image the homogeneity of the microwave response is thermometry. This method was proven very successful in the investigation of anomalous loss mechanisms in low-temperature superconducting accelerating cavities ([62], [33] and references therein). The basic principle is the transformation of dissipated microwave power into a heat flux that can be detected by means of suitable thermometers (see also Sect. 4.1). The use of 24 Pt resistors was reported in [63] to visualize inhomogeneous microwave losses in YBCO disk resonators. Despite the simplicity of this approach, its applicability is limited by the modest spatial resolution (given by the geometric extension of and the separation between the resistors), and by the necessity of homogeneous thermal and direct mechanical contacts between sensors and film. An elegant method of thermal imaging was reported in [38]. Instead of using discrete thermometers, a thin (60–100 nm) film composed of a certain Eu chelate was deposited onto a YBCO microstrip filter.

Upon activating the indicator film by exposure to UV radiation (280–320 nm), the quantum yield of its fluorescence produced a temperature image of the microwave device with μm spatial resolution, and a temperature resolution below 0.1 K. It was possible in this way to detect local hot spots developing at structural defects under application of elevated microwave power. Besides the required instrumentation, a disadvantage of this method was the irreversible modification of the sample due to the deposition of the indicator film.

A SQUID microscope was applied in [39] to detect the magnetic flux in the near-field of a microwave circuit at frequencies between 1 MHz and 1 GHz. The SQUID sensor operated like a high-sensitivity rectifier, since only the magnitude of the microwave flux modified the response characteristic. The spatial resolution reached 20 μm, mainly limited by the size of the SQUID loop. The close proximity between sensor and sample required both to be cooled to 77 K. The applicability of this method was limited to small flux values. Furthermore, the distribution of magnetic fields in resonant structures is mainly determined by the geometric boundary conditions, but is hardly affected by the surface impedance of the conductor.

Active Imaging

Various groups developed miniaturized microwave circuits to excite a superconducting sample locally and to obtain spatially resolved Z_s data. Open Fabry–Perot-type resonators were reported for low-power investigations of HTS films up to ⌀4″ [22, 40–42]. In order to optimize the spatial resolution on the mm-scale, high operating frequencies between 90 and 150 GHz were chosen. A remarkable advantage of this method is the possibility of exciting linearly polarized modes, thus providing a valuable means of investigating anisotropy effects in conductors and dielectrics. Due to the sensitivity of open resonators to parasitic diffraction losses, the achieved R_s resolution was limited to the mΩ range.

Another approach was based on dielectric resonators. The field confinement to the dielectric (Fig. 2.7) provides a natural way to investigate the surface impedance with mm resolution. Sapphire ($\varepsilon_r \approx 10$, see [26] and Sect. 2.1.3) and rutile (TiO_2, $\varepsilon_r \approx 100$, [43]) dielectric resonators operating at 19 and 10 GHz were consecutively mounted at different positions above the HTS samples. The data revealed strongly inhomogeneous power-handling capabilities (Chap. 4). An advanced quasi-scanning method was demonstrated in [44] with a rutile puck resonator at 10 GHz. The TiO_2 cylinder was mounted into a quartz cylinder, both contained in a Cu shielding. Proper dimensioning minimized the loss contributions from the shield, yielding loaded Q values of the TE_{011} mode up to 3×10^4 at 77 K. The whole resonator was spring-loaded to the HTS film and immersed in liquid nitrogen. The spatial resolution was limited by the guided wavelength to about 1–2 mm. It was anticipated that the use of $SrTiO_3$ ($\varepsilon_r \approx 2000$) might improve this feature to about 0.5 mm, given that the high and strongly temperature dependent loss-tangent did not deteriorate the measurement accuracy.

The coaxial near-field antennas described in the previous paragraph could also be operated in an active mode, with similar spatial and electrical resolutions as before. The SQUID sensor described above could be biased to induce Josephson radiation above 1 GHz (corresponding to a voltage drop of 2.06 µV across each of the two Josephson junctions). The radiation induced a microwave current in the superconducting sample, leading to an inductive coupling between sample and SQUID loop. At sufficiently reduced background fields, this method could therefore provide high-sensitivity local relative measurements of the penetration depth up to frequencies corresponding to the critical voltage of the junctions (Sect. 1.1.2).

Perturbation Scanning

The local measurement methods described so far had in common the use of miniaturized probes, acting as sensor and/or as source. Many of the approaches required careful adjustment of the sample parallel and in close proximity to the probe, thus giving rise to a high susceptibility to noise and undesired background signals. There was hardly any experimental parameter freely adjustable other than the power level of the microwave generator.

In distinction to these features, the scanning of a local perturbation across a sample introduces an additional degree of freedom, namely the "strength" of the perturbation. However, this type of mapping requires open systems giving the external perturbation access to the microwave circuit. So far, focused electron [45, 46] and laser beams [47–49] were employed for the local investigation of microwave devices. The absorption of the electronic or photonic energy in the superconducting sample leads to an increase of the local temperature. The energy density of the excitation is usually adjusted in a reversible regime where the superconductor remains in equilibrium. A typical response signal δS has the form

$$\delta S(\bm{r}_0) = \int_{\text{surface}} D(\bm{r},\bm{r}_0)\delta Z(\bm{r})H^2(\bm{r})\mathrm{d}^2\bm{r} , \qquad (2.35\text{a})$$

where the integral extends over the investigated surface. $D(\bm{r})$ represents the spatial weight of the electron or laser beam, which can often be described by a Gaussian distribution. δZ is the induced change of the measured quantity (e.g., the surface resistance or the surface reactance) which, in the case of thermal effects, is given by

$$\delta Z = \frac{\partial Z}{\partial T}\delta T. \qquad (2.35\text{b})$$

The derivative in (2.35b) contains the desired information on the local behavior of Z. Equation (2.35) reveals the main features of local perturbation techniques. First, the response is the result of a convolution of the desired quantity Z and the microwave field (or current) distribution. Conclusive data acquisition therefore requires reference measurements or appropriate numerical deconvolution procedures. Secondly, the spatial resolution is determined by the effective thermal spot size D of the beam. This depends on the energy density of the beam, the bias conditions (temperature, microwave power, frequency etc.), and on the thermal properties of the film and the substrate. Finally, the Z resolution is roughly given by the product $(\delta A/A)(\delta Z/Z)$, where δA and A are the effective area of the spot and of the microwave field distribution. In order to optimize the sensitivity to Z, the bias conditions (e.g., operating temperature and microwave field amplitude) and the strength of the perturbation (e.g., laser power density) can be adjusted such that the film is locally driven across a critical value (e.g., T_c or B_{c1}).

A 2-mW He–Ne laser was adopted in [47] to visualize the longitudinal and transverse current density distributions of YBCO microstrip and coplanar resonators on a μm scale. The current distribution for different microwave input power levels was inferred from monitoring changes of the resonant frequency as a function of the position of the laser spot. The laser beam was chopped at a frequency around 100 kHz in order to keep the thermal spot size small. The ΔX_s resolution was maximized by adjusting the operating frequency at the steepest position of $|S_{12}(f)|$ (see (2.19)). A similar setup was developed with a 750-mW infrared laser diode for the investigation of ⌀2″ YBCO disk resonators ([49] and Sect. 4.1.3).

2.2 Surface Impedance of Nb$_3$Sn Films on Sapphire

While Sect. 2.1 covered the methodological aspects of surface impedance measurements, this and the next section summarize recent experimental data on Z_s of high-temperature superconducting films at sufficiently low excitation levels to assure linear response.

In contrast to the HTS compounds, the strong-coupling A15 superconductor Nb$_3$Sn displays well-resolved microwave properties: a large electron–phonon coupling constant $\lambda_{\text{coupl}} \approx 1.8$ [64], which corresponds to $k_B T_c/\hbar\tilde{\Omega}$ ≈ 0.15 ([65, 66], c.f. Sect. 1.3.3), a large energy gap ($\Delta/kT_c \leq 2.5$) [67] and low surface impedance [68]. The short coherence length ($\xi_0 \approx 5$–7 nm, [69]) and the much larger penetration depth ($\lambda \approx 100$ nm, [69]) leads to an extreme type-II nature of Nb$_3$Sn. The ξ_0 value is still significantly larger than in HTS ($\xi_0 \approx 2.5$ nm, [70]), resulting in a reduced sensitivity to defects due to proximity coupling across them. In resemblance to HTS, Nb$_3$Sn displays (though slight) anisotropic electronic properties [71] reflecting the tetragonal crystal symmetry (Sect. 5.1). The normal-state resistivity $\rho(T_c)$ depends sensitively on the ordering of the Nb chains [72, 73], which is reminiscent of the CuO chains in YBCO. The microwave properties of polycrystalline Nb$_3$Sn are therefore interesting to analyze in comparison to those of epitaxial YBCO films (Sect. 2.3).

2.2 Surface Impedance of Nb$_3$Sn Films on Sapphire

Furthermore, the transition temperature T_c=18 K of Nb$_3$Sn (except for T_c=23 K of Nb$_3$Ge, which is more complicated to be prepared and unstable) was the highest known before the advent of the oxide superconductors. The high T_c value and the high thermodynamic critical field ($B_{c,th}$=535 mT) made this superconductor especially attractive for high-field applications. Taking into account the rapid development of closed-cycle refrigerators, which provide cooling powers in the range of several Watts at liquid-helium (LHe) temperature (see [74], [75] and Sect. 5.3), the availability of Nb$_3$Sn thin films on dielectric substrates becomes also of practical interest for compact high-performance microwave devices. Such circuits might include stable oscillators for remote sensing systems, high-power and/or tunable filters for communication systems (Chap. 6), or detectors for radioastronomy.

Section 2.2.1 summarizes the state-of-the-art obtained with Nb$_3$Sn films on dielectric substrates. Recent progress in preparing Nb$_3$Sn films of high phase purity and large grains, attained by a novel deposition route, is described in Sect. 2.2.2. Data on the energy gap and the characteristic lengths of these films in relation to the fundamental issues treated in Chap. 1 are also presented there. Section 2.2.3 discusses the influence of film thickness on these quantities.

2.2.1 Properties of Nb$_3$Sn Thin Films on Dielectric Substrates

Brief Sketch of Prominent Deposition Routes for Nb$_3$Sn Films

The preparation of thin (0.5–3 µm) Nb$_3$Sn films on bulk Nb [76, 77] or sheet material [68, 78] was mainly motivated by the economic advantage of increasing the operating temperatures of particle accelerators above the 2 K needed for Nb cavities. Among different fabrication routes, tin vapor diffusion (TVD) at a partial pressure of 10^{-1} Pa and temperatures up to 1200°C proved capable of yielding µm-thick large-area (up to several square meters) Nb$_3$Sn films [68,76–79]. The growth of the polycrystalline films was driven by Sn-diffusion from the surface along the grain boundaries to the Nb$_3$Sn-Nb boundary [80, 81]. Further details of this process are discussed in Sect. 5.1.1. Optimized Nb$_3$Sn/Nb films at LHe temperatures and frequencies between 0.2 and 20 GHz exhibited a low surface resistance R_s, which was about a factor of 100 lower than that of pure Nb at 4.2 K and 20 GHz [78]. This improvement factor resulted mainly from the enhanced reduced energy gap $\delta = \Delta/k_B T_c$ ($\delta_{Nb3Sn}/\delta_{BCS} \approx 1.3$). The quadratic frequency scaling of R_s was in accordance with the theoretical expectation for strong-coupling superconductors (Sect. 1.3.3 and below). Regarding the small microwave penetration depth $\lambda \leq 100$ nm, thin ($\leq$ µm) films already sufficed to exploit "bulk" properties of this superconductor. Finally, high power-handling capability was demonstrated by constant R_s over a large range of microwave field amplitudes, $B_s(4.2 \text{ K}) \leq 100$ mT [18b].

Despite the promising results of the TVD process, the deposition of Nb_3Sn films onto low-loss dielectric substrates like sapphire was mainly pursued by low-temperature processes. These combine in one step the deposition of Nb and Sn as well as the conversion into the A15 phase. Prominent candidates are electron-beam coevaporation of metals [82] and magnetron sputtering [83] of stoichiometric targets. Due to the thermodynamic conditions for the formation of phase-pure Nb_3Sn and the high vapor pressure of tin, the substrate temperature T_s had to be adjusted properly and within narrow margins. It was shown that the superconductive properties of Nb_3Sn films improved with higher T_s values [82, 84]. A condensation of the A15 phase ("phase-locking") was observed for an extended stoichiometric interval above 950°C [82] (Sect. 5.1.1). However, T_s remained below 1000°C in all cases.

Superconductive Parameters of Nb_3Sn Films on Sapphire Substrates

The superconductive parameters of Nb_3Sn films on dielectric substrates were mainly concluded from surface impedance measurements, tunneling spectroscopy, or from investigations of the upper critical field near T_c. Based on the measurable slope $\partial B_{c2}(T)/\partial T$, and with supplementary data on T_c, $\rho(T_c)$ and $\Delta/k_B T_c$, a complete set of characteristic lengths can be deduced on the basis of the scaled BCS theory ([64], Sect. 1.3.3). Table 2.3 summarizes representative data on a few transport properties collected from various publications. Data obtained with the TVD films on sapphire, which are described in further detail in the following section, are also included in the table.

The structural quality of the Nb_3Sn films is related to the average grain size a. It can be concluded from Table 2.3 that the high conversion temperature applied in the TVD process is beneficial for large grains. TVD films are therefore expected to suffer less from granular effects (see (1.39) and (1.40) in Sect. 1.2.2), which is confirmed by the high critical current densities. However, the impact of granular effects on the surface impedance was discussed controversely, reaching from "negligible" [80] to "Josephson coupling" [10, 83, 85].

The scatter of the T_c values might reflect the different metal stoichiometries, which resulted from the different deposition temperatures, and/or internal strain of the lattice structure. The value of the normal-state resistivity ρ depends on the density-of-states at the Fermi energy, on the electron-phonon coupling strength (and is thus related to $\Delta/k_B T_c$) and on the mean free path ℓ of the electrons. The ρ values (respectively ℓ and $\Delta/k_B T_c$ values) are found to decrease (increase) from sputtered to coevaporated and further to TVD films, indicating increasing degrees of phase-purity, ordering and coupling strength. The high resistivity of the sputtered Nb_3Sn films might hint of an additional influence of weakly coupled grain boundaries [89].

Values for the coherence length ξ_0 were deduced for the films on sapphire from $B_{c2}(T)$ analyses and for the Nb_3Sn/Nb films from numerical Z_s computations. Similar values were stated for all investigated deposition techniques.

2.2 Surface Impedance of Nb$_3$Sn Films on Sapphire

Table 2.3. Comparison of selected transport parameters of Nb$_3$Sn films prepared by different techniques. The R_s(4.2 K) values were scaled with f^2 to 87 GHz.

Quantity	Nb$_3$Sn/sapphire sputtering [83, 86]	Nb$_3$Sn/sapphire coevaporation [64,82]	Nb$_3$Sn/sapphire TVD [87, 88] (Sect. 2.2.2)	Nb$_3$Sn/Nb TVD [68, 76, 78]
Grain size (nm)	5–250	≤ 730	≤ 3000	≤ 3000
T_c (K)	16.5–17.9	16.0–17.9	18	18.2
J_c(4.2 K) (MA/cm^2)	≥ 0.5	—	5.0–6.5	—
$\rho(T_c)$ (μΩcm)	45	9–36	7–10	10–12
ℓ (nm)	—	2–10	9–12	—
$\Delta/k_B T_c$	1.9	1.2–2.1	1.8–2.3	2.1–2.3
ξ_0 (nm)	—	5.7–7.7	7.0	6.0
$\lambda_L(T=0)$ (nm)	600	90–100	65–80	40–60
R_s(4.2 K) (mΩ)	7	2	<0.4	<0.08

As indicated by the ratio ξ_0/ℓ, the electromagnetic response of Nb$_3$Sn falls between the clean and the dirty limit. The London penetration depth varied by one order of magnitude for the differently prepared films. Since the data were corrected for finite mean free path effects, the large λ value of the sputtered films again reflects the influence of grain boundaries. Regarding the inequality $\lambda_L \gg \xi_0, \ell$, the local limit is appropriate for the description of the current–field relation (Table 1.1 in Sect. 1.1.1), independent of the route of preparation. Except for the coevaporated films, the λ_L data were deduced from microwave measurements as described below. Such an analysis revealed absolute values which are by about 20 to 30% lower than indicated by $B_{c2}(T)$ measurements (Table 2.4 below). Within this range, the λ_L results for the coevaporated and the TVD films are comparable.

The data on the surface resistance at 4.2 K were scaled quadratically to 87 GHz. (Numerous TVD films on sapphire were characterized at this frequency.) The finite values found for sputtered and coevaporated films indicate the influence of impurities and Josephson-coupled grain boundaries. The R_s data obtained with Nb$_3$Sn/sapphire were limited by the resolution of the measurement system (Sect. 2.1.3).

2.2.2 Superconductive Properties of the Nb$_3$Sn Films Prepared by Tin Diffusion

Phenomenological Results on the Surface Impedance

As discussed along with Table 2.3, the application of tin vapor diffusion to the preparation of Nb$_3$Sn films on sapphire yielded excellent microwave properties. This process required the combination of two deposition techniques.

Nb precursor films were first magnetron sputtered at room temperature onto the dielectric substrates, and characterized in terms of their superconducting transport properties. In a second step, the Nb films were converted into Nb_3Sn by tin diffusion. This procedure resembled closely the preparation of Nb_3Sn on Nb cavities. Further details of the TVD process are described in Sect. 5.1.1.

The superconductive properties of the Nb precursors and of the converted Nb_3Sn films were characterized in two ways. Inductive measurements based on the detection of the third harmonic of an ac excitation signal at 1 kHz allowed a quick determination of the transition temperature T_c, the transition width ΔT_c and the critical current density J_c (4.2 K). This method, originally developed for the testing of HTS films [90], was modified for application at 4.2 K by immersing a ⌀12-mm sensor tube through the fitting of a standard LHe dewar. The surface impedance was investigated at 87 GHz using the computer-controlled Cu cavity system and at 19 GHz using the Nb-shielded sapphire resonator, both described in Sect. 2.1.3.

Typical Nb precursor films displayed $T_c = 7.8$–9.3 K, $\Delta T_c = 0.05$–0.30 K and J_c (4.2 K) = 0.8–2.8 MA/cm^2. Higher (respectively lower) T_c values and smaller (larger) transition widths correlated with higher (lower) J_c values, reflecting the influence of granularity [91] and of interstitial impurities like oxygen [92]. The corresponding parameters of the same films after conversion into Nb_3Sn were $T_c = 18.0$ K, $\Delta T_c = 0.1$ K and J_c (4.2 K) = 5–6.5 MA/cm^2.

Figure 2.8 shows a comparative set of current–voltage characteristics of selected samples before (i.e., as-sputtered Nb) and after conversion into Nb_3Sn

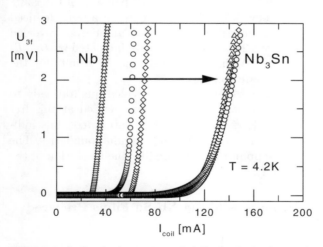

Fig. 2.8. Inductively measured AC current–voltage characteristics, used for the determination of J_c (4.2 K) from the onset of the third-harmonic voltage U_{3f}, for selected Nb precursor (*different symbols*), and for the same films after conversion into Nb_3Sn (*same symbols*) [93].

[93]. The scatter of the J_c values of the Nb precursors was strongly reduced for the A15 films. A similar improved reproducibility was observed for the surface resistance $R_s(87\,\text{GHz})$ [87]. Figure 2.9 displays some R_s data, which were corrected for the finite film thickness (Sect. 1.1.3). The temperature dependence of the surface impedance of the Nb films agreed with the expectation from BCS theory (Sect. 1.3.2). Quantitative analysis yielded $R_s(4.2\,\text{K}) = 0.5$–$2.5\,\text{m}\Omega$ and $\lambda = 37$–$155\,\text{nm}$. While the lower values correspond to data which are typical for bulk Nb, granularity and interstitial impurities deteriorated these parameters. The sensitivity of the Nb films to defects could especially be concluded from the resistivity $\rho(T_c)$ which varied by almost two orders of magnitude between 1 and $35\,\mu\Omega\text{cm}$. In accordance with [94], $\rho(T_c)$ was correlated with the transition temperature T_c. In contrast to the larger scatter for the Nb precursors, reproducibly low $\rho(T_c)$ values of 5–$10\,\mu\Omega\text{cm}$ were concluded for the Nb$_3$Sn films, indicating a high degree of ordering of the Nb chains of the A15 unit cells [72,73] (Sect. 5.1.1).

The high reproducibility of the Nb$_3$Sn films, in spite of the large scatter of the precursor quality, indicates a complete conversion into the A15 phase. Even so, there seem to be minor differences in $R_s(T)$ of the Nb$_3$Sn films below 9 K. Residual contributions from the Nb precursor film can be excluded since $R_s(T)$ is continuous at the transition temperature of Nb. As analyzed further in Sect. 2.2.3, $R_s(T \leq T_c/2)$ tends to increase slightly with increasing film thickness (MP1, 3 µm; MPZ18, 0.56 µm). However, more detailed studies of the low-temperature behavior of R_s require measurements with improved resolution: The sensitivity limit of the 87 GHz system was higher than typical R_s data achieved with optimized Nb$_3$Sn films on Nb cavities ([68], Table 2.3).

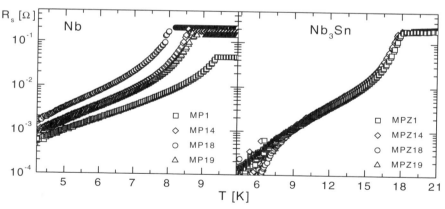

Fig. 2.9. Temperature dependent surface resistance at 87 GHz for selected Nb precursor films (indicated by the numbering in the legend, *left*), and for the corresponding converted Nb$_3$Sn films (same numbering, *right*) [87,93].

On the other hand, it was already lower than the R_s data of Nb$_3$Sn/sapphire prepared by one-step processes [82, 83].

Analysis of the Temperature Dependence of the Surface Impedance

Measurements of the surface resistance always approach a nonvanishing value as the temperature is decreased. Possible physical sources for a finite residual resistance are granularity and gapless superconductivity (Sects. 1.2.2 and 1.3.3). However, care has to be taken to generally relate the experimental data to the intrinsic properties of the material. It is well known from conventional superconductors like Pb or Nb that nonstoichiometric segregations as well as surface contaminants or roughness cause residual power dissipation, thus mimicking enhanced R_s values. Furthermore, power losses resulting from unidentified sources (see the parasitic term in (2.16)) lead to the same misinterpretation. The residual value of R_s at low temperatures is therefore usually subtracted from the $R_s(T)$ data before comparisons with theoretical expectations are performed [95]. Such a procedure makes it a challenging task to properly extract the low-temperature behavior of R_s and to relate it to physical mechanisms.

The measurable relative change of the penetration depth with temperature (Sect. 2.1.1) is usually converted onto an absolute scale by assuming an appropriate model for $\lambda(T)$. Fitting the measured data $\Delta\lambda(T)$, corrected for finite film thickness, at temperatures $T > T_c/2$ to $\lambda(T)/\lambda(0)$ values tabulated for isotropic clean BCS superconductors [96] was found to be an empirically reliable procedure, even for data obtained with HTS films ([22, 23], Sect. 2.3). However, care has to be taken to interpret the absolute values in the context of theoretical explanations.

Having these facts in mind, the temperature dependences of R_s and $\Delta\lambda$ of the Nb$_3$Sn films could be simulated numerically in the framework of the BCS theory [97]. Figure 2.10 displays a representative comparison between measured and calculated Z_s data. Good agreement was found with the low-temperature exponentials $[R_s(T) - R_s(0)] \propto \exp[(-\Delta/kT_c)T_c/T]$ and $[\lambda(T) - \lambda(0)] \propto \exp[(-\Delta/kT_c)T_c/T]$, which stresses the applicability of the scaled BCS theory. There is a trend of the measured compared to the simulated $\lambda(T)$ data to display a weaker temperature dependence below $T_c/2$. This behavior is illustrated in Fig. 2.11 in terms of the superfluid fraction $\lambda^{-2}(T)/\lambda^{-2}(0) = \sigma_2(T)/\sigma_2(0) = x_s(T)/x_s(0)$. While the data are satisfactorily reproduced by the scaled BCS theory, the described deviations indicate the corrections expected for strong pair coupling (Sect. 1.3.3). Quantitative analysis yielded for various Nb$_3$Sn films reduced energy gap values Δ/kT_c=1.8–2.3 and penetration depth values λ_L=65–80 nm. Similar gap functions fitted $R_s(T)$ equally well, namely the dependences

$$R_s(T) = R_1 \exp[(-\Delta/kT_c)T_c/T] + R_{\text{res},1} \tag{2.36a}$$

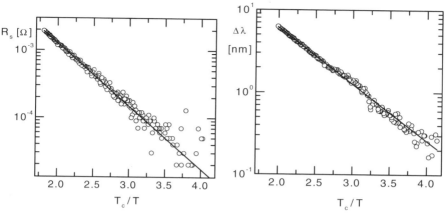

Fig. 2.10. Comparison of measured (*circles*) and calculated (*lines*) $R_s(T)$ (*left*) and $\lambda(T)$ data (*right*) at 87 GHz for a typical TVD Nb$_3$Sn film on sapphire [93].

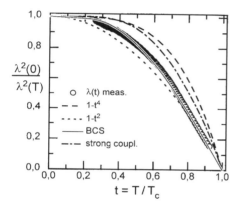

Fig. 2.11. Temperature dependence of $\sigma_2(T)/\sigma_2(0)$ resulting from the data in Fig. 2.10, compared with the expectation from BCS and strong-coupling theories ($\lambda_{\text{coupl}} \approx 3$; Sects. 1.3.2 and 1.3.3) and with the phenomenological approaches $1 - (T/T_c)^2$ and $1 - (T/T_c)^4$.

and

$$R_s(T) = R_2 T_c/T \exp[(-\Delta/kT_c)T_c/T] + R_{\text{res},2} \quad (2.36\text{b})$$

The Δ/kT_c values of the latter expression were about 20% higher than those of the former.

Once absolute λ values are available, the quasi-particle conductivity $\sigma_1(T)$ can be evaluated from the $Z_s(T)$ data according to (1.43) in Sect. 1.2.3. The σ_1 data corresponding to the surface impedance displayed in Fig. 2.10 are shown in Fig. 2.12 together with the result expected from weak-coupling BCS theory in the low-frequency limit assuming a smearing of the energy gap by 2% ([98], Sect. 1.3.2). The extremal behavior can obviously be identified with coherent scattering of the quasi-particles ("coherence peak"). The good agreement regarding the high measurement frequency $hf/\Delta \approx 0.1$ [99] and the uncertainties related to the absolute values of λ has to be considered

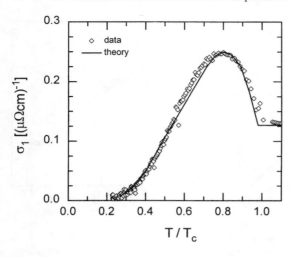

Fig. 2.12. Temperature dependent quasi-particle conductivity $\sigma_1(T)/\sigma_1(T_c)$, deduced from the data at 87 GHz in Fig. 2.10 after subtraction of the residual surface resistance, compared with the expectation from BCS theory for low frequencies [98] (see Fig. 1.7 in Sect. 1.3.2).

with care. At temperatures below the peak, $\sigma_1(T)$ decreases exponentially as expected for the existence of a finite energy gap.

Frequency Dependence of the Surface Resistance

According to the data summarized in Table 2.3, the R_s values obtained with the best Nb$_3$Sn/sapphire agree with those of Nb$_3$Sn on Nb. It is therefore tempting to assume that the same is true for the frequency dependence of R_s. Figure 2.13 displays data obtained with Nb$_3$Sn/Nb at two different temperatures [78] (circles, 4.2 K; squares, 9 K), and data measured for Nb$_3$Sn/sapphire at 9 K and 19 and 87 GHz (diamonds, [87, 88]). The error bars of the 4.2-K data reflect the influence of the residual resistance. The TVD films on sapphire fit well to the quadratic frequency dependence, which is expected for strong-coupling superconductors. The absence of deviations from $R_s \propto f^2$ up to 87 GHz reflects the high gap frequency $\Delta/h \approx 850$ GHz. With respect to granular effects (Sects. 1.2.2 and 2.3.2), it also indicates that the conductivity ratio of the grain boundaries be $y \ll 1$. If Josephson-coupled at all, the $I_c R_J$ product must be at 9 K well above $\Phi_0 f \approx 0.2$ mV.

Deduction of Characteristic Lengths from $Z_s(T)$ and $B_{c2}(T)$ Data

The $Z_s(T)$ data of the Nb$_3$Sn films yielded reliable information about the reduced energy gap and the penetration depth. More care had to be taken in the deduction of ξ_0 and ℓ because the numerical simulation of $Z_s(T)$ relied on the weak-coupling limit. An important supplement to find the characteristic lengths in a consistent way is based on the analysis of the upper critical field $B_{c2}(T)$ close to T_c. The slope $(\partial B_{c2}/\partial T)|_{T_c}$ is related to the coefficient of the normal-state electronic specific heat γ^* [100], renormalized with respect

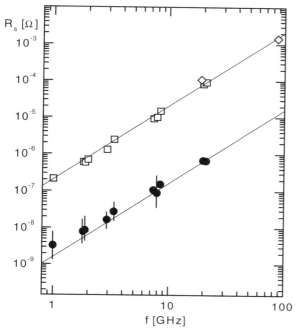

Fig. 2.13. Frequency dependence of R_s at 4.2 K (*filled*) and 9 K (*open symbols*), obtained with Nb$_3$Sn/Nb (*circles and squares*, [78]) and with Nb$_3$Sn/sapphire (*diamonds*, [87,88]). The straight *lines* illustrate the quadratic frequency dependence.

to the strong electron–phonon coupling, and to the normal-state resistivity $\rho(T_c)$ by [64]

$$-\frac{\partial B_{c2}(T)}{\partial T} = \frac{\eta_{Bc2}}{R(x)} \left[4.722 \times 10^{-8} \gamma^{*2} T_c + 5.26 \times 10^3 \gamma^* \rho(T_c) \right] . \quad (2.37)$$

The coefficient η_{Bc2} denotes the strong-coupling correction to B_{c2} within the scaled BCS theory, as introduced in (1.71). Its value can be inferred from the reduced energy gap. The function $R(x) \approx 1$ is related to the Gorkov function, evaluated at $x = 0.9\xi_0/\ell$. Once $\rho(T_c)$ is known, e.g., from the normal-state Z_s data, measurements of $(\partial B_{c2}/\partial T)|_{T_c}$ yield γ^*. Based on the formulas collected in [64], it is then straightforward to deduce ξ_0 and ℓ.

Figure 2.14 displays typical results for $B_{c2}(T)$, which were deduced from inductively measured T_c curves in a DC magnetic field applied parallel to the 1 cm × 1 cm Nb$_3$Sn films. Also shown are curves expected for the clean and dirty limits [101]. The experimental data fit well to the theoretical expectations, but did not allow one to distinguish between both limits in the accessible field range. Table 2.4 contains a detailed comparison of various superconductive parameters between two TVD and two coevaporated Nb$_3$Sn films [64]. Increasing phase purity is reflected in increasing (decreasing) values of ℓ and $|(\partial B_{c2}/\partial T)|_{T_c}$ [$\rho(T_c)$, κ, $\lambda_{GL} = \kappa \cdot \xi_{GL}$ and $B_{c2}(0)$], while ξ_0, λ_L and ξ_{GL} change little. In comparison with the microwave data for λ_L (Table 2.3), the B_{c2} analysis yields values that are about 25% higher. This discrepancy is of similar magnitude as the correction $\eta_{B_{c2}}$ and might reflect the systematic error introduced by fitting the microwave $\Delta\lambda(T)$ data to weak-coupling BCS expressions.

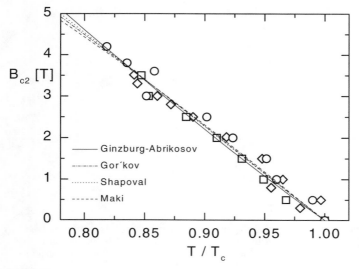

Fig. 2.14. B_{c2} values measured close to T_c for three 1 cm × 1 cm Nb$_3$Sn films on sapphire (*symbols*), and the theoretical expectation for the clean and the dirty limits (*curves*) [93].

2.2.3 Effect of Film Thickness and Grain Size on the Surface Impedance

The previous results described the typical performance of the TVD Nb$_3$Sn films on sapphire. In more detail, slight but systematic differences of the transport properties of these films were found as a function of film thickness and therefore of grain size.

Structural Properties

Figure 2.15 shows SEM images of the surfaces of three Nb$_3$Sn films on sapphire with different thickness (left, 0.6 µm; middle, 1.2 µm; right, 2.2 µm). Except for the thickness of the Nb precursor, the films were prepared under identical conditions. The average grain size a is seen to increase with increasing film thickness d (about 0.7, 1.0 and 1.3 µm in the sequence of increasing

Table 2.4. Selected superconductive parameters, deduced from $B_{c2}(T)$ data for two TVD ("MP1", $d_F \leq 3\,\mu m$ and "MPZ19", $d_F = 0.56\,\mu m$, [87]) and two coevaporated (#1 and #2, [64]) Nb$_3$Sn films on sapphire. Data on $\Delta/k_B T_c$ and $\rho(T_c)$ were inferred from transport measurements. The indices "GL" refer to the Ginzburg–Landau quantities (see text and Sects. 1.1 and 3.1).

Quantity	MP1	MPZ19	#1	#2
$-(\partial B_{c2}/\partial T)\|_{T_c}$ (T/K)	1.2	1.22	2.6	1.8
T_c (K)	18.0	18.0	16.1	17.9
$\Delta/k_B T_c$	1.9	2.2	1.9	2.1
η_{Bc2}	1.05	1.21	1.05	1.17
$\rho(T_c)$ ($\mu\Omega$cm)	10.0	7.7	33.3	9.1
ℓ (nm)	8.9	12	2.6	10
ξ_0 (nm)	7.0	7.0	5.7	7.7
λ_L (nm)	89	88	100	90
ξ_{GL} (nm)	3.9	3.9	3.2	3.2
λ_{GL} (nm)	75	54	93	64
κ	19	14	33	20
$B_{c2}(0)$ (T)	15	11	25	15

film thickness). A systematic study revealed the empirical relation $a \propto d^{1/2}$ up to $d = 3\,\mu m$ at which $a \approx 1.2\,\mu m$ [102]. The grain size thus seems to be limited by the tin diffusion during the conversion process, rather than by the film thickness. Figure 2.15 also demonstrates that the column-like growth mode of the thin film turns to grain lamellae in the thick film, which might be indicative of enhanced texture [81, 103].

Fig. 2.15. SEM photographs of the surfaces of Nb$_3$Sn films of different thickness (*left*, 0.6 μm; *middle*, 1.2 μm; *right*, 2.2 μm) prepared by tin-vapor diffusion. The magnification is marked by the white bars (length of 4 μm) in the lower part of the pictures [88].

80   2. Measurements of the Surface Impedance at Linear Response

Superconductive Properties

Figure 2.16 shows (from bottom to top) the film thickness dependences of the microwave penetration depth λ, the resistivity $\rho(T_c)$, the critical shielding current density J_c (4.2 K) and the reduced energy gap $\Delta/k_B T_c$. For thick films the $\lambda(d_F)$ trend approaches a saturation level at about 65 nm that can be identified with the London value λ_L, and which agrees well with the data deduced in the same way for Nb_3Sn/Nb. Enhanced values $\lambda > \lambda_L$ are ob-

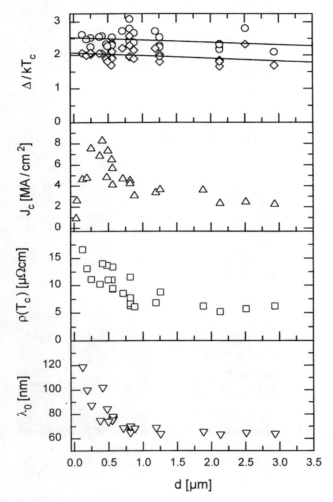

Fig. 2.16. Dependences on the film thickness d_F (from *bottom* to *top*) of the microwave penetration depth λ, the resistivity $\rho(T_c)$, the critical shielding current density $J_c(4.2$ K) and the reduced energy gap $\Delta/k_B T_c$ [from (2.36)]. The lines are only visual guides [93, 102].

served for thin films at $d_F \leq 700$ nm. The normal-state resistivity was also found to increase with decreasing film thickness in this interval, leading to the correlation between λ and $\rho(T_c)$ displayed in Fig. 2.17. If both quantities were affected by the same effect, namely quasi-particle scattering, this correlation should follow the local-limit scaling law (Sect. 1.1.1)

$$\lambda(\ell) \approx \lambda_L \sqrt{1 + \frac{\xi_0}{\ell}}, \qquad (2.38)$$

where ℓ is inversely proportional to $\rho(T_c)$. Such a behavior can indeed be resolved for $\rho \geq 11\,\mu\Omega$cm, as indicated by the solid line in Fig. 2.17. The increase (decrease) of the penetration depth (mean free path) with decreasing film thickness and thus grain size could be provoked by scattering at grain boundaries.

It is remarkable, though, that the same functional dependence is also expected for weakly Josephson-coupled grains [91]. However, the latter interpretation seems to be in contradiction with the observed increase of J_c with decreasing film thickness at $d_F \geq 300$ nm (Fig. 2.16). Obviously, in this range smaller grains introduce additional scattering centers that act at the same time as pinning centers. Pinning at grain boundaries was also reported in [81, 104] for Nb$_3$Sn filaments. $J_c(a)$ was observed to decrease like $1/a$ (and thus approximately like $d^{-1/2}$ in our case), in accordance with the present observation. At d_F below about 300 nm, the reduction of $J_c(d_F)$ indicates the existence of weakly coupled grains and/or nonstoichiometric Nb-Sn phases, which could have formed at the surface of the films during tin conversion.

Figure 2.17 reveals that $\lambda \approx \lambda_L$ stays constant at resistivities below $10\,\mu\Omega$cm. This apparent contrast to (2.38) can be resolved by accounting for the very strong sensitivity of $\rho(T_c)$ on the ordering of the Nb$_3$Sn unit cells [72], thus making it essentially independent of ℓ close to perfect ordering (see Fig. 5.3 in Sect. 5.1). The crossover value of the resistivity around

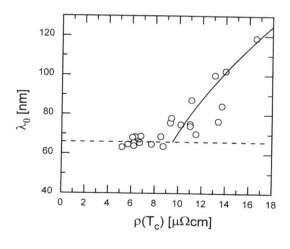

Fig. 2.17. Film thickness-induced correlation between the microwave penetration depth λ and the normal-state resistivity $\rho(T_c)$, according to the data shown in Fig. 2.16. The *solid curve* illustrates the relation (2.38) for $\rho > 9.5\,\mu\Omega$cm using $\rho = 3.3\,\mu\Omega\,\text{cm} \times \frac{\xi_0}{\ell}$. The *dashed line* indicates the saturation of λ at the London value [93, 102].

10 μΩcm agrees well with earlier results on high-quality Nb_3Sn films [64, 72, 78, 82].

The uppermost diagram of Fig. 2.16 displays the reduced energy gap values δ obtained from the $R_s(T)$ data of differently thick TVD Nb_3Sn films on sapphire. The different symbols refer to the two fitting procedures of (2.36) (diamonds (2.36a) and circles (2.36b)). The scatter of the gap values reflects the reproducibility of the film quality. There seems to be a weak trend of δ to decrease with increasing film thickness, as indicated by the lines in the diagram. A corresponding feature was already mentioned in relation to Fig. 2.9, which showed $R_s(T)$ at low temperatures to increase with d_F. Such a behavior cannot be explained by granularity for which $R_s \propto a^{-1}$ and accordingly $R_s \propto d_F^{-1/2}$ is expected. Mean-free-path effects also appear unlikely since the thick films tend to be already in the clean limit (Fig. 2.17). Rather, the effect of film thickness (or grain size) on the reduced energy gap δ and on $R_s(T \leq T_c/2)$ might be indicative of a slight anisotropy of the energy gap in large-grained Nb_3Sn. In accordance with the SEM photographs of Fig. 2.15, anisotropic properties of the polycrystalline material are expected to show up only in textured grain structures. Furthermore, according to Anderson's theorem [105], the observability of anisotropic features requires sufficiently long mean free paths along the Fermi surface to avoid averaging. Qualitative conclusions for an anisotropic energy gap in Nb_3Sn were reported also in [64, 106, 107].

2.3 Surface Impedance of Oxide Superconductor Films

The high-frequency properties of HTS single crystals and epitaxial films prepared by many different techniques (e.g., sputtering, pulsed laser deposition and physical or chemical evaporation techniques) have been the subject of numerous publications (e.g., [3, 4, 12a] and references therein). The basic features of the different deposition methods are discussed in Sect. 5.1.2. Since very high quality films can be obtained by each of them, only general features of the HTS films at microwave frequencies are treated in this section.

Before going into details, the different HTS compounds are worth a few remarks. Like for LTS, it is favorable to apply HTS at temperatures below about $T_c/2$ in order to assure advanced pair condensation (Sect. 1.3). HTS compounds with a high transition temperature provide, therefore, higher possible operating temperatures or, alternatively, lower surface impedance at a given temperature (e.g., 77 K). But according to (1.13), high transition temperatures imply short coherence lengths and large penetration depths, and thus low critical field levels that might limit the benefits of these materials (Chap. 3). To mimimize the number of nonpairing charge carriers, phase purity and high crystalline quality are required. The most prominent HTS compound, which provides lower losses than copper or gold at 77 K up to some 100 GHz, is ReBa$_2$Cu$_3$O$_{7-x}$ ($T_c = 93$ K [108], referred to as Re-123 or ReBCO, where Re is a rare-earth like Y, Yb, Gd, Nd or Eu [109]). There are about 40 different HTS compounds total, which have been analyzed from a chemical point of view [110]. Most of them are less suitable for microwave applications than Y-123 [111]. To the frequently cited HTS materials belong the Bi$_2$Sr$_2$Ca$_n$Cu$_{n+1}$O$_y$ series with $n = 1$ or 2 (referred to as Bi-2212 or Bi-2223) and the related Tl$_2$Ba$_2$Ca$_n$Cu$_{n+1}$O$_y$ (TBCCO or Tl-2212, Tl-2223) compounds. Unfortunately, the Bi system is a complex multiphase system, and the Tl superconductors demand careful handling because of the high vapor pressure of the poisonous Tl oxides. Nevertheless, numerous laboratories developed successful preparation routes for the in-situ growth of epitaxial Bi films [112–115] and the ex-situ growth of Tl films [116–121]. Bi films exhibit high R_s values due to phase impurities or granularity. The Tl films provide, between 40 K and 70 K, R_s values about a factor of 2 lower than YBCO at the same temperature [116, 117, 122, 123], but suffer from granular effects which show up at elevated magnetic field levels. Altogether, Y-123 still presents the most thoroughly investigated and most frequently adopted HTS compound for microwave applications, and Tl-2212 is a well-established alternative. Consequently, the following analysis is restricted to these two materials.

Section 2.3.1 recalls the basic features of granular YBCO layers. The main part, Sect. 2.3.2, focuses on the temperature and frequency dependent surface impedance of epitaxial HTS films. Effects of varying the cation and oxygen composition are summarized in Sect. 2.3.3. The chapter is concluded, in Sect. 2.3.4, with an analysis of $Z_s(T)$ data of YBCO films in the contrasting frameworks of the two-band model and the d-wave approach introduced in Sect. 1.3.3.

2.3.1 Granular YBa$_2$Cu$_3$O$_{7-x}$ Films

Polycrystalline bulk and thick film samples of arbitrary shape belonged to the first types of HTS material that became available for microwave investigations [124, 125]. After suitable techniques for the deposition of epitaxial films on dielectric single crystals were developed, these replaced the highly granular "precursors" for obvious reasons, though at the expense of losing the "third dimension". It is nevertheless instructive to recall the basic aspects of granular samples, especially in relation to the status-quo achieved with the surface impedance of epitaxial HTS films (Sect. 2.3.2).

Soon after the discovery of YBCO it was observed that, due to the short coherence length, boundaries between adjacent grains in polycrystalline material displayed Josephson behavior [126]. The critical Josephson current was found to increase strongly with decreasing misorientation angle between the grains [127]. The Josephson behavior was also reflected in the dependence of the surface impedance $Z_s(T, f)$ of polycrystalline samples on the average grain size and on the degree of texture [128–133]. Upon cooling below T_c,

Z_s dropped steeply due to the expulsion of the magnetic field, i.e., due to the reduction of penetration depth from the skin-depth to the London value, as typical for any superconductor. This steep transition turned into a much more gradual decrease of $Z_s(T)$ at a temperature $T_c^* \leq T_c$ only a few degrees below T_c. Here, T_c^* is the temperature at which the thermal energy $k_B T_c^*$ becomes comparable to the Josephson coupling energy $\Phi_0 I_c/2\pi$ ([134–136], Sect. 1.2.2). Below T_c^*, the surface impedance was dominated by the electrodynamic properties of the grain boundary Josephson junctions (e.g., [89]).

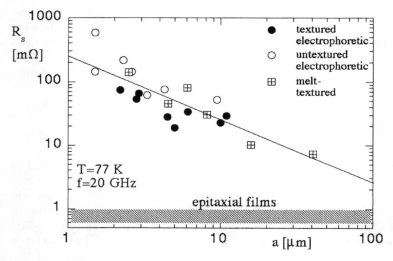

Fig. 2.18. Dependence of R_s (77 K, 20 GHz) of polycrystalline YBCO samples on the average grain dimension a, for (un-)textured electrophoretic layers [131] and melt-textured samples [129,130]. The line represents the dependence $R_s \propto a^{-1}$ expected from granular Z_s models (1.39). The dashed area indicates the quality typical of epitaxial YBCO films (Sect. 2.3.2).

Figure 2.18 shows a typical example for the dependence of $R_s(77\text{ K})$ at 20 GHz of polycrystalline YBCO layers on the average grain dimension a. Open (filled) circles stand for untextured and textured electrophoretic layers [131], crossed squares denote data from melt-textured films [129, 130]. The clearly resolved overall trend $R_s \propto 1/a$ agrees well with the expected surface resistance of granular superconductors (1.39). Furthermore, the benefit of texturing, which reflects the anisotropy of the Y-123 material [137], is also obvious from the comparison between untextured and textured samples of similar grain sizes. Indicated in Fig. 2.18 by the dashed area is the present state of the art of R_s achieved with epitaxial films at 77 K and 20 GHz. The difference between granular and epitaxial films can be interpreted as the consequence of an effectively enhanced grain size (the linear extrapolation in

Fig. 2.18 points to mm dimensions) as well as of a strongly improved structural orientation of the individual growth nuclei.

The influence of granularity on the frequency dependence of Z_s was discussed, e.g., in [133]. The main result was that the Josephson coupling between the grains yielded an enhanced conductivity ratio $y = \sigma_1/\sigma_2 \propto f/\rho$, where ρ is the normal-state resistivity (see Sect. 1.2 and below). Values of $y \approx 0.2$–0.4 (respectively $y \approx 0.8$–1.2) were concluded for electrophoretic layers at 21.5 GHz and 4.2 K (77 K) [131]. The frequency exponent $b_- = \partial \log(R_s)/\partial \log(f)$ amounted to about 2.0 (1.5). At frequencies above 30 GHz, b_- decreased with further increasing frequency. This behavior described well numerous published data on polycrystalline YBCO samples ([133] and references therein). A quantitative comparison of $R_s(f)$ between granular and epitaxial YBCO films is postponed until Fig. 2.23 in the following section.

2.3.2 Epitaxial YBa$_2$Cu$_3$O$_{7-x}$ and Tl-Ba-Ca-Cu-O Films

Temperature Dependence of the Surface Resistance

Figure 2.19 displays typical temperature dependences of the surface resistance of two epitaxial Y-123 films at 87 GHz, plotted in an Arrhenius plot [22, 23]. Three different temperature regions can be distinguished from this figure. Close to T_c, the steep decrease of R_s upon cooling results from the reduction of the penetration depth from the skin depth δ to the London penetration depth. Within the two-fluid model, the ratio λ/δ scales with temperature and frequency like $y^{1/2}$ (see (1.18c) and (1.18d)).

Below about 0.8 T_c, a nonmonotonic behavior of $R_s(T)$ is often observed for high-quality samples. At a temperature T_0, which increases with frequency (e.g., $T_0 \approx 30$ K at 2 GHz [138], 40 to 50 K at 8 GHz [139], 50 K at 87 GHz [22, 23] and 70 K at 600 GHz [140]), R_s passes through a broad maximum. This behavior was attributed in [138] to a strongly temperature dependent scattering rate: while the unpaired fraction x_n of quasi-particles decreases with decreasing temperature, $\tau(T)$ increases and causes the observed maximum. Single-crystal measurements revealed an almost exponential drop $\tau^{-1}(T) \propto \tau^{-1}(T_c) \exp\{\alpha[1-(T_c/T)^\beta]\}$ from 10^{13} Hz at T_c to 10^{11} Hz at helium temperature [138]. Similar results were reported also for epitaxial films. Such a strong scattering, which is unique to the HTS materials, must be taken into account in the theoretical analysis of their surface impedance (Sect. 2.3.4).

Figure 2.20 displays the results for $\sigma_1(T)$ which were deduced from Z_s data of epitaxial YBCO films comparable to those in Fig. 2.19 (circles and squares). Shown for comparison are the $\sigma_1(T)$ data of a typical TVD-Nb$_3$Sn film (diamonds, Fig. 2.12) and the coherence peak expected for the low-frequency limit (solid line, see discussion in Sect. 2.2). The extremal behavior of $\sigma_1(T)$ induced by incoherent quasi-particle scattering needs obviously to be distinguished from the BCS coherence effects. Phenomenologically, the increase of the quasi-particle conductivity below T_c is much more gradual in HTS than in Nb$_3$Sn, and almost independent of film quality. Furthermore, the maximum of $\sigma_1(T)$ occurs at significantly lower temperatures of about $0.5\, T_c$. The low-temperature behavior of $\sigma_1(T)$ is greatly affected by the residual resistance (see below). Finally, as argued from the theoretical point of view (e.g., [99, 141]), the coherence peak is suppressed in strong-coupling superconductors, and should be absent in HTS.

The surface resistance (Fig. 2.19) saturates at low temperatures $T \leq T_0$ at a finite level R_{res}, which is discussed separately in the following paragraph. Experimental observation of an exponential $R_s(T)$ dependence at $T < T_c/2$ with a reduced energy gap of $\delta = 0.8$ was reported for well oxygenated sputtered YBCO films, which were post-annealed and slowly cooled in plasma-activated oxygen [142]. The exponential $R_s(T)$ disappeared after a few days, presumably due to oxygen loss (Sect. 2.3.3). Further support for an exponential temperature dependence of R_s of different YBCO films and single-crystals was reported with $\delta = 0.8$–1.3 in [143–145]. The data in Fig. 2.19 yielded reduced energy gap values for the two films of $\delta = 0.7$–0.8 after a residual resistance around 1 mΩ was subtracted [146].

Also shown in Fig. 2.19 are typical R_s data of Nb$_3$Sn, which could be fitted to a well-defined reduced energy gap $\delta = 2.0$–2.2. The low-temperature R_s value was limited by the measurement accuracy of the resonator (Sect. 2.1.3). The temperature dependent surface resistance of optimized Tl-2212 films, finally, resembled closely that of Y-123 at the same reduced temperature [6], including similarly high residual resistance values at 4.2 K [147].

Residual Resistance

In comparison with the metallic high-temperature superconductor Nb$_3$Sn, even the best epitaxially grown Y-123 films reveal high residual microwave losses, which limit the surface resistance at low temperatures to R_{res} (Fig. 2.19 and [3, 4]). This limitation could result in general from any kind of (undefined) parasitic dissipation in the measurement system, from an inherent finite quasi particle conductivity $\sigma_1(T=0)$, or from the existence of a small energy gap (Sect. 1.2.2, see also comments on R_{res} in Sect. 2.2.2). However, reproducible R_{res} values at 20 K as high as 20 (1000) $\mu\Omega$ at 10 (87) GHz [23, 145, 148–150] seem typical for Y-123, independent of the measurement systems and of the depositon techniques. Similar values were reported for Tl-2212 [147]. In contrast, the lowest reported R_{res} values of Nb$_3$Sn [151], Nb [152] and Pb [153] were 170 nΩ at 21.5 GHz, 2.6 nΩ at 1.3 GHz and 30 nΩ at 10 GHz.

It is important to note that a finite residual resistance strongly affects the analysis of the quasi-particle conductivity and of the scattering time. The relation (1.43) between the complex conductivity and the surface impedance can be simplified at low temperatures $T \ll T_c$, where $R_s \ll X_s$ is a justified approximation:

2.3 Surface Impedance of Oxide Superconductor Films   87

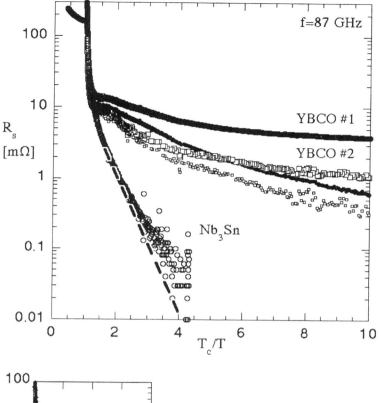

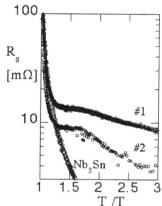

Fig. 2.19. Temperature dependence of the surface resistance at 87 GHz of two epitaxial Y-123 films [22, 23], plotted logarithmically versus T_c/T as-measured (*large symbols*) and after subtraction of the residual resistance (*small symbols, dashed line*). Data for a TVD-Nb$_3$Sn film are shown for comparison [87]. The small diagram magnifies the region $1 \leq T_c/T \leq 3$.

$$\sigma_1 - i\sigma_2 \approx \frac{1}{\omega\mu_0\lambda^2}\left(\frac{2R_\mathrm{s}}{X_\mathrm{s}} - i\right). \tag{2.39}$$

Since σ_1 is proportional to R_s, the uncertainty in determining R_res makes any theoretical conclusion about $\sigma_{1,\mathrm{res}}$ equally uncertain. Furthermore, in

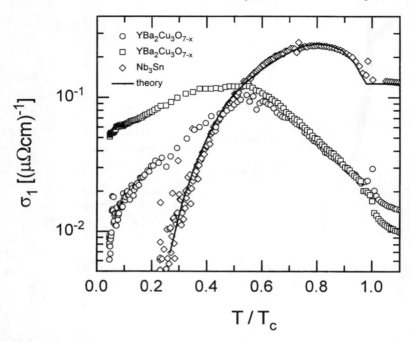

Fig. 2.20. Quasi-particle conductivity $\sigma_1(T)$ versus T/T_c for two epitaxial YBCO films (*circles* and *squares*), and a Nb_3Sn film on sapphire prepared by TVD (Fig. 2.12). Also shown is the low-frequency expectation for the coherence peak at 2% smearing of the energy gap (*solid line*, see Fig. 1.7).

the weak-scattering limit $\omega\tau \ll 1$, the relation (1.10) between the scattering time τ and the conductivity σ_1 reads:

$$\omega\tau \approx \frac{\sigma_1/\sigma_0}{1 - \sigma_2/\sigma_0}, \tag{2.40}$$

with σ_0 defined in Sect. 1.1.1 and related to the plasma wavelength λ_p by $\sigma_0^{-2} = \omega\mu_0\lambda_p^2$. Two important conclusions can be drawn from (2.40). First, absolute values of τ are as reliable as the absolute levels of the penetration depth. Second, subtraction of a finite residual resistance will lower $\sigma_{1,\mathrm{res}}$ and hence the residual value of τ. Low-temperature values of the scattering time $\tau \approx 0.5 \times 10^{-12}$ s were reported after subtraction of sample-dependent R_res values for different YBCO films, measured at different frequencies and with different systems [140, 154].

The existence of a finite residual resistance in HTS is one of the basic features which gave rise to speculation about unconventional pair interactions in these materials. It is therefore very important to gather as much empirical information on R_res as possible. One way is the analysis of correlations between R_s values obtained at different temperatures. Figure 2.21 displays the variation of the residual resistance (abscissa, identified with the 4.2-K level of R_s) versus the corresponding variation of the difference $\Delta R_s = R_s$ (77 K) $- R_s$ (4.2 K) for a variety of granular YBCO layers (triangles) and epitaxial films (filled circles) measured at 21.5 GHz [133] and scaled with f^2 to 87 GHz. The open circles denote a representative selection of data at 87 GHz, measured with epitaxial films of different quality and prepared with different techniques [22].

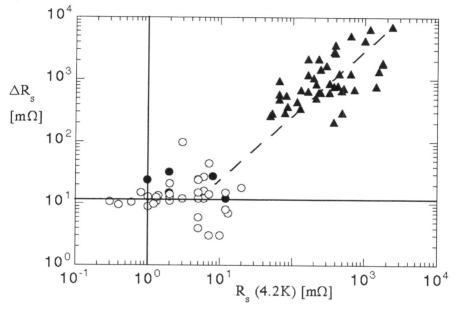

Fig. 2.21. Empirical correlation between the temperature induced increase of the surface resistance, $\Delta R_s = R_s(0.85\ T_c) - R_s$ (4.2 K), and its residual value R_s (4.2 K). Data were obtained with granular YBCO samples (*triangles*) and epitaxial YBCO films (*filled circles*) at 21.5 GHz [133] and at 87 GHz (*open circles*) [22]. The 21-GHz data were f^2-scaled to 87 GHz. The *vertical line* denotes the measurement limit. The *horizontal line* indicates the saturation of ΔR_s, the *dashed line* illustrates the overall trend.

The data on granular and epitaxial YBCO samples are clearly distinguishable. For the first group there is a clear correlation between ΔR_s and R_s (4.2 K) indicating that the Josephson coupling between the grains determines both the absolute level of R_s and its temperature dependence (dashed line). In contrast, the ΔR_s data of the epitaxial films seem to saturate around 10 mΩ, which is generally agreed to reflect the intrinsic surface resistance of YBCO at 77 K (horizontal line). Even so, the corresponding R_{res} values extend over two orders of magnitude. This result is a clear indication that absolute level and temperature dependence of R_s of high-quality YBCO films are determined by several, independent, mechanisms. Moreover, granularity can be ruled out as a dominant source of residual microwave losses for epitaxial YBCO films. It shall be noted that the sensitivity limit of the 87-GHz measurement system limits the validity of our interpretation to $R_{\text{res}} \geq 1$ mΩ (vertical line).

However, there is still an ongoing debate about whether the unusually high residual resistance reflects extrinsic or intrinsic properties. BCS estimations of R_s using material parameters for currents in the CuO_2 planes yielded R_s(10 GHz, 20 K)$\approx$10 nΩ [155]. The calculations demonstrated that the advantage of a larger gap frequency in oxide compared to metallic superconductors is partially compensated by the much larger ratio λ_L/ξ_0. More realistic R_s values can be derived, e.g., from the two-band model. The small finite, or even vanishing, value of the energy gap associated with the CuO chains depends on the magnetic impurity concentration (Sect. 1.3.3). A depressed chain gap Δ_{chain} immediately leads to a finite value of x_n and hence of R_s. The expected sensitivity of Δ_{chain} on content and order of the chain-oxygen agrees with the experimental situation: a much stronger spread is found for R_s at 20 K than at 77 K [22, 156]. The concept of an intrinsic T_c value (Sect. 1.3.3) leads to an inherently high residual resistance which should be independent of the deposition technique. This expectation is in accordance with the apparent reproducibility of R_{res} of the best Y-123 and Tl-2212 films. A complete comparison of experimental $Z_s(T)$ data with the two-band model is discussed separately in Sect. 2.3.4.

For d-wave superconductors, an intrinsic finite R_{res} value can be estimated from (1.78) assuming strong scattering. Using a reduced scattering rate $\hbar\Gamma(T_c)/\Delta_0 \approx 0.5$ [23, 157], and a normal-state resistivity $1/\sigma_n \approx 100$ μΩcm leads to $1/\sigma_{1,\text{min}} \approx 150$ μΩcm and, with $\lambda_L = 140$ nm, to $R_{s,\text{min}} \approx 1.5$ μΩ (respectively, 100 μΩ) at 10 GHz (87 GHz) (see (1.18d)). This value is significantly below the majority of measured values as judged from Fig. 2.21. However, the assumption of strong scattering inherent to the concept of a residual conductivity might be invalid for YBCO, as discussed in [23b] and Sect. 2.3.4.

Temperature Dependence of the Penetration Depth of YBCO

Figure 2.22 displays results on $\lambda(T)$ obtained at 87 GHz for two representative YBCO films [22]. The pair conductivity $\sigma_2(T)$, normalized to its value at zero-temperature, is plotted versus reduced temperature. In terms of the two-fluid formalism, this scaling reproduces the fraction $x_s = 1 - x_n$ of paired charge carriers. Except for the change of the slope $\partial x_s/\partial T$ at $T \approx T_c/2$, the data are consistent with an empirical power law $x_s(t) = 1 - t^2$. Such a scaling was also found in [138, 158] and appears to be a universal feature describing $\lambda(T)$ in YBCO films. The original Gorter-Casimir power law, $x_n = T/T_c^4$, is represented in Fig. 2.22 by the thin solid curve. The results for weakly and strongly coupled isotropic single-gap BCS-like superconductors are given by the dash-dotted and dotted curves.

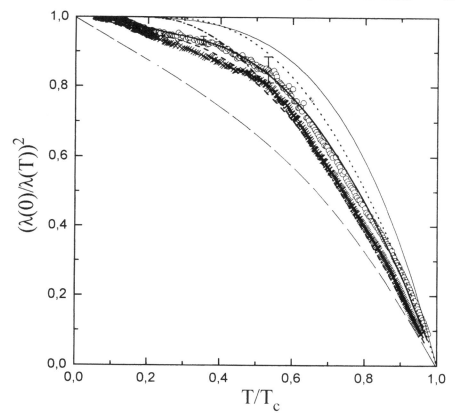

Fig. 2.22. Typical temperature dependence at 87 GHz of the pair conductivity $\sigma_2(T)/\sigma_2(0)$ for two Y-123 films [22]. Also shown are the results expected from the two-fluid model (*thin solid curve*), for a weakly (*dash-dotted*, $\lambda_{\text{coupl}} \ll 1$) and strongly coupled (*dotted*, $\lambda_{\text{coupl}}=3$) isotropic single-gap BCS-like superconductor, and for a strongly coupled d-wave superconductor [159] (*long-dashed*). The *bold solid* and *dashed* curves are fits to the two-band model [160] with $\lambda_L=160$ nm and $\delta_1=5$ (4), $\delta_2=1$ (0.5) for the two films (Sect. 2.3.4).

In contrast to this theoretical behavior, the measured $x_s(T)$ displays stronger temperature dependences at the two limits $T/T_c \ll 1$ and $T/T_c \approx 1$. This behavior was reported by several groups (e.g., [66, 148, 149]), and interpreted sometimes as indicative of a d-wave pair state. The calculations of [159] for a strongly coupled d-wave superconductor (cubic $d_{x^2+y^2-z^2}$ symmetry, vector potential $\boldsymbol{A}$ parallel to the principal axis $\boldsymbol{G}$ of the gap symmetry) are also shown in Fig. 2.22 (long dashes). The case $\boldsymbol{A} \perp \boldsymbol{G}$ and the results for weak coupling and tetragonal $d_{x^2-y^2}$ symmetry show similar or even worse deviations from the experimental data. Finally, the bold solid and dashed curves (partially covered by the experimental data in Fig. 2.22) represent calculations of $x_s(T)$ within the two-band model ([160] and Sect. 2.3.4).

The low-temperature behavior of λ (or σ_2) was compared in a recent review for a number of different measurement methods [161]. In the absence of a convergent picture it was stressed that the occurrence of a power-law behavior $\lambda(T) \propto T^a$ in a layered superconductor is not sufficient to conclude a d-wave symmetry of the pair state. It was also impossible to describe the data with a single energy gap. Similar conclusions could be derived from the theoretical analysis in Sect. 1.3.3. In relation to the discussion of local effects in Sect. 2.1.4 it is finally worth mentioning that spatially resolved measurements of the penetration depth of YBCO films revealed a variety of $\lambda(T)$ dependences within a single sample [53]. Inhomogeneity therefore has to be taken into account in the analysis of integral measurements of the surface impedance.

Frequency Dependence of the Surface Impedance

Besides reflecting the DOS of a superconductor, knowledge of the frequency dependence of the surface impedance is especially important to evaluate potential microwave applications of HTS. Shortly after the first reports on Z_s of granular HTS samples (Sect. 2.3.1), a double-logarithmic plot of R_s versus frequency emerged [162, 163] and was subsequently updated [3, 4, 12a, 111]. Figure 2.23 is a reproduction of such a plot depicting selected data for granular and high-quality epitaxial YBCO films at $T=77$ K. The advantage of epitaxial films (filled circles) compared to Cu (dashed line) or polycrystalline Y-123 (open circles) is clearly resolved. Y-123 displays a lower surface resistance compared to Cu up to about 200 GHz. Data for Tl-2212 fall, at 77 K, about a factor of 2 below the data on single-crystalline Y-123 due to the higher T_c of 106 K [117]. The overall trend of $R_s(f)$ of HTS resembles that of the metallic high-temperature superconductor Nb_3Sn at the corresponding reduced temperature. In the absence of measured $R_s(f)$ data at 15.4 K, Fig. 2.23 displays experimental results for the frequency dependent surface resistance of Nb_3Sn films at 0.5 T_c (i.e., 9 K).

Care has to be taken in scaling the surface resistance of HTS generally quadratic in frequency. Deviations from the power-law $R_s \propto f^{b-(y)}$ with $b_-(y) \approx 2 - 1.25y^2$ result from a finite value of the conductivity ratio $y > 0$ (Sect. 1.2.3). Adapting the measured data in Fig. 2.23 to the results at $f = 87$ GHz and $T = 77$ K [22], a y value of y (87 GHz)=0.16 could be derived from (1.43). Taking into account the explicit frequency dependence of y, and the complete formula $b_-(y)$, (1.45), the data in Fig. 2.23 are quantitatively reproduced (solid line). The resulting frequency exponent stayed around 2 up to 100 GHz. The data on Nb_3Sn displayed a clear f^2 dependence up to 90 GHz, which is in accordance with y (87 GHz) = 0.05. A similar procedure was also applied to the granular films at 21.5 GHz [133], as indicated in the figure. The analysis assumed a constant offset value of y_{off} (21 GHz) = 0.88

due to granularity, and a frequency dependent contribution y_{epi} (21 GHz) = 0.04 in accordance with the 87-GHz result for epitaxial films. The two-fluid formalism was capable of describing all measured data over 3 decades in f. Typical frequency exponents at 77 K were 1.44, 1.43 and 1.31 at 1, 10 and 100 GHz.

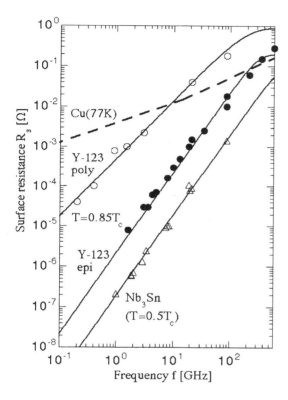

Fig. 2.23. Frequency dependence of R_s(77 K) of high-quality Y-123 epitaxial films (*filled circles*) compared to polycrystalline samples (*open circles*) and Cu (*dashed line*) [12c]. The data at $0.5\,T_c$ on Nb$_3$Sn are shown for comparison (Fig. 2.13). The *solid lines* denote the frequency dependences resulting from the two-fluid model (Sect. 1.2.3). The curves were adapted to the data at 87 GHz [22, 87] and 21.5 GHz [133]. Further explanations are given in the text.

It shall finally be noted that the b_- values were tendentially smaller for films displaying a pronounced $R_s(T)$ shoulder (Fig. 2.19) than for those which did not. This behavior reflects the influence of the scattering time τ on y [164]. The temperature dependence of the frequency exponent was reported to be weak: $|b_-(0.85\,T_c) - b_-(4.2\text{ K})|/b_-(4.2\text{ K}) \leq 20\%$ [22, 154]. The penetration depth of YBCO films was found to be independent of frequency below T_c up to frequencies of about 200 GHz [165, 166].

2.3.3 Effects of Composition

Oxygen Stoichiometry

The previous analysis indicated the important role of the oxygen stoichiometry in YBCO for the absolute level and the temperature dependence of R_s. This relevance is further illustrated with the analysis of $Z_s(87\,\text{GHz})$ of a series of YBCO samples which were subject to different oxygen loading and annealing steps [146]. Figure 2.24 reproduces a set of $R_s(T)$ curves obtained for a sputtered epitaxial YBCO film on MgO in different states of oxygenation. The sample "ERL 10", which corresponds to the as-deposited initial state, indicated a high film quality. The oxygen content of the samples "ERL11"–"ERL14" was varied by adjusting the oxygen partial pressure and the annealing temperature. The variation of the oxygen deficiency x (in $YBa_2Cu_3O_{7-x}$) was judged from the length change of the c-axis lattice parameter to be Δx=0.09, 0.12, 0.16 and 0.25 with "ERL10" serving as the reference. Figure 2.24 reveals a strong and nonmonotonic variation of R_{res} with the deficiency x. Only in the initial state was there no clear indication for a residual resistance. Other data on YBCO films with controlled oxygen content confirmed the increase of R_{res} with decreasing oxygen content, and showed furthermore an increase of the low-temperature value of the penetration depth [167].

The results displayed in Fig. 2.24 proved not only the strong impact of the oxygen content on R_{res}, but also of the ordering of the oxygen along the chains. This was confirmed by an additional set of experiments, in which the film "ERL10" was subject to different annealing conditions. An almost temperature independent R_{res} of 70 mΩ was found after quench cooling ($< 20\,\text{s}$), but a much smaller $R_s(20\,\text{K})$ of only a few mΩ after slow cooling ($< 1\,\text{h}$) of the same film, even after reducing its oxygen content (T_c=80 K). This observation was interpreted in terms of the two-band model by assuming that the different levels of oxygenation led to different values of the Cu-O chain gap.

Cation Stoichiometry

The quasi-ternary chemical composition of YBCO asks for careful stoichiometric control, as well as investigations of deviations from the correct 1:2:3 metal stoichiometry and of the impact on film morphology and surface impedance. This is all the more important (and all the more difficult), the more elements are contained in the HTS compound, e.g., in Tl-2223. Large deviations from the 1:2:3 composition obviously result in multi-phase material and non-stoichiometric segregations (Sect. 5.1.2), and hence in undesired microwave losses. However, the effect of small variations of the cation stoichiometry is less obvious.

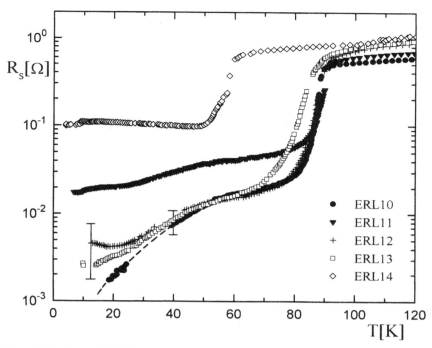

Fig. 2.24. Set of $R_s(T)$ curves at 87 GHz, obtained with YBCO films subject to different states of oxygenation [146]. (With permission from Elsevier Science.)

The effect of different metal stoichiometries of electron-beam coevaporated Y-123 films on R_s (87 GHz) was analyzed in [23b, 167]. As shown in Fig. 2.25, slight (about 3%) variations of the Cu content off the 1:2:3 composition resulted in significant modifications of the film morphology. Excess copper (part a) caused Cu-rich segregations, Cu-deficiency (part c) led to the formation of holes. The background morphology of the films (e.g., parts b and d) was mainly determined by the Ba/Y ratio with Y-rich films being extremely smooth. Figure 2.25d is an image of films close to optimum composition.

Despite these remarkable differences, only slight variations of R_s were observed around the optimum 1:2:3 composition. Additional surface resistance of similar magnitude resulted from different oxygenation levels and, possibly, different scattering rates. In fact, the different morphologies observed for films of different compositions were supposed to affect R_s dominantly by different conditions for oxygen diffusion [167]. The smoothest films were most difficult to be completely oxygenated. The analysis revealed $R_s(T)$ to be determined mainly by oxygen stoichiometry and order, rather than by granular effects. This interpretation is further in accordance with the observation of aging effects which were reported to be more pronounced the more structured the film morphology was.

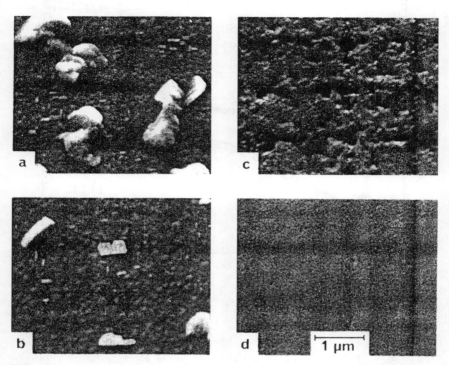

Fig. 2.25. Scanning electron microscope images of three films with varying Cu content. (**a**) 0.515, (**b**) 0.5, (**c**) 0.485 atomic fraction of the metals. (**d**) An image of a film close to optimum composition [167b].

2.3.4 Theoretical Description

While theoretical studies so far have mainly concerned the surface impedance of single crystals [168–170], two groups analyzed the measured $Z_s(T)$ of HTS films over the whole temperature range from 4.2 K up to T_c theoretically. The models and some of the results were indicated previously (Sects. 1.3.3 and 2.3.2), and shall be described in more detail in the following. It shall be stressed that the experimental 87-GHz data discussed in the contrasting frameworks of the two-band model and the d-wave pair mechanism resulted in both cases from closely comparable, partially even identical, electron-beam coevaporated Y-123 films.

2.3 Surface Impedance of Oxide Superconductor Films

Two-Band Model

The pair conductivity $\sigma_2(T)$ of the two epitaxial YBCO films discussed along with Fig. 2.22 was analyzed in the framework of the two-band model for an anisotropic strong-coupling superconductor with s-wave pairing in band "p" (CuO$_2$ planes, $\lambda_{\text{coupl}}^{\text{pp}}=3$) and induced superconductivity in the inherently normalconducting band "c" (CuO chains, $\lambda_{\text{coupl}}^{\text{cc}}=0$) [160, 171]. Based on numerical solutions of the Eliashberg equations, the interband coupling was adjusted to $T_c=90$ K ($\lambda_{\text{coupl}}^{\text{pc}}=\lambda_{\text{coupl}}^{\text{cp}}=0.2$). This parameter set agrees well with the results of other authors [172]. Assuming weak interband scattering (Γ^{pc}, $\Gamma^{\text{cp}} \ll T_c$; in obvious notation with $\hbar = k_B = 1$), and magnetic spin flip scattering (index "m") only in band "c" (i.e., $\Gamma_m^{\text{pc}} = \Gamma_m^{\text{pc}} = \Gamma_m^{\text{pp}} = 0$), the impurity scattering rates Γ^{pp}, Γ^{cc} and the magnetic scattering rate Γ_m^{cc} were left adjustable. While the first two parameters were determined by the normal-state resistivity $\rho(T_c) = 80$ μΩcm, Γ_m^{cc} remained the only free parameter.

The bold solid and dashed curves in Fig. 2.22 are the resulting fits to the 87-GHz data yielding $\lambda_L=160$ nm and reduced energy gaps $\Delta^{\text{p}}/k_B T_c = 5$ (respectively: 4), $\Delta^{\text{c}}/k_B T_c = 1$ (0.5) for the two films. Figure 2.26 shows data extracted for one of the films ($\Gamma^{\text{pp}} = \Gamma^{\text{cc}} = 32$ meV, $\Gamma^{\text{pc}} = 0.8$ meV and $\Gamma_m^{\text{cc}}=2.4$ meV) illustrating the agreement between measurement and calculation over the whole temperature range. The inset magnifies the low-temperature behavior of $\lambda(T)$ on a linear scale. The deviations from the weak-coupling BCS ($\lambda_{\text{coupl}} \ll 1$) and from isotropic strong-coupling calculations ($\lambda_{\text{coupl}}=3$) are obvious. Recently, penetration depth, surface resistance and quasi-particle conductivity of an epitaxial YBCO film were modeled with a single parameter set [174]. Figure 2.27 reproduces initial results for the quasi-particle conductivity and for the surface impedance with $\Gamma_m^{\text{cc}}=0$ and $\Gamma^{\text{pp}} = \Gamma^{\text{cc}}=6$ meV.

Antiferromagnetic Spin Fluctuations

Analysis of the surface impedance [23b] was performed in the framework of a magnetic pairing mechanism leading to a d-wave symmetry of the order parameter (see (1.75)). $Z_s(T)$ was calculated within linear response theory assuming weak pair coupling, with T_c treated as a parameter. In this approach, the momentum dependence of scattering events and of the quasi-particle velocity at the Fermi level could be neglected, leaving the plasma wavelength $\mu_0 \lambda_p^2 = (e^2 N(0) v_F)^{-1}$ an important input parameter. In order to describe the low-temperature behavior of Z_s it was found essential to assume strong elastic quasi-particle scattering, yet not strong enough to suppress the critical temperature. The elastic scattering rate in the normal state was given by ($\hbar = k_B = 1$):

$$\Gamma_{\text{el}}^{\text{N}} = \frac{n_{\text{imp}}}{\pi N(0)} \sin^2 \theta_{\text{N}}. \tag{2.41}$$

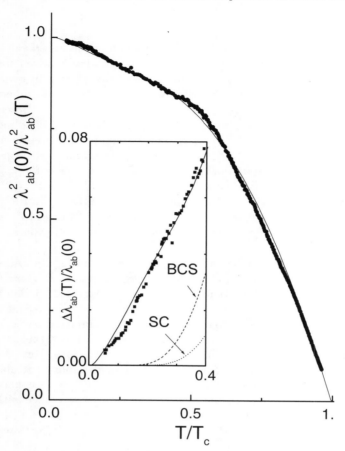

Fig. 2.26. Normalized pair conductivity $\sigma_2(T)/\sigma_2(0)$, measured [160] and calculated within the two-band model [171]. Fit parameters are given in the text. *Inset:* Low-temperature region of $\Delta\lambda_{ab}(T)$ vs. T/T_c. The *dashed* and *dotted lines* denote the results calculated for the standard weak-coupling BCS model and for isotropic strong-coupling with $\lambda_{coupl} = 3$.

$N(0)$ denotes the density of states for one spin at the Fermi level, and n_{imp} is the impurity concentration. The scattering angle θ_N turned out to be an essential parameter that determined the behavior of R_s and $\Delta\lambda$ between 0 K and the lowest measured temperatures around 5 K. The strong scattering limit $\theta_N = \pi/2$ (unitary limit), leading to the minimum conductivity given by (1.78), as well as the Born limit of weak scattering ($\theta_N = 0$) could not reproduce the data. Rather, θ_N had to be adjusted as a free parameter in order to cope with the high measurement frequency $\omega \approx \Gamma_{el}^N$. The normal-state resistivity at T_c and the extremal behavior of $\sigma_1(T)$ below T_c were

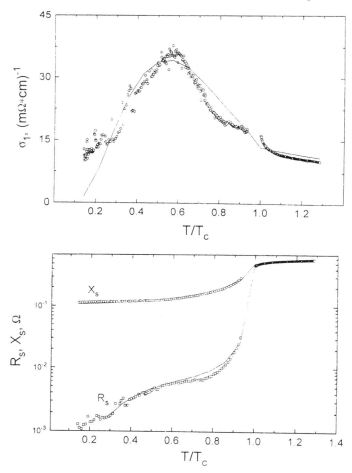

Fig. 2.27. *Top:* Quasi-particle conductivity $\sigma_1(T)/\sigma_1(T_c)$, deduced from 87-GHz data (*symbols*, [173]), and calculated within the two-band model with $\Gamma_m^{cc} = 0$, $\Gamma^{pp} = \Gamma^{cc} = 6$ meV (*line*, [174]). *Bottom:* Corresponding comparison between measured and calculated surface impedance.

simulated by taking into account strongly temperature dependent inelastic scattering:

$$\Gamma_{\text{inel}}(T) = \Gamma_{\text{inel}}(T_c) f(t) \,. \tag{2.42a}$$

The function $f(t)$ depended on the reduced temperature $t = T/T_c$, simulated either by a phenomenological approximation, or as a result of the "nested Fermi liquid" model [175]. The first approach yielded a better description of the data, where

$$f(t) = at^3 + (1-a) \exp\left[b\left(1 - t^{-c}\right)\right] \tag{2.42b}$$

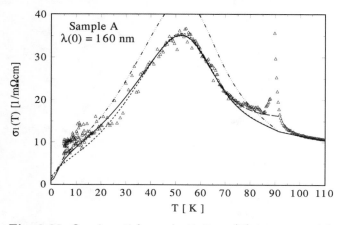

Fig. 2.28. Quasi-particle conductivity $\sigma_1(T)$ as measured (*symbols*) and as calculated (*lines*) [23b]. The parameters corresponding to the *solid line* are given in the text. (© 1999 by the American Physical Society.)

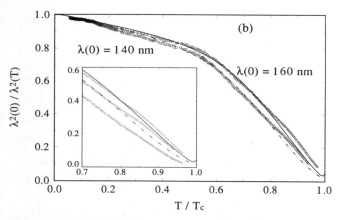

Fig. 2.29. Normalized pair conductivity $\sigma_2(T)/\sigma_2(0)$ as measured (*symbols*) and as calculated [23b]. The *solid line* denotes the same fit parameters as in Fig. 2.28. (© 1999 by the American Physical Society.)

was used with a, b, c as adjustable parameters. It is worth mentioning that the form (2.42b) implies an exponential drop of the scattering rate below T_c in accordance with experimental observations (Sect. 2.3.2). With a total of about eight parameters it was possible to obtain good agreement between the measured and the calculated $\sigma_1(T)$ data. This is demonstrated in Fig. 2.28 by the solid line using the values $\lambda(0) = 160$ nm, $\Gamma_{el}^N = 0.43$ meV, $\theta_N = 0.4\pi$, $\Gamma_{inel}(T_c) = 8.25$ meV, $a = 0.15$, $b = 5.5$, $c = 5$, $2\Delta_0/k_BT_c = 6.8$ and $\lambda_p = 155$ nm. The choice of parameters was severely restricted by fitting $\sigma_1(T)$, $\Delta\lambda(T)$ and $R_s(T)$ simultaneously. The dotted, dashed and dash-dotted lines represent slightly different parameter sets [23b]. The results for $\Delta\lambda(T)$ and $R_{\rm s}(T)$ for the fit parameters listed before are shown in Figs. 2.29 and 2.30. The change of the slope of $\sigma_2(T)$ around 50 K, which was found to be a characteristic feature in numerous measurements, could roughly be approximated with a single gap parameter (Fig. 2.29). It could be reproduced better with a recent two-band d-wave approach [176]. Furthermore, the measured slope of $\lambda(T)$ close to $T_{\rm c}$ seems to be overestimated by the model. The steep drop of $R_{\rm s}$ below $T_{\rm c}$ (Fig. 2.30) reflected the high value of the energy gap, while the large residual value $R_{\rm s}(4.2\,{\rm K})$ was indicative of a small gap. This discrepancy, which was accounted for in the two-band models by two different gap values, could be solved in the single-gap d-wave approach by adjustment of the scattering angle $\theta_{\rm N}$ without needing to account for an extrinsic residual resistance $R_{\rm res}$. The same analysis was also applied to data obtained with another, apparently better, epitaxial YBCO film. However, the agreement between experiment and calculation was less satisfactory.

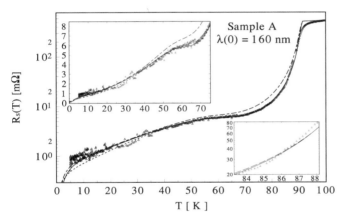

Fig. 2.30. Surface resistance $R_{\rm s}(T)$ as measured (*symbols*) and as calculated [23b]. The *solid line* denotes the same fit parameters as in Fig. 2.28. The *insets* provide magnified views of the low-T behavior in linear scales (*left*) and the behavior close to $T_{\rm c}$ in a semilog plot (*right*). (© 1999 by the American Physical Society.)

It was concluded that fitting the measured $Z_{\rm s}(T)$ data did not provide unambiguous information about the exact symmetry of the order parameter. However, the analysis was incompatible with an order parameter having a nonvanishing average over the Fermi surface. In analogy to the results from the two-band model, measurement of the surface impedance below 4.2 K with sufficiently sensitive systems was stated to be a possible experimental key to further distinguish between different parameter sets or different models.

Conclusions

In summary, both theoretical approaches reproduced the experimental data for $\sigma_1(T)$, $\Delta\lambda(T)$ and $R_s(T)$ well, though on completely different theoretical bases, and with different numbers of parameters. The latter aspect makes it difficult to differentiate between the two approaches. While the structural and electronic properties of the HTS compounds (including YBCO as a special member) are realistically considered in the two-band model, the d-wave approach provides an appealing combination of normal- and superconducting properties within the "nested Fermi liquid" model. However, it presently faces the problems of explaining the high critical temperature and the strong coupling. The kink in $\sigma_2(T)$ around $T_c/2$ and the simultaneous occurrence of a steep drop of R_s close to T_c and a high residual value at low temperature seem to find more natural explanations in the two-band s-wave model. Furthermore, this model takes into account the existence of magnetic impurities in HTS. It provides, at the same time, a consistent explanation of the experimental results for the electron-doped NdCeCuO-compound, which indicate a conventional pair state (see, e.g., [84] in Chap. 1).

Recent experiments on c-axis tunneling in HTS break junctions [177] and between HTS crystals and conventional superconductors revealed a nonvanishing pair contribution displaying s-wave symmetry [178]. Such a feature was accounted for by theoretical approaches considering the coexistence of a dominant pairing with d-wave symmetry and a subdominant s-wave contribution (e.g., [179–181]). There is thus increasing evidence for the existence of two bands displaying different gap values and, possibly, symmetries. Final answers to the multiplicity and symmetry and, independently, to the mechanism of high-temperature superconductivity are still lacking. Improved understanding could be achieved, e.g., by extending measurements of the surface impedance to temperatures below 4.2 K with sufficiently sensitive methods, and with HTS films of carefully adjusted doping levels.

3. Field and Power-Dependent Surface Impedance

> *Power makes things nonlinear: the more it has already changed, the more it longs to change further. But saturation slips away.*

Chapters 1 and 2 treated the surface impedance in the small-signal limit, i.e., assumed that the levels of the magnetic field amplitude and of the microwave power were low enough to prevent Z_s from becoming implicitly (or even explicitly) dependent on these quantities. While this condition can be met experimentally by proper adjustment of the operating conditions, it reveals an incomplete picture of the surface impedance of superconductors. Application of elevated field or power levels is important for various reasons:

1. to investigate basic field and power limitations which are characteristic for the investigated superconductor,
2. to distinguish extrinsic from intrinsic types of microwave response, and
3. to characterize the accessible range of parameters which can be exploited for the different types of applications (see Chap. 6).

Chapters 3 and 4 therefore contain an analysis of the various mechanisms leading to a nonlinear surface impedance $Z_s = R_s + iX_s$. Referring to the physical meaning of the surface resistance R_s and the surface reactance X_s in the linear regime (Sect. 1.1), we could derive a general formulation (see also (2.34) in Sect. 2.1) like:

$$\langle Z_s(\bm{r})\rangle_{\bm{r}} = \int_{\mathrm{surface}} [R_s(\bm{r}) + iX_s(\bm{r})] \frac{|\bm{H}_s(\bm{r})|^2}{|\overline{\bm{H}}|^2} \mathrm{d}^2\bm{r}, \tag{3.1}$$

where R_s and X_s could be considered to be independent of the surface magnetic field amplitude H_s or, equivalently, the circulating power $P_{\mathrm{circ}} \propto H_{s,\mathrm{max}}^2$ (Sect. 2.1). Also assumed was the independence of the dissipated power P_{diss} or, equivalently, the absence of Joule heating. Equation (3.1) thus defines a spatially averaged surface impedance. In contrast, the nonlinear regime is characterized by

$$\langle Z_{\mathrm{s}}(\boldsymbol{H}_{\mathrm{s}},T,\boldsymbol{r})\rangle_{H,T,\boldsymbol{r}} = \int_{\mathrm{surface}} Z_{\mathrm{s}}\left[\boldsymbol{H}_{\mathrm{s}}(\boldsymbol{r}),T[\boldsymbol{H}_{\mathrm{s}}(\boldsymbol{r})],\boldsymbol{r}\right] \frac{|\boldsymbol{H}_{\mathrm{s}}(\boldsymbol{r})|^2}{|\overline{\boldsymbol{H}}^2|} \mathrm{d}^2\boldsymbol{r},$$
(3.2)

where Z_s now must be expected to depend on the field amplitude $|H_\mathrm{s}|$, the local field-dependent temperature T (via the dissipated power), and on the position $\boldsymbol{r}$. Unless $Z_\mathrm{s}(H_\mathrm{s})$ can be deconvoluted from (3.2) by assuming thermal and spatial homogeneity, or measurements with spatial resolution are performed such that $H_\mathrm{s}(\boldsymbol{r})$ can be considered constant, the measurable surface impedance is always an average over the inhomogeneous microwave field distribution.

Equation (3.2) leads us to consider three cases, related to magnetic and thermal mechanisms (electric field effects on Z_s were discussed in [1]). The first case, if the excitation frequency is lower than the relaxation rates in the superconductor ($\omega\tau \ll 1$), is the adiabatic field dependence of $Z_\mathrm{s}(H_\mathrm{s})$ (Sect. 3.1). If such a microwave field and an external DC field can be considered to affect Z_s in the same way, the study of $Z_\mathrm{s}(H_\mathrm{DC})$ yields supplementary insight into the nonlinear reponse, since H_DC can be made homogeneous over the superconducting surface (Sect. 4.2). The second case (Sect. 3.2), which is also adiabatic, deals with the dependence of Z_s on (slow) heating effects. The third case finally concerns dynamic nonlinear effects evoked by rapidly varying magnetic fields or heat fluxes (Sect. 3.3).

The main objective of this chapter is to analyze possible limitations of the linear microwave response of superconducting films. The treatment is therefore limited to small or intermediate field (or power) levels. The surface impedance of superconductors in the mixed state forms a separate, very rich, issue which is worth being studied especially for the oxide superconductors. However, the reader is referred to reviews and detailed articles that extensively covered this area [2–7].

3.1 Critical Field Levels and Corresponding Surface Impedance

The shielding of magnetic fields in superconductors is characterized by a number of critical field levels. Section 3.1.1 provides a brief overview of this field, based on a schematic classification of the length and time scales involved. The analysis of the thermodynamic properties of superconductors, e.g., within Ginzburg–Landau (GL) theory [8], yields the concept of the thermodynamic critical field $H_\mathrm{c,th}$ which is equally well applicable to superconductors of first and second kind. Section 3.1.2 recalls the major aspects of this well-known approach and its impact on the surface impedance. A specific consequence of the thermodynamic properties of superconductors is the existence of supercooled and superheated regimes where the Meissner state constitutes a local (rather than the global) minimum of the free energy. Section 3.1.3 focuses on adiabatic superheating and summarizes the resulting Ginzburg–Landau phase diagram of superconductors. The lower critical fields of homogeneous superconductors and isolated Josephson junctions are discussed in Sect. 3.1.4. The last section summarizes some implications resulting from the geometry of superconducting films (demagnetizing effects and finite film thickness), and from possible unconventional pairing.

3.1.1 Overview

Classification of Critical Fields

The sources of nonlinear microwave field (and power) dependence of the surface impedance can be divided into magnetic or thermal effects, depending on the prevailing nature of the excitation energy. (In principle, in specific measurement arrangements where dielectric materials and/or gaps between the field-occupied volume and the superconductor are involved, electric field effects must also be accounted for [1].) This phenomenological distinction is well known, e.g., from the study of low-temperature superconductors for particle accelerators [9]. The field-limiting mechanisms can further be classified along with the presence (or the absence) of defects and their size r_{def}, evaluated relative to the range of pair coupling, the coherence length ξ. Based on these aspects, Table 3.1 provides a general classification of the nondynamic microwave field limitations in superconducting films. Also given are the basic parameters involved, and typical ranges in (mT) (orders of magnitude). More specific quantitative results for Nb_3Sn and Y-123 are summarized in Sect. 3.4.3 at the end of this chapter.

The magnetic limitation processes (row A in Table 3.1) result in temperature dependent critical fields $H_c(T)$, depending on the London and the Josephson penetration depths λ_L and λ_J (Sect. 1.1.2) and, via the Ginzburg–Landau parameter $\kappa = \lambda_{\text{GL}}/\xi_{\text{GL}}$ (Sect. 3.1.2), on the coherence length ξ. (The variable "ξ" without index is understood as being a general expression, referring either to the electrodynamic value ξ_0 or the thermodynamic value ξ_{GL}, Sect. 1.1.) In row B heating effects that correspond to the same physical categories as the magnetic mechanisms in row A are listed. The breakdown fields are determined by critical temperatures $T_c(H)$, which are actually lowered in non-zero field. Relevant parameters are the thermal conductance $\sigma_{\text{th,eff}}$ of the HTS film-substrate sandwich, the temperature rise $\Delta T = T_{c,\text{eff}} - T$ between the film and the coolant and, in the presence of defects (cells 2B, 3B), their surface resistance R_{def} and size r_{def} [11, 12]. A global microwave response is expected for the defect-free or proximity-coupled cases $r_{\text{def}} \ll \xi$ (column 1). In contrast, in the majority of cases in type-II superconductors, defects with $r_{\text{def}} \geq \xi$ (being intrinsic or extrinsic) induce nonlinear effects locally (columns 2 and 3).

For defect-free planar surfaces, and in the absence of thermodynamic fluctuations, a superheated Meissner state (cell 1A) can exist up to the field

Table 3.1. Classification of nondynamic microwave field limitations. From top to bottom, each cell contains a brief description of the mechanism, the resulting expressions in terms of the main parameters, and a typical order of magnitude (in mT) [10].

Process	1 $r_{\text{def}} \ll \xi$ no defects	2 $r_{\text{def}} \approx \xi$ weak links	3 $r_{\text{def}} \gg \xi$ nonsuperconducting defects
Row A magnetic limitation $H_c(T)$	adiabatic superheating $H_{\text{sh}} = Fct(\kappa)H_{\text{cth}}$ κ, T 10^3	nucleation of Josephson vortices $\mu_0 H_{\text{c1J}} = \Phi_0/2\pi\lambda_J\lambda_L$ $\lambda_L, J_{\text{cJ}}, T$ $\leq 10^1$	nucleation of Abrikosov vortices $\mu_0 H_{\text{c1}} = \Phi_0/4\pi\lambda_{\text{GL}}^2$ $(\ln \kappa + \varepsilon_{\text{self}})$ $\lambda_L, \xi_0, \kappa, T$ 10^2
Row B thermal limitation $T_c(H)$	global heat balance $H_q^2 = 2(T_c - T)/$ $(R_s \cdot R_{\text{th}})$ R_s, R_{th}, T, f $10^2\text{--}10^3$	Josephson phase slippage $H_{\text{TAPS}}^2 = 4\sigma_{\text{th}}(T_{\text{cJ}} - T)/$ $(r_{\text{def}}R_J)$ $r_{\text{def}}, J_{\text{cJ}}, R_J, \sigma_{\text{th}}, T, f$ $10^2\text{--}10^3$	local heat balance $H_{q,\text{loc}}^2 = 4\sigma_{\text{th}}(T_{c,\text{loc}} - T)/$ $(r_{\text{def}}R_{\text{def}})$ $r_{\text{def}}, R_{\text{def}}, \sigma_{\text{th}}, T, f$ $10^1\text{--}10^2$

H_{sh}, which scales with the thermodynamic critical field $H_{c,\text{th}} \approx 10^3$ mT. For $\kappa \gg 1$, $H_{\text{sh}} = 0.75\, H_{c,\text{th}}$ was calculated (e.g., [13]). Structural irregularities (like protrusions, particulates etc.) lead to local field enhancement, and are hence expected to lower this threshold to the bulk (cell 3A) or Josephson (2A) lower critical fields H_{c1} and H_{c1J}, respectively [2, 14, 15].

The lower critical field of Josephson junctions H_{c1J} (cell 2A) is defined in terms of the Josephson critical current density J_{cJ} and λ_J [2]:

$$H_{c1J} = 2J_{cJ}\lambda_J. \tag{3.3}$$

The definition of H_{c1J} is applicable and useful unless J_{cJ} approaches the critical current density J_c of the bulk superconductor. The concept of a "weak" link then becomes subtle, and remains justified merely in relation to the dynamic response [15]. The Josephson lower critical field can be interpreted in analogy to the bulk value H_{c1} only in the case of "long" junctions, where λ_J is smaller than the extension of the junction transverse to the magnetic field [16]. It then represents the breakdown of Meissner shielding and the onset of the penetration of Josephon fluxons, i.e., regions along which the phase difference across the junction varies by 2π (or integer multiples of it).

3.1 Critical Field Levels and Corresponding Surface Impedance

The bulk lower critical field (cell 3A) denotes the onset of the penetration of Abrikosov fluxons, i.e., circulating supercurrents which enclose an integer number of flux quanta. H_{c1} is related to the Ginzburg–Landau penetration depth λ_{GL}, which in the clean limit approaches the London value λ_L (Table 1.1, and [10] in Sect. 1.1.1). For anisotropic superconductors, the λ value has to be taken for that plane whithin which the shielding currents flow. For instance, for fields parallel to the c-axis of the layered HTS materials λ_{GL} is to be identified with λ_{ab}. The self energy $\varepsilon_{self} \approx 0.5$ of an Abrikosov vortex is expected to add only a small contribution to the magnitude of H_{c1} in extreme type-II superconductors like Nb_3Sn or HTS where $\ln \kappa \geq 2.5$.

Row B lists field limitations caused by heating the film above an effective critical temperature $T_{c,eff}$, thus inducing a thermal "avalanche". The critical fields are obtained by equating the dissipated microwave power density $\partial P_{diss}/\partial A$ to the heat flux $(T_{c,eff} - T)/R_{th}$ over the relevant area [11, 12]. The thermal resistance $R_{th} = d_F/\sigma_{th,F} + R_{bd} + d_S/\sigma_{th,S}$ (cell 1B) depends on the heat conductivities $\sigma_{th,F}$ and $\sigma_{th,S}$ of film and substrate, averaged over the temperature interval $\{T, T_{c,eff}\}$. It depends further on the thicknesses d_F and d_S of film and substrate, and on the thermal boundary resistance R_{bd} at their interface [17]. In the presence of defects (cell 3B), the surface resistance of the defect $R_{def} \gg R_s$ enters as the main parameter, since the total heat causing the field breakdown is created predominantly at the defect which extends over an area πr_{def}^2.

Finally, phase coherence across a Josephson junction (cell 2B) ceases at $k_B T_{cJ} = \Phi_0 \pi r_{def}^2 J_{cJ}$, where the thermal energy exceeds the Josephson coupling energy (Sect. 2.3.1).

In summary, the critical field levels terminating the linear microwave response of superconductors should be clearly distinguishable in terms of their magnitudes. According to Table 1.1, the Josephson and bulk lower critical fields as well as microwave heating at defects are considered the most relevant limiting mechanisms for the A15 and HT superconductors since the corresponding H_{crit} values are the lowest.

Equilibrium (Adiabatic) vs. Nonequilibrium (Dynamic): Characteristic Time Scales

Each source leading to nonlinear microwave response can be ascribed a characteristic time constant τ_c. As long as a microwave cycle $1/\omega$ is long compared to τ_c, the superconductor responds adiabatically, i.e., it passes through (time-varying) equilibrium states. In contrast, nonequilibrium effects are induced if the response times are long or the excitation frequency is high, such that $\omega \tau_c \gg 1$. While both limits contribute to an apparent magnetic field dependence of the surface impedance, predominantly the "fast" nonlinear mechanisms give rise to dynamic effects like nonequilibrium conductivity, harmonic generation and frequency intermodulation.

108   3. Field and Power-Dependent Surface Impedance

Prominent microwave-induced nonlinear effects in superconductors are pair breaking, quasi-particle recombination, Josephson or Abrikosov vortex nucleation, transition from the superconducting (S) to the normal (N) state, and the propagation of normal zones. Many of these mechansims were reviewed in [18] with emphasis on HTS. While a general quantitative theoretical analysis of the related time scales is complicated and today still incomplete, rough estimates can be given for the above examples.

The basic time scale of a superconductor is given by the energy gap: $\tau_{\text{pair}} = \hbar/\pi\Delta$. Taking into account the relation between ξ_0 and Δ, τ_{pair} is given by ξ_0/v_{F}. The physical meaning of this ratio is roughly the time required for the two quasi-particles constituting a Cooper pair to exchange a virtual phonon [19]. It is thus referred to as the "pairing time". Regarding the large gap energies of HTS ($\Delta \approx 30\,\text{meV}$) and Nb_3Sn ($\Delta \approx 4\,\text{meV}$), the Cooper pair formation can therefore safely be expected to be spontaneous up to frequencies of about 10^3 and $10^2\,\text{GHz}$, respectively.

Microwave irradiation of superconductors might introduce nonequilibrium states in the form of an energy imbalance of the Cooper pair system [19, 20]. The superconducting properties (T_c, $H_{c,\text{th}}$, Δ, etc.) in a nonequilibrium state can be weakened as well as enhanced. The corresponding relaxation time τ_Δ of the order parameter was found to be determined by the inelastic quasi-particle scattering time τ_{in}. Strong electron–phonon coupling as in A15 and HTS is therefore expected to reduce τ_Δ. Analytical expressions for τ_Δ were derived from Ginzburg–Landau and BCS theory, and evaluated mainly near T_c, $1 - T/T_c \ll 1$, often being valid only for dirty or gapless superconductors. However, the scaling of τ_Δ with τ_{in} proved to be rather general [19]:

$$\tau_\Delta \propto \frac{k_{\text{B}} T_c}{\Delta(T)} \tau_{\text{in}} \qquad (3.4)$$

It is tempting to assume from (3.4) that the order parameter relaxation becomes faster with increasing pair coupling. It was also noted in [19] that additional relaxation channels like scattering between quasi-particles would further reduce τ_Δ. Similar arguments were reported for superconductors displaying an anisotropic gap and/or magnetic impurities. Altogether, experimental data on τ_{in} seem more reliable than theoretical estimates to judge about order parameter relaxation. The results for quasi-particle scattering in HTS (Sect. 2.3) indicate response times τ_Δ of ps order. While corresponding data on Nb_3Sn are hardly available, much longer relaxation times ($\leq 100\,\text{ps}$) should be expected from the lower gap value and the weaker pair coupling.

The time needed for a resistively shunted Josephson junction (Sect. 1.2.2) to switch between the S and the N state is determined by the critical voltage: $\tau_{\text{cJ}} = \Phi_0/I_{\text{cJ}} R_{\text{J}}$. For ideal tunnel barriers, $I_{\text{cJ}} R_{\text{J}}$ is of comparable order of magnitude as the gap voltage (see (1.32)). In contrast, the charge transport across HTS junctions revealed many localized states, thus depressing the critical voltage much below the gap voltage. Therefore comparable τ_{cJ} values of the order of ps are expected for both Nb_3Sn and HTS.

There are no definite times predictable for the nucleation of Abrikosov vortices and for S–N transitions. Rather, they strongly decrease with field H (power P) via an increasing overcriticality factor: $\tau_c(E) \propto \tau_\Delta (E/E_c - 1)^{-1}$, where $E = H$ or $E = P$ denotes the excitation amplitude, and E_c is the relevant critical value [18, 21]. A nonequilibrium response should therefore prevail at excitation levels close to the critical values, while spontaneous transitions are induced at high overcriticality $E \gg E_c$. As a specific result, nonlinear microwave losses introduced by exceeding the lower critical field H_{c1} are expected to be rather smooth.

Finally, the time scale τ_{heat} of heating effects is determined by the thermal parameters and by the geometric arrangement of superconductor, substrate and thermal interfaces between them. As discussed in Sect. 3.2, time constants between ns and ms can occur, depending on the substrate material, the operating temperature and the overcriticality.

In summary, the different types of nonlinear microwave response should also be clearly distinguishable in terms of their temporal behavior. According to the previous analysis, heating and flux penetration are the slowest mechanisms. They are thus considered to cause specific dynamic responses, e.g., in terms of frequency intermodulation phenomena (Sect. 3.3). In contrast, pair breaking and Josephson switching are expected to be spontaneous in the microwave range for both Nb_3Sn and YBCO.

Differential Loss Tangent: A Conceptual Aid

The different sources of nonlinear microwave response become distinguishable from the associated critical field (or power) levels, and from the characteristic time constants or corresponding frequency dependences. Moreover, the effect of the different mechanisms on absolute level and functional dependence of the surface impedance is in general different. As pointed out, e.g., in [1, 2, 22–25], the analysis of the differential loss tangent (Sect. 2.1)

$$r(E) = \frac{\partial \text{Im}\{Z_s\}/\partial E}{\partial \text{Re}\{Z_s\}/\partial E} = \frac{\partial X_s(E)}{\partial R_s(E)} \tag{3.5}$$

is a valuable empirical tool to obtain unambiguous information about the type of nonlinearity. It is, however, important to note that the results depend strongly on the variable E with respect to which $r(E)$ is evaluated. In general, different results will be obtained for $E =$ temperature, frequency or magnetic field. The main advantage of the r value analysis still is the elimination of explicit dependences $Z_s(E)$, which are usually complicated to find and might contain many unknown parameters [15]. Care has to be taken to pay attention exclusively to $r(E)$, since it does not contain information of the absolute levels of R_s and X_s at linear response.

3.1.2 Thermodynamic and Upper Critical Fields

Stationary Ginzburg–Landau Equations

The thermodynamic and magnetic properties of superconductors were comprehensively described by the phenomenological Ginzburg–Landau (GL) theory [8], well before the BCS theory was developed. The major extension of the general GL framework compared to the London model is the consideration of a spatially varying superfluid density. As proven later by Gorkov through rigorous calculation [26], the GL theory is consistent with the BCS theory under two assumptions (see, e.g., [27]): 1. the order parameter is small, and 2. the electrodynamics are local. The first condition implies that the theory is strictly applicable only to weak-coupling superconductors, with the option of getting intuitive access to strong pair coupling by means of an appropriate renormalization procedure (Sect. 1.3.3). In the case of dirty superconductors the two conditions lead to the conclusion that the GL theory is valid at all field levels $H \leq H_{c2}$ for temperatures close to T_c, and at all temperatures $T \leq T_c$ for fields close to H_{c2}. The temperature region in which the GL theory can be expected to apply scales with the κ-parameter, leading to an extended range of validity in the high-κ A15 and HTS materials.

Since the GL theory provides the basic framework to derive expressions for the critical fields of superconductors, to understand the existence of supercooling and superheating fields, and to describe the implication of finite film thickness, some relevant relations are summarized in the following. The idea of the GL theory was based on the Landau theory of second-order phase transitions (e.g., [28]). Such transitions can be characterized by an order parameter (OP) which quantifies the continuously increasing ordering in a solid when cooling below the critical temperature (e.g., spin orientation in a ferromagnet, or pairing in a superconductor). For a superconductor, Ginzburg and Landau identified the OP with the macroscopic wave function $\Psi(\mathbf{r})$, which is in general complex and varies in space. Gorkov showed later that $\Psi(\mathbf{r})$ is strictly proportional to the order parameter $\Delta(\mathbf{r})$, which was introduced in Sect. 1.3.3 in relation to gapless superconductors. The basic assumption of GL is that the Helmholtz free energy $F = U - TS$, or its volume density $f = \partial F / \partial V$, can be expanded in even powers of the OP:

$$f(\mathbf{r}) = f_n + \alpha(T)|\Psi(\mathbf{r})|^2 + \frac{\beta}{2}|\Psi(\mathbf{r})|^4 + \frac{\hbar^2}{2m}|\nabla\Psi(\mathbf{r})|^2. \tag{3.6}$$

Here, $\alpha(T) = -A(T_c - T)$ and $A, m, \beta > 0$ are real constants. In comparison with the microscopic picture, m can be identified with the electron mass. (It should be mentioned that $m^* = 2m$ is used instead of m in some textbooks, which implies a different normalization of $|\Psi|^2$). The first term on the right-hand side of (3.6) denotes the volume density of the normal-state free energy. The last term results from the spatial variation of the OP, and can be regarded as a "kinetic energy" contribution to the total energy. Obviously,

3.1 Critical Field Levels and Corresponding Surface Impedance

steep variations of $\Psi(\mathbf{r})$ would enhance the free energy, and are thus relatively suppressed by an amount

$$\xi_{\rm GL}(T) = \sqrt{\frac{\hbar^2}{2m|\alpha(T)|}}, \tag{3.7}$$

which corresponds to the ratio of the coefficients of the second and the last term in (3.6). The length scale defined by (3.7) is called the Ginzburg–Landau coherence length. It denotes the minimum length over which the OP can vary significantly (e.g., across a flux line), and is related to the low-temperature BCS value ξ_0 as indicated in Table 1.1 (Sect. 1.1.1).

The interaction of the superconductor with a vector potential $\mathbf{A}$ is well-described in terms of the Gibbs free energy $G = U - TS - \mathbf{H}_{\rm ext}\mathbf{M}$ (or its volume density $g = \partial G/\partial V$), where $\mathbf{H}_{\rm ext}$ denotes an externally applied magnetic field. The resulting expression for $g(\mathbf{r})$ contains the two unknown quantities $\Psi(\mathbf{r})$ and $\mathbf{A}(\mathbf{r})$. In order to find the equilibrium state, $g(\mathbf{r})$ has to be minimized with respect to variations of both quantities. This procedure yields the two GL equations (in the gauge where div $\mathbf{A} = 0$):

$$\frac{1}{2m}(-i\hbar\nabla - 2e\mathbf{A})^2\Psi + \alpha\Psi + \beta|\Psi|^2\Psi = 0 \tag{3.8a}$$

and

$$\mathbf{J}_{\rm s} = -\frac{ie\hbar}{m}(\Psi^*\nabla\Psi - \Psi\nabla\Psi^*) - \frac{4e^2}{m}|\Psi|^2\mathbf{A}, \tag{3.8b}$$

where e is the electronic unit charge. The last equation is a generalized form of the London equation, (1.9), providing a local relation between the supercurrent $\mathbf{J}_{\rm s}$ and the vector potential $\mathbf{A}$. In comparing (3.8b) with (1.9), the physical meaning of the OP becomes obvious: $|\Psi|^2 = n_{\rm s}(T)$. The squared modulus of the OP obviously quantifies the number density of Cooper pairs. Equation (3.8b) reduces exactly to (1.9) for a bulk superconductor, if Ψ is constant and assumes its equilibrium value $|\Psi_0|^2 = -\alpha/\beta$. This assumption leads to the definition of the GL penetration depth:

$$\lambda_{\rm GL} = \sqrt{\frac{m\beta}{4e^2\mu_0|\alpha|}}. \tag{3.9}$$

The thermodynamic critical field is obtained by equating the Gibbs free energies of the superconducting and the normal state, with the result

$$\mu_0 H_{\rm c,th} = \sqrt{\frac{\mu_0\alpha^2}{\beta}} = \frac{1}{\sqrt{2}}\frac{\Phi_0}{2\pi}\frac{1}{\xi_{\rm GL}\lambda_{\rm GL}}. \tag{3.10}$$

This field describes the lowering of the free energy by the amount $(\mu_0/2)H_{\rm c,th}^2$ due to the pair condensation. It also corresponds to a maximum thermodynamic critical current density, which can be observed, e.g., in a thin film ($d_{\rm F} \ll \xi_{\rm GL}$), where $|\Psi|^2$ is spatially constant [29]:

$$J_{c,th} = \frac{2\sqrt{2}}{3\sqrt{3}} \frac{H_{c,th}}{\lambda_{GL}}. \qquad (3.11)$$

At this level of current density, the superfluid density $x_s = |\Psi|^2/|\Psi_0|^2$ has decreased to the value $x_{c,th} = 2/3$, and the corresponding critical momentum is $p_{c,th}(T \approx T_c) = 3^{-1/2}\hbar/\xi_{GL}$. For $p > p_{c,th}$, order parameter and superfluid current density decrease continuously to zero. In contrast, the zero-temperature limit obtained from the BCS theory yields the sharp limiting value $p_c(0) = \hbar/\pi\xi_0$, above which energy gap and supercurrent density steeply drop to zero [30].

Further analysis of the GL equations (3.8) yields the upper critical field H_{c2}, at which the type-II superconductor undergoes a second-order phase transition into the normal state:

$$\mu_0 H_{c2} = \frac{m|\alpha|}{e\hbar} = \frac{\Phi_0}{2\pi} \frac{1}{\xi_{GL}^2} = \sqrt{2}\kappa\mu_0 H_{c,th}. \qquad (3.12)$$

The second equality ($\mu_0 H_{c2} = \Phi_0/\pi r_A^2$) makes obvious that the effective radius r_A of an Abrikosov vortex equals approximately $2^{1/2}\xi_{GL}$. This view is justified by a theoretical analysis of the low-lying quasi-particle excitations in the mixed state ($r_A \approx \xi$, [31]), and of the magnetic induction of an isolated vortex ($r_A \approx 2^{1/2}\xi$, [32]). The last equality in (3.12) relates H_{c2} to the thermodynamic field in terms of the GL parameter $\kappa = \lambda_{GL}/\xi_{GL}$. The threshold value $\kappa = 2^{-1/2}$ (or, equivalently, $H_{c2} = H_{c,th}$) leads to the important distinction between type-II and type-I superconductors. The physical understanding of $H_{c2} < H_{c,th}$ for the latter type is discussed in Sect. 3.1.3.

In analogy to the ratio of electrodynamic lengths (Chap. 1), the GL parameter κ helps to distinguish between local ($\kappa \gg 1$) and nonlocal ($\kappa \ll 1$) current–field relations. Since λ_{GL} and ξ_{GL} both vary with temperature close to T_c like $(T_c - T)^{-1/2}$, κ is almost constant in the framework of the GL theory. The dirty and clean limits are defined, as usual, by the ratio of the electronic mean free path and the GL coherence length. The clean local limit of thin films therefore implies $d_F/\xi_{GL} \ll 1$. Regarding typical film thicknesses of a few 100 nm, this limit can be met merely with type-I superconductors, or at temperatures close to T_c. It might finally be worth noting that the thermodynamic critical field $H_{c\parallel}$ is enhanced in thin films with $d_F \leq \lambda_{GL} \ll \xi_{GL}$ by roughly a factor λ_{GL}/d_F compared to (3.10), if the field is applied parallel to the film. The superfluid density is then found to decrease like $x_s = 1 - H^2/H_{c\parallel}^2$ [29].

Time Dependent GL Equations

The evaluation of the dynamic features inherent in the GL concept requires the analysis of the time dependent formulation of (3.8). According to the review in [19], a suitable form is

$$\frac{2|\alpha|}{\hbar}\tau_{\text{relax}}\left(\hbar\frac{\partial}{\partial t}-2ie\Phi_{\text{el}}\right)\Psi=-\frac{1}{2m}(-i\hbar\nabla-2e\boldsymbol{A})\Psi-\alpha\Psi-\beta|\Psi|^2\Psi,$$
(3.13)

where Φ_{el} is an electrostatic potential and $\Psi = \Psi(\boldsymbol{r},t)$. The GL relaxation time $\tau_{\text{relax}} = h/[32k_B(T_c - T)]$ describes the exponential decay $\exp(-t/\tau_{\text{relax}})$ of deviations of Ψ from the equilibrium value $(|\alpha|/\beta)^{1/2}$. The time scale τ_{relax} has to be replaced by τ_Δ (Sect. 3.1.1) for superconductors with a finite gap value. By comparing the left-hand side of (3.8a) with the right-hand side of (3.13), one can conclude that the rate of change of Ψ is proportional to

$$\dot\Psi \propto \frac{1}{\tau_{\text{relax}}}\left(-\frac{\partial G}{\partial \Psi^*}\right),$$
(3.14)

where $\partial G/\partial \Psi^*$ is the total derivative of the Gibbs free energy with respect to the order parameter in the framework of the Euler–Lagrange variational calculus [27]. Equation (3.14) thus tells us that nonequilibrium states decay faster the farther the deviation was away from equilibrium (i.e., the larger the overcriticality). This is the result mentioned previously in Sect. 3.1.1.

Field Dependent Surface Impedance in the GL Framework

The surface impedance $Z_{\text{s,GL}}$ has been considered within framework of the GL theory for semi-infinite superconductos [33, 34] and microstriplines [35]. While exact results require numerical computation, intuition can be gained by expressing Z_s in terms of the normal fraction x_n (Sect. 1.2.3, (1.44)). Assuming complete pairing, i.e., $x_n + x_s = 1$, and generalizing the previous finding that $x_s = 1 - (H/H_{c,\text{th}})^2$ [18, 36], we obtain in the weak-scattering limit $\omega\tau \ll 1$ in terms of the reduced field $h = H/H_{c,\text{th}}$:

$$\frac{Z_{\text{s,GL}}}{R_n} \approx \sqrt{\frac{(\omega\tau)^3}{2}\frac{h^4}{(1-h^2)^3}} + i\sqrt{\frac{2\omega\tau}{1-h^2}}.$$
(3.15a)

Since for type-II superconductors the scaling field $H_{c,\text{th}}$ is much higher than typical experimental field levels, (3.15a) can be further expanded according to $h \ll 1$, yielding to order h^2:

$$\frac{Z_{\text{s,GL}}}{R_n} \approx \sqrt{\frac{(\omega\tau)^3}{2}}h^2 + i\sqrt{2\omega\tau}(1+h^2/2).$$
(3.15b)

The resulting differential loss tangent $r_{\text{GL}}(h)$ (see (3.5) in Sect. 3.1.1) is given by

$$r_{\text{GL}}(h) \approx \frac{1}{\omega\tau}(1-3h^2) \gg 1.$$
(3.15c)

Equation (3.15) is in accordance with the summary in [1], except that the dissipative field coefficient is larger than 3, which results from density-of-state effects in high-κ superconductors [33]. The high and approximately constant value of $r_{\mathrm{GL}}(h)$ reflects that both $R_{\mathrm{s,GL}}$ and $X_{\mathrm{s,GL}}$ increase quadratically with field, with the increase of $X_{\mathrm{s,GL}}$ being much stronger than that of $R_{\mathrm{s,GL}}$.

The field coefficients in (3.15) were derived in the GL limit, i.e., are strictly valid only at temperatures close to T_c. In relation to a comparison between s-wave and d-wave superconductors (Sect. 3.1.5), the field dependence (respectively current density dependence) of the superfluid fraction x_s was calculated in [30, 37] from the London equation for clean superconductors. The result could be expressed as

$$\frac{1}{\lambda(T,H)} = \frac{1}{\lambda(T)}\left[1 - \alpha(T)\left(\frac{H}{H_0(T)}\right)^2\right], \qquad (3.16)$$

where the scaling field $H_0(T)$ is of the order of $H_{\mathrm{c,th}}(T)$. The temperature dependent coefficient $\alpha(T)$ presents an important extension of the previous approximation for $Z_s(H)$. For BCS-like superconductors, $\alpha(T)$ vanishes at low temperatures as $\exp(-\Delta/k_\mathrm{B}T)$.

3.1.3 Adiabatic Superheating

Physical Concept and Bean–Livingston Surface Barrier

The transition from the superconducting to the normal state occurs in type-I superconductors at $H = H_{\mathrm{c,th}}$. This transition is known to be of first order, since the order parameter drops abruptly from its value in the S state to zero in the N state. Similarly, thorough analysis reveals that the transition from the Meissner state into the Shubnikov state at $H = H_{\mathrm{c1}}$ in type-II superconductors is not strictly of second order [18, 38]. Both transitions therefore imply a latent heat which, in the absence of perturbations, can cause superheating of the S state up to field levels which exceed $H_{\mathrm{c,th}}$ and H_{c1}, respectively (the lower critical field H_{c1} is discussed in Sect. 3.1.4). Similarly, supercooling of the N state can be observed down to fields below $H_{\mathrm{c,th}}$ and H_{c2}. These effects were originally considered already by Ginzburg on the basis of the stationary GL equations [39].

While the equilibrium S state constitutes a global minimum of the Gibbs free energy below the relevant critical fields, the superheated state still presents a local minimum. It is therefore a metastable state which can be observed only in the absence of nucleation effects. Nucleation of the N-state is inferred at defects or geometric irregularities in the superconducting sample, and by thermodynamic or electrodynamic fluctuations. As reviewed in [13] for several elementary superconductors close to T_c, superheating was experimentally confirmed for DC magnetic fields. Recent measurements of the surface impedance of high-purity polished Nb cavities at 1.3 GHz [40] showed linear response up to a microwave field amplitude $\approx H_{\mathrm{c,th}} > H_{\mathrm{c1}}$. Assuming the GL concept to stay qualitatively valid at low temperatures $T \leq T_c/2$, this result was interpreted as being an indication of superheating

3.1 Critical Field Levels and Corresponding Surface Impedance

in microwave fields. In contrast, the superheated state of metallic and oxide high-temperature superconductors has not yet been reported.

As pointed out in [13], the occurrence of a superheated S state is related, though not equivalent, to the existence of an energetical surface barrier for flux entry in type-II superconductors, as introduced by Bean and Livingston (BL) [14]. This effect, which is sometimes referred to as surface pinning, also occurs only in the absence of defects or surface irregularities. The critical field level for flux penetration is again reduced to H_{c1} if such structural fluctuations are present. Since the concept of Bean and Livingston is intuitive, its rationale will be repeated in the following. It might serve also as a basis to visualize the time constants for flux penetration in oscillating magnetic fields (Sect. 3.3). We assume, as in most references, a semi-infinite isotropic superconductor. The coordinate system is such that the surface of the superconductor extends over the y-z plane, and the magnetic field is oriented along the z direction. According to the BL analysis for $\kappa \gg 1$, the delayed flux entry results from the superposition of two competitive forces acting on a flux line. The total energy per unit length $W'(x)$ can be described by

$$W'(x) = \frac{\Phi_0}{4\pi} H_{c1} \left(1 - \frac{1}{\ln \kappa} K_0(2x/\lambda_{GL}) + \frac{H}{H_{c1}} e^{-x/\lambda_{GL}}\right). \qquad (3.17)$$

The first term represents the energy per unit length of an isolated flux line (Sect. 3.1.4). The second term approximates the magnetic field near the vortex core, which falls off at $x \gg \xi_{GL}$ like $\exp(-2x/\lambda_{GL})$. K_0 is the zeroth-order modified Bessel function of the second kind. Both terms represent the image force experienced by the flux line, which is required to cancel currents flowing normal to the surface of the superconductor. This force attracts the flux line to the surface, $x \geq 0$. In contrast, the last contribution to (3.17), which arises from the interaction of the flux line with the magnetic field H penetrating into the superconductor, dominates the energy balance in the interior at $x/\lambda_{GL} \gg 1$. Figure 3.1 shows (3.17) for $\kappa = 10$ and different field levels $h = H/H_{c1}$. The flux line remains "pinned" at the surface up to the field $H_{BL} > H_{c1}$. Inspection of (3.17) reveals $H_{BL} = H_{c,th}/2^{1/2}$ ($H_{BL}/H_{c1} \approx 3.57$ for $\kappa = 10$), which is lower by a factor of $2^{1/2}$ than the superheating field H_{sh} calculated from the GL equations (see below).

Ginzburg–Landau Superheating Field

The Ginzburg–Landau superheating field H_{sh} can be derived from (3.8) by evaluating the highest magnetic field that still yields a local minimum of the Gibbs free energy with respect to the order parameter Ψ [13]. This analysis was performed by various groups (e.g., [13, 39, 41–43] and review [44]). Recent improvements of the mathematical techniques resulted in approximations of H_{sh} for the two limiting cases [45]:

$$\frac{H_{sh}}{H_{c,th}} \approx 2^{-1/4} \kappa^{-1/2} \cdot \frac{1 + 4.6825\kappa + 3.3478\kappa^2}{1 + 4.0196\kappa + 1.0006\kappa^2}, \quad \text{for } \kappa \ll 1, \qquad (3.18a)$$

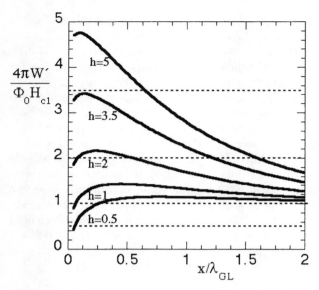

Fig. 3.1. Reduced vortex energy $4\pi W'/\Phi_0 H_{c1}$ (from (3.17)) versus normalized depth x/λ_{GL} in a superconductor for different normalized field levels $h = H/H_{c1}$ (After [14]).

and

$$\frac{H_{sh}}{H_{c,th}} = 1 + 0.461\kappa^{-4/3} + O(\kappa^{-6/3}), \quad \text{for } \kappa \gg 1. \tag{3.18b}$$

The small-κ limit presents physically relevant solutions that are stable against fluctuations. The analytical expression (3.18a) agreed with numerical data within about 1% up to $\kappa = 1$. As originally discussed in [46] and analyzed further in [13], the large-κ limit represents a mathematical formulation of the ultimate field up to which the nucleation of the N state can be delayed. However, the Meissner state becomes physically unstable in the presence of fluctuations already at a lower field $H_{sh}^* = (5^{1/2}/3)\, H_{c,th} \approx 0.745\, H_{sh}$. At field levels between H_{sh}^* and H_{sh}, the occurrence of a new metastable S-state was concluded, characterized by a layer of vortices at the surface of the superconductor. These alternating current loops were concluded to be distinctly different from Abrikosov vortices in that they had no singularity of the superfluid velocity at their center and did not carry quantized flux. Below $\kappa \leq 0.5$, no such instability was found, leaving H_{sh} according to (3.18) the relevant superheating field, above which the Meissner state becomes physically unstable.

The effect of finite film thickness d_F of a superconducting slab on H_{sh} was analyzed in [43] in terms of the reduced quantity d_F/λ_{GL}. As a representative result of the numerical computations, H_{sh} was found to increase with decreasing d_F/λ_{GL}, and to be for $\kappa = 1.5\, H_{sh} \approx 1.12\, H_{c,th}$ at $d_F/\lambda_{GL} = 10$ and $H_{sh} \approx 1.30\, H_{c,th}$ at $d_F/\lambda_{GL} = 2.5$.

While there are no theoretical treatments of the surface impedance at fields $H \geq H_{sh}$, it is tempting to assume that the global nucleation of the N state causes a steep increase of both the surface resistance and the penetration depth. Experimental investigations of the time constant of the transition from the superheated to the normal state were limited by the damping of the eddy currents in the sample. An upper limit of $\tau_{sh} < 4 \times 10^{-11}$ s (corresponding to $f_{sh} > 25$ GHz) was reported in [44], even for long-coherence length type-I superconductors. As reported in [40], increasing a microwave field amplitude above H_{sh} led to an oscillatory reduction of the energy stored in the resonator, indicating a fast ($\leq$ μs) global heating of the superconducting surface.

Complete Ginzburg–Landau Phase Diagram of Superconductors

The consideration of the critical fields H_{c1}, H_{c2}, $H_{c,th}$ and H_{sh} defines the complete phase diagram of superconductors in the framework of the Ginzburg–Landau theory. The results obtained in (3.10), (3.12) and (3.18), normalized to $H_{c,th}$, are displayed in Fig. 3.2 on double-logarithmic scales as a function of the GL parameter. The functional dependence of H_{c1} on κ (Sect. 3.1.4) is anticipated to be $H_{c1}(\kappa)/H_{c,th} \approx (\ln \kappa + 0.5)/2^{1/2}\kappa$.

Type-I and type-II superconductors occupy the regimes $\kappa < 2^{-1/2}$ and $\kappa > 2^{-1/2}$, respectively, as indicated by the vertical dashed line. The equilibrium S-state is limited at $H_{c,th}$ and H_{c1} for the two types (solid lines). The superheating fields H_{sh} and H_{sh}^* mark the ultimate field levels above which the normal and the Shubnikov phases are inevitably nucleated. The critical field H_{c2} (short-dashed line) denotes the second-order phase transition of type-II superconductors from the Shubnikov phase to the N-state. For type-I superconductors, H_{c2} (long-dashed) is expected to denote the supercooling field H_{sc} in the range $1/1.695 < 2^{1/2}\kappa < 1$ [27, 29]. At lower κ values, the supercooling field in parallel fields is identical to the surface critical field $H_{c3} \approx [\pi/(\pi-2)]^{1/2}\, H_{c2} \leq H_{c,th}$, since the nucleation of the S state in a surface layer of thickness ξ_{GL} will serve to initiate the transformation to the bulk of the sample [29].

In dealing with the microwave field dependence of the surface impedance of the extreme type-II superconductors Nb$_3$Sn ($\kappa \geq 5$) and YBa$_2$Cu$_3$O$_{7-x}$ or Tl$_2$Ba$_2$Ca$_{1,2}$Cu$_{2,3}$O$_y$ ($\kappa \geq 50$) in Chap. 4, the linear microwave response is expected to become metastable and completely unstable at H_{c1} and H_{sh}^*, respectively.

3.1.4 Lower Critical Fields

Bulk Lower Critical Field

The magnetic properties of type-II superconductors were analyzed originally by Abrikosov on the basis of the GL theory [47]. (Due to the important con-

118   3. Field and Power-Dependent Surface Impedance

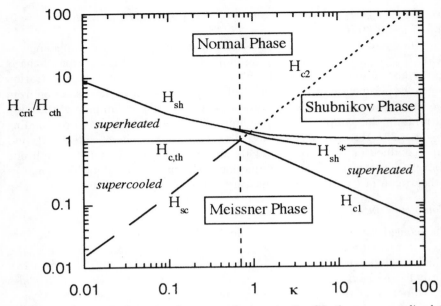

Fig. 3.2. Critical field levels of superconductors in the GL theory, normalized to the thermodynamic field $H_{c,th}$, versus the GL parameter κ.

tributions of Abrikosov and Gorkov (Sect. 3.1.2), the theory is often referred to as GLAG theory.) The thermodynamic and the upper critical fields $H_{c,th}$ and H_{c2} were deduced from the GL equations with boundary conditions describing the interface between a superconductor and an insulator [29, 38]. In contrast, the lower critical field is obtained from evaluating the properties of an S–N interface. The surface energy of such an interface is given by the difference in Gibbs free energies per unit area:

$$\sigma_{NS} = \int_{-\infty}^{\infty} dz \{G_s(H) - G_n\}. \tag{3.19a}$$

Equation (3.19a) can be expressed in terms of the order paramter Ψ and the microscopic induction B:

$$\sigma_{NS} = \int_{-\infty}^{\infty} dz \left[-\frac{\beta}{2}|\Psi|^4 + \frac{\mu_0}{2}\left(\frac{B}{\mu_0} - H_{c,th}\right)^2 \right]. \tag{3.19b}$$

According to (3.7) and (3.9), the parameter β is proportional to κ^2. General solutions of (3.19b) must be found numerically. However, σ_{NS} changes sign at the exact value $\kappa = 2^{-1/2}$, thus giving the physical basis to distinguish between type-I and type-II superconductors: σ_{NS} is positive (respectively, negative) for type-I (type-II). For the latter it is favorable to form as many S–N interfaces as possible, in order to maximally profit from the negative surface energy. Detailed analysis shows that the equilibrium state is attained

3.1 Critical Field Levels and Corresponding Surface Impedance 119

by the formation of flux lines, each carrying a single flux quantum and being arranged in a triangular lattice (see also recent comprehensive review in [48]). The formation of this "mixed" state (or Shubnikov phase) sets in at the field level H_{c1}, which corresponds to an energy W_1 per unit length of an isolated flux line (see below). Precise computations of W_1 and H_{c1} need to be performed numerically. However, for purposes of illustration, the approximative deduction within the London model is briefly summarized in the following. The exact results are stated at the end of this paragraph.

In an extreme type-II superconductor at fields well below H_{c2}, the total area occupied by the cores of the flux lines contributes only a very small amount to the total superconducting area, thus leaving $|\Psi|^2$, on average, essentially constant. If the sample contains a single flux line, oriented parallel to the z axis at the origin, the two GL equations merge into a single one, which can be viewed as an extended London equation relating the supercurrent $\boldsymbol{J}_s$ to $\boldsymbol{B}$:

$$\mu_0 \lambda_{\mathrm{GL}}^2 \mathrm{curl} \boldsymbol{J}_s + \boldsymbol{B} = \hat{z}\Phi_0 \delta_2(\boldsymbol{r}), \tag{3.20}$$

where $\delta_2(\boldsymbol{r})$ is a two-dimensional delta function at the position of the vortex. Using the Maxwell equation curl $\boldsymbol{B} = \mu_0 \boldsymbol{J}_s$, and considering that div $\boldsymbol{B} = 0$, leads to

$$\nabla^2 \boldsymbol{B} - \frac{1}{\lambda_{\mathrm{GL}}^2}\boldsymbol{B} = -\frac{\Phi_0}{\lambda_{\mathrm{GL}}^2}\hat{z}\delta_2(\boldsymbol{r}). \tag{3.21}$$

The solution of (3.21) is given by the zero-order modified Bessel function of second kind $K_0(r/\lambda_{\mathrm{GL}})$, which diverges logarithmically for zero argument and vanishes like $\exp(-r/\lambda_{\mathrm{GL}})$ at large arguments:

$$\boldsymbol{B}(\boldsymbol{r}) = \frac{\Phi_0}{2\pi\lambda_{\mathrm{GL}}^2}\hat{z}K_0(r/\lambda_{\mathrm{GL}}), \tag{3.22a}$$

with the approximate solution in the interval $\xi_{\mathrm{GL}} \ll r \ll \lambda_{\mathrm{GL}}$

$$\boldsymbol{B}(\boldsymbol{r}) \approx \frac{\Phi_0}{2\pi\lambda_{\mathrm{GL}}^2}\hat{z}\ln(\lambda_{\mathrm{GL}}/r). \tag{3.22b}$$

The energy $W_{1,\mathrm{L}}$ per unit length of a single vortex in the London model (index "L") is given by the difference between the free energy per unit length of a superconductor containing a single flux line and that describing the Meissner state. According to the notation introduced in Sect. 3.1.2, $W_{1,\mathrm{L}}$ is found to be

$$W_{1,\mathrm{L}} = \int \mathrm{d}^2 \boldsymbol{r} \left(\frac{1}{2\mu_0}B^2 + \frac{1}{2m}|(-i\hbar\nabla - 2e\boldsymbol{A})\Psi|^2 \right). \tag{3.23}$$

Since the order parameter is approximately constant in space, its value can vary only through its phase. The second term under the integral of (3.23) is then related to the supercurrent $\boldsymbol{J}_s$ (see (3.8b)). As a result, $W_{1,\mathrm{L}}$ can be regarded as being the sum of the magnetic field energy and the shielding kinetic energy. Using Maxwell's equation curl$\boldsymbol{B} = \mu_0 \boldsymbol{J}_s$ leads to

$$W_{1,\mathrm{L}} = \frac{1}{2\mu_0} \int \mathrm{d}^2 r \left[B^2 + \lambda_{\mathrm{GL}}^2 (\mathrm{curl}\boldsymbol{B})^2 \right]. \tag{3.24}$$

The logarithmic divergence of the second term at the origin (see (3.22b)) is a specific consequence of the London approximation near the core. The integration is therefore usually cut off at $r = \xi_{\mathrm{GL}}$. Integration by parts of (3.24) and using (3.21) and (3.22b) finally yields

$$W_{1,\mathrm{L}} = \frac{1}{\mu_0} \frac{\Phi_0^2}{4\pi \lambda_{\mathrm{GL}}^2} \ln \left(\frac{\lambda_{\mathrm{GL}}}{\xi_{\mathrm{GL}}} \right). \tag{3.25}$$

By definition, the lower critical field $H_{\mathrm{c1,L}}$ is given as that field at which the Gibbs free energy of the mixed state with a uniform low density of flux lines equals the Gibbs free energy of the Meissner state. This definition implies that W_1 and H_{c1} are related by

$$\Phi_0 H_{\mathrm{c1}} = W_1. \tag{3.26}$$

The index "L" in (3.26) has been omitted to account for the general validity of the expression. Combining the last two equations leads to the final result

$$\mu_0 H_{\mathrm{c1,L}} = \frac{\Phi_0}{4\pi \lambda_{\mathrm{GL}}^2} \ln \kappa. \tag{3.27}$$

This expression is valid only for $\kappa \gg 1$. Detailed analysis of the exact GL expressions adds another term to the line energy $W_{1,\mathrm{L}}$ in (3.23) (e.g., [49]):

$$W_{1,\mathrm{GL}} = W_{1,\mathrm{L}} + \int \mathrm{d}^2 r \left[\alpha |\Psi_0|^2 (x_{\mathrm{s}} - 1) + \frac{\beta}{2} |\Psi_0|^4 (x_{\mathrm{s}}^2 - 1) \right]. \tag{3.28}$$

The additional term represents the condensation energy per unit length of the vortex. Its evaluation on the basis of numerical computation was reported in [50], yielding

$$\mu_0 H_{\mathrm{c1,GL}} = \frac{\Phi_0}{4\pi \lambda_{\mathrm{GL}}^2} (\ln \kappa + 0.497). \tag{3.29}$$

It is interesting to note that Abrikosov reported the additive number to be 0.08 instead of 0.497 [38, 47]. Part of the numerical analysis of H_{c1} for arbitrary κ [51] is summarized in Table 3.2.

Flux-Flow Surface Impedance at $H \geq H_{\mathrm{c1}}$

The surface impedance of type-II superconductors in the mixed state was the subject of numerous publications, starting with early work by Gittleman, Rosenblum and Cardona [52]. Comprehensive reviews and summaries of their and related results can be found, e.g., in [2, 15, 53]. The following treatment focuses on the basic features needed to obtain valuable expressions for Z_{s} in terms of the amplitude of the microwave excitation.

In the presence of a (microwave) current density $\boldsymbol{J}$, a flux line containing one flux quantum Φ_0 experiences the force density

3.1 Critical Field Levels and Corresponding Surface Impedance

Table 3.2. Comparison of numerically computed data on the reduced lower critical field $H_{c1}/H_{c,th}$ [51] with the approximative result from (3.25b): $H_{c1}/H_{c,th} = 2^{-1/2}(\ln\kappa + 0.497)/\kappa$.

GL parameter κ	computed $H_{c1}/H_{c,th}$	calculated $H_{c1}/H_{c,th}$
0.3	1.68	–
$2^{-1/2}$	1.00	0.150
1	0.817	0.351
2	0.547	0.421
5	0.315	0.298
10	0.201	0.198
20	0.124	0.123
50	0.0622	0.0624

$$\boldsymbol{f} = \boldsymbol{J} \times \hat{z}\Phi_0. \tag{3.30}$$

We choose the coordinate system such that the superconductor extends over the y-z plane, the static flux density B and the microwave field H are along the z direction, and the x direction points into the superconductor. The equation of motion for the displacement u of the vortex is then

$$\mu_{\mathrm{em}}\ddot{u} + \eta\dot{u} + \chi u = \Phi_0 J. \tag{3.31}$$

Here, μ_{em} is the "electromagnetic" mass of the vortex per unit length quantifying its inertia, η is the viscosity per unit length, and χ is the spring constant of the potential which binds the vortices to an equilibrium position. The effective mass μ_{em} was calculated by Stephen and Suhl [54] with a time dependent form of the Ginzburg–Landau theory from the electromagnetic kinetic energy in the mixed state. They obtained the result

$$\mu_{\mathrm{em}} = \frac{2\pi\mu_0}{c^2}H_{c,\mathrm{th}}^2\xi_{\mathrm{GL}}^2\left(\frac{\lambda_{\mathrm{GL}}}{\lambda_{\mathrm{E}}}\right)^2, \tag{3.32a}$$

where λ_{E} is the shielding length of the electric field, which is of the order of a few times the inverse wave number k_{F}^{-1} at the Fermi level.

Based on the experimental analysis of the flux line viscosity η in [55], Bardeen and Stephen derived an elementary model yielding the analytical expression [56]:

$$\eta = \frac{\Phi_0}{\rho_{\mathrm{n}}}\mu_0 H_{c2}. \tag{3.32b}$$

Basic to this model is the assumption that the dissipative processes can be attributed to the normalconducting core of the flux line, and that the electrodynamics can be described by Ohm's law inside the vortex and by the London equation in the superconducting matrix. More sophisticated approaches are discussed in [29]. For fields slightly above H_{c1}, where the vortex density is low and the flux lines do not interact, η of (3.32b) was found to be enhanced by a factor of 2.62 in a superconductor containing many magnetic impurities [50].

The motion of a flux line induces an electric field $E = B\dot{u}$, where B is the spatial average of the microscopic flux density. Although care has to be taken in generally equating $B = \mu_0 H$ [29], we will usually neglect this distinction. Assuming $u = u_0 \exp(i\omega t)$, (3.31) can be solved and yields a relation between electric field and current:

$$J = \frac{1}{\rho_f}\left[1 - i\frac{\omega_{\text{dep}}}{\omega}\left(1 + \frac{\omega^2}{\omega_{\text{dep}}\omega_{\text{inert}}}\right)\right] E, \tag{3.33a}$$

where ρ_f is the flux flow resistivity [56]:

$$\rho_f = \frac{\Phi_0 B}{\eta} \approx \rho_n \frac{B}{\mu_0 H_{c2}}, \tag{3.33b}$$

and ω_{dep} is the "depinning" frequency [57]:

$$\omega_{\text{dep}} = \frac{\chi}{\eta}, \tag{3.33c}$$

which marks the transition between the limit where the flux motion is mainly reactive ($\omega \ll \omega_{\text{dep}}$), and the contrary where the nature of the flux motion is predominantly viscous, i.e., dissipative. The frequency $\omega_{\text{inert}} = \eta/\mu_{\text{em}}$ quantifies the inertial relaxation of a free vortex [54]:

$$\omega_{\text{inert}} = \frac{\eta}{\mu_{\text{em}}} = \frac{2\mu_0 c^2}{\rho_n} \kappa^2 \left(\frac{\lambda_E}{\lambda_{\text{GL}}}\right)^2, \tag{3.33d}$$

where the results of (3.32) were used. For $\kappa = 1$, $\rho_n = 100\,\mu\Omega\text{cm}$ and $\lambda_E/\lambda_{\text{GL}} \approx 10^{-3}$, (3.33d) gives $\omega_{\text{inert}} \approx 2 \times 10^{11}$ Hz, increasing quadratically with κ for $\kappa > 1$. For high-T_c superconductors, $\omega_{\text{dep}} \ll \omega_{\text{inert}}$ is therefore a realistic approximation.

The total surface impedance follows in the local limit from $Z_{s,\text{ff}} = [i\omega\mu_0(1/\sigma_{00} + 1/\sigma_{\text{ff}})]^{1/2}$ (index "ff" for "flux flow", $\sigma_{00} = \sigma_{01} - i\sigma_{02}$ denotes the intrinsic conductivity in the absence of magnetic field). At low fields and frequencies, where $\sigma_{\text{ff}}/\sigma_{02} \gg 1$, $Z_{s,\text{ff}}$ is given by ($\sigma_{01} \approx 0$):

$$Z_{s,\text{ff}} \approx i\omega\mu_0\lambda_L\left[1 + \frac{\lambda_c^2}{2\lambda_L^2}\left(1 - i\frac{\omega}{\omega_{\text{dep}}}\right)\right], \tag{3.34a}$$

with $\lambda_c = (\Phi_0 B/\mu_0\chi)^{1/2}$ the Campbell penetration depth [58]. $R_{s,\text{ff}}$ and $X_{s,\text{ff}}$ increase in this limit linearly with magnetic field. In the opposite limit, i.e., at fields far above H_{c1}, where $\sigma_{\text{ff}}/\sigma_{00} \ll 1$, the flux-flow surface impedance is determined by the conductivity ratio $y_{\text{ff}} = \text{Re}\{\sigma_{\text{ff}}\}/\text{Im}\{\sigma_{\text{ff}}\} = \omega/\omega_{\text{dep}}$. For $y_{\text{ff}} \ll 1$ it is:

$$Z_{s,\text{ff}} = R_n\sqrt{\frac{B}{\mu_0 H_{c2}}}\left(\sqrt{(\omega/\omega_{\text{dep}})^3/2} + i\sqrt{2(\omega/\omega_{\text{dep}})}\right) \tag{3.34b}$$

with $R_\mathrm{n} \propto \omega^{1/2}$ the normal-state surface resistance. We find the usual frequency dependences $R_\mathrm{s,ff} \propto \omega^2$ and $X_\mathrm{s,ff} \propto \omega$. For $y_\mathrm{ff} \gg 1$, we obtain

$$Z_\mathrm{s,ff} = R_\mathrm{n}\sqrt{\frac{B}{\mu_0 H_\mathrm{c2}}}(1+\mathrm{i}), \quad \omega/\omega_\mathrm{inert} \ll 1 \tag{3.34c}$$

and

$$Z_\mathrm{s,ff} = R_\mathrm{n}\sqrt{\frac{B}{\mu_0 H_\mathrm{c2}}}\left(\sqrt{(\omega_\mathrm{inert}/\omega)^3/2} + \mathrm{i}\sqrt{2(\omega_\mathrm{inert}/\omega)}\right), \quad \omega/\omega_\mathrm{inert} \gg 1. \tag{3.34d}$$

The unusual frequency dependences $R_\mathrm{s,ff} \propto \omega^{-1}$ and $X_\mathrm{s,ff} \propto \omega^0$ are reminiscent of the capacitive effect of the vortex inertia. The differential loss tangent $r_\mathrm{ff}(B)$ turns out to be independent of field in all of the four cases covered by (3.34):

$$r_\mathrm{ff}(B) = \frac{2\omega_\mathrm{dep}}{\omega} = \frac{2}{y_\mathrm{ff}} \gg 1 \quad \text{at } \omega \ll \omega_\mathrm{dep}, \tag{3.35a}$$

$$r_\mathrm{ff}(B) = 1 \quad \text{at } \omega \gg \omega_\mathrm{dep},\ \omega \ll \omega_\mathrm{inert}, \tag{3.35b}$$

$$r_\mathrm{ff}(B) = \frac{2\omega}{\omega_\mathrm{inert}} \gg 1 \quad \text{at } \omega \gg \omega_\mathrm{inert}. \tag{3.35c}$$

Equations (3.34) and (3.35) present limiting cases, which might not reproduce the experimental situation. A complete calculation of $Z_\mathrm{s,ff}$ needs to consider finite values of $y_0 = \sigma_{01}/\sigma_{02}$ and of $y_\mathrm{ff} = \omega/\omega_\mathrm{dep}$, leading to more complicated expressions like (1.18). This is demonstrated in Sect. 4.2.2 for a specific example that shows that $Z_\mathrm{s,ff}(B)$ varies between the above limits.

A comprehensive study of the surface impedance in the mixed state including thermally activated flux flow ("flux creep") can be found in [2, 3, 59, 60]. The field dependence $Z_\mathrm{s,ff}(B)$ remains essentially unchanged, while the slow process of flux creep introduces an additional characteristic frequency ($\ll \omega_\mathrm{dep}$) which affects $Z_\mathrm{s,ff}(f)$. The neglect of flux creep leads to an over- (respectively, under-) estimation of the flux-flow go viscosity η (depinning frequency ω_dep).

Finally, it is worth noting that for extreme type-II superconductors at field levels H slightly above the onset of the flux-flow regime it is $H/H_\mathrm{c2} \approx H_\mathrm{c1}/H_\mathrm{c2} \approx \ln\kappa/2\kappa^2 \ll 1$. Therefore, the flux-flow surface impedance presents only a very small fraction of R_n, and the field-induced nonlinearity is expected to increase rather smoothly above the low-field level $Z_\mathrm{s}(H \to 0)$.

Critical-State Surface Impedance at $H \geq H_\mathrm{c1}$

The result for the flux-flow surface impedance assumed a reversible displacement of the vortices in the superconductor. This assumption starts to fail if the flux-lines are pinned to their equilibrium position, causing a hysteretic field cycling. Pinning is supported by defects on the length scales of λ_{GL} and ξ_{GL}, which cause variations of the line energy W_1 [29]. It was recently shown that thickness variations in superconducting films also present effective pinning centers [49]. The surface impedance of superconductors in the presence of strong pinning can be evaluated in terms of Bean's critical-state model [2, 61]. The rationale of this model is that the flux lines remain pinned until the corresponding field gradient creates a critical magnetic pressure [29]. Defect-induced pinning and magnetic-pressure-induced depinning constitute the critical state, in which the field gradient is related to the critical transport current density by

$$-\mu_0 \frac{\partial B}{\partial x} = J_c. \tag{3.36}$$

Bean assumed the critical current to be independent of field, but field dependent forms of $J_c(B)$ can be implemented as well. The surface resistance $R_{s,cs}$ (index "cs" for "critical state") is found from the area of a full hysteresis loop of the magnetization curve [2, 61]. For field levels smaller than required for complete penetration into the sample, the result for $Z_{s,cs}$ is (with H_{c1} set to 0):

$$R_{s,cs} = \frac{2}{3\pi}\omega\mu_0 \frac{H}{J_c} \approx \frac{2}{3\pi} X_s \frac{H}{H_{pin}} \tag{3.37a}$$

and

$$X_{s,cs} = \frac{1}{2}\omega\mu_0 \frac{H}{J_c} \approx \frac{1}{2} X_s \frac{H}{H_{pin}}, \tag{3.37b}$$

with the linear field and frequency dependences of $R_{s,cs}$ being specific features of this mechanism. The second part of (3.37) introduced the magnetic scaling field $H_{pin} \approx J_c \lambda$. It represents the pinning of flux lines. It is to be clearly distinguished from the previous thermodynamic critical fields H_{c1}, $H_{c,th}$ and H_{c2}.

For a plate of thickness d_F which is penetrated from one side, the surface resistance passes through a maximum at $H = H^* = J_c d_F$ and decreases at higher field levels like H^*/H [61]. For critical current densities $J_c = 10^6 - 10^7$ A/cm^2 and a film thickness of 0.5 µm, the penetration field amounts to $H^* = 5$–50 kA/cm (or $\mu_0 H^* = 6.3$–63 mT). These values are within experimental reach and hence provide an important way of identifying critical-state effects from the observation of a maximum in $R_s(H)$.

The differential loss tangent $r_{cs}(H)$ follows from (3.37) as the constant number

$$r_{cs}(H) = \frac{3\pi}{4}. \tag{3.38}$$

The critical-state model was also applied to the calculation of the surface impedance of striplines with elliptic and rectangular cross-sectional geometries of the center conductor [62, 63]. The lower critical field H_{c1} was set to zero with respect to the strong demagnetizing effects (see below). Neglecting the current enhancement at the edges of the stripline, and assuming a field independent critical current I_c, the surface resistance R_1 and the surface reactance X_1 per unit length were deduced in a quasi-static approach from a Fourier expansion of Faraday's law. The different strip geometries lead to markedly different fronts of the penetrating flux, which are reflected in the dependence of Z_1 on the reduced microwave current $\iota = I/I_c$. While the complete expressions are lengthy and will not be repeated here, it is illustrative to compare the leading-order behavior of the surface impedance for the two geometries as given in [63]. The result valid in the limit $\iota \ll 1$ is for an elliptically shaped center conductor

$$Z_1 = \frac{\omega \mu_0}{6\pi^2} \iota \left(1 + i\frac{3\pi}{4}\right) + iX_0, \qquad (3.39)$$

and for a rectangular center conductor

$$Z_1 = \frac{\omega \mu_0}{6\pi^2} \iota^2 \left(1 + i\frac{15\pi}{32}\right) + iX_0. \qquad (3.40)$$

In both equations, X_0 denotes the geometric reactance which is independent of current. It is important to note that the nonlinear contributions to Z_1 increase linearly (elliptic case) and quadratically (rectangular) with the microwave current amplitude ι. Comparing (3.39) and (3.40) with (3.37) reveals the ambiguity introduced by the film geometry to the power-law behavior of the nonlinear Z_s. The r values corresponding to (3.39) and (3.40) are constant in accordance with (3.38) and amount to $3\pi/4 \approx 2.36$ and $15\pi/32 \approx 1.47$, respectively.

Nonlinear Microwave Response of Josephson Junctions

The granular nature of high-κ superconductors often induces Josephson behavior. It is therefore instructive to supplement the previous analysis by the description of the microwave response of a Josephson junction. Some of the resulting models applied to the oxide superconductors will be discussed in more detail in Sect. 3.4.1.

The electrodynamics of a single small Josephson junction are described by (1.34)–(1.36) in Sect. 1.2.2. The Josephson inductance L_J can accordingly be expressed at small excitation currents $\iota_J = I/I_{cJ} < 1$ [64]:

$$L_J = \frac{\Phi_0}{2\pi I_{cJ}} \cdot \frac{1}{\sqrt{1 - \iota_J^2}}. \qquad (3.41)$$

The same dependence on ι_J is transferred to the conductivity ratio $y_J = \omega L_J / R_J$ since R_J is considered constant. In the limit of small $y_J \ll 1$, where the inductive nature of the Josephson junction prevails, the surface impedance $Z_s(y_J)$ can be expanded in terms of the reduced current:

$$\frac{Z_{s,J}}{R_J} = \sqrt{\frac{y_0^3}{2}}\left(1 + \frac{3}{4}\iota_J^2\right) + i\sqrt{2y_0}\left(1 + \frac{1}{4}\iota_J^2\right) \quad \text{at } \iota_J \ll 1. \tag{3.42a}$$

Here, y_0 is given in terms of the critical voltage $V_c = I_{cJ}R_J$ and the measurement frequency, $y_0 = \Phi_0 f/V_c$. The corresponding r value is

$$r_J(I) = \frac{2}{3y_0}\sqrt{1 - \iota_J^2} \gg 1. \tag{3.42b}$$

The surface impedance of junctions with very low critical voltage follows the "normal skin effect" behavior $R_{s,J} = X_{s,J} = R_J$ with $r_J(I) = 1$. The apparent similarity between the small-signal limit of Josephson junctions and the Ginzburg–Landau response of superconductors (especially (3.42) and (3.15)) was emphasized in [1, 15]. However, the nature of the excitations (current versus field) and the scaling parameters (I_{cJ} versus $H_{c,\text{th}}$) are distinctly different.

At current levels $\iota_J > 1$ above the critical current, the microwave response of the junction can be described in the framework of the RSJ model, as long as $y_0 \ll 1$ remains valid. The surface resistance in this limit can be approximated by [65–67]:

$$R_{s,J} \approx R_J\sqrt{1 - \iota_J^2} \quad \text{at } \iota_J > 1. \tag{3.43}$$

The surface reactance oscillates between inductive and capacitive nature, and approaches zero for high bias currents $\iota_J \gg 1$. The same behavior is reflected in the differential loss tangent.

In the case of long Josephson junctions, the microwave response can be expressed in terms of the lower critical field $\mu_0 H_{c1J} = \Phi_0/(2\pi\lambda_L\lambda_J)$ (Sect. 3.1.1). The expressions for the surface impedance in this case are analogous to those for the flux-flow and the hysteretic mechanisms outlined above, with the substitution of H_{c1} by the much smaller value H_{c1J}, and by noting that Josephson fluxons rather than Abrikosov vortices are involved [15].

3.1.5 Implications of Film Thickness, Anisotropy and Unconventional Pairing

This section summarizes some important refinements of the GL theory, especially in relation to the oxide superconductors. Early approaches to account for strong pair coupling were reviewed, e.g., in [38]. However, there are no general results, and numerical computations are required to analyze specific problems. A similar statement holds for the consideration of two electronic conduction bands [68]. The GL analysis in this case requires two κ parameters. In contrast, the effects of finite film thickness, anisotropy and demagnetization were analytically approached, as sketched in the following. Furthermore, there has been some work on the GL theory of d-wave superconductors, which is briefly mentioned at the end of this section.

Altogether, the analysis of the magnetic field dependent transport properties of HTS in the frameworks of the conventional theories by Ginzburg–Landau and Josephson presents a valuable starting point. Quantitative conclusions, however, have to be deduced with care. This is especially true for the analysis of H_{c1} values since these are neither accurately calculable nor measurable, theoretically due to the problems listed before, and experimentally due to the influence of pinning at the surface and/or at defects, due to demagnetization, finite film thickness and granularity.

Lower Critical Field of Thin Superconducting Films

The lower critical field of a thin ($\xi_{GL} \ll d_F \leq \lambda_{GL}$) superconducting slab in a parallel DC magnetic field was first analzyed by Abrikosov [69]. The problem was treated for extreme type-II superconductors ($\kappa \gg 1$) within the framework of the Ginzburg–Landau theory in analogy to the derivation of the bulk lower critical field (Sect. 3.1.4). Since the derivation is lengthy, we merely reproduce the final result for $H_{c1}(d_F)$, relative to the bulk value $H_{c1,\infty}$, in terms of the reduced film thickness $d^* = d_F/\lambda_{GL}$:

$$\frac{H_{c1}(d_F)}{H_{c1,\infty}} = \frac{1}{1 - \cosh^{-1}(d^*/2)} \left[1 + \frac{2}{\ln \kappa + \varepsilon} \sum_{m=1}^{\infty} (-1)^m K_0(md^*) \right] \tag{3.44a}$$

with $\varepsilon = 0.497$ (see (3.29)). Equation (3.44a) shows that the lower critical field of a thin film is enhanced over that of an infinite superconductor. This enhancement reflects the reduced gain of magnetic energy by forming an S–N interface, since the external field can penetrate the film without significant shielding. The first factor in the equation, which approaches $8\lambda_{GL}^2/d_F^2$ for $d^* \ll 1$, is mainly responsible for $H_{c1}(d_F)$ exceeding $H_{c1,\infty}$. Typical values for $H_{c1}(d_F)/H_{c1,\infty}$ are, for $\kappa = 50$, $H_{c1}(d_F)/H_{c1,\infty} = 281$, 23, 7.5, 1.19 and 1.01 at $d_F/\lambda_{GL} = 0.1$, 0.5, 1, 5 and 10.

The hyperbolic cosine function in (3.44a) reflects the spatial variation of the magnetic field in the film in the Meissner state:

$$H_M(z) = H_0 \frac{\cosh(x/\lambda_{GL})}{\cosh(d^*/2)}. \tag{3.44b}$$

Here, the film is again assumed to extend over the y-z plane. The upper and lower surface are located at $x = +d_F/2$ and $x = -d_F/2$, respectively. The external field is oriented within the y-z plane, and penetrates exponentially into the superconductor from both sides. Thus, (3.44b) is a solution of (3.21) in the homogeneous case (no flux lines present, $\boldsymbol{B} = \mu_0 \boldsymbol{H}$).

As pointed out in [70], the infinite sum in (3.44a) can be understood to result from image vortices above and below the film at separations $\pm m d_F$. In analogy to the Bean-Livingston surface barrier (Sect. 3.1.3), the consideration of image vortices is required in order to cancel the components of the shielding currents vertical to the film at its two surfaces. The thin-film limit $d^* \ll 1$ of (3.44b) was evaluated in [70] for the London limit (i.e., $\varepsilon = 0$):

$$\mu_0 H_{c1}(d_F) \approx \frac{2\Phi_0}{\pi d_F^2} \left(\ln\left(\frac{d_F}{\xi_{GL}}\right) - 0.458 \right) \quad \text{for } d_F \ll \lambda_{GL} \,. \tag{3.45}$$

This expression resembles, except for the numerical factors, the bulk value of $H_{c1,\infty}$ in (3.27), with the film thickness replaced by the penetration depth. As the film thickness decreases further, and eventually becomes comparable to the Ginzburg–Landau coherence length ξ_{GL}, $H_{c1}(d_F)$ increases and approaches the upper critical field $H_{c2}(d_F)$. At even smaller thickness, the mixed state is no longer energetically favorable, and the superconducting film experiences a magnetic phase transition similar to that of type-I superconductors [69].

In contrast to the large enhancement of $H_{c1}(d_F)$ for fields penetrating the superconducting film from both sides, the effect of finite film thickness on the penetration of a magnetic field from a single surface (e.g., in the case of a superconducting film exposed to the microwave fields of a host resonator) is strongly reduced. Assuming as before that the energetically most favorable position for the nucleation of the first vortex is at $x = 0$ [70], the term $\cosh^{-1}(d^*/2)$ in the denominator of (3.44a) is to be replaced by $[\cosh^{-1}(d^*/2) + \exp^{-1}(d^*/2)]/2$. Typical values for $H_{c1}(d_F)/H_{c1,\infty}$ are then 14, 5.6, 3.4, 1.14 and 1.01 at $\kappa = 50$ and $d_F/\lambda_{GL} = 0.1, 0.5, 1, 5$ and 10.

Lower Critical Field of Anisotropic Superconductors

The lower critical field of anisotropic layered superconductors in the GL framework was analyzed in [71, 72]. It was assumed that the perpendicular coherence length extended over many layers, corresponding to bulk behavior. An alternative approach is the consideration of Josephson-coupled layers, treated by Lawrence and Doniach [73]. For the sake of clarity, we focus here on the former work which is expected to be valid at least close to T_c. Anisotropy was introduced into the GL equations in the form of an anisotropic mass tensor. By proper scale transformations, the equations could be rewritten in the classical form (Sect. 3.1.4) with an effective angle dependent κ parameter. It was noted as being an important consequence of the anisotropy that the directions of external magnetic field and of the microscopic magnetic induction are in general not aligned. For highly anisotropic superconductors, the vortices prefer to lie in the easy direction, for which H_{c1} is smallest, unless H is close to perpendicular to this direction. If the magnetic field is applied at an angle other than 0 or 90° with respect to the principle axes of symmetry, vortices with different orientations might even coexist [48, 74].

For shielding currents parallel and perpendicular to the layers, the lower critical field of anisotropic superconductors is parallel to the magnetic induction, and can be expressed in the large-κ limit in closed form (after [71]):

$$\mu_0 H_{c1,\|} = \frac{\Phi_0}{4\pi\lambda_\|^2}\left[\ln\left(\frac{\lambda_\|}{\xi_\|}\right) + 0.497\right] \qquad (3.46a)$$

and

$$\mu_0 H_{c1,\perp} = \frac{\Phi_0}{4\pi\lambda_\|\lambda_\perp}\left[\frac{1}{2}\ln\left(\frac{\lambda_\|\lambda_\perp}{\xi_\|\xi_\perp}\right) + 0.497\right] \qquad (3.46b)$$

in obvious notation. It is worth mentioning that the high sensitivity of penetration depth and intralayer coherence length of the oxide superconductors on the oxygen content necessarily leads to significant scatter of the H_{c1} data, especially since H_{c1} scales like λ^{-2} [10].

The expressions for H_{c1} in the thin-film limit result from a straightforward combination of (3.46) and (3.44) [70].

Effect of Demagnetization on the Lower Critical Field

The magnitude of the lower critical field, according to (3.27) and (3.46), can be expected to apply to experimental data only if the magnetic field is applied parallel to the surface of films. More precisely, the shape of a sample should be approximated by, e.g., an oblate spheroid with the demagnetization coefficients $D_\|$ and $D_\perp$, where $2D_\| + D_\perp = 1$ [75]. For an isotropic superconductor with the magnetic field applied at an angle θ_H with respect to the z axis this leads to a lower critical field [72]:

$$H_{c1}(\theta_H) = H_{c1,0}\left(\frac{\sin^2\theta_H}{(1-D_\|)^2} + \frac{\cos^2\theta_H}{(1-D_\perp)^2}\right)^{-1/2}, \qquad (3.47)$$

where $H_{c1,0}$ denotes the value at zero demagnetization. In case of a flat disk of thickness $2d_F$ and radius R, we have approximately $D_\| = 0$ and $D_\perp = 1 - (\pi/2)d_F/R$ [38]. For homogeneous, singly connected, thin films we expect $D_\perp \approx 1$ and therefore $H_{c1}(90°) \approx 0$.

In the presence of demagnetizing effects, the lower critical field of an anisotropic superconductor is characterized by θ_H as before, and in addition by the angle θ_B between the microscopic induction and the z axis of the sample [72, 76]. The general result is a lengthy expression that must be solved for each fixed angle θ_H of the applied field. For the easy directions, a combination of the two previous expressions is obtained, where $H_{c1,0}$ in (3.47) must be substituted by either $H_{c1,\|}$ or $H_{c1,\perp}$ from (3.46). For arbitrary angles θ_H, θ_B qualitatively remains small up to a critical value θ_H^*, and jumps to angles close to 90° at larger θ_H values. The resulting angular dependence $H_{c1}(\theta_H)$ can exhibit a discontinuity, a kink, or be monotonic, depending on the magnitude of the κ parameter and the degree of anisotropy [77].

Implications of d-Wave Symmetry

The thermodynamic implications of a pairing interaction leading to a d-wave symmetry of the order parameter were investigated theoretically and numerically, e.g., in [78–80]. The interpretation of the vortex core as a normalconducting region of radius ξ_{GL} remains valid. However, the symmetry of the order parameter changes from s wave near the core to d wave in the bulk superconductor. The microscopic description of pinning and flux-flow viscosity are therefore different for s- and d-wave superconductors [3].

The magnetic-field dependence of the pair fraction x_s and thus of the microwave penetration depth λ were analyzed in [30]. The existence of nodes of the energy gap along certain directions on the Fermi surface leads to two major modifications compared to conventional superconductors. First, the linear energy dependence of the density of states of a d-wave superconductor (Sect. 1.3.3) translates into a linear field dependence of the penetration depth at very low temperatures. This feature should be observable in a narrow interval of field levels which is limited by impurity scattering at the lower end, and by H_{c1} at the upper end. Secondly and more importantly, as further studied in [37], the low-lying excitations in a d-wave superconductor result in a much stronger and anisotropic pair breaking. In analogy to (3.16), the field dependence of the penetration depth was found to be

$$\frac{1}{\lambda(T,H)} = \frac{1}{\lambda(T)}\left(1 - \alpha(\theta_\Delta, T)\left(\frac{H}{H_0(T)}\right)^2\right). \tag{3.48}$$

(The notation in (3.48) differs slightly from that in [37], following the one reported in [30]). The coefficient α introduces anisotropy through the angle θ_Δ between the direction of current flow and the direction of the antinodes ($\theta_\Delta = 0°$ and $90°$) of the order parameter $\Delta(\theta_\Delta) = \Delta_0(T)\cos(2\theta_\Delta)$. Furthermore, the nodes of $\Delta(\theta_\Delta)$ lead at low temperatures to enhanced pair breaking $[\alpha(T) \propto \Delta_0(0)/k_B T]$ as compared to s-wave pairing. An example of $\alpha(\theta_\Delta, T)$ for $2\Delta_0(0)/k_B T_c = 6$ in the clean limit is reproduced in Fig. 3.3 from [37]. According to this result, the effective scaling field for pair breaking $H_0/\alpha^{1/2}$ should become extremely small (and comparable to the value at $T \geq 0.8 T_c$) at temperatures $T < 0.1 T_c$. Observation of such an anomalous temperature dependence would be a striking indication for d-wave symmetry of the order parameter in HTS (which has not yet been verified, Chap. 4).

3.2 Microwave Heating

The previous section summarized the different possible magnetic mechanisms which evoke a nonlinear microwave response of superconductors. These effects can be induced either by application of a DC magnetic field or by increasing the magnetic microwave field amplitude (or, equivalently, the circulating power). In contrast, this section deals with nonlinear effects caused by the dissipation of microwave power.

In contrast to the infinite conductivity of superconductors with respect to DC currents, alternating and especially microwave fields dissipate heat in

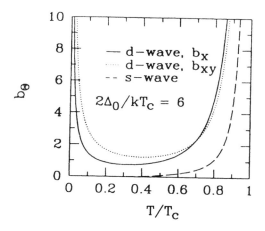

Fig. 3.3. Temperature dependence of the pair breaking coefficient $b_\theta \equiv \alpha(\theta_\Delta, T)$ of a d-wave superconductor with $2\Delta_0(0)/k_B T_c = 6$ [37] for $\theta_\Delta = n\pi/2$ (*solid line*) and $\theta_\Delta = (2n+1)\pi/4$ (*dotted line*) with $n \geq 0$ integer. Also shown is the result obtained for a BCS-like superconductor (*dashed line*).

the superconductor. Such heating effects were suggested to lead to enhanced losses at power levels which are relevant for devices [81–83]. A simple representative estimation shows that the dissipated power in a planar HTS filter may reach values above $P_\text{diss} \geq 1\,\text{W}$ for a circulating power $P_\text{circ} = 100\,\text{kW}$ (with $Q = P_\text{circ}/P_\text{diss} = 10^5$) [84, 85]. The steady-state temperature of the superconductor might hence be elevated at its surface well above the temperature of the cooling stage, especially if CW operation in an evacuated environment is concerned (Sect. 2.1.2). Microwave heating might involve the complete superconductor (global heating) as well as the introduction of heat at exceptionally lossy regions (local heating). Early analytical estimations of the latter type considered the heat dissipated in the superconductor at the normal core of isolated vortices [86], which will certainly become relevant at field levels above H_{c1}. A general treatment of microwave heating is complicated by the highly nonlinear temperature dependence of the thermal parameters of the superconducting film and the substrate. Numerical calculations seem therefore appropriate to study thermal effects under specific boundary conditions. Such approaches were proposed for bulk Nb accelerator cavities [87–89] and transferred later to the problem of YBCO films on dielectric substrates [84, 90, 91].

Section 3.2.1 describes the physical problem and sketches appropriate tools for the numerical computation of heating effects in HTS films under constant applied microwave power. Approximative results for the field levels H_q at which thermal breakdown ("quench") occurs, and for the time constants of the temperature rise are summarized in Sect. 3.2.2 in order to illustrate the problem. Exact numerical solutions are discussed in Sect. 3.2.3 on the basis of [84]. The results are studied in terms of the temperature dependent surface resistance $R_s(T)$, the thermal properties of the substrates, the properties of isolated defects, the film thickness, the operating frequency, and the cooling conditions. Dynamic effects are postponed to Sect. 3.3.1.

While the results discussed in Sect. 3.2 were deduced with material parameters appropriate for YBCO films on dielectric substrates like LaAlO$_3$ or sapphire, their validity turned out to be rather general. Qualitatively similar conclusions can therefore be drawn, e.g., for microwave heating in Nb$_3$Sn films on sapphire.

3.2.1 Description of the Problem

Physical and Mathematical Basis

The computer code developed at the University of Wuppertal for bulk Nb [88] was modified to evaluate the effect of heating on the microwave response of YBCO films [84]. The algorithm accounts for the finite thickness d_F of the HTS film, deposited onto a dielectric substrate of thickness d_S. The geometric arrangement and the discretization lattice are sketched in Fig. 3.4. The heat generated in the film at a given microwave amplitude H_s (index "s" for surface), $P_\mathrm{diss} = (1/2)[R_\mathrm{def} A_\mathrm{def} + R_\mathrm{s}(A - A_\mathrm{def})] H_\mathrm{s}^2$, is conducted to the bottom of the substrate, which is kept at the constant temperature T_B of the cold sink. R_def and $A_\mathrm{def} = \pi r_\mathrm{def}^2$ are the surface resistance of the defect and its areal extension.

The thermal power density $p_\mathrm{th}(T)$ (in W/m^3), generated by the dissipation of microwave power ∂P_diss in a volume element ∂V, is related to the

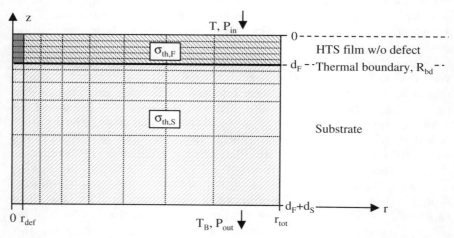

Fig. 3.4. Schematic discretization to simulate microwave heating [84]. The thickness d_F of the HTS film (*grey, top layer*) is covered by five equidistant layers. The boundary between film and substrate (thickness d_S, *light grey*) presents the thermal resistance R_bd (*bold horizontal line*). The radial mesh elements and the vertical discretization of the substrate increase away from the origin. The first uppermost column of radius r_def (*dark grey*) represents the potential position of a defect. The cylindrical arrangement is terminated at r_tot by thermal insulation.

spatial variation of the temperature $T(\mathbf{r})$ and the heat conductivity σ_{th} by [92]:

$$p_{\text{th}}(T) = -\sigma_{\text{th}}(T)\nabla^2 T = \frac{\partial P_{\text{diss}}}{\partial V} . \tag{3.49a}$$

(It is worth noting the formal analogy between the thermal transport described by (3.49a) and the electrodynamic analogue, where σ_{th} is replaced by the electrical conductivity, and where ∇T corresponds to an electric field E.) If the heat would be homogeneously conducted along the z direction, the one-dimensional simplification of (3.49a) would read

$$\dot{Q} = -\sigma_{\text{th}} A \frac{\partial T}{\partial z} = P_{\text{diss}} , \tag{3.49b}$$

where Q is the total heat flowing across the area A, and the dot denotes its time derivative. The first part of (3.49a) is solved numerically with the Gauss–Seidel relaxation method [93, 94] for each volume element (i,j) of the mesh depicted in Fig. 3.4. Here, $i \leq N$ and $j \leq M$ are the indices for the radial and the axial directions of the rotational symmetric problem, respectively. The origin of the coordinate system is adjusted such that a defect can be introduced at the origin $r = 0$, extending axially over the whole film thickness (dark grey area in Fig. 3.4). The mesh size is reduced near the origin in order to achieve maximum accuracy in the vicinity of the defect. The heat flow $\dot{Q}_{ij}$ is calculated from the sum of the heat current densities $J_{\text{th},ij}$ through all of the four boundaries A_{ij} of cell (i,j):

$$\dot{Q}_{ij} = \sum_{k=1}^{4} J_{\text{th},ij}^{(k)} A_{ij}^{(k)}, \tag{3.50a}$$

where the superscript "k" denotes the four surfaces of cell (i,j) ($k = 1, 3$, top, bottom; $k = 2, 4$, left, right). The boundary conditions of the problems are accounted for by

$$J_{\text{th},Nj}^{(2)} = J_{\text{th},Nj}^{(4)} = 0, \tag{3.50b}$$

i.e., the cylindrical film–substrate arrangement is terminated at an adjustable radius r_{tot} by a thermal insulation. Furthermore,

$$J_{\text{th},i1}^{(1)} = 0 \tag{3.50c}$$

takes into account that no heat is conducted away from the surface of the superconductor. Finally, at the interface between film and substrate at $j = L$, a finite boundary resistance R_{bd} leads to the discontinuous temperature difference ΔT (which is the thermal analog to Ohm's law):

$$J_{\text{th},iL}^{(3)} = \frac{\Delta T}{R_{\text{bd}}} . \tag{3.50d}$$

Substituting the current densities $J_{\text{th},ij}$ by temperature gradients, the first part of (3.49a) can be iteratively solved for the temperature T_{ij} of each cell.

The temperature $T_{ij}^{(n+1)}$ of each cell in step $(n+1)$ can be calculated from the temperature of the neighboring cells in the preceding iteration step n. After sufficient approximation of the temperature distribution for a given distribution of the power dissipation (set as an input), an improved P_{diss} distribution is calculated in a second iteration loop (second part of (3.49a)). The algorithm is terminated when this two-loop iteration balances the heat power P_{in} absorbed at the top of the film with the heat power P_{out} at the bottom of the substrate within an adjustable ε criterion: $(P_{\text{in}} - P_{\text{out}})/P_{\text{in}} < \varepsilon$.

If the film thickness d_{F} is comparable to the microwave penetration depth λ, the redistribution of the electromagnetic field within the film should be considered in the calculation of the dissipated power. In a first step, representative $R_{\text{s,eff}}(T)$ and $\Delta\lambda_{\text{eff}}(T)$ data measured with thin films are converted into intrinsic $R_{\text{s}}(T)$ and $X_{\text{s}}(T)$ values, according to the analysis in Sect. 1.1.3. The dissipated power in each cell (ij) is then given by [95]:

$$P_{\text{diss},ij} = \frac{1}{2} A_{ij}^{(3)} \int dz\, \sigma_1 |E(z)|^2, \tag{3.51}$$

where the average quasi-particle conductivity $\sigma_1(T)$ is derived from $R_{\text{s}}(T)$ and $X_{\text{s}}(T)$ according to (1.43). The spatial dependence of the electric field gradient $E(z)$ is obtained by considering the reflections and transmission of a plane electromagnetic wave at the two interfaces vacuum–film and film–substrate with appropriate boundary conditions.

After having achieved power balance as well as the corresponding three-dimensional steady-state temperature distribution, the power dissipated in every cell is summed up, and the result is divided by the film surface πr_{tot}^2. The magnetic field amplitude H_{s} is considered constant across this area. From this result, the thermally enhanced surface resistance of the HTS film can be recalculated for each H_{s} value, yielding a synthetic $R_{\text{s}}(H_{\text{s}})$ curve. The corresponding value of the penetration depth at the elevated temperature can be deduced from the measured $\lambda(T)$ dependence. It is important to note that H_{s} enters these considerations as a "heat load". Increasing H_{s} is equivalent to raising the temperature of the sample, and the thermally induced $R_{\text{s}}(H_{\text{s}})$ reflects the temperature dependent surface resistance $R_{\text{s}}(T)$ of the sample.

Dynamic Calculations

The dynamic features of microwave heating were taken into account on the basis of the continuity equation [92]

$$\text{div}\, \boldsymbol{J}_{\text{th}} = -C_V \dot{T} + p_{\text{th}} \tag{3.52a}$$

with C_V denoting the heat capacity per unit volume. Integration of (3.52a) and using the Gauss theorem leads to the time dependent expression ($P_{\text{diss}}(t)$ is considered constant):

$$\dot{Q} = P_{\text{diss}} - \oint_{\partial V} \boldsymbol{J}_{\text{th}}\, d\boldsymbol{A} = C_{\text{tot}}\, \dot{T} \tag{3.52b}$$

with C_tot the total heat capacity in J/K. Supplementary, transient power dissipation was also investigated [91] (see also Sect. 3.2.3), leading to implicit equations which prevent the derivation of analytical approximations.

In the steady state $T(t) = \text{const.}$, (3.52b) corresponds to (3.50a), and the surface integral here is evaluated in an analogous computation as the $\dot{Q}_{ij}$'s there. Discretization of the time derivation leads to an iterative calculation of the temperature for each cell [94]:

$$T_{ij}^{(n+1)} = T_{ij}^{(n)} + \Delta t \frac{\dot{Q}}{C_\text{tot}} \qquad (3.53)$$

The time step Δt has to be adjusted to the mesh size and can be chosen as a compromise between reasonable CPU time and a computationally stable algorithm.

Input Parameters

The input data for $R_\text{s}(T)$ and $\lambda(T)$ were taken from microwave measurements (Chap. 2). The numerical algorithm also takes into account the temperature dependences and the anisotropy of the thermal conductivities of the HTS film, $\sigma_\text{th,F}^\text{ab}(T)$ and $\sigma_\text{th,F}^\text{c}(T)$, of the substrate, $\sigma_\text{th,S}(T)$, and of the thermal boundary resistance $R_\text{bd}(T)$. In order to minimize the number of fit parameters and to imitate realistic conditions, measured rather than calculated data were collected from the literature for $\sigma_\text{th,F}(T)$ [96, 97], $R_\text{bd}(T)$ [98–100] and $\sigma_\text{th,S}(T)$ (LaAlO$_3$ (LAO, [101]), SrTiO$_3$ (STO), MgO [102], and sapphire [103]).

The temperature dependence $\sigma_\text{th,F}(T)$ of a BCS-like superconductor results in general from an electronic contribution, which vanishes as $\exp(-\Delta/k_\text{B}T)$, and from a phononic contribution, which vanishes like a power law and hence dominates $\sigma_\text{th,F}$ at low temperatures. A peak in $\sigma_\text{th,F}(T)$ is usually observed slightly below T_c, the origin of which can be attributed to the interdependence of electronic and phononic mechanisms [96]. The absolute values and the anisotropy of the heat conductivity of YBCO were found to depend strongly on the quality of the samples. Values reported for high-quality single-crystals amounted to maximum values of $\sigma_\text{th,F}^\text{ab} \approx 30\,\text{W/Km}$ and $17\,\text{W/Km}$.

The phononic heat conductivities of the dielectric substrate materials increase as T^3 at low temperatures and level off (LAO, STO) or decrease (MgO, sapphire) above 20–30 K. Sapphire and MgO display heat conductivities which are about a factor of 100 higher than LAO and STO. The boundary resistance is approximately constant above about 30–40 K. It increases at lower temperatures like T^3, as expected for the Kapitza resistance reflecting the mismatch of acoustic phonons at the film–substrate interface [104]. Additionally, the temperature dependent heat capacities $C_\text{V,F}$ and $C_\text{V,S}$ of film and substrate were implemented to simulate the dynamic evolution of the microwave heating [101, 102, 105–107].

The presence of a defect is accounted for by its areal extension and its surface resistance, with A_{def} and R_{def} being free input parameters. If none was specified, global microwave heating was simulated. Except for one case (see below), $R_{\text{def}} = (\omega \mu_0 \rho / 2)^{1/2}$ represented a normalconducting defect with resistivity $\rho(T_c) = 100\,\mu\Omega\text{cm}$. The frequency was fixed to $\omega/2\pi = 19\,\text{GHz}$.

3.2.2 Analytical Approach

Before discussing the detailed numerical results for the thermal $R_s(H_s)$ curves in Sect. 3.2.3, analytical approximations of the thermally induced breakdown field H_q are introduced. This parameter is of major interest for the analysis of nonlinear phenomena, since it presents the thermal analog of the critical magnetic fields H_{crit} (Sect. 3.1). A global quench occurs when the dissipated power suffices to induce a thermal S–N transition of the entire film. For an isotropic bulk superconductor ($\sigma_{\text{th}}^{\text{ab}} = \sigma_{\text{th}}^{c} = \sigma_{\text{th}}$) one obtains H_q from (3.49b) with $\Delta T = T_c - T_B$, where T_B is the temperature of the cold sink, and with the total thermal resistance R_{th} of the film–interface–substrate sandwich

$$R_{\text{th}} = \frac{d_F}{\sigma_{\text{th,F}}} + R_{\text{bd}} + \frac{d_S}{\sigma_{\text{th,S}}} \tag{3.54a}$$

which reflects the sum rule for a series connection of heat resistivities. The result is

$$H_q^2 = \frac{2(T_c - T_B)}{R_s R_{\text{th}}} . \tag{3.54b}$$

Equation (3.54a) means that the thermal resistance for $\sigma_{\text{th,F}} \approx \sigma_{\text{th,S}}$ (e.g., for YBCO on LAO) is dominated by the heat conductivity of the substrate due to the much larger thickness $d_S/d_F \approx 1000$. H_q in this limit is independent of the film thickness. Equation (3.54) yields $B_q = \mu_0 H_q \approx 40\,\text{mT}$ and $300\,\text{mT}$ for YBCO on LAO and on sapphire at $77\,\text{K}$ and $19\,\text{GHz}$, respectively. Comparison with the numerical data shows that the analytical result is an upper limit of H_q, because the simple derivation fails close to T_c where $R_s(T)$ becomes strongly temperature dependent [108–110]. (3.54) implies a frequency dependent quench field, which reflects the frequency dependence of R_s. Assuming $R_s \propto f^b$ (Sect. 1.2.3), the global quench field scales like $f^{-b/2}$.

Defect-Induced Quenching

A local quench occurs when the temperature of the superconducting matrix surrounding the defect exceeds the critical temperature $T_{c,\text{loc}}$, which can be lower than the bulk value T_c. The local S–N transition initiates an avalanche effect, i.e. the entire film becomes normal because of the diffusion of heat into the remainder of the superconductor. Expressions for H_q can be deduced under specific assumptions for the direction of the heat flow.

The isotropic model ("iso") assumes that the heat propagation away from the defect is semi-spheroidal in shape. Such a case can be expected, e.g., for YBCO on LAO, where $\sigma_{\text{th,F}} \approx \sigma_{\text{th,S}}$, if the thermal boundary resistance can be neglected (e.g., at elevated temperature). Equation (3.49) then yields, in analogy to the electric field contours around a point-like space charge

$$\frac{P_{\text{diss}}}{2\pi r^2} = -\sigma_{\text{th,S}}(T)\frac{\partial T(r)}{\partial r}. \tag{3.55}$$

Equation (3.55) can be integrated by approximating the geometrical situation as $d_{\text{S}}/d_{\text{F}} \to \infty$, and by setting $T(r_{\text{def}}) = T_{\text{c,loc}}$ and $T(r \to \infty) = T_{\text{B}}$. Neglecting losses in the superconducting matrix, the dissipated power is $A_{\text{def}} R_{\text{def}} H_{\text{s}}^2/2$, and we arrive at the result

$$H_{\text{q,loc}}^2(\text{iso}) = \frac{4\overline{\sigma}_{\text{th,S}}(T_{\text{c}} - T_{\text{B}})}{R_{\text{def}} r_{\text{def}}}, \quad \text{with} \tag{3.56a}$$

$$\overline{\sigma}_{\text{th,S}} = \frac{1}{T_{\text{c}} - T_{\text{B}}} \int_{T_{\text{B}}}^{T_{\text{c}}} dT \sigma_{\text{th,S}}(T) \tag{3.56b}$$

denoting the thermal conductivity averaged over the relevant temperature interval.

An alternate approximation can be obtained by assuming a purely two-dimensional (2d) heat propagation in the anisotropic film $\sigma_{\text{th,F}}^{\text{ab}} \gg \sigma_{\text{th,F}}^{\text{c}}$ (i.e., thermal insulation along the z direction). This model might be relevant in the case of thick films, and if substrates of poor thermal conductivity are employed. The term $P/2\pi r^2$ in (3.55) must then be replaced by $P/2\pi r d_{\text{F}}$, and the integration is terminated at the outer radius r_{tot}. The remaining calculation is analogous to the previous approach, with the result

$$H_{\text{q,loc}}^2(\text{2d}) = \frac{4\overline{\sigma}_{\text{th,F}}(T_{\text{c}} - T_{\text{B}})}{R_{\text{def}} r_{\text{def}}} \frac{d_{\text{F}}}{r_{\text{def}} \ln(r_{\text{tot}}/r_{\text{def}})}. \tag{3.57}$$

With $r_{\text{tot}} \gg r_{\text{def}} \geq d_{\text{F}}$ in real cases, this quench field is always lower than the isotropic value (3.56).

The contrary case of quasi-one dimensional (1d) heat conduction along the substrate (i.e., in the z direction) can be expected in the case of highly conductive substrates like sapphire. The heat equation (3.49) can be solved analytically for appropriate boundary conditions [111]. A similar problem was treated in [100] for the local illumination of a superconducting film with a laser. Adapting this approach to the present problem, one arrives at the general expression

$$H_{\text{q,loc}}^2(\text{1d}) = \frac{2(T_{\text{c}} - T_{\text{B}})}{\alpha R_i} \left(\frac{d_{\text{F}}/\alpha - \lambda}{\sigma_{\text{th,F}}} + R_{\text{bd}} + \frac{r_{\text{min}}}{\sigma_{\text{th,S}}} \right)^{-1}, \tag{3.58a}$$

where $\alpha = 1 - \exp(-d_{\text{F}}/\lambda)$, λ is the microwave penetration depth, and r_{min} denotes the minimum of $r_i/2$ and d_{S}. The index "i" refers to the two cases of global ($i = 1$) and local ($i = 2$) heating, where $R_1 = R_{\text{s}}$, $R_2 = R_{\text{def}}$, $r_1 = r_{\text{tot}}$

and $r_2 = r_{\text{def}}$. It is instructive to consider some limiting cases of (3.58a). In the thick-film limit $d_F \gg \lambda$, α reduces to unity, and the expression (3.58a) reduces to (3.54b) for global heating ($i = 1$). In the case of local heating $r_{\min} = r_{\text{def}}/2$, because we are interested in small defects $r_{\text{def}} \ll 2d_S$. The quench field becomes in this case

$$H_{\text{q,loc}}^2(1\text{d}) \approx \frac{2(T_c - T_B)}{R_{\text{def}}(d_F/\sigma_{\text{th,F}} + R_{\text{bd}} + r_{\text{def}}/2\sigma_{\text{th,S}})}. \tag{3.58b}$$

If the thermal conductivity of the substrate dominates the heat flow, the latter expression reproduces the 2d result (3.56). Finally, local heating of thin films ($d_F \ll \lambda$) leads to $\alpha = d_F/\lambda$, and the local-quench field becomes independent of the thermal conductivity of the film:

$$H_{\text{q,loc}}^2(1\text{d}) \approx \frac{2\lambda(T_c - T_B)}{R_{\text{def}}d_F(R_{\text{bd}} + r_{\text{def}}/2\sigma_{\text{th,S}})}. \tag{3.58c}$$

It is important to note that the penetration depth enters (3.58c) explicitly. The thin-film limit is expected to apply at elevated temperatures, where $\lambda(T)$ might exceed typical values of the film thickness, $d_F \leq 500$ nm.

The frequency dependence of $H_{\text{q,loc}}$ is $f^{-1/4}$ independent of the considered models, due to the assumed normal skin effect behavior of R_{def}. A quantitative comparison between the three models (3.56)–(3.58) is given in Table 3.3 for selected data. In accordance with expectation, the quench field levels decrease with increasing values of temperature and defect radius. The isotropic and the one-dimensional model yield closely comparable results, except at small defects and high temperatures. The two-dimensional model leads to much lower quench field levels and a stronger dependence of H_q on r_{def}. Realistic input parameters yield $\mu_0 H_{\text{q,loc}}$ values spanning the range between 0.1 and 150 mT, and thus being relevant for the experimental situation. The fact that the global-quench level falls below the $H_{\text{q,loc}}(77\,\text{K})$ values for $r_{\text{def}} = 1\,\mu\text{m}$ ("iso" and "1d") illustrates the limited validity of the assumptions which led to the analytical expressions. Finally we remark that the consideration of a finite boundary resistance R_{bd} will further reduce the calculated H_q values, especially at low temperatures where its contribution to R_{th} (3.54a) becomes appreciable.

Time Constant for Heat Diffusion

The dynamic evolution of microwave heating proceeds through several phases. Power dissipation leads at first to a quasi-spontaneous heating of the superconductor or of a defect. The corresponding time scale is given by the specific heat of the quasi-particles and the phonons, which is of order 10^{-12}–10^{-10} s in HTS ([112], Sect. 3.1). On a much larger time scale follows the global heating of the entire film and of the substrate. This time constant can be estimated from (3.49a) and (3.52a) assuming a freely decaying temperature gradient ($J_{\text{th}} = 0$):

Table 3.3. Comparison of local-quench field levels $\mu_0 H_{q,loc}$ in mT, calculated from (3.56) to (3.58) for YBCO on LAO at $f = 19\,\mathrm{GHz}$, with $T_c = 90\,\mathrm{K}$, $d_F = 200\,\mathrm{nm}$ and $r_{tot} = 25.4\,\mathrm{mm}$ for two values of T_B. An average thermal conductivity of 13 W/Km was assumed. The defect was assumed normalconducting with $\rho(T_c) = 100\,\mu\Omega\mathrm{cm}$. The penetration depth was estimated to be 150 nm (300 nm) at 4.2 K (77 K). The boundary resistance was neglected in the calculations.

Geometry	T_B (K)	$r_{\mathrm{def}} = 1\,\mu\mathrm{m}$	$r_{\mathrm{def}} = 10\,\mu\mathrm{m}$	$r_{\mathrm{def}} = 100\,\mu\mathrm{m}$
isotropic	4.2	160	50	16
Eq. (3.56)	77	60	19	6
2d	4.2	25	2.5	0.3
Eq. (3.57)	77	9	1.0	0.1
1d, $d_F/\lambda \gg 1$	4.2	135	50	16
Eq. (3.58b)	77	52	19	6
1d, $d_F/\lambda \ll 1$	4.2	139	44	14
Eq. (3.58c)	77	76	24	8

$$C_V \frac{\partial T}{\partial t} = -\sigma_{\mathrm{th}} \frac{\partial^2 T}{\partial z^2}. \tag{3.59}$$

A temperature profile $T \approx T_0 \exp(-z/\delta_{\mathrm{th}})$ extending over the thermal penetration depth δ_{th} decays exponentially with time like $\exp(-t/\tau_{\mathrm{th}})$, where (3.59) yields

$$\tau_{\mathrm{th}} = \frac{\delta_{\mathrm{th}}^2 C_V}{\sigma_{\mathrm{th}}}. \tag{3.60}$$

As in the stationary case, the thermal properties of the substrate dominate the decay time τ_{th} because of their major contribution to the specific heat. Estimations based on the literature data described before reveal for δ_{th}, being equal to a typical thickness $d_S = 0.5\,\mathrm{mm}$ of the substrate, as orders of magnitude: $\tau_{\mathrm{th}} \approx 10^2\,\mu\mathrm{s}$ and $10^{-1}\,\mu\mathrm{s}$ for sapphire at 77 K and 4.2 K, respectively. The corresponding time constants for LAO substrates are about $10^4\,\mu\mathrm{s}$ ($10^2\,\mu\mathrm{s}$). Taking into account typical pulse lengths of 100–200 μs (Sect. 2.1.2), heating could affect the transient microwave response (Sect. 3.3.1) markedly. As a result, the derivation of $R_s(H_s)$ dependences from the pulse shapes becomes ambiguous, and implicitly dependent on the time scales of the measurement setup.

3.2.3 Numerical Simulations

Typical Field Dependence of R_s for Global and Local Heating

Figure 3.5 displays a set of typical $R_s(B_s)$ curves simulated for representative $R_s(T)$ data of a ⌀1″ YBCO film at 19 GHz and different temperatures. (Here, as in the following, the notation for the flux density B_s is used synonymously

140   3. Field and Power-Dependent Surface Impedance

with that for the magnetic field H_s, with the conversion $B_s = \mu_0 H_s$.) The computations were performed for a film without defects ("ideal" limit) and with a defect of 1 μm radius. Due to the negligible contribution of losses in the defect to the total losses, $R_{\text{def}} r_{\text{def}}^2 \ll R_s r_{\text{tot}}^2$, identical $R_s(B_s)$ dependences are obtained for both cases. The shape of the curves is hence determined by the global heating of the entire film. At low and intermediate field levels, the functional dependence of $R_s(B_s)$ can be approximated well by

$$R_s(T, B_s) = R_s(T, 0) + a_{R,1}(T) b_s^2 + a_{R,2}(T) b_s^4, \tag{3.61}$$

where b_s is the field amplitude B_s normalized to the global quench field $b_q = \mu_0 H_q$ (see (3.54)). This dependence is indicated in Fig. 3.5 by the dashed lines at two temperatures. The only, though very significant, effect of the presence of a normalconducting spot is the strong reduction of the quench fields at intermediate and low temperatures. Global quenches occur at the ends of the curves, while local quenches are indicated by the arrows in Fig. 3.5.

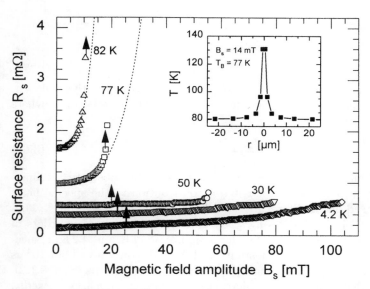

Fig. 3.5. Simulated $R_s(B_s)$ curves (*symbols*) for a YBCO film on LAO at different temperatures [84]. All curves end with a global quench. The quench fields are reduced for a 1-μm normalconducting defect (*arrows*). The curves can be approximated at low and intermediate field levels by the even power law (3.61) (*dashed lines*).

The values $B_{q,\text{loc}} = 25\,\text{mT}$ and $18\,\text{mT}$ at $4.2\,\text{K}$ and $77\,\text{K}$, respectively, are comparable to the estimates for the 2d case listed in Table 3.3. Even lower quench fields were deduced for larger defects (see below). At elevated temperatures, where $R_s(T)$ enters the transition region ($T \geq 70\,\text{K}$), local and global quench fields become comparable. A typical temperature profile near the defect is shown in the inset for $T = 77\,\mathrm{K}$. It reveals a local hot spot with a temperature well above that of the cold sink. It can also be seen that the temperature rise remains localized at the position of the defect until the entire film turns into the normal state.

Figure 3.6 represents the corresponding $R_\mathrm{s}(B_\mathrm{s})$ dependences (symbols) computed for a YBCO film on a CeO_2-buffered sapphire (CbS) substrate [84b]. The thin CeO_2 buffer layer ($d \geq 30\,\mathrm{nm}$) was neglected in the thermal simulation. According to the much higher thermal conductivity of sapphire compared to LAO, the temperature rise at the surface of the superconducting film is significantly reduced. The global-quench fields are above $900\,\mathrm{mT}$ at $4.2\,\mathrm{K}$ and around $45\,\mathrm{mT}$ at $77\,\mathrm{K}$. Assuming a defect of $1\,\mathrm{\mu m}$ radius, the thermal breakdown fields are strongly reduced (arrows in Fig. 3.6) to field levels of 44 and $26\,\mathrm{mT}$, respectively. As for LAO substrates, microwave heating in YBCO films on CbS can be described at low and intermediate field levels by the power law (see (3.61), dashed lines).

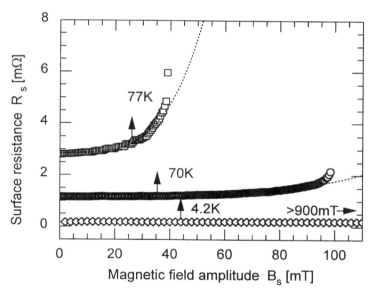

Fig. 3.6. Simulated $R_\mathrm{s}(B_\mathrm{s})$ curves (*symbols*) for a YBCO film on CeO_2-buffered sapphire (CbS) at different temperatures [84]. All curves end with a global quench. The quench fields are reduced for a 1-μm normalconducting defect (*arrows*). The curves can be approximated at low and intermediate field levels by (3.61) (*dashed lines*).

The impact of local defects on the thermal $R_\mathrm{s}(B_\mathrm{s})$ curves was investigated in detail in [84b]. The general behavior seen in Figs. 3.5 and 3.6 remained unaffected for defects up to $r_\mathrm{def} \leq 20\,\mathrm{\mu m}$, except for an increasingly large reduction of the breakdown fields. Larger defects deteriorated even the average $R_s(B_s = 0)$ value and led to a slight increase of the slope $\partial R_s/\partial B_s$. In contrast to the normalconducting defects considered so far, the case of a weakly superconducting defect ($r_{\text{def}} = 100\,\mu\text{m}$) was simulated with data for oxygen-depleted YBCO, which displayed a reduced transition temperature $T_{c,\text{loc}} \approx 77\,\text{K}$. The resulting $R_s(B_s)$ curves showed the same behavior as before, with local quenches occurring approximately when the critical temperature of the defect was exceeded.

In summary, (3.61) provides a general phenomenological approximation of $R_s(B_s)$ for microwave heating, independent of the substrate material and of the presence and the transport properties of local defects. A similar expression (with field coefficients $a_{X,1}(T)$ and $a_{X,2}(T)$) is expected to hold for the thermally induced rise of the surface reactance. As a result, the differential loss tangent $r_{\text{th}}(B_s)$ can be approximated at low and intermediate field levels by

$$r_{\text{th}}(B_s) \approx \frac{a_{X,1}(T) + 2a_{X,2}(T)b_s^2}{a_{R,1}(T) + 2a_{R,2}(T)b_s^2}. \tag{3.62}$$

In contrast to the r values for most magnetic-field-induced nonlinearities (Sect. 3.1), $r_{\text{th}}(B_s)$ depends on field and temperature. Representative simulations yielded for YBCO on LAO $r_{\text{th}} \ll 1$ at low field levels and 4.2 K and $r_{\text{th}} \gg 1$ at low field levels and 77 K.

Effect of Various Parameters on the Quench Fields

Local-quench fields were computed for film thicknesses d_F ranging from 50 to 600 nm in the presence of a 1-µm normalconducting defect. The result is shown in Fig. 3.7 for the two temperatures 4.2 K (solid squares) and 77 K (open triangles). The functional dependence $B_{q,\text{loc}} \propto d_F^{1/2}$, which is expected from (3.57), is represented by the solid line. The analytical approximation agrees roughly with the numerical results at 77 K up to $d_F = 400$ nm, while $B_q(d_F)$ at 4.2 K is better described by $B_q \propto d_F^{0.7}$. A similar exponent ($B_q \propto d_F^{0.6}$) was estimated for isotropic bulk Nb samples [113]. The use of thick films in applications where high power-handling capability is required (Chap. 6) is therefore favorable mainly at low temperatures. The saturation of the local-quench fields in Fig. 3.7 at 77 K points to a crossover from two- to one-dimensional heat conduction and/or to an increasing importance of global heating.

The numerical computations of $B_{q,\text{loc}}$ as a function of defect radius are summarized in Fig. 3.8 with r_{def} ranging from 1 to 100 µm. The squares and triangles (respectively: diamonds) denote data obtained for YBCO on LAO at 4.2 and 77 K (YBCO on CbS at 77 K). Also shown are the results from (3.56) and (3.57). The data for YBCO on LAO are well approximated by $B_q \propto r_{\text{def}}^{1/2}$ (solid line), indicating dominant contribution to the heat conduction from the substrate. For small defects ($r_{\text{def}} < 5\,\mu\text{m}$), $B_{q,\text{loc}}(r_{\text{def}})$ tends to weaken,

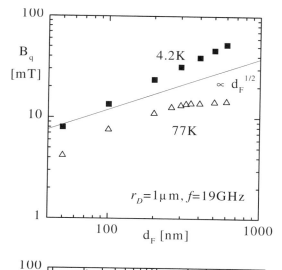

Fig. 3.7. Simulated dependence of the local-quench field $B_{q,loc}$ at 19 GHz on film thickness d_F for 4.2 K (*full squares*) and 77 K (*open triangles*) in the presence of a 1-μm normalconducting defect [84]. The *straight line* represents the expectation from (3.57).

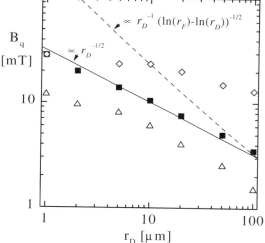

Fig. 3.8. Simulated dependence of the local-quench field $B_{q,loc}$ at 19 GHz on defect radius r_{def} for YBCO on LAO (*solid squares*: 4.2 K, *open triangles*: 77 K) and on CbS (*open diamonds*: 77 K) [84]. The solid and dashed lines represent the results from (3.56) and (3.57).

due to increasing contributions from the boundary resistance and the heat conductivity of the film, and due to global heating. Similar arguments were reported in [113]. In contrast, the 2d model (see (3.57), dashed line) cannot describe the data in the investigated range of r_{def} values, for CbS even less than for LAO. In the case of sapphire, the numerical results for $B_q(r_{def})$ are described by a small exponent: $B_{q,loc} \propto r_{def}^{-0.3}$, which can be expected from (3.58) when considering the finite boundary resistance $R_{bd} > 0$ and the high thermal conductivity of sapphire. The dissipated power is hence concluded to spread in the film over a range $\geq r_{def}$, and to be thermally "short-circuited" to the cold sink, i.e., the substrate.

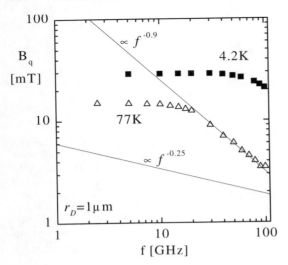

Fig. 3.9. Simulated dependence of the local-quench field $B_{q,loc}$ for $r_{def} = 1\,\mu m$ on frequency f for YBCO on LAO (solid squares: 4.2 K, open triangles: 77 K) [84]. The steep ($\propto f^{-0.9}$) and the flat lines ($\propto f^{-0.25}$) denote the results expected for global and local heating.

The $R_s(T)$ data measured at 19 GHz for the film discussed so far (YBCO on LAO) were scaled with $f^{1.8}$ to different frequencies. Assuming the presence of a defect ($r_{def} = 1\,\mu m$), the quench field $B_{q,loc}(f)$ could thus be studied between 2 and 100 GHz as displayed in Fig. 3.9. Below about 10 GHz (respectively: 40 GHz) at 77 K (4.2 K), where $R_{def} \geq 1000\,R_s$, $B_{q,loc}(f)$ is nearly constant. The reduction of the local-quench field at higher frequencies could be described by $B_{q,loc} \propto f^{-0.9}$, indicating global heating as the prevailing field limitation [as in (3.54b)].

Figure 3.10 summarizes the temperature dependence of the reduced quench field assuming the different conditions listed in Table 3.4. The solid line denotes $B_q(T) \propto (1-T/T_c)^{1/2}$, which would follow from (3.55) to (3.58b) if the difference $T_c - T$ were the only source of temperature dependent effects. According to this approximation, the curve gives only a coarse desription of the data. The results obtained with normalconducting defects are better described by the empirical dependence $[1-(T/T_c)^2]^{1/2}$ (dashed line). The three different types (without defects, normalconducting or weakly superconducting defects) can clearly be distinguished. The strongest deviation from the solid line to lower B_q values is observed for the last type. This is most likely due to the steep increase of $R_{def}(T)$ at elevated temperatures, but still well below $T/T_c = 1$. This interpretation is confirmed by the data for the ideal case, which still fall below the analytical estimate, but at a lesser degree because of the higher T_c value. In contrast, for normal defects, $B_{q,loc}(T)$ deviates from the solid curve to higher values. This behavior might reflect the dominant role of the temperature dependent thermal conductivity in determining the quench field. Finally, the choice of sapphire substrates weakens the temperature dependence of B_q, especially for extended defects. This is qualitatively in accordance with the thin-film limit in (3.58c), which becomes applicable at

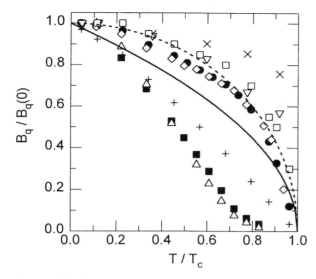

Fig. 3.10. Simulated dependence of the reduced quench field $B_q(T)/B_q(0)$ at 19 GHz on reduced temperature for different YBCO films (parameters see Table 3.4) [84]. The *solid* and *dashed* lines represent the dependences $(1 - T/T_c)^{1/2}$ and $[1 - (T/T_c)^2]^{1/2}$, respectively. Further explanations are given in the text.

elevated temperatures. This approximation yields $B_q(T) \propto [1 - (T/T_c)]^{1/4}$, if $\lambda(T)$ is scaled like $[1 - (T/T_c)^2]^{-1/2}$ (Chap. 2).

Table 3.4. Summary of conditions to simulate the temperature dependence of $B_q(T)$ in Fig. 3.10. All computations were performed at $f = 19\,\text{GHz}$, with $\rho(T_c) = 100\,\mu\Omega\text{cm}$ in the case of normalconducting (NC) defects, and with $T_{c,\text{loc}} = 76.8\,\text{K}$ for weakly conducting (WC) defects.

Symbol in Fig. 3.10	Substrate material	Type of defect	T_c/K	$B_q(T=0)$ /mT
"+" (plus signs)	LAO	none	88.2	95
solid circles	LAO	NC, 1 µm	88.2	29
solid diamonds	LAO	NC, 10 µm	88.2	10
open squares	LAO	NC, 1 µm	90.7	29.5
open triangles upside down	CbS	NC, 1 µm	84.0	50
"×" (crosses)	CbS	NC, 10 µm	84.0	31
solid squares	LAO	WC, 100 µm	90.4	87
open triangles	MgO	WC, 100 µm	90.4	540

3.3 Dynamic Nonlinear Surface Impedance

The previous sections treated various magnetic and thermal effects which evoke nonlinear response of superconducting films to an excitation at the high frequency $\omega = 2\pi f$. It was assumed so far that, as the magnetic field H_{DC}, the microwave power levels P_{circ} and/or P_{diss} exceeded critical values, the nonlinear effects showed up at a characteristic response time τ_c such that $\omega \tau_c \ll 1$. This assumption fails especially in the case of "slow" nonlinearities like magnetic superheating (Sect. 3.1.3) or Joule heating (Sect. 3.2.2), where τ_c can become considerably long compared to $1/\omega$. Sect. 3.3.1 describes the characteristic features in the time domain of these two effects, which are considered representative for "slow" nonlinear effects. Further distinction between fast and slow nonlinear effects can be obtained in the frequency domain as discussed in Sects. 3.3.2 and 3.3.3 on the basis of two-tone frequency intermodulation.

3.3.1 Dynamic (Nonequilibrium) Microwave Response

Dynamic Evolution of Microwave Heating

Numerical simulations of the dynamic phases during microwave heating of superconducting films were performed in the framework of the previously described model (Sect. 3.2, [84]). Figure 3.11 displays representative $R_s(B_s)$ results at subsequent instants upon application of a microwave signal, obtained for YBCO on LAO at 77 K and 19 GHz. Shown for comparison is the stationary case ($t \to \infty$). The presence of a NC defect ($r_{\text{def}} = 1\,\mu\text{m}$) was found not to affect the time evolution of $R_s(B_s)$, but only the absolute value of the quench field. This result supplements the previous conclusion (Sect. 3.2.3) that defects act as "nuclei" for thermal processes which, after being initiated, always involve the entire film. The inset to Fig. 3.11 shows the dynamic temperature rise at the position of the defect ($r = r_{\text{def}}$, left ordinate) and of the film at the outer contour ($r = r_{\text{tot}}$, right ordinate, c.f. Fig. 3.4). Both curves follow the relation

$$T(t) = T_\infty \left[1 - \exp\left(-\frac{t}{\tau_{\text{th,eff}}}\right)\right]. \tag{3.63}$$

The effective time constants deduced for the film were in good agreement with the analytical approximation (3.60) in Sect. 3.2.2, namely of order $10^2\,\mu\text{s}$ and $10^{-1}\,\mu\text{s}$ for sapphire at 77 K and 4.2 K, and of order $10^4\,\mu\text{s}$ and $10^2\,\mu\text{s}$ for LAO at the two temperatures. The thermal penetration depth δ_{th} in (3.58) was set equal to the thickness of the substrate ($d_S = 0.5\,\text{mm}$).

The dynamic simulations also yielded values for the time constants τ_q of the thermal breakdown at field levels $B_s \geq B_q$. As expected, the τ_q values were well below the time constants of heat diffusion. Values of τ_q of the order of $50\,\mu\text{s}$ were concluded for field amplitudes B_s close to the quench field

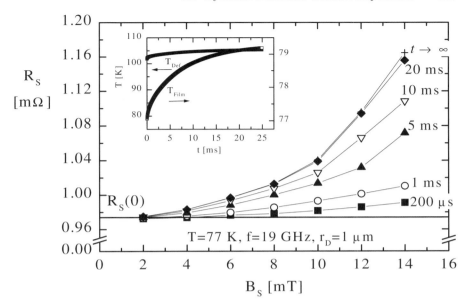

Fig. 3.11. Simulated $R_s(B_s)$ dependences at various instants during microwave heating for YBCO on LAO at 77 K and 19 GHz with $r_{\rm def} = 1\,\mu{\rm m}$ [84]. The steady-state solution is denoted by $t \to \infty$. The *inset* shows the evolution of the temperature at the defect ($r = r_{\rm def}$, *left ordinate*) and at the outer radius of the investigated film area ($r = r_{\rm tot}$, *right ordinate*).

for YBCO on LAO at 77 K. Faster breakdowns were deduced for decreasing defect size and thus increasing B_q values (see also [114]). Furthermore, the quench times decreased with increasing overcriticality $B_s/B_q > 1$.

The dynamic results demonstrate that a complete and unambiguous determination of $R_s(B_s)$ and of the relevant loss mechanisms should be achieved in pulsed operation, preferably with variable duty cycles. However, if resonant methods are applied (Sect. 2.1), the measurable microwave response becomes complicated by the coexistence of two different time scales: ω/Q for the energy relaxation in the resonator, and $\tau_{\rm th}$ for the heating process. A numerical analysis of the dynamics of both effects was reported in [90b, 91]. Figure 3.12 reproduces a result of [91a] computed at 20 K and 14 GHz for the time dependent energy charging of a two-fold coupled dielectric resonator containing two HTS films on LaAlO$_3$ (type "3a" in Fig. 2.4). With increasing power, a peak appears in the output signal. It extends up to some μs and levels off at later times to a reduced steady-state signal. The involved time scales reflect the thermal diffusion times of the substrate material at the investigated temperature. The decrease of the output signal, which leads to the formation of the peak, indicates the dynamic increase of power dissipation in the films. The subsequent saturation is due to the enhanced thermal conductivity of

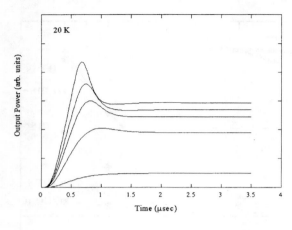

Fig. 3.12. Computed output power versus time of a two-fold-coupled dielectric resonator containing two microwave-heated HTS films on sapphire, at 20 K and 14 GHz [91a]. The input power increases as the curves shift from *bottom* to *top*.

the substrates, $\partial\sigma_{\mathrm{th,s}}/\partial T > 0$. A continuous decrease of the output signal is expected (and was observed) at elevated temperatures where no equilibrium can be attained, e.g., because of the negative slope $\partial\sigma_{\mathrm{th,s}}/\partial T$ and/or the steep increase of $R_\mathrm{s}(T)$ near T_c. Recent applications of this analysis to films on sapphire verified the benefit of its superior thermal properties by the absence of peaks in the transient signal, and revealed even an influence of the buffer layer [115].

Curves similar to those displayed in Fig. 3.12 can arise also from non-thermal nonequilibrium effects, and an unambiguous identification of the underlying mechanism requires some care. However, nonlinear transient microwave response can be monitored properly only if the "sampling rate" is faster than the relaxation rate $1/\tau$ of the sample. Regarding the broad interval of τ values expected for the different mechanisms, the optimum pulse timings have to be adjusted properly to the experimental situation. Additional information about nonlinear effects can potentially be attained by performing frequency-selective measurements as described in Sects. 3.3.2 and 3.3.3.

Dynamic Magnetic Superheating

The superheated state of a type-II superconductor presents a metastable state, i.e., a local minimum of the Ginzburg–Landau (GL) free energy with respect to the order parameter Ψ up to fields H_{sh} well above the lower critical field $H_{\mathrm{c}1}$ (Sect. 3.1). According to (3.14), long relaxation times τ_{sh} are expected in the absence of structural or electrodynamic fluctuations.

The dynamic behavior of superheating can be illustrated in terms of the Bean–Livingston surface barrier (Sect. 3.1.3). Figure 3.13 is a modified version of Fig. 3.1 for various reduced fields $h = H/H_{\mathrm{c}1}$ and $\kappa = 50$, where the BL superheating field amounts to about $11\,H_{\mathrm{c}1}$. For fields $H > H_{\mathrm{c}1}$, there is a finite quantum-statistical probability for a vortex to "tunnel" through the energy barrier of height $\Delta W = W_{\max} - W(x=0)$ and width w, as indicated in the figure. At $H = H_{c1}$, the energy per unit length of a free vortex is only asymptotically recovered as x (directed into the superconductor) grows to infinity. We find consequently $w \to \infty$ and a stable Meissner state. The presence of defects can be considered as positions of locally reduced H_{c1}, thus leading to an enhanced probability of vortex penetration. This interpretation is in accordance with the thermodynamic understanding of superheating and supercooling, which implies the absence of nucleation sites. The "lifetime" of the superheated state decreases also with increasing overcriticality $\varepsilon = h - 1 > 0$ (e.g., [18, 21]). This can be illustrated by considering that increasing ε values lead to a reduction of both the height ΔW and the width w of the pinning potential (Fig. 3.13).

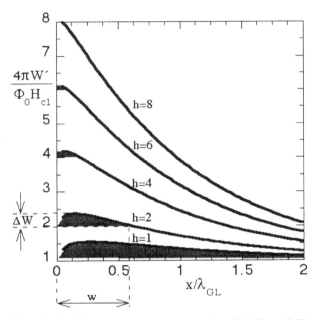

Fig. 3.13. Reduced vortex energy (see (3.17) and Fig. 3.1 in Sect. 3.1.3) versus normalized depth x/λ_{GL} in the superconductor for $\kappa = 50$ and field levels $h = H/H_{c1} = 1, 2, 4, 6$, and 8. Indicated by the gray areas are the height ΔW and the width w of the potential well, which renders the penetration of vortices improbable. The lifetime of the superheated state decreases with decreasing values of ΔW and w.

The above argumentation is merely hand-waving, but consistent computations of $\tau_{\mathrm{sh}}(\varepsilon)$ are presently lacking. Time-resolved measurements of the decay of the Meissner state are therefore required. The response time of superconductors to the penetration of vortices was studied with induction coils in pulsed magnetic fields (e.g., [116]). It was found to be of the order of μs. Based on this observation was the conclusion that dynamic superheating should generally occur at microwave frequencies $\omega \gg 1\,\mathrm{MHz}$ [117]. However, as discussed in [18, 44], the measured response times were limited by the attenuation of the induced eddy currents, rather than by the formation of vortices itself. It thus remains an experimental challenge to determine τ_{sh}, e.g., from frequency dependent microwave measurements at field amplitudes $H > H_{\mathrm{c1}}$.

3.3.2 Two-Tone Frequency Intermodulation in the Adiabatic Limit

General Framework

The nonlinear response of superconductors to microwave fields was considered so far in the time-domain at constant frequency. The preferred choice of the measurement frequency was the resonant value $\omega_0 = 2\pi f_0$ in the case of resonant systems. In the frequency domain, the nonlinear field dependence of the surface resistance gives rise to the generation of harmonics at $f_n = nf_0$, and to frequency intermodulation (IMD) at $f_{mn} = |mf_1 \pm nf_2|$ if two microwave sources are applied at frequencies f_1 and f_2 [118]. Such effects can be relevant for microwave applications, and can be either desirable (e.g., for modulators, mixers and limiters, see Sect. 6.3) or undesired (e.g., for filters [119]). Harmonic generation and IMD can also provide information about the nature of the nonlinearity. It is thus supplementary to the integral studies of $R_{\mathrm{s}}(H_{\mathrm{s}})$ at a single frequency. (The nonlinear response of the surface reactance can be treated analogously, but we restrict the discussion to the surface resistance.)

At linear response, the electric field $\boldsymbol{E}_{\mathrm{s}}$ induced parallel to the surface of a superconductor is proportional to the parallel magnetic field $\boldsymbol{H}_{\mathrm{s}}$. The proportionality constant is the surface impedance Z_{s} (Sect. 1.1.1):

$$\boldsymbol{E}_{\mathrm{s}}(\boldsymbol{r}, t) = Z_{\mathrm{s}} \boldsymbol{H}_{\mathrm{s}}(\boldsymbol{r}, t). \tag{3.64a}$$

The component of $\boldsymbol{E}_{\mathrm{s}}$ which follows the magnetic field in phase is correspondingly given by the surface resistance, $\boldsymbol{E}_{\mathrm{s,in}} = R_{\mathrm{s}} \boldsymbol{H}_{\mathrm{s}}$. In contrast, nonlinear response can be described in many cases by a power series of $E_{\mathrm{s,in}}$ in terms of H_{s}, which replaces (3.64a) (the indices are omitted for simplicity in the following):

$$E(t) = \sum_{k=1}^{\infty} \gamma_k [H(t)]^k. \tag{3.64b}$$

Equation (3.64b) implies that, at a given instant t, the electric field is determined by the value of the magnetic field at the same time. Its validity is therefore limited to nondynamic systems, i.e., systems without dispersive or dissipative relaxation ($\omega \tau_{\mathrm{c}} \ll 1$, where ω is a characteristic frequency, and τ_{c} could have the form $(LC)^{1/2}$, L/R or $1/RC$ with L, C, R an inductance, capacitance and resistance). The coefficients γ_k in (3.64b) are then real and constant. We assume these conditions to apply for the following. The contrary cases of dispersive (or time independent dynamic) and dissipative (or time dependent dynamic) nonlinearities are treated separately in Sect. 3.3.3.

Let us assume that $H(t)$ is the superposition of two harmonic microwave signals of identical amplitude H_0. In principle, consideration of a DC bias $H_{dc} \neq 0$ and of asymmetric excitation amplitudes $H_1 \neq H_2$ would reveal additional information about the nonlinear mechanism (see recent application to HTS in, e.g., [120]). However, since the mathematics become involved, the simplest approach shall suffice to illustrate the basic features. We thus have

$$H(t) = H_0(\cos \omega_1 t + \cos \omega_2 t). \tag{3.65a}$$

ω_1 and $\omega_2 > \omega_1$ are chosen symmetrically around the center value $\omega_0 = (\omega_1 + \omega_2)/2$ which could be, e.g., the resonant frequency of the measurement systems described in Chap. 2. Let $\delta\omega$ denote half the difference frequency $\delta\omega = (\omega_2 - \omega_1)/2$. If in a resonant system both fundamental frequencies fit into the passband, $\delta\omega < \omega_0/Q \ll \omega_0$ holds. Equation (3.65a) becomes

$$H(t) = 2H_0 \cos(\delta\omega t) \cos(\omega_0 t), \tag{3.65b}$$

which expresses an oscillation at carrier frequency ω_0, harmonically amplitude-modulated at the much lower frequency $\delta\omega$. The application of two harmonic signals is thus seen to introduce as an additional time scale $1/\delta\omega$. For the following discussion we assume that the response time τ_c of the nonlinearity be such that $\delta\omega\tau_c \ll 1$. The harmonic components of $E(t)$ can be evaluated from inserting (3.65a) into (3.64b) using the usual trigonometric identities. Due to the assumption of symmetric excitations, the frequency spectrum of $E(\omega)$ will be symmetric with respect to ω_0. Basic, though lengthy, calculations yield for the harmonic amplitudes of E, normalized to $\gamma_n H_0^n$, at $\omega = (n-2m)\omega_1$, where n, m are positive integers,

$$\frac{1}{2^{n-1}} \binom{n}{m} \sum_{k=0}^{m} \binom{m}{k} \binom{n-m}{k} \quad \text{for } 2m < n, \tag{3.66a}$$

at $\omega = (n - 2m + 1)\omega_1 - \omega_2$

$$\frac{m}{2^{n-1}} \binom{n}{m} \sum_{k=0}^{m-1} \binom{m-1}{k} \binom{n-m}{k} \frac{1}{k+1} \quad \text{for } 2m+1 \leq n, \tag{3.66b}$$

at $\omega = (n - 2m)\omega_1 - 2\omega_2$

$$\frac{m \cdot (n-m)}{2 \cdot 2^{n-1}} \binom{n}{m} \sum_{k=0}^{m-1} \binom{m-1}{k} \binom{n-m}{k} \frac{1}{k+2} \quad \text{for } 2m \leq n, \tag{3.66c}$$

and so on. While the expressions (3.66) appear complicated at first glance, their value is the consideration of all terms contributing to a given intermodulation frequency, which are usually neglected except for the principal term (i.e., the sums are set to unity) [121].

Third-Order Intermodulation

The order of the frequency mixing derived in the preceding paragraph is $n-2m$ in case of (3.66a) and $n-2m+2$ for (3.66b) and (3.66c). For resonant systems, the third-order IMD is of special interest because the frequencies $2\omega_2 - \omega_1 = \omega_0 + 2\delta\omega$ and $2\omega_1 - \omega_2 = \omega_0 - 2\delta\omega$ are closest to the fundamental frequencies, and thus fall also into the passband of the resonator. Their amplitudes are found from (3.66b) for $n = 3$ (order of intermodulation) and $m = 1$ (lowest term contributing to this order):

$$E_0(2\omega_1 - \omega_2) = \frac{3}{4}\gamma_3 H_0^3. \tag{3.67}$$

This value is by a factor 3 larger than that of the third-order harmonic generation $E_0(3\omega_1)$, as can be concluded from (3.66a) for $n = 3$ and $m = 0$. Figure 3.14 sketches the general dependence $E(H)$ for some frequencies, according to the approach in (3.64b).

The physical meaning of the third-order coefficient γ_3 becomes obvious from setting $E = R_s(H)H$ and recalling from Sects. 3.1 and 3.2 that many nonlinear mechanisms cause $R_s \approx R_s(H = 0) + \gamma_{\mathrm{NL}} H^2$ (index "NL" for nonlinear). We then have $E \approx R_s(0)H + \gamma_{\mathrm{NL}} H^3$ and obtain the identities $\gamma_1 = R_s(0)$ and $\gamma_3 = \gamma_{\mathrm{NL}}$. It also follows that generally $\gamma_{2m} = 0$ ($m \geq 1$) if the magnetic field dependence of the surface resistance can be described by even powers. The third-order intermodulation (3IM) then quantifies the leading term of the nonlinear behavior.

The energy $W_{3\mathrm{IM}}$ stored in a resonator at $\omega = 2\omega_1 - \omega_2$ is usually measured with a spectrum analyzer. Since this frequency is close to ω_0, the electric field with amplitude (3.67) induces the time varying magnetic field $H_{3\mathrm{IM}} = E_{3\mathrm{IM}}/R_s(\omega_0)$, and $W_{3\mathrm{IM}} \propto H_{3\mathrm{IM}}^2$. An important figure-of-merit for microwave applications is the third-order intercept (3OI). This point denotes the hypothetical field amplitude where the energy W_0 stored in the linear case equals the energy stored at the same field level at $\omega_{3\mathrm{IM}}$ (Fig. 3.14). One finds from the previous analysis

$$H_{3\mathrm{OI}} = \sqrt{\frac{4}{3} \frac{R_s(0)}{\partial R_s/\partial(H^2)}}, \tag{3.68}$$

which made use of the physical meaning of γ_1 and γ_3. Roughly speaking, the field amplitude $H_{3\mathrm{OI}}$ is shifted to higher levels for less nonlinear samples. In reality, the intercept cannot be measured directly but has to be extrapolated, since at field levels comparable to $H_{3\mathrm{OI}}$ the higher-order terms neglected in the derivation of (3.67) and (3.68) become relevant (indicated by the curvature and the dotted continuations in Fig. 3.14).

There are two further important consequences for the analysis of nonlinear surface resistance. First, as $R_s(H)$ becomes nonlinear with increasing field level H, more and more energy is converted from the fundamental frequency to higher harmonics. This "lost" energy appears in single-tone measurements

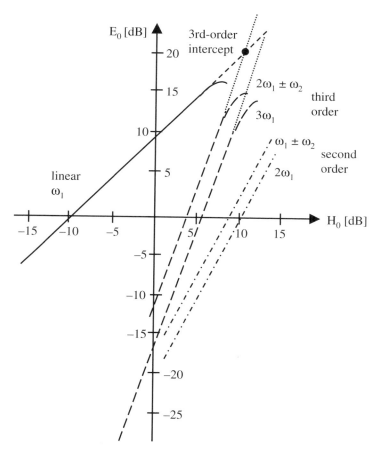

Fig. 3.14. Electric field amplitude E_0 versus magnetic field amplitude H_0 at different frequencies on logarithmic scales in dB, considering only the lowest terms in the sums of (3.66). The 3rd-order intercept denotes the coordinates of the hypothetic intersection between the fundamental ($\propto H_0$) and the 3IM ($\propto H_0^3$) signals (see (3.68)). After [118].

as an enhanced surface resistance. It thus presents a constructive feedback mechanism for $R_s(H)$ curves, which become steeper than quadratic in H. Secondly, IMD measurements are much more sensitive to the nonlinear properties of a sample than single-tone measurements. This can be seen by writing the latter response in the form

$$R_s = R_{s,\text{linear}} + R_{s,\text{nonlinear}}(H), \tag{3.69}$$

where the second term presents usually only a small fraction of the linear contribution at the threshold to nonlinearity. In contrast, the IMD response

154   3. Field and Power-Dependent Surface Impedance

is solely determined by $R_{\mathrm{s,nonlinear}}$. Even the smallest deviation from linearity contributes therefore to a nonvanishing signal at ω_{IMD}.

Deviations of the 3IM Field Dependence from the Cubic Power Law

The measurable field amplitude H_{3IM} is approximately proportional to H_0^3 at low field levels, independent of the source of nonlinearity. It is therefore not surprising that $H_{\mathrm{3IM}} \propto H_0^3$ was theoretically deduced also for superconductors with d-wave symmetry [37], especially since the pair breaking was assumed to be instantaneous. However, this cubic dependence is by no means universal, for the various reasons discussed in the following.

First, intermodulation products of odd order higher than 3 contribute additionally to the signal at the 3IM frequency. This can be seen from (3.66b) for $(n,m) = (3,1), (5,2), (7,3)$, and so forth. Taking explicitly into account the corresponding field amplitudes, one finds

$$E_0(2\omega_1 - \omega_2) = \frac{3}{4}\gamma_3 H_0^3 + \frac{25}{8}\gamma_5 H_0^5 + \frac{735}{64}\gamma_7 H_0^7 + \ldots \quad . \tag{3.70}$$

Clearly, the steeper the field dependence of $R_{\mathrm{s}}(H)$ is, the more the expectation of a cubic field dependence of the 3IM amplitude of the electric field leads to an underestimation.

Second, the power-law expansion of $E(H)$ in (3.64b) fails if the nonlinearity depends on the modulus $|H|$ of the magnetic field, e.g., in the case of strong pinning (Sect. 3.1.4). Noncubic behavior is also expected if the basic assumption of nondynamic response is not appropriate. As argued in the next section, dynamic conditions are important to be considered, e.g., in granular materials displaying Josephson behavior, or if microwave heating occurs at a time scale where $\delta\omega\tau_{\mathrm{th}} \geq 1$, leading to steeper or flatter field dependences of the intermodulation distortion.

The previous arguments concerned physical reasons for noncubic IMD behavior. These have to be carefully distinguished from the phenomenological appearance of two-tone intermodulation in experiment. Usually IMD measurements detect a signal which depends on $E_{\mathrm{3IM}}/R_{\mathrm{s}}(H_0)$ (like the stored energy measured at ω_{3IM}). Thus it is not E_{3IM} as given by (3.70) which is measured, but a more complicated quantity which, in general, must be expressed as

$$H_{\mathrm{3IM}} = \frac{E_{3IM}}{R_{\mathrm{s}}(H_0)} = \frac{(3/4)\gamma_3 H_0^3 + (25/8)\gamma_5 H_0^5 + \ldots}{\gamma_1 + \gamma_3 H_0^2 + \gamma_5 H_0^4 + \ldots}. \tag{3.71}$$

Neglecting terms of order 7 and higher, (3.71) yields at field levels $H_0 \gg (4\gamma_1/3\gamma_3)^{1/3}$ an approximate polynomial field dependence $H_{\mathrm{3IM}} \approx (3H_0/4)(1 - (\gamma_5/g_3)H_0 + \ldots)$ of order 1 to 2.

Finally, nonlinear effects are judged in many cases (especially if applications of superconducting films, e.g., for microwave filters, are concerned) from the unloaded quality factor Q_0 of a microstrip, dielectric, or cavity resonator. The relations between input power P_{inc} at the fundamental frequency ω_0 and output powers both at the fundamental and at the 3IM frequencies $\omega_0 \pm 2\delta\omega$ were calculated in terms of $Q_0(P_{\text{inc}})$ for transmission line resonators in [122] and for disc resonators in [123]. It turned out that the suppression of Q_0 at elevated power levels resulted in a deviation of $P(\omega_0)/P_{\text{inc}}$ from the linear behavior, and of $P(\omega_{3\text{IM}})/P_{\text{inc}}$ from the cubic dependence. In fact, the same formalism shows that the deviation from cubic power does not appear in $E_{3\text{IM}}(H_0)$, especially since the authors of [123] assumed a nondynamic response. This example demonstrates that conclusions drawn from measured quantities (like Q and P) might not be applicable without modifications to the underlying physical quantities (R_s and H).

3.3.3 Two-Tone Frequency Intermodulation in Dynamic Systems

Third-Order Intermodulation in Josephson Junctions

In view of granular effects in polycrystalline superconductors it is instructive to study the IMD behavior of a single Josephson junction. The microwave response of a Josephson junction, as already treated in Sects. 1.2 and 3.1.4, can be described by a parallel circuit of the constant resistance R_J and the nonlinear Josephson inductance L_J. The reactive nature causes the storage of energy. The simple power-law expansion with constant real coefficients (see (3.64b)) hence fails, as does the cubic power-law behavior of the 3IM component. This was confirmed by modeling the junction as a switch between the superconducting and the normal state [124]. However, the 3IM response of a Josephson junction can be deduced in a more rigorous way [125], which resembles that of the derivation of Shapiro steps [16].

In contrast to a homogeneous superconductor, the microwave response of a Josephson junction depends on the flux Φ threading it, rather than on the magnetic field. The basic expression relating $\Phi(t)$ to the total time dependent current $I(t)$ through the junction is given in terms of the reduced variable $\varphi(t) = (2\pi/\Phi_0)\Phi(t)$ by

$$I(t) = I_{cJ}\sin(\varphi(t) + \varphi_{\text{off}}), \qquad (3.72)$$

where φ_{off} stands for a DC bias flux or a constant phase difference between the superconducting electrodes on both sides of the junction (Sect. 6.3). Applying a two-tone harmonic excitation analogous to (3.65), the current through the junction becomes

$$I(t) = I_{cJ}\{\sin[\varphi_1(t)]\cos[\varphi_2(t)] + \sin[\varphi_2(t)]\cos[\varphi_1(t)]\}. \qquad (3.73)$$

The harmonic components of $I(t)$ can be evaluated in terms of Fourier series. The trigonometric functions sin and cos lead to mth and nth order Bessel functions of the first kind as the coefficients of this expansion:

$$I(t) = I_{\text{cJ}} \sum_{m=-\infty}^{\infty} \sum_{n=-\infty}^{\infty} J_m(\varphi_{\text{HF}}) J_n(\varphi_{\text{HF}}) \sin[(m\omega_1 + n\omega_2)t + \varphi_{\text{off}}], \quad (3.74)$$

where φ_{HF} denotes the microwave flux amplitude. The 3IM component of $I(t)$ is given, e.g., by $m = 2$ and $n = -1$, where $J_{-1}(x) = -J_1(x)$. Results for $I(\varphi_{\text{HF}})$ are displayed in Fig. 3.15 for $\varphi_{\text{off}} = 0$. Due to the asymptotic behavior of the Bessel functions, $J_n(x) \propto x^n/(2^n n!)$ at $x < 1$, the IMD signal of Josephson junctions starts with a cubic dependence. This regime corresponds to flux amplitudes where $\sin(\varphi_{\text{HF}}) \approx \varphi_{\text{HF}} + \ldots$ holds, and the 3IM product should be cubic. As the flux amplitude increases, however, oscillations develop, in marked contrast with the power-law behavior of a homogeneous adiabatically responding superconductor.

The response of a granular superconductor is expected under realistic conditions to resemble a superposition of junctions with different critical currents. As a result, the oscillations of the IMD response are washed out, and the average 3IM signal vanishes at large amplitudes.

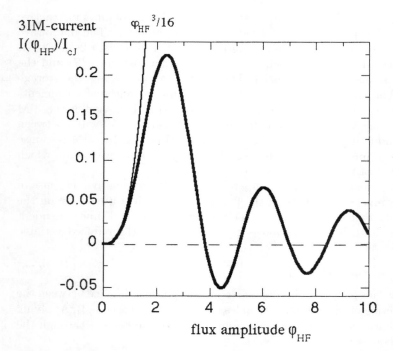

Fig. 3.15. 3IM component of the normalized current $\iota(\varphi_{\text{HF}}) = I(\varphi_{\text{HF}})/I_{\text{cJ}}$ through a Josephson junction versus the microwave flux amplitude φ_{HF} according to (3.74). The small-signal cubic behavior $\iota(\varphi_{\text{HF}}) \propto \varphi_{\text{HF}}^3/16$ is shown for comparison.

Third-Order Intermodulation Resulting from a Time Dependent Nonlinear Resistance

Another realistic example of a dynamic system is a time dependent nonlinear surface resistance. R_s can become time dependent through, e.g., the time dependence of the temperature of the superconducting film (see (3.63)). If the temperature rise remains modest, it is sufficient to interpolate $R_s(T)$ linearly: $\Delta R_s(T(t)) \propto \Delta T(t) \propto P_{\text{diss}}(t)$. The second proportionality expresses the dissipated power as the source of heating. In order to reproduce the exponential growth of R_s with time, which was identified in (3.63), one finds

$$R_{s,\text{th}}(t) = R_0 \left[1 + 2\beta \int_{-\infty}^{t} dt' \exp\left(-\frac{t-t'}{\tau}\right) \cdot P_{\text{diss}}(t') \right]. \quad (3.75a)$$

The index "th" indicates the thermal origin of R_s, R_0 describes a nonthermal offset resistance, and $\beta > 0$ is a constant. The effect of (3.75a) can be checked by setting $P_{\text{diss}}(t') = 0$ from the past until the time $t = 0$ ($-\infty < t' \leq 0$), and $P_{\text{diss}}(t') = P_0$ afterwards ($t' > 0$). The integral can then be easily solved, with the result $R_{s,\text{th}}(t) = R_0\{1 + 2\beta\tau P_0[1 - \exp(-t/\tau)]\}$ as desired. The meaning of the factor β follows from setting $P_{\text{diss}} = P_0$ at ($-\infty < t' \leq +\infty$). Equation (3.75a) then yields

$$R_{s,\text{th}} = \frac{R_0}{\eta - 1} \left\{ \eta \exp\left[(\eta-1)\frac{t}{\tau}\right] - 1 \right\} \quad (3.75b)$$

with $\eta = 2\beta\tau P_0/R_0$. Equation (3.75b) leads for $\eta < 1$ to the steady-state solution $R_{s,\text{th}} \to R_0/(1-\eta) > R_0$, while $R_{s,\text{th}}$ increases exponentially with time for $\eta > 1$. The condition $\eta = 1$ obviously describes the situation when P_{diss} has reached a critical value P_c above which the steady-state solution becomes unstable. From this we can identify $(\beta\tau)^{-1/2} = 2P_c/R_0$ to be a critical current.

Equation (3.75b) was refined in [126] by writing for the dissipated power $P_{\text{diss}}(t) = [R_{\text{inst}}(t) + R_{\text{th}}(t)]H^2(t)/2$, where $R_{\text{inst}} = \alpha H^2$ is the surface resistance of the superconductor which follows the microwave excitation instantaneously. This term was needed to account for the experimentally observed coexistence of a "fast" ($< 30\,\text{ps}$) and a "slow" ($10\,\text{ns}$) response of a YBCO coplanar line to short microwave pulses [127]. The slow response was found to be thermal in origin while the fast response was attributed to a nonequilibrium quasi-particle population (see also [112]). Putting all the pieces together, the time dependent surface resistance due to microwave heating can be modeled as

$$R_{s,\text{th}} = \beta \int_{-\infty}^{t} dt' \exp\left(-\frac{t-t'}{\tau}\right) H^2(t') \left[R_{s,\text{inst}}(t') + R_{s,\text{th}}(t')\right]. \quad (3.76)$$

The solution of (3.76) for arbitrary $H(t)$ requires numerical computation. Figure 3.16 reproduces results for the 3IM signal evaluated from (3.76) for the two-tone excitation (3.65). The parameters (see figure caption) were chosen for optimum agreement between simulation and experimental results. The four curves were simulated for two different values $\omega_0 \tau \gg 1$ (24π for curves 1 and 3 and 200π for curves 2 and 4). Curves 1 and 2 (respectively: 3 and 4) describe situations where $\delta\omega\tau < 1$ (> 1). It can be seen (curve 1) that for $\delta\omega\tau = 0.012\pi \ll 1$ the third-order intermodulation comes close to the result expected for nondynamic systems (Sect. 3.3.2, dashed line). As $\delta\omega\tau$ approaches unity ($\delta\omega\tau = 0.1\pi$, curve 2), the 3IM-signal still starts with a cubic behavior at low fields, but increases stronger than cubic at intermediate field levels and saturates at high fields. Curve 3 represents the case $1 \ll \delta\omega\tau = 2.4\pi \ll \omega_0\tau$ which directly leads from cubic behavior to saturation. For even larger values of $\delta\omega\tau$ ($\delta\omega\tau = 20\pi$, curve 4), nonmonotonic behavior of the 3IM signal shows up.

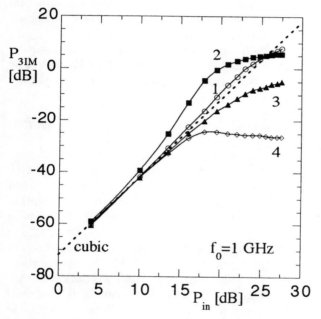

Fig. 3.16. Third-order intermodulation signal, obtained from (3.76) at $f_0 = 1\,\text{GHz}$ using $\alpha = 800\,\Omega\text{m}^2/\text{A}^2$ [126]. The curves represent the parameters $\delta\omega\tau = 0.012\pi$ (1, *circles*), 0.1π (2, *squares*), 2.4π (3, *triangles*), and 20π (4, *diamonds*). Curves 1 and 3 (2 and 4) were simulated with $\omega_0\tau = 24\pi$ (200π) and $\beta = 6.7 \times 10^9\,\text{A}^{-2}\text{s}^{-1}$ ($4.5 \times 10^9\,\text{A}^{-2}\text{s}^{-1}$).

3.4 Summary: Consequences for Nb_3Sn and $YBa_2Cu_3O_{7-x}$

We conclude the discussion of the nonlinear surface impedance of superconductors with a summary of features that are specific to the high-κ compounds Nb_3Sn and $YBa_2Cu_3O_{7-x}$. Various research groups observed a decrease of the surface resistance, the penetration depth, or both, in increasing DC and/or HF magnetic fields. Since such a decrease appears contrary to the mechanisms discussed so far, it is reasonable to refer to this phenomenon as an "anomalous" field effect. Different models were proposed to explain this anomaly, based on granularity or scattering. Both approaches are briefly reviewed in Sects. 3.4.1 and 3.4.2, with some emphasis on the chronological evolution of the coupled-grain models. The different scales of the critical magnetic fields and of the corresponding relaxation times introduced in the previous sections are summarized in Sect. 3.4.3 for both superconductors. The aim of this summary is to provide a reliable basis for the discussion of the experimental results described in Chap. 4.

3.4.1 Anomalous Field-Effect in Superconductors: Coupled-Grain Models

Most of the magnetic and thermal nonlinear effects considered in Sects. 3.1 and 3.2 affected the surface resistance and the surface reactance simultaneously. In many cases, a power-law behavior of the type $Z_s(H) \propto H^p$ with $p = 0.5$, 1 or 2, could approximate the analytical or numerical results, with the field slopes $\partial Z_s/\partial H \propto pH^{p-1}$ being positive. The differential loss tangent $r(H) > 0$ was also positive, and in some cases even constant. This conclusion applies to the conventional picture as well as to the d-wave scenario presented in Sect. 3.1.5. However, it contrasts with the experimental observation of negative field slopes, and sometimes negative r-values, reported by several groups for HTS films at different frequencies and temperatures [128–134]. There are presently two approaches to model an anomalous field dependence of $Z_s(H)$, which are discussed in some detail in the following.

Chronological Evolution of Coupled-Grain Models

The granular nature of the high-κ (i.e., short coherence length) superconductors Nb_3Sn and, more serious, the HTS compounds initiated an evolution of coupled-grain models which accounted for various microwave properties. Instead of repeating the details of the numerous approaches, this section aims to give a brief chronological review in order to help in differentiating between them in terms of assumptions and limitations. The various proposals have been accompanied by many elaborate publications concerning the comparison between models and experimental results. It is impossible, and therefore not attempted, to cite all of these, although they contributed significantly to the present understanding of $Z_s(T, f, H)$.

Basic work on the high-frequency response of a single Josephson junction to microwave and light irradiation was reported in 1972/73 by Auracher and van Duzer [64] (Sect. 3.1.4) and Russer [135], respectively. The former group modeled $Z_{s,J}$ for currents I below and above the critical current, and for extreme limits of the reduced frequency $\Omega = \Phi_0 f / I_{cJ} R_J$. The surface impedance was found to increase quadratically at low currents, and to diverge as I approached the critical value. The reactance was inductive in nature. At currents above I_{cJ}, the surface resistance approached the normal-state value R_J, and the reactance oscillated between inductive ($X_{s,J} > 0$) and capacitive ($X_{s,J} < 0$) behavior.

In 1988/89 Hylton and Beasley [136] developed an effective-medium model of granular HTS films (Sect. 1.2.2). The surface impedance was calculated in terms of an effective complex conductivity, which could be described by an equivalent circuit connecting grains and grain boundaries in series. The areal contribution of the junctions to the total impedance was weighted by the factor $2\lambda_g/a$, with λ_g the penetration depth of the grains and a the average grain size. Results derived for low field (or current) levels were sufficient to describe the frequency dependence of $Z_{s,J}$. A characteristic feature of the effective-medium approach was the assumption that the average behavior of the network of junctions would show up as that of a single junction with an effective critical current and an effective normal resistance.

A combination of both these works was reported in 1990 by Attanasio et al. [137] used to analyze the field/current dependence of the surface impedance of granular superconductors. The discussion was again tailored to an effective medium, and to field amplitudes well below the critical value. It was noted that a realistic description of granular superconductors would imply the consideration of distributed junction parameters.

Based on the data for temperature dependent transport properties of natural and engineered HTS Josephson junctions collected since 1990, Halbritter composed the various elements into a comprehensive description of the dependences of Z_s on temperature, frequency and field [22]. The implications of Josephson coupling between the grains were generalized, and applied to the low- and high-frequency transport properties in the superconducting and in the normal state. The discussion was again restricted to the effective parameters of a single junction (Sect. 1.2.2).

As the quality of the HTS films improved it was recognized that an effective-medium approach could be inappropriate to model the surface impedance. Portis and Cooke developed in 1992/93 a transmission line model to account for heterogeneous penetration of the microwave fields into the grains and the boundary junctions, respectively ([2, 138], Sect. 1.2.2). Inductive shunting of the junctions was further found essential to describe the data measured with HTS films in different laboratories [23]. The analysis was capable of describing $Z_s(T, f, H)$ in terms of effective junction parameters, and in relation to different degrees of granularity.

In view of applications of HTS films to microwave applications, the field dependence of the surface impedance became a major concern. Numerous systematic experiments on $Z_s(H)$ of stripline resonators at various temperatures between 4.2 K and T_c and frequencies between 1.5 and 20 GHz often revealed a quadratic increase of $R_s(H)$ and $\Delta\lambda(H)$ [139]. This observation led, in 1993, to a further extension of the previous coupled-grain models [136, 137] by Nguyen et al. in terms of microwave power dependence [140]. The quadratic field dependence of Z_s, which follows from the two Josephson equations (Sect. 1.2.2), advanced to the major characteristic of the coupled-grain model. In view of the present understanding, which revealed numerous mechanisms to display also a quadratic field dependence of R_s (Sects. 3.1.2, 3.1.4, and 3.2.3), such a generalization is certainly not justified. However, the MIT work [139, 140] demonstrated impressively that many experimental data could be well described by the extended version of the coupled-grain model, leaving only a few adjustable parameters like the critical current, normal resistance, and the geometrical dimensions of the junctions. As in the previous approaches, effective junction parameters were assumed to model the grain–junction network.

Experimental studies of the surface impedance of engineered HTS grain boundary junctions have been reported since 1994 ([65–67, 141], Sect. 3.1.4). The motivation for this work was to better understand the transport properties of real junctions, to extract measured parameters for I_{cJ}, R_J etc., and to justify the wide-spread application of the coupled-grain models. The studies also renewed interest in simulations of the electrodynamics of short junctions in terms of the RSJ model (Sect. 1.2.2), and of long (in)homogeneous junctions. The main output of these studies was a confirmation of the analysis [64], and the realization that the surface impedance of Josephson junctions increases step-like rather than smoothly. This is illustrated in Fig. 3.17 for the simulated RSJ surface impedance compared with experimental data at 6 GHz [65a]. The decreasing oscillations of the reactance around zero at currents above I_{cJ} turned out to be important for an explanation of the anomalous field effect.

Recent extensions of these studies revealed that small-angle grain boundaries ($\Theta < 10°$) are unlikely to be responsible for nonlinear microwave response in HTS films [66a].

Anomalous Field-Effect in Networks of Josephson Junctions

Another extension of the coupled-grain model was the consideration of a distribution of short Josephson junctions with different parameters, their entity being simulated in terms of the RSJ model [129, 142]. This approach provided two new conclusions. First, field dependences of $Z_s(H)$ steeper than H^2 (e.g., H^4) could be translated into an appropriate probability distribution

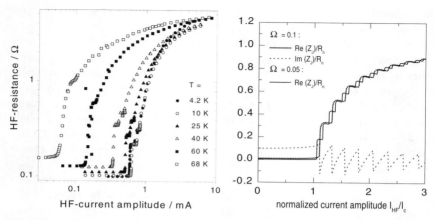

Fig. 3.17. Measured surface impedance of an engineered YBCO Josephson junction (*left*), incorporated into a microstrip line resonating at 6 GHz. The response was simulated within the RSJ model (*right*) for different reduced frequencies Ω [65a].

$p(I_{cJ}R_J)$ of $I_{cJ}R_J$ values. It was claimed an advantage to model the body of experimental data within a single approach, and to avoid the necessity of considering other mechanisms such as vortex penetration, flux-flow and hysteresis [6]. Since the latter effects inevitably occur at sufficiently high field levels, this advantage can be doubted. Furthermore, the explanation of $R_s(H)$ through $p(I_{cJ}R_J)$ is not a reduction of parameters but a transcription. The physical significance of this interpretation hence requires an independent way of determining a realistic $p(I_{cJ}R_J)$. The second novel result of the network model was the simulation of negative r values. Since the Josephson inductance oscillates with increasing field level (Fig. 3.17), an appropriate tailoring of the model parameters allows one to transfer this effect to the total effective surface reactance of the network. However, this procedure yields $r(H) < 0$ only for selected frequencies, and an anomalous field effect for $R_s(H)$, namely $\partial R_s(H)/\partial H < 0$, is inherently absent in this approach.

It shall be noted that a nonmonotonic power dependence of the microwave losses was observed also with low-T_c low-κ superconductors [143, 144]. $R_s(P)$ increased at low power levels as expected, but decreased at higher power levels. Azlamazov and Larkin proposed a model based on inhomogeneous Josephson contacts between the grains, which could reproduce such an extremal behavior [145]. The crossover of $\partial R_s/\partial P$ from positive to negative values occured for junctions with a length L smaller than the coherence length $\xi(T)$. The applicability of this approach to high-κ superconductors seems therefore unrealistic.

Finally, the coupled-grain model was recently re-analyzed in terms of inductive shunting across the grain boundaries [146, 147]. By proper dimensioning of the involved circuit elements and their electromagnetic properties it was possible to simulate a current redistribution, including negative field slopes of $R_s(H)$ and $X_s(H)$. However, the anomalous field dependence of the surface impedance of shunted grain boundaries affected predominantly the losses rather than the penetration depth. A physical validation of this approach remains, therefore, presently lacking.

3.4.2 Anomalous Field-Effect in Superconductors: Pair Breaking Effects

Nonequilibrium Quasi-particle States

Sridhar and Mercereau observed a quadratic dependence of the surface resistance of thin Sn and In films in an applied DC field H_{DC}: $R_s(H_{DC}) = R_s(0) + \alpha(T) H_{DC}^2$ [148]. The static field was oriented parallel to the microwave field. The experimentally determined coefficient $\alpha(T)$ was positive near T_c and negative at low temperatures. The temperature dependent field-induced enhancement of superconductivity was analyzed in terms of a generalized current–field relation in the local limit (Chap. 1):

$$J_{tot} = \frac{1}{\mu_0} K(q, \omega, T)(a_{DC} + a_\omega). \qquad (3.77)$$

The index "ω" denotes the high-frequency contribution of the Fourier component a of the vector potential A, and q is the wave vector. The local limit implies the evaluation of (3.77) at $q = 0$. The surface impedance was considered in the thin-film limit: $Z_s = 1/\sigma d_F$ [52], where $\sigma = \sigma_1 - i\sigma_2 = -iK/\omega\mu_0$ is the complex conductivity (see (1.59a)). The kernel K was evaluated in terms of a microscopic theory for dirty (type-I) superconductors in magnetic fields [149].

The superposition of the DC and the microwave field induces two pair-breaking parameters (Sect. 1.3.3): $\Gamma_{DC} \propto A_{DC}^2$ and $\Gamma_\omega \propto A_{DC} A_\omega$. The first parameter represents the suppression of the order parameter due to the static field, while the second indicates an interference between the static and the microwave fields, which leads to a dynamic perturbation of the pair state. This term is absent if both fields are oriented perpendicular to each other. The effect of the static pair breaking is analogous to that of paramagnetic impurities: the energy gap is reduced and the number of quasi-particles increased. The resulting increase of the surface impedance, $\Delta R_s/R_s > 0$ and $\Delta X_s/X_s > 0$, is proportional to H_{DC}^2. The numerical evaluation of the field coefficient $\alpha(T)$ agreed with the Ginzburg–Landau result derived in Sect. 3.1.2.

The dynamic pair breaking was found to cause oscillations of the order parameter Δ_{OP} and of the density of states (and thus of the quasi-particle number density N_{qp}) at the driving frequency ω. All quantities follow the excitation adiabatically as long as $\omega \ll 1/\tau_\Delta$ and $\omega \ll 1/\tau_{in}$ (Sect. 3.1.1). The effect of Γ_ω is then similar to that of Γ_{DC}. Here, it is important to note that the relaxation time τ_Δ is in general different from that of quasi-particle excitations, $\tau_{in} < \tau_\Delta$. At sufficiently high frequencies $\omega\tau_{in} \gg 1$, N_{qp} cannot adjust to the fast perturbation but approaches a constant value. The quasi-particle distribution function develops instead a nonthermal high-frequency component $f_\omega(E)$ which oscillates out of phase with the excitation A_ω. As a result, the microwave field experiences "fewer" quasi-particles than in the equilibrium state, and the superconducting properties are recovered: $\Delta R_s/R_s < 0$ and $\Delta X_s/X_s < 0$. Since the relaxation time of the quasi-particles decreases with increasing temperature, this model could explain the experimental observation of a temperature dependent sign-change of the coefficient α. Nonequilibrium effects determined the field dependent surface resistance at low temperatures ($\alpha < 0$) while "Ginzburg–Landau" pair breaking dominated at elevated temperatures ($\alpha > 0$).

Two important remarks need to be added. First, it was noted in [148] that the type of nonequilibrium state introduced by the static field was different from photon-enhanced superconductivity, since the high-frequency response remained linear (i.e., R_s remained independent of microwave field amplitude). The effect of the static field was to increase the supercurrent via the fraction x_s of paired charge carriers and, independently, to decrease the normal current via x_n. Both effects were "sensed" by the high-frequency field. In contrast, photon-induced nonequilibrium affects x_s alone, and the microwave signal excites and senses the energy imbalance of the pair-state at the same time. (This difference should be reflected by a differential loss tangent that is smaller in the former case than in the latter.) Secondly, the occurrence of the non-equilibrium effect required local electrodynamics as well as the high-frequency limit $\omega\tau_{in} \gg 1$. Both conditions could be met with sufficiently thin films such that $\ell < \xi(T)$, and with low-κ material where the basic time scale corresponds to GHz frequencies. While local electrodynamics are applicable also to the high-κ strong-coupling superconductors Nb$_3$Sn and YBCO ($\ell < \lambda$), this type of nonequilibrium might hardly be excited in the microwave range due to the very short relaxation times $\tau_{in} \approx (T\text{Hz})^{-1}$ (Sect. 2.3).

Magnetic Field Dependence of the Scattering Rate

According to the previous analysis, nonequilibrium effects appear unlikely as an explanation of negative field coefficients of Z_s in HTS compounds. Still, there is another, adiabatic, mechanism causing a recovery of superconductivity. It is related to the presence of magnetic scattering centers in the oxide superconductors. The basic manifestation of this recovery effect was the observation of an anomalous increase of the upper critical field $H_{c2}(T)$ at low temperatures (e.g., [150] and references therein). In conventional superconductors, $\partial H_{c2}(T)/\partial T$ vanishes as the temperature approaches $T = 0\,\text{K}$, and becomes constant as T approaches T_c. In contrast, measurements of various doped HTS compounds revealed positive slope over the whole temperature

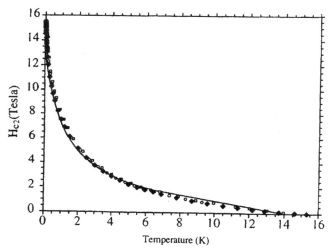

Fig. 3.18. Temperature dependence of the upper critical field $H_{c2}(T)$ for $Tl_2Ba_2CuO_6$ [150]. *Symbols* represent experimental data, the *solid line* denotes the theoretical result. (© 1999 by the American Physical Society.)

range, as reproduced in Fig. 3.18 (symbols). Also shown in the figure is the result of a theoretical analysis (line) which describes the data quantitatively.

The basic features of the theory are not limited to the temperature as parameter but also apply to magnetic fields (see, e.g., [151]). The arguments given in [149] are therefore presented in the following in terms of the generalized variable ("order parameter") Ψ. Increasing Ψ values are regarded to represent decreasing temperature ($\Psi \propto 1/T$) or increasing magnetic field ($\Psi \propto H$).

The presence of magnetic impurities induce spin-flip scattering of the quasi-particles at a rate Γ (see corresponding Hamiltonian in Sect. 1.3.3). As a result, Cooper pairs are broken, and the superconducting state is weakened (Figs. 1.9 and 1.10). The probability for such scattering events is high if the impurity spins are unconstrained, since the total spin is then conserved. Spin-flip scattering is favored if the impurity spins are uncorrelated, namely, at low levels of Ψ (e.g., at elevated temperatures). However, as Ψ increases, mutual interactions between the impurities become important. As a result, the spin-flip scattering rate Γ (Ψ) becomes dependent on the ordering parameter Ψ. In effect, $\Gamma(\Psi)$ decreases as Ψ increases, and the superconducting pair state is recovered. The appearance of this mechanism resembles that of photon-induced superconductivity, except that the superconductor remains in equilibrium. The theoretical fit to the data in Fig. 3.18 was obtained by putting $\Gamma(\Psi)$ into the form

$$\Gamma(\Psi) = \Gamma_0 \frac{1 + \beta \Psi^*/\Psi}{1 + \Psi^*/\Psi} \tag{3.78}$$

with $\beta > 1$ and Ψ^* the "cross-over" value of the ordering parameter, above which Γ decreases markedly. It was further assumed that Ψ^* was small compared to the critical value Ψ_c (corresponding to $1/T_c$ or H_{crit}). The optimum fit to the data in Fig. 3.18 was obtained with $T^* = 1\,\mathrm{K}$ and $\beta = 1.26$.

As pointed out in [133, 150], the ordering of impurity spins recovers all superconducting parameters, including the penetration depth. This can qualitatively be verified with (1.66), which was derived for a BCS superconductor in the dirty limit (Sect. 1.3.3):

$$\frac{\lambda_{\mathrm{L}}^2}{\lambda^2(\Gamma)} = \int_0^\infty \mathrm{d}\varepsilon \frac{\Delta_{\mathrm{OP}}^2}{(\varepsilon^2 + \Delta_{\mathrm{OP}}^2)[(\varepsilon^2 + \Delta_{\mathrm{OP}}^2)^{1/2} + \hbar\Gamma_2]}$$
$$\times \left(1 - \frac{\hbar\Gamma\Delta_{\mathrm{OP}}^2}{(\varepsilon^2 + \Delta_{\mathrm{OP}}^2)^{3/2}}\right). \tag{3.79}$$

With Γ given, e.g., by (3.78), this expression yields a reduction of $\lambda[\Gamma(\Psi)]$ with increasing impurity ordering Ψ. If this ordering is induced by application of a magnetic field, (3.79) provides the basis for an anomalous field effect in HTS films. It is finally noted that, as long as the driving frequency is below the relaxation rate of the impurity spins, the effects of DC and microwave fields should remain indistinguishable. A quantitative theoretical analysis of the oxide superconductors as indicated in [151] is presently in progress [152].

3.4.3 Typical Field Ranges for Nb$_3$Sn and YBCO

Summary of Critical Fields

This section concludes Chap. 3 with a summary of the nonlinear mechanisms. For the sake of consistency, all physical types of nonlinear microwave response are represented by critical fields H_{crit}, although some have currents as the physical origin. Table 3.5 is a collection of the various definitions together with the basic dependences $H_{\mathrm{crit}}(T)$, $R_{\mathrm{s}}(H)$, and $r(H)$ derived in the preceding sections. The sources of anomalous field effects were not considered in the table because a detailed understanding is presently lacking.

The absolute values of the various critical fields as well as their temperature dependences turn out to be the most conclusive features to distinguish between them, whereas the field dependences of R_{s} and r reveal ambiguities which may be difficult to resolve experimentally. Figure 3.19 is a graphic representation of $\mu_0 H_{\mathrm{crit}}(T)$ from the data in Table 3.5. The pinning field was omitted in the figure, because it presents a scaling field rather than a critical field. The thermodynamic limitation is represented by the superheating field, which in the large-κ limit is $H_{\mathrm{sh}} \approx 0.745\,H_{\mathrm{c,th}}$. The absolute values of the different critical fields were adjusted to published data on YBCO. Table 3.6 summarizes these data in comparison with results for Nb$_3$Sn. The temperature dependences of all H_{crit} values were expressed in terms of those of the penetration depth, and of the bulk and Josephson critical current densities.

3.4 Summary: Consequences for Nb$_3$Sn and YBa$_2$Cu$_3$O$_{7-x}$ 167

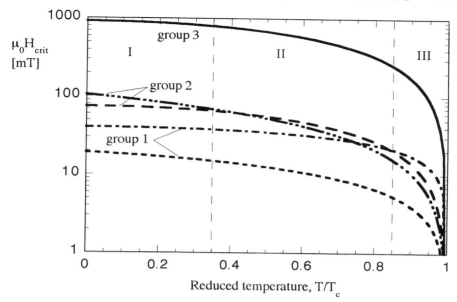

Fig. 3.19. Comparison of the temperature dependent critical fields listed in Table 3.5: H_{sh} (*solid line*), H_{c1} (*long-dashed*), H_{c1J} (*short-dashed*), H_q (*dot-dot-dashed*) and $H_{q,loc}$ ("iso") (*dot-dashed*). The thermal quench fields were evaluated for YBCO on LAO and $f = 19$ GHz (Sect. 3.2). Further explanations are given in the text.

According to the results derived in Sect. 2.3, $\lambda(T) \propto [1 - (T/T_c)^2]^{-1/2}$ was assumed for the whole temperature range instead of using the GL approximation, which is valid only close to T_c. Further, the empirical dependences $H_q(T) \propto [1 - (T/T_c)]$ ("+" in Fig. 3.10) and $J_{cJ} \propto [1 - (T/T_c)]$ (Sect. 1.2.2) were adopted. The figure can provide only a qualitative illustration, since the assumed absolute levels and temperature dependences involve numerous parameters, and are partially subject to ongoing discussion. However, several important conclusions of general validity can be drawn.

First, the five fields shown in Fig. 3.19 form three groups. Josephson coupling puts the strongest restrictions on the linear range of $Z_s(H)$, followed at $T < 0.7\,T_c$ by heating at defects (group 1). The threshold levels for vortex penetration (H_{c1}) and global heating fall close together and constitute group 2. Group 3 is given by the thermodynamic superheating field. Three different temperature regions can be distinguished in Fig. 3.19. In the absence of weakly superconducting defects (Josephson coupling) and normalconducting defects (local heating), the linear microwave response is limited by H_{c1} at temperatures $T \leq 0.35\,T_c$ (region I: $H_{c1} \leq H_q$). In contrast, local and global heating lead to the lowest critical fields up to about $T \approx 0.85T_c$ (region II: $H_{q,loc}, H_q \leq H_{c1}$). At higher temperatures, H_q and H_{c1} are of comparable

Table 3.5. Summary of the critical fields H_{crit} characterizing nonlinear microwave response. Included are the expected temperature dependences of H_{crit}, and the approximate field dependences of R_s and r at field levels below ($<$) and above ($>$) H_{crit}. The distinction in λ between the electromagnetic and the Ginzburg–Landau quantities was neglected. In brackets are given the references to the corresponding equations. The field dependences are expressed in normalized units, with the scaling fields identifiable from the indices, e.g, $h_{c2} \equiv H/H_{c2}$.

Mechanism	H_{crit}	Definition of $\mu_0 H_{\text{crit}}$	Temperature dependence $H_{\text{crit}}(T) \propto$		Field dependence $R_s(H) \propto$		Field dependence $r(H) \propto$
GL depairing s wave	$H_{\text{c,th}}$	$\Phi_0\kappa/(2^{3/2}\pi\lambda^2)$ (3.10)	$\lambda^{-2}(T)$	$<$: $>$:	$h_{c,\text{th}}^2$ normal (3.15)	$<$: $>$:	$1/\omega\tau \gg 1$ $= 1$
GL depairing d wave	$H_{\text{c,th}}$	$\Phi_0\kappa/(2^{3/2}\pi\lambda^2)$ (3.10)	$\lambda^{-2}(T)$	$<$: $>$:	$\alpha(T)h_{c,\text{th}}^2$ normal (3.48)	$<$: $>$:	$1/\omega\tau \gg 1$ $= 1$
superheating $\kappa \gg 1$	H_{sh}	$5^{1/2}/3 H_{c,\text{th}}$ (3.18b)	$\lambda^{-2}(T)$	$<$: $>$:	$h_{c,\text{th}}^2$ normal	$<$: $>$:	$1/\omega\tau \gg 1$ $= 1$
bulk lower critical field ($\kappa \gg 1$)	H_{c1}	$\Phi_0\ln\kappa/(4\pi\lambda^2)$ (3.29)	$\lambda^{-2}(T)$	$<$: $>$:	h_{c2}^0 $h_{c2}^{1/2}$ (3.34)	$<$: $>$:	$\sigma_2/\sigma_1 \gg 1$ $\omega_{\text{dep}}/\omega \gg 1$
Vortex depinning field	H_{pin}	$\approx J_c\lambda$ (3.37)	$\approx (1-T/T_c)^{1.2}$	$<$: $>$:	h_p^0, h_p^1, h_p^2 h_p^1 (3.37, 39, 40)	$<$: $>$:	$\sigma_2/\sigma_1 \gg 1$ $= 3\pi/4$
Josephson lower critical field	H_{c1J}	$2J_{cJ}\lambda_J$ (3.3)	$[I_{cJ}(T)/\lambda(T)]^{1/2}$	$<$: $>$:	$1 + 3\iota_J^2/4$ ι_J^0 (3.42, 43)	$<$: $>$:	$2V_c/3\Phi_0 f \gg 1$ oscillating $\to 0$
global-quench field	H_q	$(2\Delta T/R_s R_{\text{th}})^{1/2}$ (3.54)	various $\approx (1-T/T_c)$	$<$: $>$:	$1 + \alpha_R b_q^2$ normal (3.60)	$<$: $>$:	various $= 1$
local-quench field, isotropic	$H_{q,\text{loc}}$	$(4\sigma_{\text{th}}\Delta T/R_d r_d)^{1/2}$ (3.54)	$\approx [1 - (T/T_c)^2]^{1/2}$	$<$: $>$:	$1 + \alpha_R b_{q,\text{loc}}^2$ normal	$<$: $>$:	various $= 1$

magnitude (region III: $H_{\text{q,loc}} \geq H_{\text{c1}}$), with vortex penetration again limiting the linear microwave response of YBCO under optimum operating conditions.

Secondly, the field dependence of R_{s} (and correspondingly of X_{s} and r) above the lowest critical field at a given temperature (e.g., along the dashed vertical line at $T/T_{\text{c}} = 0.35$ in Fig. 3.19) will in a real sample not display the "pure" character listed in Table 3.5. It will rather be a superposition of field dependences related to the different mechanisms involved. In contrast, in an ideal material (group 2), the slope $\partial R_{\text{s}}(H)/\partial H$ at H_{c1} scales like $2H_{\text{c1}}/H_{\text{c,th}}^2 \propto \ln\kappa/\kappa^2$. Especially in high-$\kappa$ materials, R_{s} is therefore expected to be constant in the Meissner regime.

Finally, there are numerous positive feedback mechanisms which enhance the slope $\partial R_{\text{s}}/\partial H$ in the nonlinear regime (see also Sect. 4.3): Joule heating becomes more and more pronounced as $R_{\text{s}}(H)$ increases, thus "accelerating" this increase. Also, the scaling fields decrease with increasing temperature, giving rise to a similar effect. Induced by a temperature rise, coherence length and penetration depth increase, too. While a precise prediction of the related effects on $R_{\text{s}}(H)$ is complicated, a weakening of the superconducting state is the most likely consequence. In the case of fast nonlinearities, frequency conversion will transfer power from the fundamental frequency f_0 to higher harmonics, which appears as an additional loss mechanism at f_0. A dynamically delayed response (e.g., at H_{c1}) of the superconductor will also become faster (and thus steeper) as the penetration into the nonlinear regime proceeds.

Critical Fields for Nb_3Sn and YBCO

Table 3.6 summarizes the absolute values at zero temperature of the characteristic lengths and the critical magnetic fields. Literature data are mainly available for the bulk and Josephson current densities J_{c} and J_{cJ}, and for H_{c1}. Data on the higher critical fields are hardly measurable in a direct way, and are usually derived from the characteristic lengths.

Published data on $\mu_0 H_{\text{c1}}$ of Nb_3Sn scatter by almost one order of magnitude in the range 20–140 mT [153–155]. Corresponding data for YBCO were reported for films and single-crystals between 40 and 130 mT [6b, 65c, 156–159]. This scatter reflects the difficulties in proper determination of the lower critical field of superconducting films, which is complicated by the quality of the samples, e.g., in terms of different mean free paths. Surface- and defect-induced pinning usually lead to an overestimation of the lower critical field. Anisotropy, demagnetizing effects and thin-film corrections have to be considered also. The Ginzburg–Landau critical fields listed in Table 3.6 were therefore estimated from ξ_{GL} and λ_{GL} data, which display less uncertainties. The data for YBCO correspond to fields applied perpendicular to the $CuO_2(ab)$ planes. For parallel fields, λ_{L}^2 has to be replaced by $\lambda_{\text{ab}}\lambda_{\text{c}}$, leading to a lower critical field reduced by about a factor of 3–5 (Sect. 3.1.5). The upper critical field is enhanced by the same factor in parallel orientation.

The resulting values in Table 3.6 are in good agreement with the body of literature data, thus yielding a consistent framework.

Values for the bulk critical current density were taken from transport measurements, yielding $J_c = 1 \times 10^7$ A/cm^2 (respectively: 6×10^7 A/cm^2) for Nb$_3$Sn (YBCO). The Josephson critical current density J_{cJ} depends on the crystallinity and the phase purity of the samples. A resonable upper limit to sustain the concept of "weak-coupling" was set by $J_{cJ} = 0.1 J_c$. The thermal quench fields were omitted in Table 3.6 because of their complicated dependence on the types of defects, substrate materials and thermal conductivities. Analytical estimates for a few cases are listed in Table 3.6 and shown in Fig. 3.10 (Sect. 3.2).

Table 3.6. Characteristic lengths and critical field values of the bulk, strong-coupling, and extreme type-II superconductors Nb$_3$Sn and YBa$_2$Cu$_3$O$_{7-x}$ at $T = 0$ K. Quench fields of thermal effects are listed in Table 3.3. Remarks: (a) Average value for TVD films from Table 2.3 (Sect. 2.2). (b) Average value for TVD films from Table 2.4 (Sect. 2.2). (c) Evaluated as derived in Sect. 3.1, and partially listed in Table 3.5. (d) A transport critical current density of $J_c = 1 \times 10^7$ A/cm^2 was assumed (c.f., Table 2.3). (e) The maximum critical Josephson current density was set to 0.1 J_c. (f) Results according to Sects. 1.3.3 and 2.3.2. (g) The GL quantities were derived for the clean limit (Sect. 1.1): $\xi_{GL} = 0.74\xi_0$, $\lambda_{GL} = 0.71\lambda_L$. (h) A transport critical current density of $J_c = 6 \times 10^7$ A/cm^2 was assumed.

Characteristic parameters	Nb$_3$Sn	Remarks	YBa$_2$Cu$_3$O$_{7-x}$	Remarks
ξ_0 (nm)	7	a	2.5	f
λ_L (nm)	90	a	140	f
ξ_{GL} (nm)	4	b	1.9	g
λ_{GL} (nm)	60	b	100	g
$\kappa = \lambda_{GL}/\xi_{GL}$	15	c	54	c
$\mu_0 H_{c2}$ (mT)	20500	c	91000	c
$\mu_0 H_{c,th}$ (mT)	970	c	1200	c
$\mu_0 H_{sh}$ (mT)	720	c	900	c
$\mu_0 H_{c1}$ (mT)	140	c	75	c
$\mu_0 H_{pin}$ (mT)	10	d	100	h
$\mu_0 H_{c1J}$ (mT)	≤ 10	e	≤ 20	e

4. Measurements of the Surface Impedance at Nonlinear Response

> *Empirical perception proceeds piecewise. Every single one of the pieces available is therefore worth consideration. Only then might the true picture dawn.*

Chapter 3 dealt with the various theoretical aspects of nonlinear response (i.e., field dependent surface impedance) of superconductors. No explicit distinction was made between DC and microwave magnetic fields, except for the consideration of microwave heating which is absent in DC fields. In contrast to this apparent symmetry, the body of experimental data reported in the literature can be divided into the two groups $Z_s(H_s)$ and $Z_s(H_{DC})$ (with the index "s" for microwave surface field and "DC" for static field). As mentioned, e. g., in [1] and discussed below, the effects of H_s and H_{DC} on the surface impedance are empirically not equivalent. They are considered, therefore separately in Sects. 4.1 and 4.2, with a comparison given in Sect. 4.3. It is impossible, and therefore not attempted, to refer to every single one of the numerous publications in this field, but it is hoped that the selected references will provide a representative frame of the present state of the art on $Z_s(H_s, H_{DC})$ for $YBa_2Cu_3O_{7-x}$ and Nb_3Sn. For further reading the interested reader is referred to previous books and reviews [1–5].

Results on the nonlinear surface impedance of HTS and A15 superconductors are discussed in parallel, since the mechanisms are similarly applicable to both materials. The discussion begins with the oxide materials for which much more data are available than for Nb_3Sn films on dielectric substrates.

4.1 Microwave Field Dependence of the Surface Impedance

This section aims to give a comprehensive overview of the phenomenology of $Z_s(H_s)$. Many groups have investigated HTS films (mainly Y-123 or Tl-2212, and Tl-2223) prepared by different methods (Chap. 5), and measured with different instrumental setups (Chap. 2). Accordingly, there is a variety of nonlinear behavior observed. Before trying to draw conclusions about the underlying mechanisms, it seems therefore advisable to describe in some detail the empirical status. This is done in Sects. 4.1.1 and 4.1.2 for YBCO and TBCCO, and in Sect. 4.1.4 for Nb_3Sn. Section 4.1.3 contains remarks on the local character of the microwave field dependence of Z_s, which constrains possible interpretations. The analysis of and the distinction between magnetic and thermal nonlinearities are treated in Sect. 4.3.

4.1.1 Phenomenological Picture

Measurement Methods and Typical Results for Nonlinear Microwave Response

The nonlinear microwave properties of HTS have most commonly been studied by investigating the power dependence of the surface impedance of unpatterned or patterned films. Other, and more sensitive, methods can be applied as well, which mainly sense the dynamic nature of the nonlinearity. Such measurements might address the performance issues of useful devices more realistically. One can distinguish five different methods in terms of the multiplicity, the power level and the modulation of the driving signal(s) [6]: harmonic generation, intermodulation, saturation and desensitization, cross modulation, and conversion of amplitude- to phase-modulation. In addition, nonlinear effects can be studied either in the frequency domain or in the time domain where the dynamic response of the test objects to excitation pulses of variable amplitude and duration is investigated. Many of these methods have been applied to study nonlinear effects in unpatterned [2, 7–13] or patterned coplanar [6, 14–18], microstrip [19–21] or stripline arrangements [22] using Y-123 or Tl-2212.

Figure 4.1 shows a representative example of the temperature and microwave field dependent surface resistance of two YBCO films on LAO at

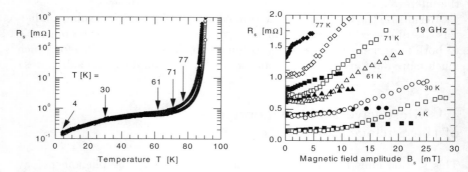

Fig. 4.1. Temperature dependence at 19 GHz (*left part*) of the surface resistance a ⌀2″ laser-ablated film (*diamonds*, "L49", [23]) and a ⌀1″ DC-sputtered film (*circles*, "S145", [24]). The *right part* displays the field dependences $R_s(B_s)$ of both films at the temperatures indicated by *arrows* in the left part. *Filled (open) symbols* refer to the laser-ablated (sputtered) films.

19 GHz [23]. The diamonds denote data for the $\varnothing 2''$ laser-ablated film ("L49", $d_\mathrm{F} = 260$ nm, $T_\mathrm{c} = 90.1$ K), [24]), the circles stand for the $\varnothing 1''$-DC-sputtered film ("S145", $d_\mathrm{F} = 220$ nm, $T_\mathrm{c} = 89.8$ K, [25]). The left part of the figure demonstrates that both samples reveal within 10% identical R_s values at low field levels ($B_\mathrm{s} = \mu_0 H_\mathrm{s} \ll 1$ mT) at all temperatures. Slight differences occur in the transition regime $70\,\mathrm{K} \leq T \leq T_\mathrm{c}$. In contrast, the two films display obvious differences at elevated fields [right part: open (respectively: filled) symbols refer to "S145" ("L49")], which become more pronounced as the temperature increases. Similar observations were reported also for the surface reactance $X_\mathrm{s}(T, B_\mathrm{s})$ [26]. The results demonstrate that the field dependence of Z_s provides information on the microwave response of the superconducting films which is supplementary, and in many cases uncorrelated, to the linear regime (Chap. 2). A comprehensive understanding of Z_s therefore requires careful analysis of its field dependence at various temperatures.

Representative Results for Two-Tone Frequency Intermodulation

Figure 4.2 illustrates the two-tone frequency intermodulation observed for a non-optimized laser-ablated $\varnothing 3''$ YBCO film on sapphire [27] (upper part) in comparison with $R_\mathrm{s}(H_\mathrm{s})$ (lower part, note the logarithmic scales for B_s in the figure) [28, 29]. Two microwave signals of identical power level, separated in frequency by $\Delta f = f_2 - f_1 = 600$ Hz, were applied to the Nb-shielded sapphire resonator (Sect. 2.1.3) around the center frequency $(f_1 + f_2)/2 = 19$ GHz. Third-, fifth- and seventh-order intermodulation (IMD, Sect. 3.3.2) were monitored at the frequencies $2f_1 - f_2(f_{1-})$ and $2f_2 - f_1(f_{1+})$, $3f_1 - 2f_2(f_{3-})$ and $3f_2 - 2f_1(f_{3+})$, $4f_1 - 3f_2(f_{5-})$ and $4f_2 - 3f_1(f_{5+})$ with a spectrum analyzer. Although not representative for optimized YBCO films, the diagram illustrates several characteristic features. First, the IMD signals are symmetric with respect to the center frequency, i.e., the signals at $f_{1\pm}$, $f_{3\pm}$ and $f_{5\pm}$ form three pairs of curves (open and filled symbols). Secondly, the signal levels rise at low field levels with increasing power $P_\mathrm{inc} \propto B_\mathrm{s}^2$ almost as expected for the order of the intermodulation, i.e., $\propto B_\mathrm{s}^{n/2}$ for the nth order (indicated by the dashed lines). The third-order signal is the strongest one, starting at about 50 dB separation from the linear signal, and at field levels where $R_\mathrm{s}(B_\mathrm{s})$ was still flat. This feature characterizes the high sensitivity of intermodulation measurements. The figure also reveals a crossover of the IMD signals from the power $n/2$ to $1/2$ (linear power dependence) at field levels above about 0.1 mT. Since slopes change already at modest field levels, one might conclude the nonlinear mechanism not to be in equilibrium with the microwave excitation, even if the difference frequency $\delta\omega = 2\pi\delta f$ is very small. Dissipative or reactive hystereses are possible explanations for such a behavior, as discussed in Sects. 3.3.2 and 3.3.3. Moreover, the large sensing time-constant ($\approx 1/\delta f \approx$ ms) leaves the possibility for a thermal cause of the IMD.

174   4. Measurements of the Surface Impedance at Nonlinear Response

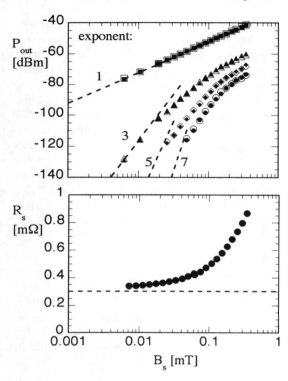

Fig. 4.2. *Top*: Signal power versus microwave field amplitude B_s at 4.2 K for a $\varnothing 3''$ laser-ablated YBCO film on sapphire at the fundamental frequencies f_1 and $f_2 > f_1$ (*open and filled squares*), and at the third (*triangles*), fifth (*diamonds*) and seventh (*circles*) order intermodulation. *Open* and *filled symbols* denote the IMD components above f_2 and below f_1. The *dashed lines* indicated the power-law behavior corresponding to the order of the IMD. *Bottom*: Microwave field dependence of R_s (note the logarithmic B_s scale) [29].

The behavior displayed in Fig. 4.2 presents only one example of a variety of two-tone IMD observed for different films and at different temperatures. There is a trend of weaker frequency intermodulation at elevated temperatures (but still sufficiently below T_c), which supports the potential relevance of microwave heating. While the detailed understanding of these effects is presently not complete, the example illustrates the dynamic nature of the nonlinearities, which cannot be judged from the $Z_s(B_s)$ curves alone.

Impact of Nonlinear Microwave Response on Time and Frequency Domain Signals

Figure 4.3 illustrates the implications of nonlinear surface resistance (Fig. 4.3a), as determined from the transmitted power in the time domain (pulsed operation, Fig. 4.3b), in comparison with frequency domain data (sweep-mode, Fig. 4.3c) [23, 29, 30]. The behavior is typical for many published data, and is qualitatively well understood (e.g. Sect. 2.1.2, [1–4, 11]).

Figure 4.3b displays results for the nonlinear microwave response of the $\varnothing 2''$ laser-ablated YBCO film "L49", measured in the time domain. The data were obtained by adjusting the 19-GHz resonator above a bad spot of the film

4.1 Microwave Field Dependence of the Surface Impedance 175

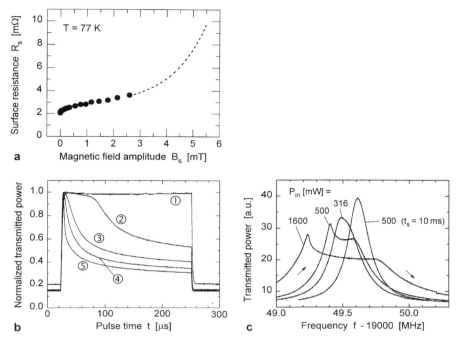

Fig. 4.3. (a) $R_s(B_s)$ of the ⌀2″ YBCO film "L49" at 19 GHz and 77 K. The *symbols* denote the region of undistorted pulse shapes (c.f., b). The *dashed line* indicates the trend of $R_s(B_s)$ when evaluating the maximum transmitted power of distorted pulses. (b) Normalized transmitted power versus time during a 225-µs square pulse at different maximum field levels B_s in the nonlinear regime of $R_s(B_s)$ (see a): B_s (mT) = 2.6 (curve 1), 3.1 (2), 3.6 (3), 4.1 (4), and 4.6 (5). (c) Transmitted power versus frequency, swept across the resonance (sweep interval = 1.3 MHz) with a sweep time t_s = 200 ms at different levels of input power P_{in}, and with t_s = 10 ms at P_{in} = 500 mW. *Arrows* indicate the sweep direction. After [23, 30].

(Sect. 4.1.3). The transmitted power was normalized to its value $P_2(t_{on})$ at the onset of the signal [$P_2(t_{on})$ (mW) = 4.2 (curve 1), 5.9 (2), 7.8 (3), 10.3 (4), and 13.0 (5)]. The film responds in equilibrium to the microwave signal up to B_s = 2.6 mT during the full 225-µs period (symbols in Fig. 4.3a). At higher field levels, the transmitted signal breaks down on a time scale which decreases with increasing overcriticality. The breakdown is not complete ($P_2 > 0$ at the end of the pulse), but continues decaying slowly until the microwave power is switched off. While the fast signal decay can be due to magnetic as well as thermal effects, the slow part of the response points to heating (c.f., Fig. 3.12). Results like those shown in Fig. 4.3b have been reported frequently in the literature. They are considered representative of locally induced nonlinear losses. The onset field level for pulse deformation was reported to decrease with increasing temperature like $1 - (T/T_c)^m$, with m between 2 and 4.

Nonresonant time-resolved measurements of the dynamic nonlinear response of Y-123 films indicated two different loss mechanisms (Sect. 3.3.3). The impulse response of thin coplanar transmission lines to ps-pulses revealed a "slow" effect with a time constant τ_c of the order of 10 ns, and a nonthermal "fast" effect ($\tau_c < 30$ ps) [17]. The slow response was attributed to local heating effects while the physical origin of the fast nonlinearity remained unclear. Similar time scales were observed for the photoresponse of thin current-biased HTS films to picosecond laser pulses at wavelengths of 1.06 and 0.80 µm [31, 32]. The authors attributed the widths of the photoresponse pulses of about 30 ps to nonequilibrium mechanisms (Sects. 3.1.1 and 3.4.2). Response times of 100–200 nm thick Y-123 films of as low as 1.5 ps were reported at 79 K for 150-fs laser pulses at 0.79 µm [33]. Due to phonon relaxation, the fast signals were followed by a much slower thermal response ($\tau_c \approx 10$ ns).

Since the transmitted power was not constant during the microwave pulse (Fig. 4.3b), the field dependent surface resistance could not be deduced without ambiguities. The dashed line in Fig. 4.3a indicates the results for $R_s(B_s)$ if evaluated from the maximum transmitted power. However, the absolute level of P_2 depends in general on the different time scales involved in the measurement (load time ω/Q of the resonator, relaxation time τ_c of the nonlinear mechanism, response time of the microwave detector, etc.). The physical significance of $R_s(B_s)$ curves derived from distorted pulses must hence be considered with great care.

Figure 4.3c illustrates the effect of nonlinear microwave losses on the transmitted power monitored during a frequency sweep Δf across the resonance for two different sweep times t_s. The measurement conditions were otherwise the same as in Figs. 4.3a and b. Above an input power about 310 mW, the frequency characteristic deviated from the expected Lorentzian shape (Sect. 2.1.1). As a result, the determination of a loaded quality factor from the FWHM becomes ill-defined. Values for R_s and B_s can merely be evaluated from the maximum signal, with resulting uncertainties depending additionally on the chosen sweep velocity $\Delta f/t_s$.

It was reported that the field dependence of R_s degraded irreversibly in some cases, after a certain microwave field amplitude was exceeded (e.g., [34]). The deteriorated performance turned out to be reproducible, even after warming of the film above T_c and subsequent cooldown. Such observations were attributed to local field-induced film damages. However, microscopic investigations failed in many cases to identify regions of structural degradation.

4.1.2 Critical Field Levels and Power-Law Behavior

Experimental results on the field dependent surface impedance are usually published as $R_s(B_s)$ curves. Notification of the measurement conditions (quality of the sample, resolution of the measurement system, mode of operation, involved time constants, cooling conditions etc.) is often incomplete, and one is left with an attempt to derive at least a phenomenological picture. Due to the number of groups involved, different notations (typical indices being "s", "rf", "ω") and units have been used for the microwave field amplitude. They reflect individual considerations but do not in general indicate different physical mechanisms. The conversion between magnetic field H (or surface current) and flux density B is given by $B = \mu_0 H$ with $\mu_0 = 4\pi \times 10^{-7}$ Vs/Am, unless otherwise noted. For instance, 1 Oerstedt $= 4\pi 10^{-3}$ A/m, 10^3 A/m correspond to 0.4π mT ≈ 1.25 mT, and 1 Gauss $= 0.1$ mT.

Parametrization of $Z_s(B_s)$

Two regions can be roughly distinguished for $Z_s(B_s)$ (c.f., Fig. 4.1): linear behavior at field strengths below a critical value $b_s \equiv B_s/B_{crit} < 1$, where R_s and X_s are constant, and nonlinear behavior at $b_s > 1$, where $Z_s = R_s + iX_s$ can often be approximated by a power law with temperature dependent coefficients:

$$Z_s(T, b_s) \approx Z_{s,0}(T) \quad \text{at } b_s < 1, \tag{4.1a}$$

and

$$Z_s(T, b_s) \approx Z_{s,0}(T) + \sum_{m>0}[a_{R,m}(T) + ia_{X,m}(T)](b_s - 1)^m \text{ at } b_s \geq 1. \tag{4.1b}$$

For example, the thermal and magnetic mechanisms discussed in Chap. 3 yielded the limiting cases $m = 0, 0.5, 1$ or 2. A graphic representation of the real part of (4.1) is sketched in Fig. 4.4. At field levels well above B_{crit}, $Z_s(B_s)$ often increases steeply with power larger than 2, this increase likely indicating a transition from the superconducting to the normal state.

The linear region $Z_s(b_s) \approx Z_{s,0}$ is of crucial importance for the construction of devices which need to sustain high microwave field levels at optimum performance (Sect. 6.1). In this case, the critical field B_{crit} should be as high as possible. B_{crit} denotes the local maximum field level to which the superconducting film is exposed. It can be derived numerically (in some cases analytically) for given boundary conditions (Sect. 2.1.1). Compared to unpatterned films or cavity resonators, B_s takes enhanced values at the center conductors of patterned devices, e.g., in coplanar, stripline or microstrip transmission line devices (Fig. 2.1). Additional field enhancement at the edges of these lines may result from the cross-sectional geometries in such devices, which potentially lead to enhanced powers m in (4.1) [35], thus obscuring the physical nature of the nonlinear effect (Sect. 3.1.4). In the opposite case, nonlinear effects [e.g., $a_{R,X} > 0$, $m \gg 1$ in (4.1)] may be desired for some types of signal processing like mixing, limiting or switching [5]. Linear structures might then be favorable to benefit from the field enhancement. Furthermore, Josephson elements rather than homogeneous films could be employed to induce a specific, periodic nonlinear behavior (Sect. 6.3).

178   4. Measurements of the Surface Impedance at Nonlinear Response

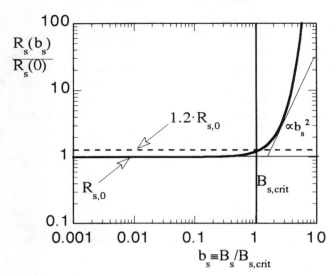

Fig. 4.4. Schematic representation of the field dependent surface resistance of superconductors. Indicated are the low-field value $R_{s,0}$ and the critical field $B_{s,\mathrm{crit}}$.

Determination of the Critical Field

The nonlinear phenomena need to be understood, or at least be reproducible and empirically controllable, in order to optimize either type of applications. Yet in the absence of a comprehensive understanding, a comparative analysis of the reported $Z_s(B_s)$ data provides important information about the underlying mechanisms. Such an analysis relies on the accurate determination of the temperature dependent level $Z_{s,0}(T)$ of the surface impedance in the linear regime (Chap. 2), of the critical field $B_{s,\mathrm{crit}}$ and its temperature dependence, and of the differential loss tangent $r(B_s) = \partial X_s(B_s)/\partial R_s(B_s)$ at a given temperature. However, accurate $Z_{s,0}(T)$ and $B_{\mathrm{crit}}(T)$ measurements are challenging for the two following reasons.

First, the local microwave field amplitude has to be known quantitatively. Since the field distribution in linear resonators depends on the penetration depth, which by itself can be nonlinear, the involved numerical algorithms cause systematic uncertainties in B_s of about $\pm 30\%$. Much better accuracy ($\leq 10\%$) can be obtained with the two- and three-dimensional resonant techniques described in Sect. 2.1, where the field-to-power conversion coefficient depends only on the geometry of the resonators. Additional uncertainty comes, in any case, from the determination of the coupling strengths $Q_{1,2}$, which induce systematic errors of up to $\pm 20\%$ in absolute B_s values. In contrast, relative variations of the field level can be measured much more precisely with a typical accuracy of better than 5%.

The second challenge relates to the determination of the surface impedance. The deduction of absolute X_s values requires, in most cases, curve-fitting to model $\Delta X_s(T)$ curves, inducing uncertainties which are difficult to quantify. We therefore focus on the analysis of the surface resistance. Even in this case, parasitic effects like radiation losses or the finite resolution of the measurement system might mask the "true" onset of nonlinearity. The nonlinear penetration depth $\Delta\lambda(B_s)$ drives the superconductor eventually into the thin-film limit which leads to enhanced $R_{s,\text{eff}}$ values (Sect. 1.1.3). The apparent critical field is then an underestimate of the bulk nature of the nonlinearity. In analogy to the determination of a critical current from current–voltage characteristics, it is therefore necessary to define a criterion for $B_{s,\text{crit}}$ based on a measurable, though arbitrary, increase of $R_s(B_s)$ over its low-field level $R_{s,0}$. $B_{s,\text{crit}}$ is defined in this work according to

$$\frac{R_s(T, B_{s,\text{crit}}) - R_s(T,0)}{R_s(T,0)} = +20\% \ . \tag{4.2}$$

Misinterpretations due to nonlinear signal distortions are avoided by such a conservative definition. Figure 4.4 illustrates $B_{s,\text{crit}}$ in terms of the nonlinear surface resistance according to (4.1).

Review of $Z_s(B_s)$ and Emerging Phenomenology

Many different loss mechanisms have been proposed for the nonlinear microwave response of HTS in relation to specific data from individual groups. Mainly magnetic mechanisms like intrinsic pair breaking [36], Josephson behavior (coupled grain models, Sect. 3.4.1), bulk vortex dynamics [37], combinations of these [1, 22, 38] and critical state effects [1, 19] were considered. Many authors reported a quadratic increase of $R_s(B_s)$ above 0.1 mT, with a quartic or even stronger dependence dominating above a frequency and temperature dependent field level [19, 20, 22]. For an appreciable period of time, this behavior was believed intrinsic to HTS. The nonlinear response at low and intermediate field levels $B_s \leq B_{s,\text{crit}}$ was attributed to the presence of intragranular weak links, whereas at higher field levels bulk vortex dynamics were assumed to dictate the nonlinear $R_s(B_s)$ ([22, 39], Sect. 3.4.1). The situation recently became more diverse, as more data on high-quality unpatterned samples became available, which displayed constant R_s at low field levels.

Table 4.1 summarizes the status achieved with different samples at different frequencies and temperatures. The listing is an upgrade of a previously published version [5], which contains improved or novel data for unpatterned Y-123 films on MgO, LAO and CbS. According to the experimental observations, the $R_s(B_s)$ characteristics are divided into three categories. Type A describes data which clearly reveal constant R_s at $B_s < B_{s,\text{crit}}$, with a strong increase at higher fields. Type B represents quadratic behavior of $R_s(B_s)$, and type C summarizes curves with intermediate field exponents

$\partial \log(R_s)/\partial \log(B_s)$ around 1. Except for the data at 19 GHz, which were obtained in pulsed measurements (duty cycle $\approx$ 200 μs : 1 s), the field dependence of Z_s was usually deduced from slow ($\geq$ 10 ms) frequency sweeps across the resonant frequency. The cooling conditions also differed for the listed experiments.

The listed data cover a range of film thicknesses d_F from 200 to 1000 nm, and of frequencies from 1.5 to 19 GHz. Correspondingly, the $R_{s,0}$ data reveal a large scatter which, when scaled by f^2 to 10 GHz, spread from 40 to 500 μΩ at low temperature ($T \leq 20$ K). No correlations are visible, neither between d_F and $R_{s,0}$, nor between $R_{s,0}$ and $B_{s,crit}$, nor between $R_{s,0}$ or $B_{s,crit}$ and the type of behavior. Also, no general frequency dependence of $B_{s,crit}$ could be established between 1.5 and 15 GHz [14, 22]. Some groups reported correlations between the slope $\partial R_s/\partial B_s$ and the critical transport current density during an early stage of HTS technology [7, 8, 15]. Interestingly, the maximum field levels around 50–100 mT were obtained with patterned devices [41] (see also Fig. 4.5). This fact might reflect the field enhancement in such samples which allows easier excitation of strong fields at the edges of the signal lines than in unpatterned films. Furthermore, these data render the damaging of films, which could have been induced during the patterning process, to be unlikely. A similar conclusion was drawn from a set of measurements performed with one sample before and after patterning [42].

The quadratic behavior (type B) was observed mainly in patterned devices, while type A prevailed in unpatterned films. This difference might also be related to geometric field enhancement [35]. Furthermore, the small dimensions of linear resonators favor a rapid diffusion of heat through the superconductor, thus facilitating thermal effects. Finally, type C is predominantly found for granular samples (e.g., Tl-2223) and/or at elevated temperatures.

Footnotes and Remarks to Table 4.1: [a] Y = Y-123, T1 = Tl-2212, T2 = Tl-2223. [b] cp = coplanar resonator, ms = microstrip, sl = stripline, up = unpatterned. [c] The values in this column present the threshold fields above which R_s increases strongly in type A behavior. For type B, the data denote the field level at which quadratic behavior changes into a quartic or stronger dependence.

Specific Remarks 1. Data obtained at the fundamental, first and second harmonic mode [14b]. 2. Data correspond to variously prepared films [26]. The film with higher $R_{s,0}$ showed a slight increase of R_s already above 0.1 mT. 3. The data comprise results of variously prepared films. Type B was reported for the film with the lower $B_{s,crit}$ [18]. 4. The authors note that the high residual resistance is likely to be affected by the teflon spacer between signal line and ground plate [20c]. 5. A weak increase of R_s was observed above $B_{s,crit}$, extending at 4.2 K up to 80 mT [20a]. 6. Quadratic behavior was observed at all field levels above 0.05 mT [22b]. 7. The data revealed a weak linear increase from the lowest levels [40]. 8. Both films were at least $\varnothing 1''$ in diameter. The data in the upper (lower) row were obtained with a laser-ablated (thermally coevaporated) film [11]. 9. $R_s(B_s)$ increased linearly with field at low field levels [9b]. 10. The data were obtained on $\varnothing 2''$ films using a sapphire resonator [2]. 11. The data were measured on a similarly prepared film at another institute using a parallel-plate resonator with teflon spacer [10].

4.1 Microwave Field Dependence of the Surface Impedance   181

Table 4.1. Summary of the $R_s(B_s)$ behavior reported for Y-123, Tl-2212 and Tl-2223 samples. For footnotes see previous page.

HTS/substrate[a]	Film thickness (nm)	Type of device[b]	f (GHz)	T (K)	$R_{s,0}$ (μΩ)	$B_{s,crit}$[c] (mT)	Type of $R_s(B_{rf})$[c]	Remarks/Ref.
Y/LAO	350	cp	5.2, 10.4, 15.6	15	25, 100, 230	9–19	A	1
Y/MgO	350	cp	8	15	25, 33	0.7, 12.5	A, A, C	2
Y/MgO	800	cp	5.7	4.2	≥ 35	0.1–4	A,B	3
Y/LAO	200	cp	6.2	50, 50	60, 75	3.5, 5	B,C, B,C	[15]
Y/LAO	380	ms ring	3.7	70–77	not given	7.5–10	B	[19]
Y/LAO	200–300	ms meander	3	7	50–100	2	B	4
Y/LAO	300	ms meander	2	4, 47, 75	20, 55, 850	≥ 40, 40, 11	A, A, A	5
Y/LAO	not given	ms meander	2.2	4.2	not given	< 0.1	C	[20b]
Y/LAO	700	sl meander	1.5	4.2, 77	4.3, 15	28, 5	B, B	6
Y/LAO	700	sl meander	1.5	4.2, 77	6, 20	≥ 50, ≥ 15	B, B	[41]
Y/LAO	400	ms disk	1.6	45, 76	not given	20, > 7	A, A	[23]
Y/MgO	350	up	19	4.2, 77	110, 1450	> 50, ≈ 2	C, C	7

Table 4.1. (continued)

HTS/substrate[a]	Film thickness (nm)	Type of device[b]	f (GHz)	T (K)	$R_{s,0}$ (μΩ)	$B^c_{s,\text{crit}}$ (mT)	Type of $R_s(B_{rf})$[c]	Remarks/Ref.
Y/LAO	330	up	19	4.2	200	≥ 27	A	[8b]
				50	500	≥ 23	C	
				77	750	14	C	
Y/LAO	350	up	19	4.2	200	> 43	A	[23]
				77	1100	6.5	A	
Y/LAO	not given	up	14.4	20	500	6.5	A	8
Y/LAO	not given	up	14.4	10	200	18.5	A	
Y/LAO	300	up	5.6	50	not given	13	C	9
Y/CbS	400	up	9.5	10	55	15	A	[9a]
				52	175	11	A	
				66.5	350	10	A	
Y/CbS	330	up	8.5	77	430	15	A	[23]
Tl/LAO	700	up	19	4.2	100	12	A	[40]
				77	550	7.5	A	
Tl/LAO	700	up	5.6	4.2	18	0.5	A	
				80	60	0.3	A	10
				21	70	≥ 6	A	
				80	105	≈ 2	A	11
T2/LAO	1000	up	19	4.2	170	< 0.1	C	[40]
				77	600	< 0.1	C	

4.1 Microwave Field Dependence of the Surface Impedance 183

Figure 4.5 is a graphic representation of the $B_{\rm s,crit}(T)$ values for unpatterned (squares and circles) and patterned YBCO films (triangles and diamonds), resulting from the data in Table 4.1 and supplemented by further data [43]. Also shown are the maximum levels $B_{\rm s,max}(T)$ achieved in the different experiments (open symbols). The symbols stand for different substrates and resonator structures. For clarity, the data corresponding to different categories were offset in temperature, which introduces systematic errors especially at temperatures close to $T_{\rm c}$.

The large scatter reflects the variety of test methods and the continuing trend of improved deposition techniques. The highest field levels at $T \leq 60\,{\rm K}$ were reported for patterned YBCO films, where the measurements with unpatterned films were usually limited by the available power. Above 60 K, the $B_{\rm s,max}$ data were often limited by field quenches, hinting at the proximity of $B_{\rm c1}(T)$ and the importance of the positive nonlinear feedback mechanisms (Sect. 4.3.3). Despite the lattice mismatch between sapphire and YBCO and the necessity of depositing a CeO_2 buffer, films on CbS display similar power handling as on LAO and MgO substrates. The majority of $B_{\rm s,crit}$ data can be enveloped by the empirical relation $B_{\rm s,crit}(T) \leq 30\,({\rm mT})[1 - (T/T_{\rm c})^2]$ (dashed line in Fig. 4.5). Similarly, the maximum field levels are limited by $B_{\rm s,max}(T) \leq 75\,({\rm mT})[1 - (T/T_{\rm c})^2]$ (solid curve). This curve coincides well with the expected level and temperature dependence of $B_{\rm c1}$ (Sect. 3.4.3). The average of the best data is indicated by the dash-dotted curve: $\overline{B}_{\rm s,max} \approx 50\,({\rm mT})[1 - (T/T_{\rm c})^2]$.

In summary, a constant surface resistance at a low absolute level is found to indicate high film quality, and thus intrinsic properties of epitaxial HTS films. The highest $B_{\rm s,crit}$ values reported to date are $\geq 50\,{\rm mT}$ at 4.2 K and $\approx 20\,{\rm mT}$ at 77 K. In contrast, the absence of constant $R_{\rm s}$ at sufficiently low $B_{\rm s}$ is typical for films having defects such as weakly coupled grains, multiphase material, particle contaminations or cracks. The large scatter and the absence of correlations (Tab. 4.1) lead to the conclusion that the nonlinear microwave response of HTS films could be affected by systematic uncertainties related to the measurement systems (e.g., due to unspecified cooling conditions) or, and more likely, by defects in the variously prepared films. According to the separation between the solid and dashed lines in Fig. 4.5, there is still significant potential to improve the power handling of HTS films. $B_{\rm s,crit}$ seems to be limited by $B_{\rm c1}$ at elevated temperatures, while imperfect film quality and/or cooling terminate the linear microwave response below about 60 K. This interpretation is in accordance with the theoretical conclusions of Chap. 3 (Fig. 4.19). Further improvements of $R_{\rm s}(B_{\rm s})$ concern therefore the optimization of the deposition, oxygenation and technical handling of the films, as well as of cooling the film and the film–substrate, substrate–housing and housing–coolant interfaces (Chap. 5). Applications involving high circulating powers might need to exploit the temperature dependence of $B_{\rm s,crit}$, i.e., profit from a reduced operating temperature (Chap. 6). As mentioned in

184   4. Measurements of the Surface Impedance at Nonlinear Response

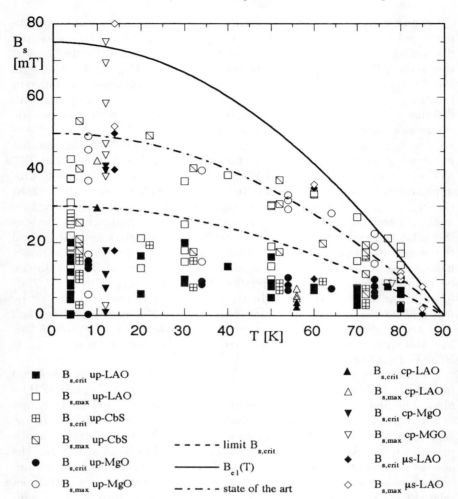

Fig. 4.5. Survey of the field amplitude $B_{s,\mathrm{crit}}$ as defined by (4.2), and of the maximum achieved fields $B_{s,\mathrm{max}}$ for unpatterned YBCO films on LAO (*simple squares*), CbS (*marked squares*, offset by +2K) and MgO (*circles*, offset by +4K). Data obtained with coplanar (stripline) devices are marked with *triangles* and *diamonds*, and offset by +6K and +8K. The *dashed*, *dash-dotted* and *solid lines* represent the functions $B_0[1 - (T/90\,\mathrm{K})^2]$ with $B_0 = 30$, 50, and 75 mT [43].

[6], the lower B_{c1} value of Tl-2212 compared to Y-123 films (about 5–8 mT at 77 K) might be a serious drawback for such applications.

4.1.3 Local Measurements of the Nonlinear Surface Impedance

General Considerations

Measurements of the surface impedance (the discussion is focused on the resistance) integrate over the entire part of the superconducting surface which is exposed to the microwave field. If this surface contains isolated defects of surface resistance R_def and areal extension A_def, embedded in a homogeneous matrix of constant R_s, the total dissipated power is

$$P_\mathrm{diss} = \frac{1}{2} R_\mathrm{s} \int_A |H_\mathrm{s}(\boldsymbol{r})|^2 \mathrm{d}^2\boldsymbol{r} + \frac{1}{2} \sum_k R_{\mathrm{def},k} A_{\mathrm{def},k} H_\mathrm{s}^2(\boldsymbol{r}_k) \,. \tag{4.3}$$

The integration has to be performed over the complete superconductor surface except for the degraded areas. The effective surface resistance, derived from an integral measurement, is found from (4.3) to be

$$R_{\mathrm{s,eff}} = R_\mathrm{s} \left(1 + \sum_k \frac{R_{\mathrm{def},k} A_{\mathrm{def},k}}{R_\mathrm{s} \tilde{A}} \frac{H_{\mathrm{s},k}^2}{H_{\mathrm{s,ave}}^2} \right) \tag{4.4a}$$

The field levels $H_{\mathrm{s},k}$ at the positions $\boldsymbol{r}_k$ of the defects can be assumed to be comparable to the average value $H_{\mathrm{s,ave}}$, because defects at positions of low field do not contribute much to the dissipation. The H_s^2 ratio in the sum of (4.4a) is therefore near unity and can be neglected for the following discussion. We then obtain in dimensionless units $\rho_k = R_{\mathrm{def},k}/R_\mathrm{s}$ and $a_k = A_{\mathrm{def},k}/A$:

$$R_{\mathrm{s,eff}} \approx R_\mathrm{s} \left(1 + \sum_k \rho_k a_k \right) \,. \tag{4.4b}$$

Equation (4.4) allows us to draw the following important conclusions.

First, defects lead to an enhanced surface resistance only if the local products of surface resistance times area, summed over all defects, add a measurable fraction to R_s times the total area. It was illustrated in Sect. 3.2 in relation to microwave heating at defects that a single normalconducting spot in a ⌀1″ film remains "invisible" at 19 GHz and 77 K up to a radius of $r_\mathrm{def} \approx 20\,\mathrm{\mu m}$. The value of the sum $\sum \rho_k a_k$ in (4.4b) was then of order 10^{-4}. The sensitivity of $R_{\mathrm{s,eff}}$ to defects increases at a given film size with decreasing frequency (since $\rho_k \propto f^{-3/2}$) and with decreasing temperature [since $R_\mathrm{s}(T)$ usually decreases more strongly than $R_{\mathrm{def},k}(T)$]. Furthermore, due to the low dimensionality of microstrip resonators, the sum $\sum \rho_k a_k$ is enhanced by a factor of about ℓ/w compared to unpatterned films, where ℓ is the length of the microstrip line (or the diameter of the corresponding film) and w the linewidth.

Second, integral R_s measurements are not suited to determine the number k of defects, nor the individual values of ρ_k and a_k. Such a diversification requires local measurements where the investigated spot area becomes comparable to or even smaller than the defect size.

186   4. Measurements of the Surface Impedance at Nonlinear Response

Third, if the formation of defects happens statistically, rather than being determined by the applied deposition technique, the number k is expected to increase with increasing film size. Consequently, absolute value and field dependence of the effective surface resistance of large-area films can be easily inferior to those of smaller samples.

Finally, even if defects do not show up at low field levels (because $\sum \rho_k a_k$ is below the sensitivity limit), they might still be effective in limiting the linear response. The reason is that ρ_k as well as a_k are expected to increase at elevated field amplitudes. In addition, the critical fields discussed in Chap. 3 are usually reduced at the positions of weakened superconductivity. The identification and prevention or elimination of defects is therefore crucial for applications of superconductors in high-frequency fields or pulsed currents.

The local character of nonlinear effects in superconducting films has been observed in numerous different experiments (e.g., [8c] and references therein; see also Sect. 2.1.4). The potential and the limitations of various approved types of spatially resolved measurements are illustrated in the following in terms of some recent examples.

Multimode Analysis

Figure 4.6 illustrates the microwave field dependence of the unloaded quality factor at 19 GHz for the individual sides of two $\varnothing 1''$ double-sided patterned YBCO films on LAO (open symbols) and at 3.5 GHz in the TM_{010} mode of the corresponding disk resonators (Sect. 2.1.3, filled symbols) [44]. The Q_0 values were normalized to the value at low fields to facilitate the comparison of the field dependences. The upper diagram presents data on the film "S236" ($d_F = 350$ nm, $T_c = 87.3$ and 87.7 K). The two film sides can clearly be distinguished in terms of absolute value and field dependence of $Q_0(B_s)$. Moreover, the worse film side could be identified to limit the properties of the disk resonator. Regarding the different investigated areas (about $\varnothing 7$ mm at 19 GHz and $\varnothing 21$ mm at 3.5 GHz) and field distributions, the results indicate an extended area ($\varnothing \leq 20$ mm) of reduced film quality rather than the existence of few localized defects. The situation was different for sample "S237" ($d_F = 350$ nm, $T_c = 86.9$ and 87.1 K, lower part of Fig. 4.6). Both film sides (open symbols) were of comparable quality as judged from the data at 19 GHz. Still, the $Q_0(B_s)$ performance of the disk resonator turned out to be significantly worse. The conclusion was that the damaged film regions were smaller ($\varnothing \leq 10$ mm) or subject to stronger heating.

The multimode analysis based on the comparison of modes in different resonators at different frequencies senses different surface areas. The resulting spatial resolution is poor. However, as pointed out in Sect. 2.1.3 in relation to the 87-GHz resonator, different high-Q modes (e.g., TE_{013} and TE_{021}) can also be excited in a single setup. Basic conditions are a proper dimensioning of the resonator and an appropriate frequency separation of the modes in terms of measurement resolution and bandwidth of the available instrumentation.

4.1 Microwave Field Dependence of the Surface Impedance   187

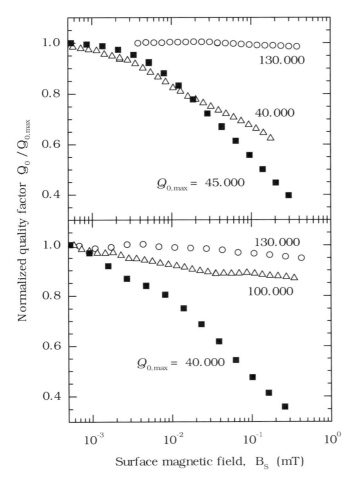

Fig. 4.6. Microwave field dependence of Q_0 at 77 K, measured for the individual sides of two different $\varnothing 1''$ double-sided patterned YBCO films on LAO at 19 GHz (*open symbols*) and at 3.5 GHz in the TM_{010} mode of the corresponding disk resonators (*solid symbols*) [44]. The *circles* (*triangles*) represent the disk side (ground plane) of the disk resonator. (© 1999 IEEE).

Microwave Scanning

A Fabry-Perot resonator operating at 145 GHz [45] was used to scan across a $\varnothing 3''$ double-sided YBCO film deposited onto CbS [46]. The R_s resolution was about $\pm(5$–$10)$ mΩ, and the spatial resolution was around 6 mm. Figure 4.7 shows first results obtained with this method, which indicate an inhomogeneity of the effective surface resistance (i.e., $R_{s,\text{eff}}$ was not corrected for the finite film thickness) of a factor of 3–5. The lower diagram compares the temperature dependence of $R_{s,\text{eff}}$ at the two positions A and B. Differ-

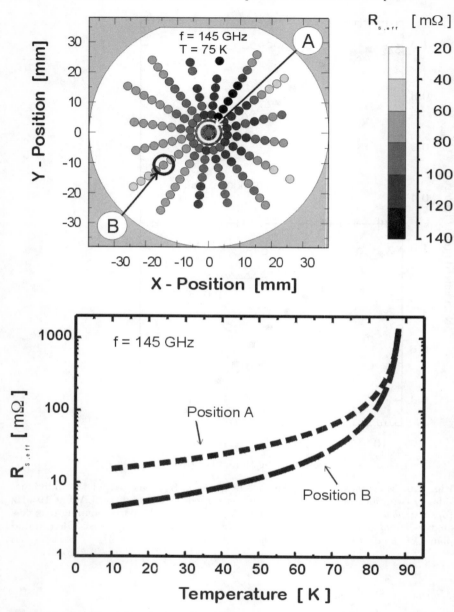

Fig. 4.7. *Top:* Spatial variation of the effective surface resistance $R_{s,\text{eff}}$ at 145 GHz and 75 K across a ⌀3″ YBCO film on CbS [45,46]. *Bottom:* Temperature dependent $R_{s,\text{eff}}$ at the two positions marked by "A" and "B" in the upper diagram.

ences in film quality can be attributed to the two halves of one film side. This method thus provides important information on optimizing the deposition process for such large-area samples, as recently illustrated for laser-ablated films on buffered sapphire substrates [47]. An exact analysis of bulk R_s values requires knowledge of the local film thickness and of the local penetration depth (Sects. 1.1.3 and 2.1.4). In addition, due to the high measurement frequency, the method is suitable to image inhomogeneities of R_s only at low microwave field amplitudes.

Thermal Imaging

Imaging of temperature distributions in superconducting films was already discussed in Sect. 2.1.4. The most elegant approach has been the deposition of a thermo-optic indicator film onto the superconductor, which provided images with high temperature, time and spatial resolution of local and global heating in patterned YBCO devices at variable field levels. The simplest, though efficient, approach has been the observation of local boiling above the surface of HTS films and devices, which were immersed into a liquid cryogen and exposed to high microwave power [44, 48, 49]. A method of intermediate complexity employed 24 standard platinum resistors (sensing area about 2 mm × 2.5 mm, sensitivity 400 mΩ/K above 77 K), arranged on three radii on the surface of a disk resonator [23, 40, 50]. Figure 4.8 shows the temperature map obtained at 1.8 GHz in liquid nitrogen at 77 K at a maximum field amplitude $B_s = 1$ mT, which was well below the quench field of about 3 mT. The resonator was operated in the CW mode. Positions of enhanced temperature are clearly revealed, which point to an inhomogeneous surface resistance. Similar temperature maps were also obtained at lower field. R_s was thus concluded to be enhanced but field independent, due to, e.g., locally reduced critical temperatures.

Although this method is capable of imaging inhomogeneous microwave properties of HTS films at elevated field levels, the extraction of quantitative values for the variation of R_s along the positions of the sensors is complicated. According to the analysis in Sect. 3.2, the temperature rise $\Delta T(\mathbf{r})$ at a position $\mathbf{r}$ is related to the local surface resistance $R_s(\mathbf{r})$ and to the local field amplitude $H_s(\mathbf{r})$ through the thermal boundary resistance $R_{\mathrm{th}}(\mathbf{r})$, which in general also depends on position:

$$\Delta T(\mathbf{r}) \approx \frac{1}{2} R_s(\mathbf{r}) \overline{H}_s^2(\mathbf{r}) R_{\mathrm{th}}(\mathbf{r}) . \tag{4.5}$$

The bar over H_s denotes averaging over the sensing area of the thermometer. If the thermal contacts between the sensors and the film surface were all identical, the measurement yielded information about the local value of the product $R_s H_s^2$. The signals measured along one radius (such that $H_s(|\mathbf{r}|) = $ constant) were then proportional to the azimuthal variation of R_s. Obviously,

190   4. Measurements of the Surface Impedance at Nonlinear Response

the simple approach illustrated in Fig. 4.8 was not yet sensitive enough to resolve the field distribution in the disk resonator, as H_s^2 should be distinctly different for the three radii, being highest for the intermediate ring of sensors. This weakness was attributed to spatial variations of R_{th} for the individual sensors in the presence of the liquid cryogen. Further disadvantages are the large dimensions of the thermometers, as well as their fixed positions which are required to prevent mechanical damage. Finally, the microwave field distribution is potentially modified through the proximity of the sensors and the open environment.

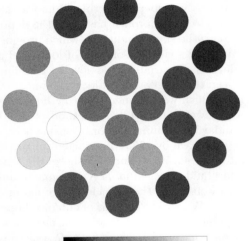

ΔT: 60 mK   440 mK

Fig. 4.8. Grey-coded temperature map of a disk resonator at 1.8 GHz and 77 K at a maximum field amplitude $B_s = 1\,\text{mT}$ [23, 50].

Laser Scanning

A promising approach which avoids several of the previously noted problems is the investigation of the field dependent microwave response of a superconducting film when exposed to local thermal perturbation introduced, e.g., by a focused laser spot (Sect. 2.1.4, [51, 52]). According to the previous analysis [see (4.3)], the modulation of the dissipated power and of the energy stored in the superconductor in a small region δA can be approximated by

$$\frac{\delta P_{\text{diss}}}{P_{\text{diss}}} + \mathrm{i}\frac{\delta W_{\text{S}}}{W_{\text{S}}} \approx \frac{\delta A}{A}\left(\frac{\delta R_{\text{s}}}{R_{\text{s}}} + \mathrm{i}\frac{\delta X_{\text{s}}}{X_{\text{s}}}\right)\,. \tag{4.6}$$

Changes of P_{diss} and/or W_S can be measured, e.g., by monitoring the modulation of the coefficient $S_{12}^2(f)$ of the transmitted power (Sect. 2.1.2). If the modulation signal results mainly from changes of R_s, the highest sensitivity is obtained by evaluating $S_{12}(f)$ at the resonant frequency $f = f_0$. In the contrary case of X_s variations prevailing, $S_{12}(f)$ should be evaluated at the steepest parts of the resonance curve [see (2.19)], i.e., at $f_\pm = f_0 \pm \delta f$, where $f_0/\delta f = 4 \times 3^{1/2} Q$ and $S_{12}^2(f_0 \pm \delta f) = (3/4)|S_{12}(f_0)|^2$.

The top of Fig. 4.9 displays recent data measured by scanning a modulated 750-mW laser beam (optical wavelength $\lambda_{\text{opt}} = 805$ nm, modulation frequency $f_{\text{mod}} = 15$ Hz) across the patterned side of a YBCO/LAO disk resonator, operated at 40 K and 2.216 GHz in the TM$_{310}$ mode [23,52]. The measurement was sensitive to changes of the resonant frequency, and thus to spatial variations of the penetration depth. The experimental data clearly display the threefold rotational symmetry, which is characteristic of the investigated mode (first mode index = 3). The radial distribution of the azimuthal and the radial magnetic field components $H_\varphi(r,\varphi)$ and $H_r(r,\varphi)$ expected in terms of a magnetic wall model, i.e., neglecting field enhancement at the edges, are displayed in the bottom of Fig. 4.9 (dotted and solid lines). Both field components can be distinguished from the radii of maximum field, occurring at $r/R = 0.63$ and 0.86, respectively, with R the radius of the disk. As shown by the dashed line, the measured H_φ data agreed better with the idealized model when an effective $R_{\text{eff}} = 1.15\,R$ was assumed, indicating edge field enhancement [52]. The local laser-induced temperature rise as derived from a comparison between measured data and the model was found to be about 1 K. The achieved spatial resolution was estimated from the signal tails at $r > R$ to be around 2–3 mm, limited by the low modulation frequency. Finally, the systematic difference in signal heights on the two halves of the disk indicated an inhomogeneous effective microwave penetration depth. This conclusion was in accordance with the thermometric analysis of the same disk resonator (Fig. 4.8).

As this and other results [51] demonstrate the potential of laser scanning for microwave measurements of HTS films and devices with high spatial and signal resolution, it is worth investigating how to optimize the spatial resolution and the signal-to-noise ratio. According to the analysis in [52] and Sect. 2.1, the frequency variation $|S_{12}(f)|^2$ is given by

$$|S_{12}^2(f)| = \frac{\alpha Q^2}{1 + 4Q^2[(f/f_0)^2 - 1]} \tag{4.7a}$$

with $\alpha = 4/Q_1 Q_2 \ll 1$. The laser-induced signal for modulations in f_0, monitored at the inflection points $f_\pm$ of the resonance curve, follows from differentiating (4.7a):

$$\delta|S_{12}^2| = \frac{9}{2\alpha}|S_{21}(f_0)|^4 \frac{\partial f_0}{f_0}. \tag{4.7b}$$

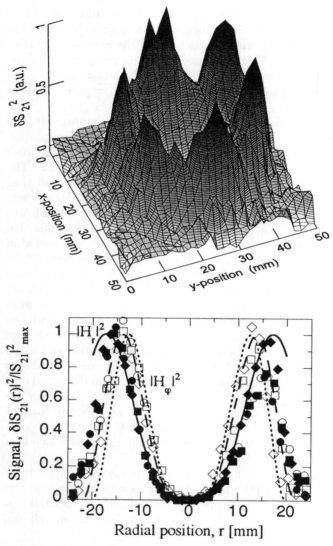

Fig. 4.9. *Top:* Measured modulation of the transmission coefficient $\delta|S_{21}|^2$ of a YBCO/LAO disk resonator operated at 40 K and 2.21 GHz in the TM$_{310}$ mode upon laser scanning across the disk side [52]. The laser was modulated at 15 Hz. *Bottom:* Measured distribution of the azimuthal ($|H_\varphi|^2$, *open*) and the radial field components ($|H_r|^2$, *filled symbols*) along the three independent axes of symmetry (*circles, squares, diamonds*). The data sets were normalized to their average maxima. Also shown are the field profiles calculated for H_φ (*dotted*) and H_r (*solid line*) using the geometric disk radius, and an enhanced radius 1.15 R (*dashed line*).

4.1 Microwave Field Dependence of the Surface Impedance   193

According to Slater's theorem [see (2.11a)], the changes of f_0 can be attributed to changes of the penetration depth $\lambda(\mathbf{r})$ induced by the laser spot, $\delta\lambda(\mathbf{r}) = (\partial\lambda(\mathbf{r})/\partial T)\delta T(\mathbf{r})$:

$$\frac{\partial f_0}{f_0} \propto -\frac{|H(\mathbf{r})|^2}{W_S}\delta A \frac{\partial \lambda(\mathbf{r})}{\partial T}\delta T(\mathbf{r}) \,. \tag{4.7c}$$

In typical experiments the effective spot size of the laser beam is determined by the heat diffusion in the film–substrate sandwich, rather than by the optical wavelength [53]. The area δA of the temperature rise is therefore determined by the thermal conductivity $\sigma_{\rm th}$ of the sample, its heat capacity per volume C_V, and by the relaxation rate $f_{\rm mod}$: $\delta A \approx \sigma_{\rm th}/(C_V f_{\rm mod})$. In order to achieve a spatial resolution of $\delta_{\rm th} = (\delta A)^{1/2} \approx 100\,\mu{\rm m}$ (respectively: $10\,\mu{\rm m}$) in YBCO on LAO at 77 K (c.f., $\tau_{\rm th}$ data in Sect. 3.2.2), the modulation frequency $f_{\rm mod} = 1/\tau_{\rm th}$ has to be adjusted around 2.5 kHz (250 kHz). Vice versa, at a given modulation frequency, the spatial resolution is reduced for, e.g., sapphire instead of LAO.

The temperature rise δT scales with the absorbed laser power density $p_{\rm las}$. Rearranging all previous terms, (4.7a–c) yield

$$\delta|S_{12}^2(\mathbf{r}, T)| \propto -\alpha Q^4(T)\frac{\sigma_{\rm th}(T)}{C_V}\frac{p_{\rm las}}{f_{\rm mod}}\frac{|H(\mathbf{r})|^2}{W_S}\frac{\partial \lambda(\mathbf{r},T)}{\partial T} \,. \tag{4.7d}$$

The detected signal is a convolution of the microwave field distribution and the local effective penetration depth. High values of Q, $\sigma_{\rm th}/C_V$, and $p_{\rm las}$ are required for a good signal-to-noise ratio. The thermally limited spatial resolution $\delta_{\rm th}$ of the technique improves as $f_{\rm mod}$ is increased. Equation (4.7d) hence implies a payoff between sensitivity and spatial resolution.

Similarly to the real and/or imaginary part of the surface impedance, laser scanning can also be applied to imaging the spatial variation of the critical microwave field $B_{\rm s,crit}$. This mode requires biasing the sample at field amplitudes $B_{\rm s,0}$ slightly below $B_{\rm s,crit}$. The laser-induced temperature rise is then expected to provoke a signal $\delta Z_{\rm s}$ proportional to $(\partial Z_{\rm s}/\partial B_{\rm s})\delta B_{\rm s}$, such that the reduction $\delta B_{\rm s,crit} = |\partial B_{\rm s,crit}/\partial T|\delta T > B_{\rm s,crit}(T_0) - B_{\rm s,0}$ exceeds the critical field of the illuminated region.

Recent applications of laser scanning to YBCO [23] and Nb$_3$Sn films [54], which formed one endplate of a dielectric resonator (Sect. 2.1.3), revealed even the possibility of imaging the spatial distribution of the surface impedance of single-sided unpatterned superconducting films.

Concluding Remarks on the Field Dependent Surface Impedance of HTS

In conclusion, the detailed understanding of the mechanisms limiting the linear microwave response of HTS films requires further studies of the spatial distribution of $R_{\rm s}(B_{\rm s})$ in relation to microstructure and stoichiometry. The achievement of high $B_{\rm s,crit}$ values relies on optimized film deposition and handling. This is especially true for patterned devices, the surface impedance of which displays a high sensitivity to the presence of defects.

The best YBCO films presently available indicate field limitations which are comparable to the temperature dependent lower critical field. At field levels below $B_{\mathrm{s,crit}}(T)$, the surface impedance is constant on a low absolute level, which is dictated by the residual fraction of unpaired charge carriers at the investigated temperature, $x_{\mathrm{n0}}(T)$ (Sect. 1.2.3). Third-order intermodulation is expected, in the absence of global heating, to follow a cubic power law up to frequencies of the order of the gap frequency. Since magnetic pair breaking [see (3.15b), Sect. 3.1] remains the only effective nonlinear mechanism in high-quality films, the ultimate third-order intercept field B_{3OI} [(3.68), Sect. 3.3] should scale with the thermodynamic critical field like $B_{\mathrm{3OI}}(T) = B_{\mathrm{c,th}}(T)\,[(4/3)x_{\mathrm{n0}}(T)]^{1/2}$. This result is more than sufficient for many high-frequency applications (Sect. 6.2), regarding the high critical temperature and the high thermodynamic critical fields of HTS.

4.1.4 Survey of Data on Nb$_3$Sn Films

In contrast to the numerous data on the microwave field dependent surface resistance of Y-123 and Tl-2212 or Tl-2223, far fewer results are currently available for $R_{\mathrm{s}}(B_{\mathrm{s}})$ of Nb$_3$Sn films on dielectric substrates [55, 56]. In order to focus the discussion on the basic field limitations, the following results refer mainly to unpatterned Nb$_3$Sn films on sapphire of different film thickness, which were prepared by tin vapor diffusion (Sects. 2.2 and 5.1.1) [56].

Temperature Dependence of $R_{\mathrm{s}}(B_{\mathrm{s}})$ of Nb$_3$Sn in Comparison with YBCO

Figure 4.10 displays typical results measured with a ⌀1″-film at 19 GHz. The temperature dependent nonlinear behavior resembles qualitatively that of the YBCO film "S145" in Fig. 4.1 at the corresponding reduced temperatures T/T_{c}. Moreover, comparable values of the critical fields are observed, exceeding 25 mT at 4.2 K and dropping rapidly at elevated temperatures. This similarity appears at first glance rather surprising when regarding the strikingly different linear microwave response of A15 and HTS (Sects. 2.2 and 2.3). The onset of nonlinear behavior occurs well below the expected B_{c1} values (Table 3.6 in Sect. 3.4.3), thus indicating mechanisms that are not intrinsic to the material. Above about $0.7\,T_{\mathrm{c}}$, a pronounced linear increase of $R_{\mathrm{s}}(B_{\mathrm{s}})$ develops, which has similarly been observed in YBCO films of medium quality. The arrows in Fig. 4.10 indicate the threshold fields B^* above which the time-domain signals, recorded during 400-μs microwave pulses, became distorted (Sect. 4.1.1 and Fig. 4.11 below). The response at fields $B < B^*$, including the saturation behavior, is thus in equilibrium with the excitation. The trend of $R_{\mathrm{s}}(B_{\mathrm{s}})$ above B^* is indicated by the dashed lines, as evaluated from the maximum transmitted power. The inset to Fig. 4.10 displays the temperature dependence of B^*, which could be described well by the power law $1 - (T/T_c)^4$ over an extended temperature range, and for numerous films. The fourth power is reminiscent of, though not unambiguously proving, the Gorter–Casimir temperature dependence of the superfluid fraction x_s (Chap. 1). The field level B^* was found to be only slightly lower than the critical field $B_{s,crit}$ as defined in (4.2).

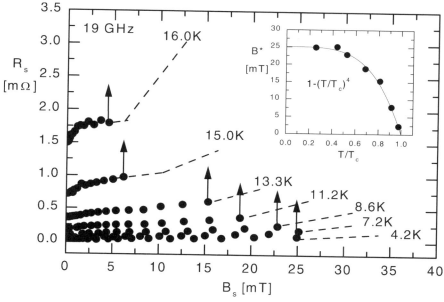

Fig. 4.10. $R_s(B_s)$ at 19 GHz of a ⌀1″ Nb$_3$Sn film on sapphire [56]. The *arrows* indicate the threshold fields B^* above which the time-domain signals became distorted (*dashed lines*, see also Figs. 4.3a and 4.11). The *inset* displays the temperature dependence $B^*(T) \propto [1 - (T/T_c)^4]$.

Figure 4.11 displays a typical set of pulse shapes monitored for a Nb$_3$Sn sample at different field strengths. Below $B^* = 17$ mT, the response remained adiabatic, while at B^* the transmitted power started to decrease slightly during the pulse. At even higher field amplitudes, an oscillatory dissipative mechanism developed with time-decaying amplitudes. The period of the oscillations decreased with increasing overcriticality as expected, but increased slightly during the pulses. The oscillatory behavior, which was rarely reported for YBCO films, indicates a fast recovery of the dissipative source, while the decaying amplitudes point to heating. Field- or power-induced switching between the superconducting and a lossy state is the most likely explanation of this phenomenon [57]. Interpreting the decay time of the order of 100 μs as an effective thermal relaxation time, the diffusion of heat appears to be

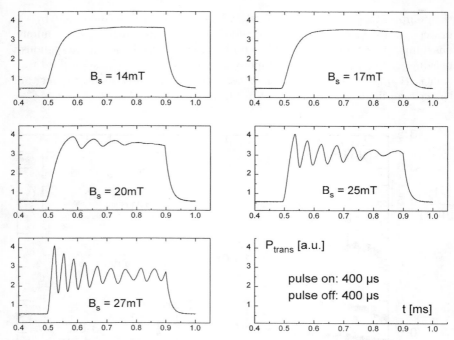

Fig. 4.11. Sequence of pulse shapes monitored at 19 GHz with a ⌀1″ Nb$_3$Sn film on sapphire at increasing field levels [56].

strongly suppressed by the low thermal conductivity of the Nb$_3$Sn film [58] and/or the thermal boundary resistance between the metallic film and the dielectric substrate. As discussed in Sect. 3.2, the heat was expected to diffuse along the 0.5-mm-thick sapphire substrate at least within some µs. The relevance of the low thermal conductivity $\sigma_{\text{th,F}}$ can be verified by estimating the local-quench field according to (3.58b) in Sect. 3.2.2. Assuming a normal-state resistance of 87 mΩ at 19 GHz (which corresponds to a resistivity of 10 µΩcm), a film thickness $d_\text{F} = 1\,\text{µm}$ and an average thermal conductivity of Nb$_3$Sn $\sigma_{\text{th,F}} = 0.5$–$1\,\text{W/Km}$ [58], and neglecting the thermal boundary resistance, yields $B_{\text{q,loc}}(1\text{d}) \approx 16$–$23\,\text{mT}$. This value is much lower than for YBCO (Table 3.3 in Sect. 3.2.2), mainly due to the lower T_c and $\sigma_{\text{th,F}}$ values. The estimated quench fields agree favorably with the measured data $B_{\text{s,crit}} = 15$–$25\,\text{mT}$ [56]. Furthermore, the dependence $B_{\text{q,loc}} \propto d_\text{F}^{-1/2}$ expected for defect-induced heating could be verified with Nb$_3$Sn films of thickness ranging between 1 and 3 µm.

In contrast, for films of thickness below about 1 µm, the onset of nonlinear response decreased with decreasing d_F values, identifying the above $B_{\text{s,crit}}$ values as a maximum of $B^*(d_\text{F})$. Since with decreasing film thickness the grain size and the critical current density were reduced (Sect. 2.2.3), the nonlinear response of the thin films could be attributed to field dependent losses at the grain boundaries.

Comparison of $R_s(B_s)$ of Nb_3Sn Films on Different Substrates

Figure 4.12 compares the field dependent surface resistance at 4.2 K of Nb_3Sn films deposited onto different substrates. The open symbols denote films prepared in the two-step procedure as briefly discussed in Sect. 2.2. The two films on sapphire (circles and diamonds) had different film thicknesses (2.1 and 1.2 µm, respectively). The triangles show data for a Nb_3Sn film which was converted into the A15 phase after a Nb layer was sputter-deposited onto a bulk Nb disk, thus imitating as closely as possible the preparation of films on sapphire substrates. The solid symbols represent data obtained with a thin ($\approx$ 1.5–2 µm) TVD film deposited onto a 3-GHz high-purity Nb accelerating cavity [59], and scaled to 19 GHz by f^2. The figure reveals the trend of decreasing nonlinearity (i.e., larger $B_{s,\mathrm{crit}}$ values and smaller slopes $\partial R_s/\partial B_s$) with decreasing film thickness for the dielectric substrates, thus confirming the relevance of microwave heating as the limiting mechanism. However, the superior performance of the film on bulk Nb points to the thermal boundary resistance as a further parameter affecting the nonlinear surface resistance of Nb_3Sn films.

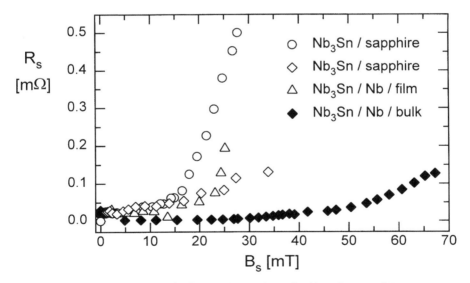

Fig. 4.12. Comparison of $R_s(B_s)$ at 19 GHz of Nb_3Sn films deposited in a two-step process onto sapphire (*open symbols*, [56]) and a TVD film on a high-purity bulk Nb accelerating cavity (*solid symbols*, [59]). The latter R_s data were scaled with f^2 from 3 to 19 GHz.

Similar conclusions were drawn from systematic studies on Nb_3Sn/Nb accelerating cavities with different thermal conductivities [60]. The field slope $\partial R_s/\partial B_s$ was found to decrease with increasing $\sigma_{th,S}$. The local character of the nonlinear dissipation was confirmed by thermometry (Sects. 2.1.4 and 4.1.3). Figure 4.13 shows a temperature map obtained with a 3-GHz 5-cell cavity. The heat flow $\dot{Q}$ is plotted versus the axial (z) and azimuthal (ϕ) position along the outer surface of the cells as indicated. At a field amplitude of 10.7 mT (Fig. 4.13a) only global heating is visible from the heat flow modulation along the z direction. However, at 10.9 mT (b) a local defect induced a hot spot. Another local defect was activated at 16.3 mT (c).

Field-induced degradation of the surface resistance was reported for Nb_3Sn, as for YBCO (see Sect. 4.1.1). However, warming the sample above T_c and cooling it down again was sufficient to recover the previous state. In comparison with the irreversible degradation of YBCO it is therefore tempting to assume that local spots could have heated the HTS films locally so much that the superconducting phase was destroyed (e.g., by the loss of oxygen). Such an assumption is in accordance with the stated absence of visible damaged regions.

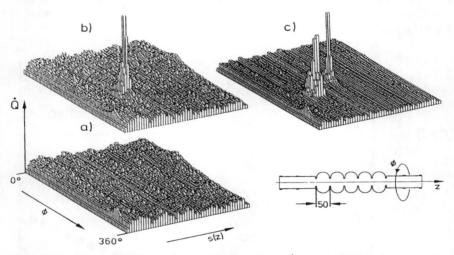

Fig. 4.13. Temperature map at 4.2 K (heat flow $\dot{Q}$ versus position) of a 3-GHz 5-cell cavity prepared from TVD Nb_3Sn on bulk Nb [60]. The thermometers were arranged as indicated. The diagrams refer to microwave field levels 10.7 mT (**a**), 10.9 mT (**b**) and 16.3 mT (**c**).

Finally it is worth noting that the maximum critical field levels reported so far for high-quality Nb_3Sn films on bulk Nb were around 100 to 120 mT [61, 62]. This range is very close to the lower critical field value expected for this material (Table 3.6 in Sect. 3.4.3).

4.2 DC Magnetic Field Dependence of the Surface Impedance

The surface impedance of the oxide superconductors in the mixed state has been investigated by many groups (see introduction to Chap. 3, and review in [63]). Since most of the microwave applications discussed later (Chap. 6) are restricted to weak external DC fields $B_{DC} < B_{c1}$, the following description of $Z_s(B_{DC})$ is focused on the "virgin state", i.e., on the microwave response upon the first increase of B_{DC}. It is nevertheless worth noting that the study of $Z_s(B_{DC})$ in the mixed state provides valuable insight into the complicated magnetic phase diagram of HTS. It is also a sensitive means of judging the presence, distribution and kind of magnetic defects in the films (Sect. 4.2.2 and [8c, 64]) and therefore presents, in principle, a valuable method of optimizing the various deposition techniques.

The results discussed in this section were obtained with the 87-GHz system described in Sect. 2.1.3 [64, 65]. An external static magnetic field up to $B_{max} = \pm 260\,\text{mT}$ was provided by a pair of superconducting NbTi Helmholtz coils. The dimensions of the coils and of the Cu cavity allowed orientation of the field either perpendicular or parallel to the film (i.e., parallel or perpendicular to the c axis of the epitaxial Y-123 films). All investigated samples were deposited on 10 mm × 10 mm substrates and were unpatterned. In order to investigate the impact of film geometry, one YBCO film was subsequently patterned into a disk of diameter ⌀8.8 mm. The cylindrical aperture of the cavity was 8 mm. The data were acquired in the TE_{013} mode, after the setup was cooled to 4.2 K in zero external field (ZFC). Supplementary to measurements of various Y-123 films, first data were recently acquired with a disk-shaped (⌀9 mm) Nb_3Sn film, prepared by tin-vapor diffusion on sapphire [56].

Section 4.2.1 contains a phenomenological description of the field dependent surface impedance $Z_s(B_{DC})$. The surface impedance in perpendicular and parallel fields is analyzed in more detail in Sects. 4.2.2 and 4.2.3. The penetration of magnetic flux into the superconducting films in perpendicular field was, in addition, investigated by magneto-optic imaging (Sect. 4.2.2).

4.2.1 Phenomenological Description of $Z_s(B_{DC})$

Typical $Z_s(B_{DC})$ Curves and Related Parameters

Figure 4.14 displays a representative set of hysteresis curves observed at 4.2 K for the surface resistance (top) and the change of the microwave penetration depth (bottom) of the electron beam coevaporated Y-123 film "EC56" ($d_F = 350\,\text{nm}$, $T_c = 90.4\,\text{K}$, [66]). The different branches of the curves were obtained by varying the magnitude and the sign of the magnetic field as indicated by the arrows and the different symbols. The scale for the magnetic flux density refers to the properties of the coils and does not account for demagnetizing effects. The surface impedance was not corrected for finite film thickness, but the general conclusions remain essentially valid also for the bulk values of Z_s since the variations of λ usually remained in the 10% regime. In order to plot the measurable changes of the penetration depth on an absolute scale, the λ data were adjusted to an offset value of 150 nm, which comes close to the value $\lambda_0 = 135$ nm deduced from the temperature variation $\lambda(T)$ (Sect. 2.3).

The small black circles represent the virgin state of the sample monitored during the first field ramping. Below a critical onset field $B_{crit,\perp}$ (the index "DC" is omitted for simplicity), which was about 20–50 mT for the YBCO samples, $Z_s(B_{DC})$ in high-quality samples stayed constant at the level $Z_{s,0}$ within measurement accuracy (see also Fig. 4.17). The $B_{crit,\perp}$ values deduced from $\Delta\lambda(B_{DC})$ were usually slightly lower than those deduced from $\Delta R_s(B_{DC})$. At $B_{DC} = B_{crit,\perp}$, surface resistance and reactance increased in proportion to the external flux density up to about 100 mT. At higher fields, the increase of Z_s became weaker. The maximum applied field $B_{max} = 260$ mT was well above the lower critical field, especially in perpendicular field orientation (as in Fig. 4.14). Demagnetizing effects lead to a strong field enhancement by about the factor $D_\perp = R/d_F$, where d_F and R are the thickness and the lateral dimension of the superconductor, respectively (Sect. 3.1.5). $D_\perp$ is typically of order 10^4 for $d_F \leq 500$ nm and $R \leq 5$ mm.

After having reached B_{max}, the field level was reduced to zero and then increased in the opposite direction up to $-B_{max}$ (open circles). The real and imaginary part of the surface impedance first decreased upon reducing the field from its maximum value, passing through a minimum at B_{min}, and then increased again as B_{DC} approached $-B_{max}$. Finally, the field magnitude was reduced from $-B_{max}$ to zero with the surface impedance displaying a minimum at $-B_{min}$ (open diamonds). The level of the surface impedance in the remanent state $Z_{s,rem}$ was higher than the starting level $Z_{s,0}$. The extremal behavior of Z_s upon field cycling is due to flux pinning (see, e.g., [1,67]). It leads to the characteristic "butterfly" shape of a complete hysteresis cycle. The area enclosed by the hysteresis loops varied for films of different quality [64]. As shown in Fig. 4.14 for $\Delta\lambda(B_{DC})$, the two branches at positive and negative field directions were for some samples subject to a shift along the ordinate, due to time dependent flux creep [64].

The DC field-induced variation of Z_s affected mainly the surface reactance: $\Delta X_s(B_{DC}) = \omega\mu_0\Delta\lambda(B_{DC})$ with $\omega\mu_0 = 0.687$ mΩ/nm at 87 GHz. A typical increase of the penetration depth was between 10 and 20 nm, corresponding to $\Delta X_s = 6\text{--}13$ mΩ. In contrast, the increase of the surface resistance $\Delta R_s \leq 2$ mΩ was much lower, with a typical average differential loss tangent $r(B_{DC}) = \Delta X_s(B_{DC})/\Delta R_s(B_{DC}) \approx 3$ (see Fig. 4.21 below).

While the data shown in Fig. 4.14 were measured in perpendicular field, the results for $Z_s(B_{DC})$ in parallel orientation were qualitatively similar (see also Figs. 4.21 and 4.24), with the absolute values of $B_{crit,DC}$ and the field

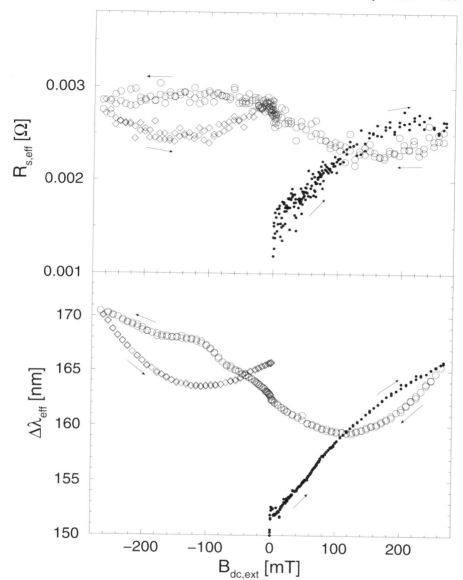

Fig. 4.14. Effective surface resistance (*top*) and change of the penetration depth (*bottom*) of the epitaxial Y-123 film "EC56" at 87 GHz and 4.2 K [64]. The *arrows* indicate the direction of the field variation, dividing the curves into three different parts (*different symbols*).

slopes $\partial Z_s/\partial B_{DC}$ being slightly lower. A detailed comparison of the two geometries is postponed until Sect. 4.2.3.

Correlations Between $Z_s(B_{DC})$ and Critical Current Density

Figure 4.15 summarizes results of $Z_s(B_{DC})$ measurements in a perpendicular field of YBCO films prepared by different techniques [64]. The data are plotted in terms of the zero-field zero-temperature penetration depth λ_0 (Sect. 2.3), the critical field $B_{crit,\perp}$ and the position B_{min} of the minimum of $\Delta\lambda(B_{DC})$, versus the critical transport current density $J_c(0.85\,T_c)$. The absolute temperatures at which J_c was evaluated (at $0.85\,T_c$) were different for the various samples due to differences in the T_c values.

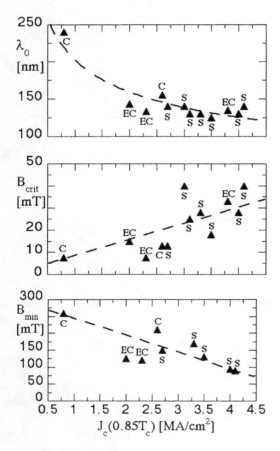

Fig. 4.15. Correlations (from top to bottom) of λ_0, $B_{crit,\perp}$, and B_{min} with the critical current density J_c at $0.85\,T_c$ [64]. The data were deduced from the $Z_s(B_{DC})$ dependences of various Y-123 films at 87 GHz and 4.2 K. The letters indicate the deposition techniques – DC sputtering: "S" [25], electron beam coevaporation: "EC" [66], metalorganic chemical vapor deposition: "C" [68]). The dashed lines indicate the average trend.

The plot illustrates that the three magnetic quantities are, on average, correlated with the critical current density. The correlation between λ_0 and J_c bears information about the pinning mechanisms in HTS. Since the penetration depth was found to be sensitive to the oxygen stoichiometry (Sect. 2.3), oxygen vacancies and/or twin domain boundaries seem to affect also the pinning potential experienced by the flux lines, and thus J_c (e.g., [69]). The correlations of $B_{\mathrm{crit},\perp}$ and $B_{\min}$ with J_c confirm that pinning phenomena are responsible for the $Z_s(B_{\mathrm{DC}})$ characteristics in perpendicular fields. $B_{\mathrm{crit},\perp}$ (respectively: $B_{\min}$) increases (decreases) with J_c at an average slope of about 7×10^{-13} Vs/A (-5×10^{-12} Vs/A). The negative sign of $B_{\min}/J_c$ indicates that the minimum of $Z_s(B_{\mathrm{DC}})$ results from flux leaving the sample as the external field is reduced below $|B_{\max}|$. The physical meaning of the slope $B_{\mathrm{crit},\perp}/J_c$ is discussed in Sect. 4.2.2 in relation to the microscopic flux penetration into the square-shaped films.

The data in Fig. 4.15 show a large scatter, which reflects the relevance of defects for the magnetic field dependence of the surface impedance. However, the role of defects in $Z_s(B_{\mathrm{DC}})$ measurements might differ from that at elevated microwave fields B_s, since the former experiment senses merely magnetic rather than thermal sources of the nonlinear microwave response.

Comparison of $Z_s(B_{\mathrm{DC}})$ of Y-123 and Nb_3Sn Films

There are presently not yet many data available on $Z_s(B_{\mathrm{DC}})$ of Nb_3Sn films on sapphire [55], and even fewer at 87 GHz. A detailed comparison between the magnetic effects in A15 superconductors and HTS must therefore await improved statistics. However, some characteristic differences can already be concluded, e.g., from the $R_s(B_{\mathrm{DC}})$ curve displayed in Fig. 4.16. The data

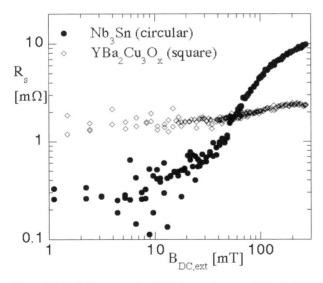

Fig. 4.16. DC magnetic field dependence of R_s at 87 GHz and 4.2 K of a 2.9-μm thick Nb_3Sn film on sapphire (*filled circles*) [56]. The field was applied perpendicular to the ⌀9-mm disk-shaped film. The virgin data of the square-shaped YBCO film EC56 (Fig. 4.14) are shown for comparison (*small diamonds*).

were obtained in perpendicular field with a disk-shaped (⌀9 mm) Nb_3Sn film prepared by tin-vapor diffusion (filled circles). First, the critical field $B_{crit,\perp}(4.2K) \leq 10\,mT$ was lower in Nb_3Sn. The initial relative increase of the surface resistance above $B \approx B_{crit,\perp}$ was more gradual than in Y-123. These features are likely to reflect mainly the geometric difference between disk- and square-shaped samples (see below). More importantly, the total variation of $\Delta R_s \approx 10\,m\Omega$ at $B_{max} = 260\,mT$ was much greater for Nb_3Sn than for the oxide films. At the same time, the penetration depth varied only slightly less than in YBCO, with $\Delta X_s \approx 6\,m\Omega$. The corresponding average loss tangent $r(B_{DC}) \approx 0.6$ was therefore about a factor of five lower than for YBCO.

4.2.2 Flux Penetration and Surface Impedance in Perpendicular Fields

In order to relate the observed field dependence of the surface impedance in perpendicular DC fields to the penetration of flux into the sample, magneto-optic images were produced for one high-quality YBCO film (EC56, Z_s data in Figs. 4.14 and 4.17) and for one oxygen-deficient YBCO film ("EC58", $T_c = 74\,K$) [8c, 70]. The following analysis develops a detailed picture of the virgin state of flux penetration.

Magneto-optic Imaging of the Flux Penetration in a High-Quality YBCO Film

Figure 4.17 displays the virgin part of Fig. 4.14 on a logarithmic field scale in the main diagrams, and on a linear scale in the insets. The arrows and letters indicate the field levels at which the flux profile was monitored with the magneto-optic Faraday effect. The results are displayed in Fig. 4.18. Light (respectively: dark) regions denote the presence (absence) of magnetic flux. Each picture covers an area of about 3 mm × 3 mm.

The low-field images in Fig. 4.18 a–d display regions near the sample edge as sketched to the right of the photographs, while the flux penetration at elevated flux densities was monitored in the center of the sample, Fig. 4.18e–h. According to Fig. 4.18a,b the flux remains pinned at the edges of the square-shaped sample up to $B_{DC} \geq 16\,mT$. The bright feature in the upper part of these diagrams indicates the enhanced penetration of magnetic flux at a defect (such as a scratch). The surface impedance stayed constant in this field range (Fig. 4.17). With B_{DC} further increasing (Fig. 4.18c,d), more flux was pressed towards the interior of the sample from the edges and from the defect. At elevated field levels ($B_{DC} \geq 39\,mT$, Fig. 4.18 e–h) the arrival of flux at the center of the sample became visible. The dark regions between adjacent flux fronts denote discontinuity lines where the shielding currents abruptly change their direction [71,72]. The observed geometrical arrangement of the

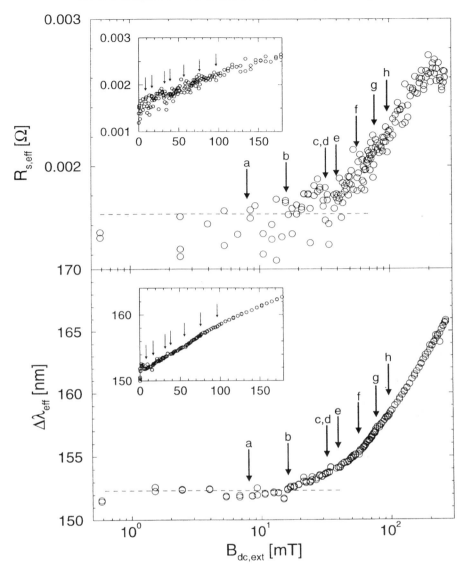

Fig. 4.17. The virgin $\Delta R_s(B_{DC})$ and $\Delta\lambda(B_{DC})$ data of sample EC56 replotted from Fig. 4.14 on logarithmic (*inset*: linear) field scales [8c,64]. The *arrows* and *letters* indicate the field levels at which magneto-optic images were taken (Fig. 4.18).

flux fronts and of the discontinuity lines is typical for square-shaped samples with only few defects. In an ideal superconducting film the discontinuity lines follow the diagonals of the square since the shielding currents flow parallel to the contour of the sample [72].

206 4. Measurements of the Surface Impedance at Nonlinear Response

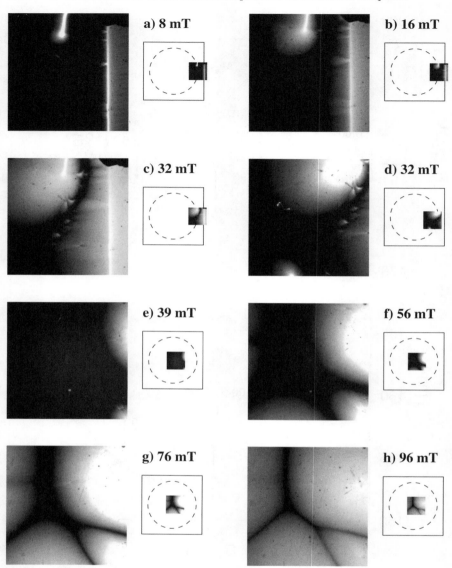

Fig. 4.18. Magneto-optic images [70] of film EC56 taken at 5 K at the field levels indicated in Fig. 4.17. The field was oriented perpendicular to the film surface. The *sketches* to the right of the photographs indicate the position of the investigated area relative to the contour of the sample (*solid square*) and to the aperture of the cavity (*dashed circle*) [64].

Comparing Figs. 4.17 and 4.18, the penetration of flux into that region of the sample which is sensed by the microwaves can be related to the onset of nonlinear microwave losses, and to the steepened increase of $\Delta\lambda(B_{\mathrm{DC}})$ around 40 mT. The increase of $\Delta\lambda(B_{\mathrm{DC}})$ at lower fields is likely to reflect the presence of Meissner shielding currents in the interior of the sample while the flux is still expelled to outer regions [73]. Between $B_{\mathrm{DC}} = 40\,\mathrm{mT}$ and $100\,\mathrm{mT}$ (Fig. 4.18 f–h), the surface impedance increased linearly with field.

Magneto-optic Imaging of the Flux Penetration in an Oxygen-Deficient YBCO Film

Figures 4.19 and 4.20 are analogous to Figs. 4.17, 4.18 but for the YBCO sample EC58. The electron-beam coevaporated film [66] was prepared under the same conditions as EC56 (Figs. 4.14, 4.17 and 4.18). The only and drastic exception was a decisive oxygen depletion that degraded the critical temperature from 90 to 74 K. In contrast to the results for EC56, Fig. 4.19 displays nonmonotonic $Z_{\mathrm{s}}(B_{\mathrm{DC}})$ behavior, with the first increase occuring already above 1 mT. The effect of magnetic field on the surface impedance became more strongly pronounced at field levels above 30 mT.

As illustrated in Fig. 4.20, the penetration of flux into this sample was very inhomogeneous. Flux was observed at localized positions already at fields below 10 mT. The steep increase of $Z_{\mathrm{s}}(B_{\mathrm{DC}})$ above 30 mT could be ascribed to the arrival of flux in the sensing area of the microwave measurement (not shown in Fig. 4.20). The geometric arrangement of the discontinuity lines was much more complicated than in the high-quality sample EC56 and reflected the superposition of flux fronts originating from the edges of the sample and from defects.

The obvious correlation between microstructure, stoichiometry and nonlinear surface impedance might bear important consequences for the microwave field dependence of Z_{s}. First, the oxygen stoichiometry of the Y-123 films as well as the related superconducting properties (energy gap, penetration depth, etc.) vary spatially (Sect. 2.3.2 and [74]). Second, the positions of large oxygen depletion tend to have lower critical fields and/or critical currents compared to fully oxygenized films. Nonlinear behavior must therefore be expected to show up at reduced field levels. Finally, the nonlinear Z_{s} behavior at low field levels, including the anomalous field dependence of $\lambda(B_{\mathrm{DC}})$ (Sects. 3.4.2 and 4.3.3), can tentatively be related to the oxygen depletion.

Perpendicular Critical Field and Depinning Frequency

The magneto-optic investigations provide a quantitative explanation of the critical field $B_{\mathrm{crit},\perp}$, that marked the penetration of flux into the microwave sensing focus. The "Bean" critical-state model [1, 75] relates the depth $d(B_{\mathrm{DC}})$, up to which a flux front can proceed in a superconductor at the field level B_{DC}, to its critical current density. Assuming a cylindrical sample and a critical current independent of field, $d(B_{\mathrm{DC}})$ is given by

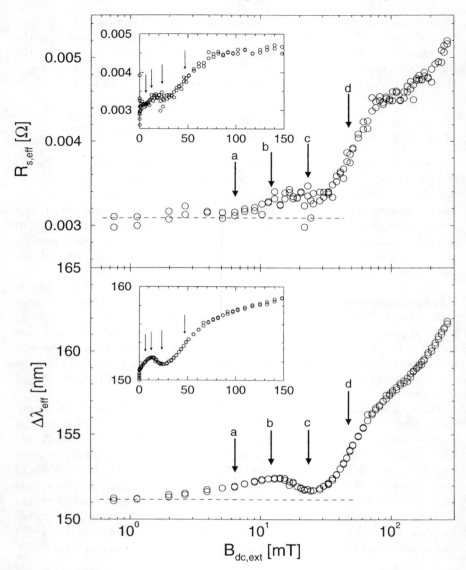

Fig. 4.19. Virgin $R_s(B_{DC})$ and $\Delta\lambda(B_{DC})$ data of the oxygen-deficient YBCO film EC58 on logarithmic (*inset*: linear) field scales [8c, 64]. *Arrows and letters* indicate the field levels at which magneto-optic images were taken (Fig. 4.20).

$$d(B_{DC}) = \frac{B_{DC}}{\mu_0 J_c} \,. \tag{4.8a}$$

As long as $d(B_{DC})$ remains smaller than the smallest difference $\Delta R = R_{\text{sample}} - R_{\text{cavity}}$ between the dimensions of the sample and of the resonator,

4.2 DC Magnetic Field Dependence of the Surface Impedance   209

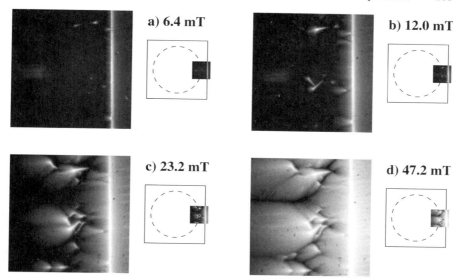

Fig. 4.20. Magneto-optic images [70] of sample EC58 taken at 5 K at the field levels indicated in Fig. 4.19. The field was oriented perpendicular to the film surface. The sketches to the right of the photographs indicate the position of the investigated area relative to the contour of the sample (*solid square*) and to the aperture of the cavity (*dashed circle*) [64].

no flux can be sensed by the microwave fields. The measurable microwave critical field is hence given by equating $d(B_{DC}) = \Delta R$, and by taking into account the demagnetizing effects in perpendicular fields:

$$B_{\text{crit},\perp} = \mu_0 \frac{\Delta R}{D_\perp} \times J_c \approx \mu_0 \frac{\Delta R}{R_{\text{sample}}} d_F \times J_c \ . \tag{4.8b}$$

Equation (4.8b) yields a linear relationship between $B_{\text{crit},\perp}$ and J_c, as was experimentally observed (Fig. 4.15). In order to extract absolute numbers for $\Delta R/D_\perp$ from the experimental $B_{\text{crit},\perp}(J_c)$ data, the $J_c(0.85\,T_c)$ values considered in Fig. 4.15 must be scaled to 4.2 K, where the magnetic measurements were performed. It has been found empirically that $J_c(4.2\,\text{K}) \approx 10\, J_c(0.85\,T_c)$ is a good approximation to published J_c data. Using the experimental result $B_{\text{crit},\perp}/J_c \approx 7 \times 10^{-13}$ Vs/A, we obtain $\Delta R \approx 0.6$ mm. This result is in fair agreement with the actual dimensions $R_{\text{sample}} \leq 5$ mm and $R_{\text{cavity}} \approx 4.1$ mm (Sect. 2.1.3), especially when regarding the potentially reduced film quality at the outermost regions of the substrate. As an important result of this analysis, the similarity between the magnitude of $B_{\text{crit},\perp}$ and the lower critical field B_{c1} (Sect. 3.4.3 and [8c]) turn out to be accidental. While B_{c1} plays an important role in parallel fields (Sect. 4.2.3 below), flux penetration in the vertical geometry is governed by pinning, and thus by the critical current density J_c.

The relevance of pinning was confirmed by comparing the onset field levels for a square-shaped sample before and after patterning it into a disk. In perpendicular field a reduced $B_{\text{crit},\perp}$ was found for the disk as expected from the reduced ΔR. Furthermore, the slopes $\partial R_{\text{s}}/\partial B_{\text{DC}}$ and $\partial X_{\text{s}}/\partial B_{\text{DC}}$ were enhanced at low field levels, in accordance with the more efficient penetration of flux due to the absence of discontinuity lines (see also Fig. 4.16 for the disk-shaped Nb$_3$Sn film).

The consistency of the above analysis can be proven further by analyzing the field level B^* at which the flux fronts reached the center of the sample. B^* was found from the magnetooptic analysis in Fig. 4.18 to be between 76 mT and 96 mT (Fig. 4.18e,f). Alternatively, B^* can be estimated from (4.8b) by equating $\Delta R = R_{\text{sample}}$ and taking $J_{\text{c}} \approx 3.8 \times 10^{11}$ A/m^2 [$J_{\text{c}}(0.85 T_{\text{c}})$ was 3.8×10^{10} A/m^2 for EC56]. Setting as before $D_{\perp} = 10^4$, we obtain $B^* = 76$ mT, which is very close to the experimental result. This agreement justifies the assumption of constant $J_{\text{c}}(B_{\text{DC}})$ and of the temperature scaling of $J_{\text{c}}(T)$.

We now turn to the analysis of the magnetic field dependence of Z_{s} above $B_{\text{crit},\perp}$. Since the microwave amplitude is small and constant in the experiments, it is straightforward to attribute the observed $Z_{\text{s}}(B_{\text{DC}})$ behavior to flux flow. According to the theoretical description in Sect. 3.1.4, the differential loss tangent $r(B_{\text{DC}})$ should then provide information on the depinning frequency of the investigated samples: $r(B_{\text{DC}}) = 2 f_{\text{dep}}/f$. Figure 4.21 shows the correlation between $X_{\text{s}}(B_{\text{DC}})$ and $R_{\text{s}}(B_{\text{DC}})$ for the virgin state at $B_{\text{DC}} \geq B_{\text{crit,DC}}$, for the DC-sputtered YBCO sample "S88" on LAO ($d_{\text{F}} = 335$ nm, $T_{\text{c}} = 91.3$ K), and for the Nb$_3$Sn sample discussed in relation to Fig. 4.16. For parallel and perpendicular fields, $r(B_{\text{DC}}) = 2.9$ and 3.2 were deduced for the YBCO sample, and $r(B_{\text{DC}}) = 0.6$ for the Nb$_3$Sn film in perpendicular field. The corresponding depinning frequencies are $f_{\text{dep}} = 120$–140 GHz for YBCO and about 26 GHz for Nb$_3$Sn. The data of sample EC56 (perpendicular field) indicated a higher value of $r(B_{\text{DC}}) \approx 7$–8, which correlated with the higher J_{c} value compared to sample S88 [$J_{\text{c}}(0.85 T_{\text{c}}) = 2.3 \times 10^6$ A/cm^2]. The range of f_{dep} for the investigated YBCO samples is slightly higher than, but still in accordance with, the results of other groups [63, 67]. The depinning frequency deduced for Nb$_3$Sn on sapphire agrees with earlier expectations from Nb$_3$Sn on Nb accelerator cavities [60b]. However, measurements at 1.5 GHz with Nb$_3$Sn films sputtered onto sapphire and patterned into meanders revealed $r(B_{\text{DC}}) \approx 5.0$ and therefore $f_{\text{dep}} \approx 4$–5 GHz [55]. Since these films were granular, the relatively low f_{dep} value might be indicative of weak pinning, since $f_{\text{dep}} = \chi/\eta$ [see (3.33c)].

Modified "Bean Model": Taking into Account the Microwave Field Distribution

Quantitative analysis of the magnetic field dependent surface impedance requires the consideration of the microwave field distribution across the superconducting sample, since Z_{s} depends on the average microscopic induction

4.2 DC Magnetic Field Dependence of the Surface Impedance   211

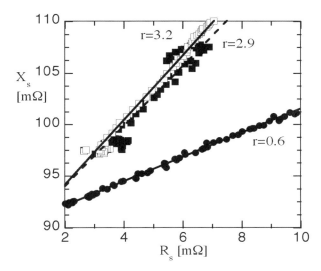

Fig. 4.21. Correlation plot to evaluate the differential loss tangent $r(B_{\mathrm{DC}})$ of the epitaxial Y-123 film "S88" on LAO in parallel (*filled*) and perpendicular field (*open squares*), and of a Nb$_3$Sn film on sapphire in perpendicular field (*circles*, see also Fig. 4.16). The X_s scale of the Nb$_3$Sn film was expanded by a factor of 2. Note the suppressed origin of the coordinate system.

$B_{\mathrm{DC}}(r)$ rather than on the externally applied field $\boldsymbol{H}(\boldsymbol{r})$. According to (3.2), the resonator can be considered to be a detector with a spatially varying "responsivity" $D(\boldsymbol{r})$. The average induction sensed by the microwaves is

$$\boldsymbol{B}(\boldsymbol{r}) = \frac{\int_{\mathrm{surface}} D(\boldsymbol{r}) \mu_0 \boldsymbol{H}(\boldsymbol{r}) \mathrm{d}^2 \boldsymbol{r}}{\int_{\mathrm{surface}} D(\boldsymbol{r}) \mathrm{d}^2 \boldsymbol{r}} \ . \tag{4.9}$$

For illustration, (4.9) is discussed within the critical-state model by assuming a step-like rotational symmetric normalized responsivity, where it is sufficient to consider the modulus $r = |\boldsymbol{r}|$:

$$D(\boldsymbol{r}) = \begin{cases} 1 & \text{if } R_- \leq r \leq R_+ \ , \\ 0 & \text{otherwise} \ , \end{cases} \tag{4.10}$$

with $R_\pm = R_0 \pm \Delta R$. R_0 is the center position of maximum sensitivity, and $2\Delta R$ is the radial width. The calculation of the numerator in (4.9) is illustrated in Fig. 4.22 for three different field levels. According to the penetration of magnetic flux from the outer contour of the sample, the radial coordinate is measured from the right (edge) to the left (center). We set $R_{\mathrm{cavity}} = R_{\mathrm{sample}} = R$ in the following for simplicity. The evaluation of (4.9) is then straight-forward. The result is:

$$B_z(r) = \begin{cases} 0 & \text{if } H_0 \leq J_c R_- , \\ \dfrac{\mu_0}{4J_c \Delta R}(H_0 - J_c R_-)^2 , & \text{if } J_c R_- \leq H_0 \leq J_c R_+ , \\ \mu_0(H_0 - J_c R_0) , & \text{if } H_0 \geq J_c R_+ , \end{cases} \quad (4.11)$$

where $H_0 = |\boldsymbol{H}(0)|$ denotes the field at the edge of the sample, and the orientation of H was assumed to be along the z direction. There is an apparent critical field $B_{\text{crit},\perp} = \mu_0 J_c R_-$, which vanishes if R_0 and ΔR are set such that $R_- = 0$. The average flux density increases quadratically with the external field above $B_{\text{crit},\perp}$, and merges into a linear dependence at $H_0 \geq J_c R_+$.

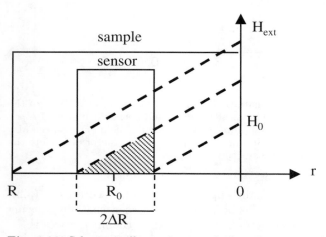

Fig. 4.22. Schematic illustration to calculate the average flux sensed by a detector with the responsivity given by (4.10). The three *dashed lines* represent the field levels at which the flux front reaches and leaves the sensitivity window, and at which it reaches the center of the sample. The *dashed area* indicates the numerator of (4.9) for one case.

While the simple example described by Fig. 4.22 and (4.11) introduces the basic features of the problem, it is not appropriate for the physical situation if the sample forms one endplate of a resonator which is excited in the TE_{013} mode. Rather, (4.9) changes into

$$B_z(r) = \mu_0 \frac{\int_0^r dr' H(r') J_1^2[\alpha(R - r')]}{\int_0^r dr' J_1^2[\alpha(R - r')]} , \quad (4.12)$$

where J_1 is the Bessel function of first kind and $\alpha = 3.8317\ldots$ is its first zero (Sect. 2.1). The radial distribution $H(r')$ in the critical-state model is given by $H(r') = H_0 - J_c r'$. The exact solution of (4.12) requires numerical computation. The result is displayed in Fig. 4.23 in the reduced units $B_z/\mu_0 J_c R$ and $H_0/J_c R$ (dots). Also shown in the figure is the result of (4.11) for comparison (solid line). The center radius $R_0/R(= 0.481\ldots)$ and the area $2\Delta R/R$

4.2 DC Magnetic Field Dependence of the Surface Impedance

($= 0.492...$) of the rectangular responsivity were adjusted to those of the Bessel function J_1^2. However, a fair description of the Bessel-type response is found only at high fields $H_0 \geq J_c R$, whereas below about $0.5 J_c R$ the linear relation $B_z/\mu_0 \approx 0.26 H_0$ yields a reasonable approximation (dashed line). Taking $J_c \approx (3–5) \times 10^{11}\,\mathrm{A/m^2}$, $R = 5\,\mathrm{mm}$ and $D_\perp = 10^4$ as before, this linear range covers external flux densities up to 200–300 mT, which are comparable to $B_{\max}$ in the experiments described before. Altogether, we arrive at the important result that the microwave fields sense an average flux density, which is proportional to the external value, as soon as the flux front has reached the edge of the resonator. Furthermore, the change of the slopes of $R_s(B_{DC})$ and $\Delta\lambda(B_{DC})$ observed in Sect. 4.2.1 turns out to be characteristic of the magnetic field induced surface impedance, rather than to be merely a geometric effect.

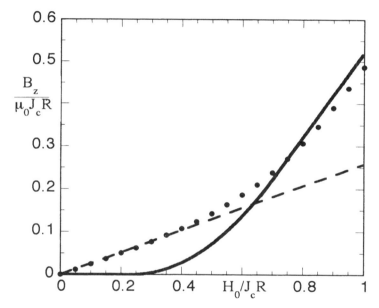

Fig. 4.23. Effective induction $B_z/\mu_0 J_c R$ versus external field $H_0/J_c R$ in the critical-state model for the square-shaped responsivity [*solid line*, (4.11)], for the Bessel-type response [*dots*, (4.12)], and its linear approximation (*dashed line*).

4.2.3 Surface Impedance in Parallel Fields

Comparison of $Z_s(B_{DC})$ for Perpendicular and Parallel Fields

The application of a magnetic field perpendicular to a superconducting film induces shielding currents parallel to the plane of the film (e.g., within the

214   4. Measurements of the Surface Impedance at Nonlinear Response

CuO$_2$ planes of HTS samples). This situation is comparable to the exposure of a sample to microwave fields, if it forms one endplate of a cavity or a dielectric resonator. However, in the DC case the lower critical field is effectively reduced to zero due to the strong demagnetization. Such an enhancement is absent in closed resonator geometries. Aside from frequency dependent effects, the microwave field dependence of the surface impedance in the Meissner state is therefore expected to correspond to that in a (hypothetical) perpendicular DC field with zero demagnetization. Such a situation ($D = 0$) can be studied experimentally with thin films only in the parallel orientation, and by taking into account the anisotropic properties of the investigated samples.

Figure 4.24 compares the change of the surface reactance of the YBCO film S88 measured in the two extreme field orientations in the virgin state (open symbols: perpendicular, filled symbols: parallel field) [76]. The data display an effective anisotropy of the critical field $B_{\mathrm{crit,DC}}$ of less than a factor of 2. In perpendicular field, demagnetizing effects and pinning were identified to be key parameters determining the $Z_{\mathrm{s}}(B_{\mathrm{DC}})$ curve. In contrast, demagnetization is absent in the parallel orientation, and the threshold for flux penetration should be related to the corresponding lower critical field.

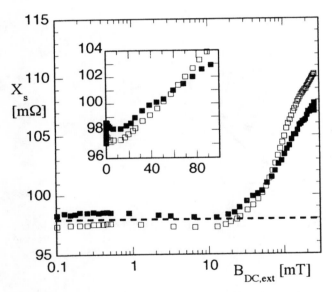

Fig. 4.24. Representative change of the surface reactance, measured with the sputtered YBCO film S88 at 87 GHz and 4.2 K in a parallel field (*filled symbols*) [76]. Shown for comparison is the corresponding result in perpendicular field (*open symbols*). The *dashed line* facilitates the estimation of $B_{\mathrm{crit,DC}}$. The *inset* magnifies the low-field regime on linear scales. Note the broken X_{s} scale.

4.2 DC Magnetic Field Dependence of the Surface Impedance

Another difference between the two curves is the slightly weaker field dependence of $\Delta\lambda(B_{\mathrm{DC}})$ in the parallel compared to the transverse case. Since the depinning frequency was comparable for both field orientations (Fig. 4.21), this feature reflects the different relationships between the external field (plotted on the abscissa) and the effective average induction (causing the increase of Z_s). Furthermore, different scaling behavior of $Z_\mathrm{s}(B_{\mathrm{DC}})$, which could follow from the largely different effective field regions, could also be responsible for this feature.

Complete Analysis of the Flux-Flow Surface Impedance

For the sake of comprehensiveness, the flux-flow impedance $Z_{\mathrm{s,ff}}$ was evaluated in Sect. 3.1.4 only for some limiting cases. Since we are now interested in the analysis of the critical parallel field $B_{\mathrm{crit},\parallel}$, the magnetic field dependence $Z_{\mathrm{s,ff}}(B_{\mathrm{DC}})$ needs to be known at arbitrary field levels. The general expression for $Z_{\mathrm{s,ff}}$ was given in (3.34):

$$Z_{\mathrm{s,ff}} = \sqrt{\mathrm{i}\omega\mu_0\left(\frac{1}{\sigma_0}+\frac{1}{\sigma_\mathrm{f}}\right)}\,, \tag{4.13}$$

where $\sigma_0 = \sigma_{01} - \mathrm{i}\sigma_{02}$ is the complex conductivity of the superconductor in the linear regime and $\sigma_\mathrm{f} = \sigma_{\mathrm{f}1} - \mathrm{i}\sigma_{\mathrm{f}2} = (1+\mathrm{i}f_{\mathrm{dep}}/f)/\rho_\mathrm{ff}$ is the complex flux-flow conductivity according to (3.33). The two associated conductivity ratios are $y_0 = \sigma_{01}/\sigma_{02}$ and $y_\mathrm{f} = \sigma_{\mathrm{f}1}/\sigma_{\mathrm{f}2}$. Introducing the new variable

$$b_{\mathrm{ff}} = \frac{\sigma_{01}}{\sigma_{\mathrm{f}1}}\frac{1+y_0^{-2}}{1+y_\mathrm{f}^{-2}} \tag{4.14}$$

enables us to express the flux-flow surface impedance in analogy to the linear case (Chap. 1):

$$Z_{\mathrm{s,ff}}(b_{\mathrm{ff}}) = R_\mathrm{c}\sqrt{(1+b_{\mathrm{ff}})\frac{1+y_{\mathrm{eff}}^{-2}}{1+y_0^{-2}}}\{\varphi_-[y_{\mathrm{eff}}(b_{\mathrm{ff}})]+\mathrm{i}\varphi_+[y_{\mathrm{eff}}(b_{\mathrm{ff}})]\}\,, \tag{4.15a}$$

with $R_\mathrm{c} = (\omega\mu_0/2\sigma_{01})^{1/2}$, $\varphi_\pm$ as defined in (1.18), and with the effective conductivity ratio

$$y_{\mathrm{eff}} = y_0\frac{1+b_{\mathrm{ff}}}{1+b_{\mathrm{ff}}\,y_0/y_\mathrm{f}}\,. \tag{4.15b}$$

Except for the root in (4.15a), the result is formally identical with the two-fluid formulation of the surface impedance at zero external field. Equation (4.15) reduces exactly to the linear case if the "flux-flow parameter" b_{ff} vanishes and y_{eff} approaches y_0. The scaling of $Z_{\mathrm{s,ff}}$ with magnetic field is thus described by b_{ff}, which also bears the desired information for the critical field.

The left part of Fig. 4.25 presents a double-logarithmic plot of the normalized flux-flow impedance $Z_{\mathrm{s,ff}}(b_{\mathrm{ff}})/R_\mathrm{c}$ using $y_0 = 0.02$ and $y_\mathrm{f} = 0.6$, which are typical for YBCO at 87 GHz (Sect. 4.2.2). Also indicated are the expected field slopes 1 and 1/2. It can be seen that $Z_{s,ff}$ stays constant up to field levels $b_{ff} \approx 0.03$. Due to the choice of parameters at 87 GHz, there is no pronounced field regime where $Z_{s,ff}/R_c \propto b_{ff}$. The expected square-root behavior at high field levels shows up at $b_{ff} \gg y_f/y_0 \approx 30$, where y_{eff} saturates at y_f [see (4.15b)]. The right-hand part of Fig. 4.25 displays the flux-flow-induced increase of $X_s(b_{ff})$ and $R_s(b_{ff})$. The loss tangent $r(b_{ff})$ (slope of the diagram) approaches the constant value $2f_{dep}/f$ only at high fields $b_{ff} \gg y_f/y_0$. At lower fields $r(b_{ff})$ is field dependent, and the evaluation $f_{dep} = r(b_{ff})f/2$ would underestimate the true depinning frequency.

We are finally interested in the evaluation of the critical field $B_{DC,\parallel}$ in terms of the b_{ff} value at the onset of nonlinear microwave response of YBCO in parallel fields. The first factor in (4.14) is found from the flux-flow analysis in Sect. 3.1.4 to be $(B-B_{c1,\parallel})/B_{c2,\parallel}$. Here, $B_{c1,\parallel}$ and $B_{c2,\parallel}$ denote the critical fields in parallel orientation (Sect. 3.1.5). The second factor in (4.14) requires knowledge of the conductivity ratios y_0 and y_f. The zero-field value y_0 can be deduced from the surface impedance $Z_{s,0}$ according to (1.43). It is $y_0 \approx 2R_{s,0}/X_{s,0} \ll 1$ in the limit $R_{s,0} \ll X_{s,0}$. Typical values for YBCO films at 87 GHz are $R_{s,0}(4.2\,\mathrm{K}) = 1\,\mathrm{m}\Omega$, $\lambda_0 = 150$ nm and thus $y_0 \approx 0.02$. Altogether, we have in terms of the reduced field $b = B_{DC}/B_{c1,\parallel}$:

$$b_{ff} \approx p_{ff,\parallel}(b-1) \tag{4.16a}$$

at $b \geq 1$, and with the prefactor $p_{ff,\parallel}$

$$p_{ff,\parallel} \approx \frac{B_{c1,\parallel}}{B_{c2,\parallel}} \frac{X_{s,0}^2}{4R_{s,0}^2} \left[1 + \left(\frac{f_{dep}}{f}\right)^2\right]^{-1}. \tag{4.16b}$$

The flux-flow ratio $y_f = f/f_{dep}$ was about 0.65, resulting in $(1+y_f^{-2})^{-1} = 0.3$. The ratio of the critical fields in (4.16b) can be evaluated in terms of the Ginzburg–Landau parameter κ (Chap. 3) and the anisotropy factor $\gamma = \lambda_c/\lambda_{ab}$:

$$\frac{B_{c1,\parallel}}{B_{c2,\parallel}} = \frac{\ln(\gamma\kappa) + 0.5}{2(\gamma\kappa)^2}. \tag{4.17}$$

Using $\gamma \approx 5$ and $\kappa = 54$, this ratio amounts to 4.2×10^{-5}, yielding a total prefactor $p_{ff,\parallel} \approx 3.2 \times 10^{-2}$. The effective critical field for nonlinear surface impedance can be related to the lower critical field by solving (4.16a) for $b = b_{crit,\parallel}$:

$$B_{crit,\parallel}(87\,\mathrm{GHz}) \approx B_{c1,\parallel}\left(1 + \frac{b_{ff,on}}{p_{ff,\parallel}}\right). \tag{4.18}$$

As judged from Fig. 4.25, the flux-flow surface impedance starts to increase markedly at $b_{ff,on} \approx 0.03$, and (4.18) yields $B_{crit,\parallel} \approx 2\,B_{c1,\parallel}$. Using $B_{c1,\perp} = 75$ mT (Table 3.6), a typical result at 87 GHz is $B_{crit,\parallel} \approx 30$ mT, in reasonable agreement with experimental data like those in Fig. 4.24.

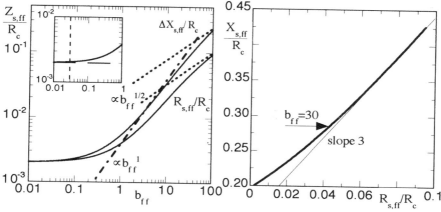

Fig. 4.25. Flux-flow surface impedance (*left*) and $X_s - R_s$ correlation (*right*) at 87 GHz, according to (4.15) with $\sigma_{01} = 1.5 \times 10^6$ $(\Omega m)^{-1}$, $y_0 = 0.02$ and $y_f = 0.65$. The *inset* is a low-field magnification of $R_{s,ff}(b_{ff})$. In the left diagram are indicated the field slopes for the limiting cases $\sigma_{01} \approx 0$ and $1/\sigma_{01} \approx 0$ (Sect. 3.1.4), in the right diagram the region of constant loss tangent $r(b_{ff})$.

Summary of $Z_s(B_{DC})$

The experimental and theoretical investigations of the flux-flow surface impedance revealed distinctly different mechanisms determining $Z_s(B_{DC})$ of epitaxial Y-123 films in perpendicular and parallel fields. Granular effects were not relevant for the interpretation of $Z_s(B_{DC})$ and of the critical fields since the investigated field levels $B_{DC} > 10$ mT were well above the characteristic fields of Josephson junctions. The microwave response was assumed to be spontaneous at the measurement frequency of 87 GHz (c.f., Sect. 3.1.1).

The data measured in perpendicular field could be explained with the critical-state model. The penetration of flux into the sample was governed by homogeneous and defect-related pinning. The low-field regime of constant $Z_s(B_{DC})$ resulted from the measurement geometry, also taking into account strong demagnetizing effects. Accordingly, much lower onset field levels, of about 2 mT, were reported for 200-μm wide microstrip lines [55] than for the 87-GHz data on 10 mm × 10 mm samples. The features of the hysteresis curves correlated with the critical current density. The differential loss tangent provided consistent information on the depinning frequency $f_{dep} \approx 130$ GHz for flux lines oriented parallel to the c axis of the YBCO unit cell.

In contrast, the field dependence of Z_s in parallel orientation could be explained by flux penetrating the films above the lower critical field $B_{c1,\parallel}$. The evaluation of the onset field was complicated by the finite quasi-particle conductivity σ_{01} and by the high depinning frequency of YBCO. A quantitative analysis required numerical analysis of $Z_{s,ff}$ as well as the consideration of sample quality and measurement frequency. The lowest possible onset field was $B_{c1,\parallel}$ [see (4.18)]. The depinning frequency for vortices aligned with the CuO_2 planes was deduced from the $Z_s(B_{DC})$ data and found comparable to the value deduced for perpendicular fields.

The apparent similarity of the $B_{crit,DC}$ values for YBCO films in both field orientations is accidental. According to (4.8b), $B_{crit,\perp}$ is of order $\mu_0 J_c \lambda \approx 100\,mT$ and independent of frequency. This value is comparable to the lower critical field $B_{c1,\perp} \approx 75\,mT$ for perpendicular fields (Table 3.6 in Sect. 3.4.3). On the other hand, $B_{crit,\parallel}$ is given by the (3–5)-fold lower value of $B_{c1,\parallel}$. However, a sample and frequency dependent enhancement factor tends to reduce this discrepancy. Samples having a higher zero-field surface resistance and/or measurements at lower frequencies would lead to higher onset fields.

Further investigations of $Z_s(B_{DC})$ in both field orientations with Nb_3Sn films, which display much less anisotropy, will be a valuable tool to confirm the flux-flow analysis. Since the pinning field strength $10\,mT \approx \mu_0 J_c \lambda \ll B_{c1} \approx 140\,mT$ is much below the lower critical field (Table 3.6), it will be worth verifying the expected inequality $B_{crit,\perp} \ll B_{crit,\parallel}$.

4.3 Identification of Magnetic and Thermal Field Limitations in YBCO Films

This section completes the analysis of the field dependent surface impedance by comparing the effects of static and microwave magnetic fields on Z_s, with the focus on the oxide superconductor Y-123. The identification of magnetically induced nonlinearities is discussed in Sect. 4.3.1. The anomalous field dependence of Z_s of YBCO films is described in Sect. 4.3.2. The identification of thermally induced nonlinearities and the resulting thermal magnetic feedback are discussed in Sect. 4.3.3. A brief summary of the nonlinear microwave response of the A15 and HTS films in Sect. 4.3.4 concludes the chapter.

4.3.1 Identification of Magnetically Induced Nonlinearities

There are basically two ways to identify whether a nonlinear microwave response is magnetic in nature. First, the critical field levels in static ($B_{crit,DC}$) and microwave fields ($B_{s,crit}$) can be compared for the appropriate field configurations. Such a comparison must take into account the characteristic time scales of the superconductor. Since these are in HTS typically of order $10^{-11}\,s$ or below, the experimentally observed critical fields should be of comparable magnitude up to about $100\,GHz$, if magnetic phenomena govern $Z_s(B)$ (the index at the variable B is omitted if both DC and microwave fields are considered at the same instant). Secondly, the influence of a superimposed DC field on the microwave field dependence $Z_s(B_s)$ can be investigated. If the nonlinear microwave response remains unaffected by application of the stationary field, the underlying mechanism can be expected to be most likely thermal in nature.

Comparison of DC and Microwave Magnetic Field Dependent Surface Impedance

Typical $R_s(B_{DC})$ data obtained with 10×10 mm^2 YBCO films at 4.2 K and 87 GHz in DC field were compared with the microwave field dependence $R_s(B_s)$ of larger ($\geq \varnothing 1''$) YBCO films at 4.2 K and 19 GHz in [76]. Both dependences were quantitatively comparable in terms of the critical magnetic field above which R_s became nonlinear. This similarity was interpreted as an indication that the linear microwave response of the best films was limited by magnetic mechanisms at this low temperature.

However, as argued in Sect. 4.2, the observed onset of nonlinear $R_s(B)$ cannot generally be identified with the lower critical field. In order to estimate the effective enhancement $B_{s,crit}/B_{c1} \geq 1$, the flux-flow analysis presented in Sect. 4.2.3 was applied to typical $R_s(B_s)$ data measured with epitaxial films at 19 GHz. The conductivity ratio $y_f = f/f_{dep}$ amounted at this frequency to $19/150 = 0.13$. Using $\lambda(4.2\,\text{K}) = 150$ nm and $R_s(4.2\,\text{K}) = 0.1$ mΩ as typical values yielded $y_0 = 8.9 \times 10^{-3}$. Since the microwave shielding currents flow within the CuO$_2$ planes of the oxides, the relevant Ginzburg–Landau critical fields are $B_{c1,\perp}$ and $B_{c2,\perp}$, with $B_{c1,\perp}/B_{c2,\perp} = (\ln \kappa + 0.5)/2\kappa^2$ and $\kappa \approx 54$. The index "$\perp$" refers, as before, to the field orientation far from the sample relative to the film surface. Taking these numbers, the reformulation of (4.18) for microwave fields at 19 GHz reads, with $p_{ff,\perp} = 0.15$:

$$B_{s,crit}(19\,\text{GHz}) \approx B_{c1,\perp}\left(1 + \frac{b_{ff,on}}{0.15}\right). \tag{4.19}$$

Numerical evaluation of the flux-flow surface impedance $Z_{s,ff}$ according to (4.15) yielded $b_{ff,on} \leq 0.05$. Flux-flow losses are therefore indeed expected to show up at field levels slightly above the lower critical field. The inevitable interaction between magnetic and thermal nonlinear loss mechanisms (Sect. 4.3.3) will bring the measurable critical field even closer to B_{c1}. This result is consistent with the interpretation of the $B_{s,crit}$ data displayed in Fig. 4.5 (Sect. 4.1.2).

The nonlinear thermal-magnetic feedback at $B_s > B_{s,crit}$ makes it difficult, if not impossible, to derive theoretical expressions for the expected field slopes $\partial R_s/\partial B$. Unless magnetic effects are the dominant source of nonlinearities (e.g., at low temperatures), the field slope is expected to increase with increasing field. This feedback mechanism is characteristic of microwave heating, and is therefore absent in DC fields.

Microwave Field Dependent Surface Impedance in Superimposed DC Fields

The investigation of the microwave field dependence $Z_s(B_s)$ in an applied DC field introduces many new features that reflect the sample quality and the measurement conditions [DC field oriented perpendicular or parallel to the films, zero-field cooled (ZFC) or field cooled (FC) measurements]. Instead of trying to discuss all possible phenomena, four examples are selected to illustrate some typical features. The discussed data result from $Z_s(B_s, B_{DC})$ measurements performed with pairs of superconducting films, arranged as the two endplates of a $\varnothing 8$ mm $\times$ 16 mm sapphire (type 3a in Fig. 2.4) [23, 30, 77]. The film–sapphire–film stack, which was resonant at 8.5 GHz, was shielded by a closed copper housing. A DC magnetic field could be applied by a pair of Helmholtz coils, either perpendicular or parallel to the $\varnothing 2''$ YBCO films. The resulting Z_s data are averages of both films, integrated over the sampling focus of 8 mm. All measurements were performed with the resonator housing immersed in liquid nitrogen, and in pulsed mode with a duty cycle below 10^{-3}. The maximum achieved microwave amplitudes $B_{s,max}$ were limited by the available power. At field levels slightly below $B_{s,max}$, an increase of the duty cycle indicated microwave heating.

Three different critical fields have to be distinguished when analyzing the $R_s(B_s, B_{DC})$ data: The application of a DC field perpendicular and parallel to the films defines the critical fields $B_{crit,\perp}$ and $B_{crit,\|}$. The onset of nonlinear losses at increasing microwave field amplitude defines $B_{s,crit}$. Typical values of the critical fields, which were estimated in terms of the analysis in Sect. 4.2, are summarized in Table 4.2. The depinning frequency f_{dep} was not known at 77 K, but scaled as $f_{dep}(4.2\,\mathrm{K})/10$ in analogy to the ratio of the critical current densities, $J_c(77\,\mathrm{K})/J_c(4.2\,\mathrm{K})$. The resulting B_{crit} values agree well with the critical fields observed in typical experiments.

Table 4.2. Definition and estimated values of the critical DC fields in perpendicular and parallel orientation, and of the critical microwave field. Remarks:
a. The value was deduced from (4.8b) with $\Delta R = R_{sample} - R_{sapphire} = (25.4-8.0)$ mΩ, $d_F = 300$ nm and $J_c(77K) = 3 \times 10^6$ A/cm^2.
b. The flux-flow prefactor $p_{ff,\|}$ in (4.16b) is for $f = 8.5$ GHz, $f_{dep} = 15$ GHz, λ (77 K) = 300 nm and $R_{s,0} = 0.4$ mΩ. The critical field $B_{c1,\|}$ was scaled to 77 K assuming a $1 - (T/T_c)^2$ dependence and $T_c = 90$ K, resulting in $B_{c1,\|}$ (77 K) = 4 mT. The onset value of the flux-flow nonlinearity was $b_{ff,on} \approx 0.03$.
c. Analogously to b, but with the critical fields $B_{c1,\perp}$ and $B_{c2,\perp}$. The onset value of the flux-flow nonlinearity was $b_{ff,on} \approx 0.02$. The critical field $B_{c1,\perp}$ was scaled to 77 K assuming a $1 - (T/T_c)^2$ dependence and $T_c = 90$ K, resulting in $B_{c1,\perp}$ (77 K) = 18 mT.

Variable	Definition	Expected value (mT)
$B_{crit,\perp}$	Onset of nonlinear $Z_s(B_{DC})$ in DC fields perpendicular to the films	8[a]
$B_{crit,\|}$	Onset of nonlinear $Z_s(B_{DC})$ in DC fields parallel to the films	≈ 10[b]
$B_{s,crit}$	Onset of nonlinear $Z_s(B_s)$ at elevated microwave field amplitudes	≈ 18[c]

4.3 Magnetic and Thermal Field Limitations in YBCO Films

As shown below, three field regions could be distinguished from the $R_\mathrm{s}(B_\mathrm{s}, B_\mathrm{DC})$ curves. Regions I and II refer to low and intermediate microwave amplitudes, respectively. Region III denotes the region of steep $R_\mathrm{s}(B_\mathrm{s})$ increase. Table 4.3 summarizes the data on the characteristic field levels and on the low-field values $R_\mathrm{s}(B_\mathrm{s} \approx 0, B_\mathrm{DC} = 0)$ of the investigated pairs of films. The samples exhibited distinctly different quality as can be judged from both the low-field R_s values and the field ranges. The relation between physical parameters and deposition methods (see footnotes for Table 4.3) are accidental. Optimized parameters usually allow for all of these techniques to yield YBCO films of similarly high quality.

Table 4.3. Low-field $R_\mathrm{s}(B_\mathrm{s} \approx 0, B_\mathrm{DC} = 0)$ data at 77 K and 8.5 GHz of various pairs of ⌀2″ YBCO films on LAO. The nomenclature of the films is the same as in Sects. 4.1, 4.2, and [77]. The three field ranges discussed in the text are limited by the two microwave crossover fields listed in the third and fourth columns. Remarks:
a. The films were deposited by pulsed laser ablation ("L") [24]. The numbers represent internal identification codes.
b. The initial "S" stands for high-pressure DC sputtering [25].
c. The data for R_s and the magnetic fields were the same for perpendicular and parallel DC fields.
d. The films were prepared by thermal coevaporation ("TC") [78].

Pairs of YBCO films	$R_\mathrm{s}(B_\mathrm{s} \approx 0, B_\mathrm{DC} = 0)(\mu\Omega)$	transition I/II (mT)	transition II/III (mT)
L48/L49 [a]	440	0.2	2.5
L49/S178 [b,c]	430	0.4	3.4
TC74/TC80 [d]	370	1.0	5.5
S232/S306 [b]	200	none	10.5

Example 1: Granular behavior in perpendicular field. Figure 4.26 displays the microwave field dependence $R_\mathrm{s}(B_\mathrm{s})$ of the pair L48/L49 at different levels of the zero-field cooled DC field (ZFC) in perpendicular orientation. The three regions I, II and III can be attributed to the different field slopes $\partial R_\mathrm{s}/\partial B_\mathrm{s}$ (Table 4.3). The right part of Fig. 4.26 displays the same data after subtracting the curve obtained in zero DC field. The data scatter in regions I and II characterizes the measurement accuracy.

The $R_\mathrm{s}(B_\mathrm{s})$ curves measured at $B_\mathrm{DC} = 0$, 3.5 and 7 mT are almost indistinguishable, while the curves corresponding to $B_\mathrm{DC} = 10.5$ and 14 mT are shifted entirely along the ordinate. The negligible effect of B_DC below 7 mT is due to the geometric arrangement of the small sapphire at the center of the large film. Despite the large demagnetization factor $D_\perp \approx 8.5 \times 10^4$, the magnetic flux stayed outside the microwave sensing focus in this field range (Table 4.2). The agreement between the zero-field and the two low-field curves hence illustrates the measurement accuracy. At superimposed DC field

$B_{DC} > B_{crit,\perp} \approx 7.5\,\mathrm{mT}$, additional losses occurred at a slope $7.5\,\mu\Omega/\mathrm{mT}$, which was almost independent of the microwave field amplitude.

The finite slopes of $R_s(B_s)$ at low and intermediate field levels ($B_s < 2.5\,\mathrm{mT}$, regions I and II) were attributed to flux-flow behavior in a network of Josephson junctions (Sect. 3.1). The transition between the regions II and III was unaffected by the DC field. The slope $\partial R_s/\partial B_s(B_{DC} = 0) \approx 230\,\mu\Omega/\mathrm{mT}$ at microwave amplitudes above $2.5\,\mathrm{mT}$ (region III) increased by only about 20% at $B_{DC} > 0$ and remained constant at higher DC fields. Such a behavior hints at a thermal cause for the nonlinear $R_s(B_s)$ at the highest applied microwave field levels.

Similar microwave field dependences of R_s were also reported from other groups (e.g., [79]), and especially for Tl-2223 films, the granular nature of which is well known.

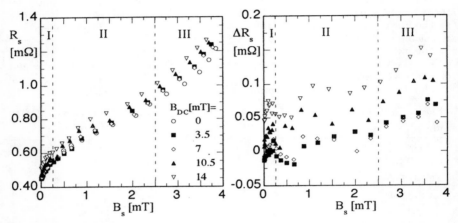

Fig. 4.26. *Left*: Microwave field dependence of R_s at 77 K and 8.5 GHz of the pair L48/L49 of laser-ablated YBCO films on LAO in perpendicular DC fields. *Right*: Microwave field dependence of the DC field-induced increase $\Delta R_s = R_s(B_{DC}) - R_s(0)$. The *dashed vertical lines* indicate B_s regions of distinguishable slopes $\partial R_s(B_s, B_{DC})/\partial B_s$.

Example 2: Granular behavior and anomalous field effect in perpendicular and parallel fields. Figure 4.27 shows the results on $R_s(B_s, B_{DC})$ for the pair L49/S178 at different DC field levels. The external field was applied perpendicular (left part) and parallel (right part) to the surface of the films. There are again three different regions visible as in the previous example, starting at comparable low-field values of the surface resistance. In contrast to the data in Fig. 4.26, $R_s(B_s)$ remained constant in region I ($B_s \leq 0.4\,\mathrm{mT}$), and was only weakly field dependent in region II. The high-field region (III, $B_s \geq 3.4\,\mathrm{mT}$) could clearly by identified from the onset of microwave field-induced losses, independent of the orientation of the DC field.

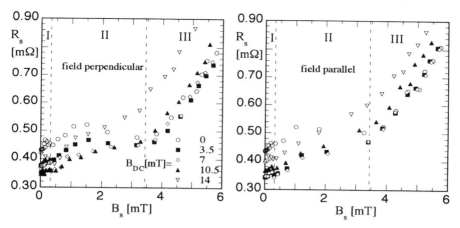

Fig. 4.27. Microwave field dependence of R_s at 77 K and 8.5 GHz of the pair L49/S178 (L = laser-ablated, S = sputtered) in perpendicular (*left*) and parallel DC fields (*right*).

In perpendicular field, $R_s(B_s)$ remained unaffected in the high-field region III up to $B_{DC} \approx 10$ mT. This value corresponds to an average critical current density of $J_c(77\,\text{K}) \approx 4 \times 10^6$ A/cm^2, which agreed with the results of transport measurements. In parallel field, $R_s(B_s)$ increased already at DC field levels between 7 and 10.5 mT. Both critical field values agree fairly well with the lower critical fields estimated in Table 4.2. The DC field-induced losses were independent of the microwave amplitude, with the average slope $\partial R_s/\partial B_s$ staying at 175 µΩ/mT. The crossover from region III to region II occurred around 3.4 mT, independent of the strength and the orientation of the DC field. The nonlinear $R_s(B_s)$ at $B_s \geq 4$ mT was therefore again attributed to thermal effects. Surprisingly, in regions II and I, $R_s(B_s)$ decreased upon increasing B_{DC}, even at the lowest value $B_{DC} = 3.5$ mT. The same behavior was also observed in parallel fields. Further analysis of this anomalous field dependence is postponed until Sect. 4.3.2.

Example 3: High critical current density in perpendicular fields, ZFC and FC.
Figure 4.28 displays results on a pair of thermally coevaporated YBCO films (TC74/ TC80) at two values of the external perpendicular field, after zero-field cooling (filled symbols) and after cooling in ambient field (open symbols). The general shape of the $R_s(B_s)$ curves in the ZFC state are comparable to those in Fig. 4.27. The critical current density could be concluded from the absence of a DC field effect up to 14 mT to be higher than 5.3×10^6 A/cm^2. Such a high value indicates a high film quality and the absence of weak Josephson junctions. The slope $\partial R_s/\partial B_s$ was not affected by the external field. At field amplitudes $B_s = 5$ mT it amounted to about 85 µΩ/mT. It was thus about a factor two lower than in the previous examples. Cooling the films below the critical temperature in an ambient field of $B_{DC} = 7$ mT caused

an enhanced, though constant, R_s value at low and intermediate microwave field levels, while the high-field behavior of $R_s(B_s)$ remained unchanged. This effect was attributed to switching local film areas of reduced lower critical field into their normal state. Cooling in $B_{DC} = 14\,\text{mT}$ introduced an additional surface resistance at a rate $12\,\mu\Omega/\text{mT}$ over the whole investigated range of B_s values.

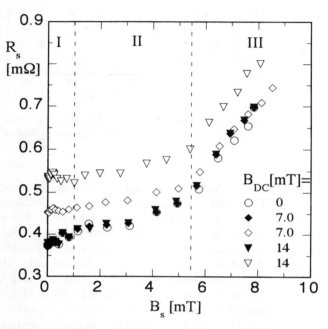

Fig. 4.28. Microwave field dependence of R_s at 77 K and 8.5 GHz of the pair TC74/TC80 (TC = thermal coevaporation) in perpendicular DC fields. *Filled* and *open symbols* refer to ZFC and FC measurements, respectively.

Example 4: DC Field dependence of $R_s(B_s)$ of high-quality films. Figure 4.29 displays the microwave field dependence of two high-quality sputtered YBCO films. The critical current density, judged from the penetration of the perpendicular DC field into the microwave sensing focus, was $(4\text{--}5) \times 10^6\,\text{A/cm}^2$. $R_s(B_s)$ remained constant in zero DC field at the very low absolute level of $0.2\,\text{m}\Omega$ up to $B_s = 10\,\text{mT}$. The slope $\partial R_s/\partial B_s$ at higher fields was only about $40\,\mu\Omega/\text{mT}$, and thus much lower than in the high-J_c samples TC74/TC80. Application of the strongest DC field $B_{DC} = 14\,\text{mT}$ still left $R_s(B_s)$ unchanged at low and intermediate amplitudes, but caused the nonlinear losses to occur at the reduced field level $B_s \approx 8\,\text{mT}$. Such a behavior indicates magnetically induced nonlinearities, especially if the close proximity of the estimated lower critical field value is taken into account (Table 4.2).

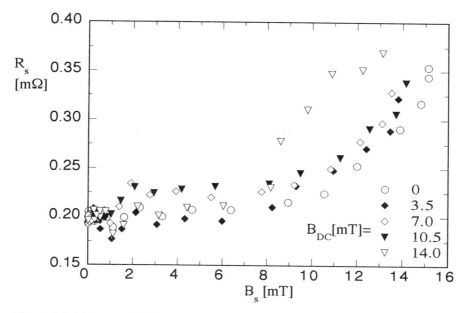

Fig. 4.29. Microwave field dependence of R_s at 77 K and 8.5 GHz of the sputtered YBCO film pair S232/S306 in perpendicular DC fields (ZFC).

In summary, defects and weak links affect the microwave field dependence of the surface resistance at 77 K typically at field levels below a few mT. They also introduce a measurable modification of $R_s(B_s)$ with applied DC fields, which is absent in high-quality films. The level at which (respectively: the steepness with which) $R_s(B_s)$ increases at elevated microwave field amplitudes increases (decreases) with increasing sample quality, i.e., shows up as an increasingly pronounced flatness of R_s at low and intermediate B_s levels. The onset of nonlinear surface resistance appears to be thermal in nature for samples of moderate quality. In contrast, it seems predominantly magnetic in nature for high-quality films with no indications of granular effects. There was no obvious correlation between the slope $\partial R_s/\partial B_s$ at high field amplitudes and the critical current density.

4.3.2 Anomalous Field Effect in YBa$_2$Cu$_3$O$_{7-x}$ Films

Figure 4.27 revealed an anomalous dependence of the surface impedance of epitaxial YBCO films on magnetic field. Similar reductions of the surface resistance and of the surface reactance were observed by various groups at increasing static as well as microwave magnetic fields (Sect. 3.4.2 and references there). Inductive and microwave measurements of patterned and unpatterned films at low (4.2 K) and high temperatures (77 K), low ($\leq$ MHz) and high frequencies ($\leq$ 87 GHz) revealed qualitatively similar effects. A recovery of the surface impedance was observed for granular YBCO samples as well as for high-quality films. The following paragraphs present a brief summary and a tentative discussion of typical results on unpatterned films [77].

DC Field-Induced Recovery of the Surface Impedance

Anomalous field effects in $Z_s(B_{DC})$ were observed in DC magnetic field for various 10 mm × 10 mm YBCO films at 87 GHz. The measurement system was described in Chap. 2, and further in Sect. 4.2. Similar field effects were also observed with pairs of ⌀2″ YBCO films using the sapphire resonator at 8.5 GHz (Sect. 4.3.1). Typical results of the DC field dependences of R_s and $\Delta\lambda$ obtained in this system at 77 K are displayed in Fig. 4.30. The recovery effect occurred already at low field levels, $B_{DC} \leq B_{crit,\perp} \approx 10$ mT, when magnetic flux was still absent from the microwave sensing area, probably except for flux nucleated at local defects (c.f., Figs. 4.18 and 4.20). However, application of the DC field induced shielding currents in the film that are of the order of the critical current. In turn, the shielding currents induce a magnetic field forcing the external field lines to align parallel to the surface of the films. The relative orientations between the magnetic field lines, the shielding currents and the HTS films are therefore expected to be comparable for the DC and the microwave fields.

Figure 4.30a displays $\Delta\lambda(B_{DC})$ obtained with various pairs of films, including the samples discussed in Sect. 4.3.1 (Table 4.3). All investigated samples showed a decrease of the penetration depth with increasing DC field, arriving at or passing through a minimum at intermediate levels ($B_{DC} \approx 10$ mT). Assuming absolute values of $\lambda(77\,\text{K}) \approx 300$ nm, the fractional reduction of the penetration depth reached up to $\Delta\lambda/\lambda \approx -3\%$. Figure 4.30b shows data on the magnetic field dependent surface resistance for the same samples. From both diagrams, correlations of the anomalous field dependence can be concluded for the changes ΔR_s and ΔX_s, while ΔX_s did not correlate with the absolute levels of the surface resistance. With increasingly pronounced effect in $\lambda(B_{DC})$, the field slope of $R_s(B_{DC})$ decreased from small positive values, and eventually changed sign. It was, in detail: $\partial\lambda/\partial B_{DC} \approx -0.4$ nm/mT and $\partial R_s/\partial B_{DC} = 0$ at low fields for L48/L49, L48/L50 and L49/L50, $\partial\lambda/\partial B_{DC} \approx -0.7$ nm/mT and $R_s \approx$ constant up to 15 mT for T74/T80, and $\partial\lambda/\partial B_{DC} \approx -1.2$ nm/mT with decreasing $R_s(B_{DC})$ for L49/S178.

Similar results were observed in $Z_s(B_{DC})$ of four small single YBCO films at 87 GHz and 4.2 K. Regarding typical values $\lambda \leq 200$ nm at $B_{DC} = 0$ and $T = 4.2$ K, the maximum fractional change of the penetration depth was close to -2%. Comparative analysis of the microstructure of the different samples revealed a correlation between the occurrence and the strength of the anomalous field effect and the relative amount of a axis grains [64].

4.3 Magnetic and Thermal Field Limitations in YBCO Films 227

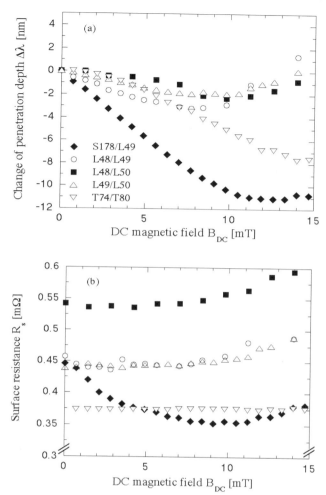

Fig. 4.30. Summary of $\Delta\lambda(B_{DC})$ (**a**) and $R_s(B_{DC})$ (**b**) data obtained at 8.5 GHz and 77 K for different pairs of $\varnothing 2''$ YBCO films on LAO [77] (see Table 4.3 for further explanation). Note the broken R_s scale in Fig. 4.30b.

Microwave Field-Induced Recovery of the Surface Resistance

A reduction of the surface resistance at elevated microwave field amplitudes could be observed for some pairs of $\varnothing 2''$ YBCO films, investigated with the 8.5-GHz system. The effect could be confirmed by $R_s(B_s)$ measurements at 19 GHz and 77 K on the individual films [77]. Due to the pulsed mode, these experiments were not sensitive enough to monitor the changes of the surface reactance (Sect. 2.1.3). Figure 4.31 presents recent results on anomalous $R_s(B_s)$ curves observed at 19 GHz and different temperatures with the two

DC-sputtered films S178 (at $T = 77\,\text{K}$) and S373 (at $T = 4.2$, 30, and 50 K). It is important to note that R_s of the sample S373 was only weakly dependent on B_s up to very high levels $B_s > 40\,\text{mT}$ at 4.2 K and still up to $B_s \geq 10\,\text{mT}$ at 50 K. Similar to the DC case, the surface resistance decreased at elevated microwave field amplitudes, namely between 1 and 10 mT, with field slopes between about -10 and $-100\,\mu\Omega/\text{mT}$. However, the relevant field scales turned out to be about a factor of 2 smaller than in the DC case (Fig. 4.30b). The recovery of $R_s(B_s)$ of film S373 decreased with increasing temperature and vanished above 50 K.

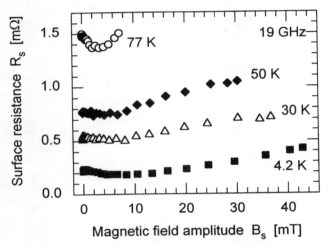

Fig. 4.31. Anomalous microwave field dependences $R_s(B_s)$ at 19 GHz for the two DC-sputtered films S178 ($T = 77\,\text{K}$, *circles*) and S373 ($T = 4.2\,\text{K}$ (*squares*), 30 K (*triangles*) and 50 K (*diamonds*)) [23].

Possible Interpretations of the Anomalous Field Effect

The extremal behavior of $\Delta\lambda(B)$ [and correspondingly of $R_s(B)$] is caused by the competition of counteracting effects. Usually expected is an increase of $Z_s(B)$, which might be the result of switching weakly superconducting regions into a lossy (e.g., the normal) state. Due to the weak coupling, this effect occurs mainly at or below mT field amplitudes. At much higher levels ($B > 10\,\text{mT}$), penetration and motion of magnetic flux (Table 4.2) and/or microwave heating (Sect. 3.2) lead to positive field slopes of Z_s (e.g., [80,81]). In contrast to the surface impedance increasing with field, the recovery of $Z_s(B)$ is an unconventional feature. The consistent observation of negative field slopes, similar in value for DC and microwave fields, indicates relaxation times of the underlying mechanism below 50 ps.

Different mechanisms were proposed to explain the recovery of microwave superconductivity (Sect. 3.4.2). A negative field slope of the penetration depth and of the surface resistance were explained in [22a, 82] by properly chosen networks of grain–boundary Josephson junctions, leading to frequency dependent current redistribution around the dissipative circuit elements. This mechanism might apply to low-quality samples, and especially to the weak coupling at the boundaries between a-axis and c-axis grains. Nevertheless, negative field slopes also occurred in films where granularity played only a minor role, if any. It is therefore questionable whether this approach can explain the coexistence of the $Z_s(B)$ recovery and of excellent microwave field handling (e.g., Fig. 4.31).

Another possible explanation of field-enhanced superconductivity was nonequilibrium quasi-particle relaxation [83]. A relation of the $Z_s(B)$ recovery in HTS films to this mechanism was proposed in [84]. However, regarding the high values of the characteristic frequencies in HTS, which are of the order of ps (Sect. 3.1), it seems unlikely that nonequilibrium effects could show up already in the low-GHz regime, at least at low temperatures.

The observed negative slopes $\partial Z_s / \partial B < 0$ were attributed in [77] to current- or magnetic field-induced ordering of free spins present in the YBCO films. Ordering of the impurity spins was found to frustrate spin-flip transitions, and thus to suppress the probability of magnetic pair breaking ([85] and Sect. 3.4.2). The presence of a-axis grains could enhance the recovery effect in two ways. First, a-axis grains contain open-ended CuO chains which are known to be susceptible to oxygen loss. Oxygen vacancies create unsaturated Cu spins and thus increase the number of pair breaking centers. This results in enhanced low-field values of R_s but, at the same time, in a more strongly pronounced recovery effect. Such a correlation is in accordance with the 87-GHz data (Sect. 4.2.2 and below). Second, the alignment of impurity spins by an external field in a-axis grains implies, in contrast to c-axis grains, a rotation from the free orientation perpendicular to the CuO_2 planes to the forced parallel orientation. This process might enforce the frustration of spin-flip transitions. A detailed theoretical treatment of the recovery effect is in progress.

Analysis of the Pair Breaking Mechanism Within the Two-Fluid Model

The suppression of pair breaking due to magnetic alignment of impurity spins could lead to a recovery of superconductivity by reducing the number density x_n of normal carriers by Δx_n and, at a correlated rate, by increasing the number density x_s of Cooper pairs: $x_n + x_s = $ constant. Whether such a correlated behavior is compatible with the previously described results can be checked by analyzing the surface impedance in terms of the two-fluid model ("TFM", Sect. 1.2.3). It is also noted that the constancy of $x_n + x_s$ is specific to the pair breaking mechanism, and is not necessarily valid in the Josephson network model nor in the nonequilibrium approach. The TFM-analysis of the anomalous field effect was presented in [77] in terms of the DC field-induced variation of the conductivity ratio $y = \sigma_1/\sigma_2 \propto x_\mathrm{n}/(1 - x_\mathrm{n})$. The quasi-particle conductivity σ_1 was considered to be constant, with the variation of y mainly reflecting that of σ_2. Such an approach is justified because $|\partial\sigma_1/\sigma_1| \ll |\partial\sigma_2/\sigma_2|$ holds if the superconductor contains a sizeable quasi-particle reservoir $x_{\mathrm{n},0}$ with only a small fraction $x_\mathrm{pb}/x_{\mathrm{n},0} \ll 1$ ("pb" = pair breaking) being affected by the field effect. Based on the same assumption, the total surface impedance can be written as the linear superposition of a field independent term $Z_{\mathrm{s},0}(T,f)$ and a small addition $Z_{\mathrm{s,pb}}(B_\mathrm{DC})$ which is susceptible to the pair breaking mechanism. This contribution can be evaluated in terms of $y(B_\mathrm{DC})$, in accordance with (1.18) in Sect. 1.1.2:

$$Z_{\mathrm{s,pb}} = R_{\mathrm{s,pb}} + \mathrm{i}X_{\mathrm{s,pb}} = R_\mathrm{c}\{\varphi_-[y(B_\mathrm{DC})] + \mathrm{i}\varphi_+[y(B_\mathrm{DC})]\} \quad (4.20)$$

with the prefactor $R_\mathrm{c} = (\omega\mu_0/2\sigma_1)$. The dependences of $\varphi_\pm$ on y are displayed in Fig. 1.5. The measured $Z_\mathrm{s}(B_\mathrm{DC})$ data were adapted to the two-fluid model by comparing the experimental slopes $\Delta X_\mathrm{s}(B_\mathrm{DC})/\Delta R_\mathrm{s}(B_\mathrm{DC})$ with the theoretical value $r_\mathrm{pb}(B_\mathrm{DC})$ resulting from (4.20):

$$r_\mathrm{pb}[y(B_\mathrm{DC})] = \frac{\sqrt{1+y^2} + (1-y^2)}{y(2+\sqrt{1+y^2})}. \quad (4.21)$$

Equating the measured $r(B_\mathrm{DC})$ values at $B_\mathrm{DC} = 0$, and at the field B_min where Z_s reached its minimum value, with the numerical expression (4.21) determined the values of $y(0)$ and $y(B_\mathrm{min})$. The total measured impedance changes, $\Delta Z_\mathrm{s} = Z_\mathrm{s}(B_\mathrm{min}) - Z_\mathrm{s}(0)$, yielded R_c by use of (4.20). The consistency of the analysis was confirmed by an unambiguous determination of R_c, which agreed to within $\pm 10\%$ for the two independent data sets ΔX_s and ΔR_s. R_c was also found to quantify the recovery effect: larger Z_s reductions correlated with larger R_c values. This behavior is in accordance with the previous arguments about the role of a-axis grains. Figure 4.32 displays the $Z_\mathrm{s,pb}$ results in terms of the measured ΔR_s and ΔX_s data, normalized to the average of the two R_c values. The different symbols denote $Z_\mathrm{s}(B_\mathrm{DC})$ data obtained at 87 GHz (circles, squares, diamonds and triangles) and at 8.5 GHz (crosses) with different YBCO films. All $Z_\mathrm{s}(B_\mathrm{DC})$ curves merged onto the TFM scaling curve [(4.20), dotted line]. The position of the data sets along the curve was fixed by the y values. Depending on the film quality, application of the DC field covered different ranges of y. The fractional change $[y(B_\mathrm{min}) - y(0)]/y(0)$ amounted to -38% to -98%. The good agreement between the experimental data and the TFM analysis confirmed the assumption that only a few charge carriers contributed to the recovery effect. It is further tempting to speculate that the large residual values $R_{\mathrm{s},0}$ and $X_{\mathrm{s},0}$ typical for the oxide superconductors could reflect a sizeable pair breaking effect due to intrinsic scatterers. This assumption fits well into the framework of the two-band model (Sects. 1.3 and 2.3).

However, recent results on anomalous $Z_\mathrm{s}(B)$ revealed complicated behavior [86], leaving a general understanding still open to discussion.

4.3 Magnetic and Thermal Field Limitations in YBCO Films   231

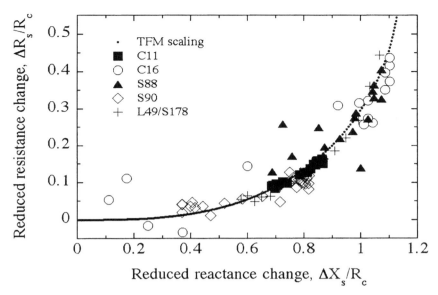

Fig. 4.32. TFM analysis of $Z_s(B_{DC})$ data of different samples (*circles, squares, diamonds* and *triangles*) that displayed a recovery effect at 87 GHz and at 8.5 GHz (*crosses*) [77]. The *dotted line* denotes the scaling of the surface impedance according to (4.20).

4.3.3 Identification of Thermally Induced Nonlinearities

The relevance of microwave heating for the nonlinear surface impedance became obvious from the $Z_s(B_s)$ results of YBCO (Sect. 4.1.1) and Nb$_3$Sn films (Sect. 4.1.4). Supplementary to this qualitative statement, a quantitative comparison between measured $Z_s(B_s)$ data and the corresponding numerical results (Sect. 3.2) might provide additional insight into the nonlinear mechanisms. Furthermore, the impact of operating conditions, like the measurement mode and the cryogenic arrangement, on $Z_s(B_s)$ at elevated field amplitudes is of crucial importance for applications of HTS films in microwave devices.

The following paragraphs describe these two aspects in terms of data obtained with Y-123 films on LAO substrates. The same arguments can be applied also to the consideration of CbS substrates. However, it has to be kept in mind that the mismatch of structural and thermal parameters between substrate and HTS renders the preparation of high-quality YBCO films on sapphire much more challenging than on LAO (Chap. 5). Frequent observations of structural defects in films on CbS have thus been complicating the analysis of the nonlinear surface impedance.

Regarding Nb$_3$Sn films on sapphire, indications for microwave heating came from the observed time constants of the response pulses, which were well above those expected for the high thermal conductivity substrate (Sect. 4.1.4). Due to the lower values of critical temperature and thermal conductivity of Nb$_3$Sn compared to YBCO, local defects are therefore expected to be much more deleterious in reducing the local-quench fields in the A15 compound. It is presently not clear in how far the finite thermal boundary resistance at the interface between superconductor and substrate enhances this problem. While these aspects affect mainly the field scales at which nonlinear effects show up, the comparison of the measured and simulated values of the field dependent surface resistance would be analogous to the discussion for YBCO.

Comparison Between Measured and Simulated $Z_s(B_s)$ Curves in Pulsed Operation

Figure 4.33 compares simlulated $R_s(B_s)$ curves with data at 19 GHz measured with the sputtered YBCO film S145 ($d_F = 220$ nm, $T_c = 89.8$ K) and the laser-ablated sample L49 ($d_F = 260$ nm, $T_c = 90.1$ K; see also Sects. 4.1.1 and 4.3.2) [23, 30]. The low-field temperature dependence of R_s, which entered the numerical calculations as input data, are displayed in the insets to the diagrams. Figure 4.1 provided a direct comparison of $R_s(T)$ and $R_s(B_s)$ of the two films. The dotted lines in Fig. 4.33 indicate the power-law approximation

$$R_s(B_s) - R_s(0) = \alpha B_s^2 + \beta B_s^4 , \tag{4.22}$$

in slightly different notation than in (3.61). The results for sample S145 indicate systematic differences between measurement and simulation. The experimental $R_s(B_s)$ data increase more steeply and at lower field amplitudes than expected for microwave heating alone. However, the introduction of a normalconducting 1-µm defect yielded local-quench fields (indicated by arrows) which come close to the quench fields measured at elevated temperatures ($T \geq 77$ K), respectively to the maximum field levels achieved below 77 K. Similar conclusions can be drawn for the $R_s(B_s)$ characteristics found for sample L49 at elevated temperatures. The maximum achieved field amplitudes were limited by the available power so that the predicted quench fields could not be verified. In distinction from sample S145, the agreement between measured and simulated data at low temperatures was better for the film L49.

Thermal–Magnetic Feedback of $R_s(T, B_s)$

Inspection of Fig. 4.33 leads to the following conclusions. First, the nonlinear mechanisms affecting the microwave field dependence of Z_s of non-optimized YBCO films at elevated temperatures is not exclusively thermal in origin. This follows from the discussion in Sect. 3.2.3 that showed that the assumed

4.3 Magnetic and Thermal Field Limitations in YBCO Films 233

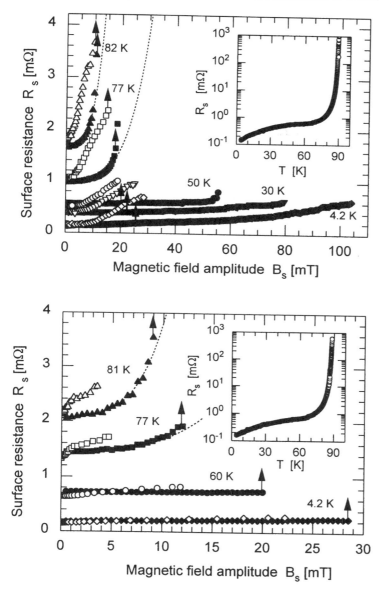

Fig. 4.33. Simulated (*solid*) and measured (*open symbols*) $R_s(B_s)$ curves at 19 GHz for the DC-sputtered YBCO film S145 (*top*) and the laser-ablated film L49 (*bottom*) [23, 30]. The *arrows* indicate the calculated local-quench fields assuming a normalconducting defect of 1 μm diameter. The *dotted lines* denote the even-power approximation of $R_s(B_s)$; (4.22). The insets display the low-field temperature dependences of R_s.

properties of the local defects (size, conductivity, T_c) could be varied significantly without modifying the resulting "thermal" $R_s(B_s)$ characteristic. Referring to the DC field dependent surface impedance discussed in Sect. 4.3.1, the onset of nonlinear effects at low fields likely reflects the presence of weakly superconductive regions.

Secondly, the observable quench fields agreed reasonably well with the predicted values. Such an agreement indicates that thermal effects become nonetheless relevant. The main effects of microwave heating are thus considered to limit $R_s(B_s)$ at elevated field amplitudes, and to modify the slope of $R_s(B_s)$. This can be illustrated by assuming a simple phenomenological field dependence of the integral surface resistance R_s:

$$R_s(T, B_s) \approx R_s(T, 0) \left[1 + \left(\frac{B_s}{B_{s,\text{crit}}(T)}\right)^m\right], \qquad (4.23a)$$

where $B_{s,\text{crit}}$ denotes the lowest critical magnetic field present in the superconductor, e.g., at a weak spot or at grain–boundary Josephson junctions. The power m in (4.23a) is typically $m = 1$. Microwave heating is reflected in an effective field dependence of the temperature at the surface of the film, which can be approximated in the global case by:

$$T(B_s) = T(0) + \frac{R_s(T, B_s) R_{\text{th}}}{2\mu_0^2} B_s^2 . \qquad (4.23b)$$

In the absence of defects or for very high $B_{s,\text{crit}}$ values (e.g., in high-quality films), the two parts of (4.23) become uncoupled, and reveal the microwave heating phenomena analyzed in Sect. 3.2. In the opposite case, if defects are present and/or $B_{s,\text{crit}}$ is low (e. g., at elevated temperatures), (4.23a) and (4.23b) constitute a single, implicit, equation. The resulting slope dR_s/dB_s reflects, according to the chain rule of differential calculus, a combination of magnetic and thermal mechanisms:

$$\frac{dR_s(T(B_s))}{dB_s} = \frac{\partial R_s}{\partial B_s} + \frac{\partial R_s}{\partial T}\frac{\partial T}{\partial B_s} . \qquad (4.23c)$$

If we assume, for simplicity, a temperature independent critical field $B_{s,\text{crit}}$, and a square-law dependence $m = 2$, then (4.23) yields

$$\frac{dR_s(T(B_s))}{dB_s} \approx \frac{2R_s B_s}{B_{s,\text{crit}}^2}\left(1 + \frac{\partial R_s}{\partial T} R_{\text{th}} \frac{B_{s,\text{crit}}^2}{2\mu_0^2}\right) . \qquad (4.24a)$$

The relative strength of the thermal-magnetic feedback is described by the second term in parentheses. It increases markedly with the critical field amplitude $B_{s,\text{crit}}$, and with the operating temperature $T(0)$, i.e., with increasing level $R_s(T)$ and steepness $\partial R_s/\partial T$. The thermal–magnetic feedback is further enhanced by a high thermal resistance R_{th} (Sect. 3.2):

$$R_{\text{th}}(T) = \frac{d_s}{\sigma_{\text{th},s}(T)}\left[1 + \frac{d_F}{d_S}\frac{\sigma_{\text{th,S}}(T)}{\sigma_{\text{th,F}}(T)}\right] + R_{\text{bd}} , \qquad (4.24b)$$

i.e., if thick and/or low-thermal conductivity substrates are employed, and if the film–substrate interface has a low phonon transmittivity.

A thermally enhanced field dependence of R_s can be identified from the temperature dependence of the quadratic coefficient $\alpha(T)$ in (4.22). Numerical simulations of a YBCO film without defects [23, 30] showed $\alpha(T)$ at 19 GHz to be constant up to about 50 K and to increase by about three orders of magnitude between 70 K and $0.94 T_c$. Quantitatively similar behavior was reported for YBCO stripline resonators at 1.5 GHz [87].

Microwave heating enhances the thermal–magnetic feedback further by the temperature dependence of the penetration depth, which drives the superconductor at sufficiently high temperatures into the thin-film limit (Sect. 1.1.3):

$$R_s(T(B_s)) \approx R_s(T, B_s) \coth\left(\frac{d_F}{\lambda(T(B_s))}\right), \qquad (4.25a)$$

where

$$\lambda(T) \approx \lambda(0) \frac{1}{\sqrt{1 - [T(B_s)/T_c]^2}}. \qquad (4.25b)$$

Finally, exact results for $R_s(T, B_s)$ derived from (4.23) require numerical computation taking into account all temperature and field dependences of the involved quantities.

Comparison Between Measured and Simulated $Z_s(B_s)$ Curves in CW Operation

The $R_s(B_s)$ characteristics discussed so far were obtained in pulsed operation. While this mode is appropriate to study the mechanisms of nonlinear surface impedance, it is not suited for many microwave applications, e.g., in filters (Chap. 6). In steady-state operation, the diffusion of heat is not limited by the pulse duration, but can proceed further. As a result, the entire sample fixture, if not sufficiently cooled, might heat up.

Figure 4.34 displays the $R_s(B_s)$ curve of the YBCO film S145 (open symbols, see also Fig. 4.33) measured with the 19-GHz sapphire resonator in CW operation. Since the samples are decisively thermally decoupled from the cooling bath in this system (Sect. 2.1.3), the thermal boundary resistance R_{bd} was rather large. It was deduced from reference measurements using a calibrated heater to be at 30 K (respectively: 77 K) $R_{bd} = 800$ Km²/W (350 Km²/W). The inset to the figure provides a direct comparison of $R_s(B_s)$ for the pulsed and the CW measurements. In contrast to the pulsed operation, and in accordance with expectation, much steeper $R_s(B_s)$ curves and reduced quench field levels were observed at all investigated temperatures.

Also shown in Fig. 4.34 are synthetic $R_s(B_s)$ curves simulated for microwave heating (filled symbols). The experimental cooling conditions were

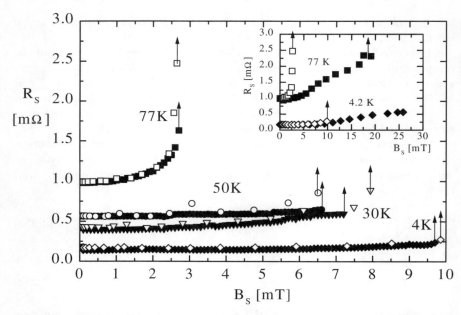

Fig. 4.34. Simulated (*filled*) and measured (*open symbols*) $R_s(B_s)$ curves at 19 GHz for the DC-sputtered YBCO film S145 in CW operation. The *arrows* indicate the calculated local-quench fields assuming a normalconducting defect of 18 µm diameter. The *inset* compares $R_s(B_s)$ for the CW (*open*) and the pulsed measurements (*filled symbols*, see also Fig. 4.33) [23, 30].

considered by taking into account a floating temperature at the bottom side of the substrate, and an additional thermal boundary resistance. The simulation agreed quantitatively with the measured $R_s(B_s)$ curves and with the quench fields at all temperatures if a normalconducting defect with a radius of 9 µm was assumed. This agreement illustrates the deleterious role of local defects on the nonlinear field dependence of R_s if microwave heating cannot be prevented.

4.3.4 Summary

The nonlinear microwave response of superconducting films was found to be caused predominantly by magnetic and thermal mechanisms. Electric field induced nonlinearities could be neglected since the discussed data were obtained in measurements where the electric field amplitudes at the surfaces of the films were weak.

Magnetic Field Induced Nonlinearities

Three cases of magnetically induced nonlinear effects had to be distinguished. The application of DC fields perpendicular to the superconducting films led to flux-flow losses as soon as the flux front reached the microwave sensing focus (Sect. 4.2.2). The proceeding flux was determined by the pinning properties of the films. The considered film geometry resulted for Y-123 in onset fields of nonlinear surface impedance around 30–50 mT at 4.2 K, which were accidentally comparable with the bulk lower critical field for this field orientation.

The application of DC fields parallel to the superconducting films led to an enhanced field dependent surface impedance at field strengths at which the flux-flow conductivity exceeded the residual low-field conductivity of the films (Sect. 4.2.3). Typical quality of Y-123 films yielded at 87 GHz critical fields of around 20–40 mT at 4.2 K. These values were about a factor of two enhanced above the bulk lower critical field for shielding currents flowing in part perpendicular to the CuO_2 planes of the HTS films.

Magnetic microwave field-induced nonlinearities could be treated in analogy to parallel DC fields, taking into account the anisotropy of B_{c1} for shielding currents flowing within the CuO_2 planes of the HTS films (Sect. 4.3.1). Typical $R_s(T)$ data of Y-123 at 8.5 and 19 GHz yielded critical field levels comparable to the bulk lower critical field, i.e., about 75 mT at 4.2 K and about 20 mT at 77 K. An enhanced Ginzburg–Landau lower critical field of thin films above the bulk value (Sect. 3.1.5) was not observed. The absence of thin-film effects could be due to the use of sufficiently thick films ($d_F > \lambda$ especially at low temperatures), to a breakdown of the London assumption of constant order parameter in the highly anisotropic layered material, or due to a competition with other λ dependent effects (e.g. the enhancement of $R_{s,eff}/R_s$). The linear field region of Z_s increased with decreasing temperature (Fig. 4.5), but the achieved field levels are still too low to verify the expectation from d-wave models (Sect. 3.1.5), which predicted a strong increase of the pair breaking coefficient at low temperatures.

Samples of different quality, including films with only few defects, displayed an anomalous field effect (Sect. 4.3.2). Surface resistance and surface reactance decreased with increasing DC or microwave field levels in the mT range. While the physical mechanisms are not yet completely understood, the high field levels, the short relaxation times and the correlative behavior between $R_s(B)$ and $X_s(B)$ point to magnetic alignment of impurity spins. The relation of the recovery effect to oxygen deficiencies provides an important support of the two-band model of high-temperature superconductivity.

Thermally Induced Nonlinearities and Thermal–Magnetic Feedback

Microwave heating as a source of nonlinear surface impedance was experimentally verified from its dynamic appearance (Sects. 4.1.1 and 4.1.4). It contributed to the field dependent surface impedance predominantly at elevated temperatures and field amplitudes. The local-quench fields predicted by the numerical simulations agreed well with data measured in films of moderate quality, ranging from about 50 mT at 4.2 K to 20 mT at 77 K for

10-μm normalconducting defects (Sect. 3.2.2). The measured $R_s(B_s)$ curves varied strongly with film quality, being approximately constant at a low frequency dependent level for the best films (Sect. 4.1.2). The maximum field levels presently achieved with unpatterned Y-123 films were limited by the available power at about 50 mT at 4.2 K, and by quenches around 20 mT at 77 K. In comparison to the epitaxial YBCO films, polycrystalline Nb_3Sn films on sapphire displayed only slightly lower quench fields, around 25 mT at 4.2 K. This limit was attributed to microwave heating, accounting for the lower values of critical temperature and thermal conductivity compared to HTS.

While reproducing $R_s(B_s)$ curves measured in CW operation, the thermal simulations could not reproduce the functional dependence $R_s(B_s)$ of typical YBCO films obtained in pulsed mode (Sect. 4.3.3). This failure was attributed to the presence of local areas of reduced critical magnetic fields. Microwave heating was nonetheless found to induce a thermal-magnetic feedback between the two nonlinear mechanisms. The resulting field slopes dR_s/dB_s were enhanced, not the least by the thin-film limit due to the temperature dependent penetration depth.

Potential for Further Improvement and Consequences for Applications

The temperature and field dependent surface impedance of Y-123 and Nb_3Sn films has approached the intrinsic limits of individual optimized films. However, the statistical view of $R_s(B_s)$ at different temperatures reveals a variety of behaviors. This reflects the inhomogeneous film quality, due to local defects or extended areas of reduced superconductivity. The technological origin and the physical nature of such damaged regions could not yet be identified, and meeting the intrinsic limits still appears to be accidental.

In analogy to the continuing improvement of low-temperature superconductors for particle accelerators [88], the remaining challenge of obtaining high-quality HTS films is the optimization of the film technology (Chap. 5). This includes a properly adjusted film thickness, oxygenation and homogeneity. It also includes the identification, elimination or even the prevention of defects. Deleterious to a linear microwave response are weakly coupled grains, normalconducting segregations, cracks and holes, and an inhomogeneous oxygen stoichiometry and critical temperatures.

The achieved state of the technology sufficed to demonstrate the enormous potential of high-temperature superconducting thin films compared to conventional technology for microwave applications [43]. Still, most of these applications envisage high operating temperatures, well above $T_c/2$. As an important consequence, the material must be developed on an industrial scale to its intrinsic limits in order to meet the desired properties, both in terms of linear and nonlinear microwave response. Promisingly, the interfaces between material science, microwave and cryogenic engineering, and marketing have proceeded to a level advanced enough to be appropriate to implement HTS devices in operational systems (Chap. 6).

5. Technology of High-Temperature Superconducting Films and Devices

Technology means to manufacture and to control. Matured technology means to understand how.

The description here of the technological processing of high-temperature superconducting films and devices aims to connect the preceding chapters, which dealt with the physics at linear and nonlinear microwave response, to the last part of the book, which is devoted to potential microwave applications. For this reason, the presentation should not be expected to be complete in every regard. Rather, it should provide a guideline to identify the technological challenges, and to help the interested reader to find references for further information.

Chapter 5 is divided into three parts that cover the preparation of Nb_3Sn and $YBa_2Cu_3O_{7-x}$ thin films on dielectric substrates (Sect. 5.1), and various aspects related to the integration of films into microwave devices (Sects. 5.2 and 5.3). Reflecting its relevance to realistic applications of superconducting devices, cryogenic refrigeration is discussed separately in Sect. 5.3, leaving to Sect. 5.2 issues such as the choice of substrates, simulation tools and packaging techniques.

5.1 Thin-Film Deposition Techniques

Most of the potential applications of superconductors in the microwave region are based on compact, and in many cases even planar, devices. One important exception is the use of cavity resonators for particle accelerators, e.g., to investigate fundamental physics at very high energies [1], or to provide diagnostic and therapeutical tools for medical purposes. This field has been well established during its 25 years of development [2], and it will not be considered in further detail here. Our discussion focuses rather on thin-film technology as the key to planar devices. The preparation of Nb_3Sn layers and films is considered first, in Sect. 5.1.1. Section 5.1.2 summarizes the most prominent deposition techniques that have proven suitable for the fabrication of epitaxial YBCO films. In allusion to the discussion of passive microwave devices based on the Josephon effect (Chap. 6), Sect. 5.1.3 contains a brief sketch of the preparation of YBCO grain-boundary Josephson junctions at substrate steps.

5.1.1 Nb$_3$Sn Films on Sapphire Substrates

Structure and Electronic Properties

The intermetallic system Nb$_3$Sn belongs to the group of cubic A15 structures, which contains more than 50 superconducting compounds [3]. Among these are the "classical high-temperature superconductors" with $T_c > 10$ K, which are binary compounds of the type A$_3$B. Usually, "A" stands for a transition element like vanadium (V) or niobium (Nb), and "B" represents aluminum (Al), silicon (Si), gallium (Ga), germanium (Ge), or tin (Sn). A record value of $T_c = 23.2$ K was found for Nb$_3$Ge [4]. However, in terms of mechanical and chemical stability, the discovery of Nb$_3$Sn ($T_c = 18$ K) [5] turned out to be of major technological relevance. Reflecting the enormous impact on the science and technology of high-T_c type-II superconductors (e.g., for high-field magnets), the structural and electronic properties of the A15 compounds have been discussed in numerous reviews [3, 6–8].

A unique feature of the binary A$_3$B compounds is the formation of a body-centered cubic lattice by the B atoms as shown in Fig. 5.1. The A atoms are arranged at the faces of the cubes, forming three orthogonal non-intersecting chains along the [001], [010] and [100] directions of the crystal. The distance between adjacent atoms along one chain (a/2 ≈ 0.265 nm in Nb$_3$Sn) is half of the lattice dimension (a = 0.529 nm). It is smaller than the separation of atoms between neighboring chains (≥ 0.324 nm in Nb$_3$Sn). As a result, the electronic wave functions extend mainly along the chains, although interactions between chains were found to contribute significantly to the electronic band structure [6].

Most of the high-T_c compounds of the A15 superconductors display a softening of the phonon frequencies with decreasing temperature. Especially in Nb$_3$Sn, a first-order phase transition from the cubic symmetry to a tetragonal distortion occurs around 43 K as illustrated in Fig. 5.2. At low temperatures, the lattice constant of the Sn cubus increases, as does the separation of the Nb atoms along the correspondingly oriented chain [7]. It is speculated that phonon softening and lattice instability are related to the high transition temperature. A theoretical approach to this electronic and structural interrelation was derived in terms of a one-dimensional model by Labbé and Friedel [8, 9]. According to their model, which neglects electronic interactions between the chains, the electronic density of states exhibits singularities arising from the overlap between the d orbitals of adjacent chain atoms. A close proximity between one of these singularities and the Fermi energy provides the basis for both a high transition temperature (Chap. 1) and a structural phase transition. While this interpretation provides a reasonable qualitative

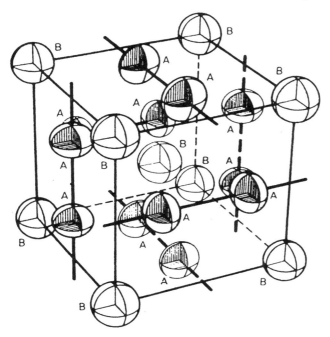

Fig. 5.1. Schematic crystal structure of a binary A15 compound of type A_3B [3].

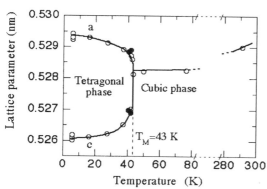

Fig. 5.2. Structural phase transition in Nb_3Sn from the cubic to the tetragonal phase at $T = 43$ K [3].

description of some features of A15 superconductors, a quantitative analysis requires more sophisticated calculations of the band structure [6].

As another property of the A15 lattice structure, the critical temperature was found to correlate with the structural order of the chain atoms (see [10] and references therein). For Nb_3Sn, T_c saturates slightly above 18 K as the stoichometric ratio 3:1 is approached [11]. The absolute value of T_c is sensitive to internal strain [12]. Even more sensitive to the atomic ordering turned out to be the resistivity $\rho(T_c)$ slightly above the critical temperature [10,

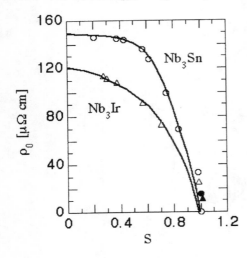

Fig. 5.3. Dependence of the resistivity $\rho(T_c)$ slightly above the critical temperature [10] on the atomic ordering represented by the parameter S ($S = 1$ denotes perfect order). (© 1999 IEEE).

11]. This is illustrated in Fig. 5.3 for completely reacted Nb_3Sn and Nb_3Ir samples (solid symbols, near $S = 1$), which were decisively disordered by nuclear irradiation (open symbols) [10]. The atomic ordering is represented by the parameter S with $S = 1$ denoting perfect order. The figure reveals a steep drop of $\rho(T_c)$ as perfect ordering is approached. This quantity therefore provides a sensitive measure of the quality of Nb_3Sn films (see below and Tables 2.3 and 2.4 in Sect. 2.2).

Thermodynamic Phase Diagram and Thin-Film Deposition Techniques

Figure 5.4 presents a section of the phase diagram of the Nb-Sn system developed by Charlesworth [13] (see also survey in [14]). It magnifies the stability region of the A15 superconductor Nb_3Sn. It especially takes into account results on the stoichiometry and the morphology of electron beam coevaporated Nb_3Sn films [15].

The A15 compound is thermodynamically stable in a stoichiometric interval between about 18 and 25 at% Sn at all temperatures. Within this stoichiometric region, the structural quality improves with increasing growth temperature from disordered material to a polycrystalline morphololgy of increasing grain size and texture. In the region of excess tin (not shown in Fig. 5.4), there exist two further Nb-Sn compounds: $Nb_3Sn+Nb_6Sn_5$ at $c_{Sn} = 25\text{--}43$ at%, $Nb_6Sn_5+NbSn_2$ at $c_{Sn} = 43\text{--}66$ at%, and $NbSn_2+Sn$ at $c_{Sn} > 66$ at% (peritectic decomposition at $(231 \pm 9)°C$ to $Nb_2Sn+melt$ and at $(845\pm7)°C$ to Nb_6Sn_5+melt). Above about 880°C, phase locking of slightly Sn-rich samples (at $c_{Sn} \geq 25$ at%) to the (Nb:Sn = 3:1) stoichiometry was observed [16]. It was stated that the re-evaporation of the excess Sn transferred the problem of exact rate control of Nb and Sn to that of adjusting the appropriate growth temperature. Above $(930 \pm 8)°C$ and below the decomposition

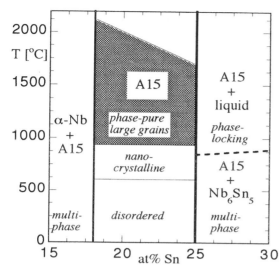

Fig. 5.4. Schematic phase diagram of the system Nb-Sn around the stability region of Nb_3Sn. The *shaded* area indicates the region of stable growth of large-grained films. After [13, 15].

temperature between about 1700°C (25 at% Sn) and 2130°C (18 at% Sn), Nb_3Sn presents the only stable Nb-Sn compound. Many preparation methods approach the A15 phase from the Sn-rich part of the phase diagram (see below). The terminating phase and its purity are then given by the relative thermodynamic stability, and the thicknesses of the resulting layers reflect the different growth kinetics [12].

Various preparation techniques were reported for Nb_3Sn as summarized in [14]. Among these are solid-state sintering of Nb and Sn powders [17], solid-state diffusion of tin from a Cu-Sn matrix into Nb filaments ("Bronze process") [12 and references therein, 18] and different liquid-phase assisted processes [10, 19]. Some of these methods were successful for producing Nb_3Sn wires and tapes with high-current densities at high magnetic fields. However, due to the presence of other solids which usually are also subject to diffusion and related contamination processes, they may not be suited to prepare thin films for microwave applications.

The deposition of phase-pure Nb_3Sn thin films was predominantly achieved by diffusion of gaseous tin into Nb substrates [20] (tin-vapor diffusion, TVD). This technique has proven capable of yielding μm-thick large-area (up to several square meters) Nb_3Sn films on cavity resonators machined from Nb bulk [21, 22] and sheet materials [23, 24]. The TVD process was recently applied also to thin Nb films sputtered onto sapphire [25] (see below). Alternatively, the Nb_3Sn films were deposited on sapphire by means of low-temperature processes, which combine in one step the deposition of Nb and Sn as well as the conversion into the A15 phase. Prominent methods are the electron-beam coevaporation of metals [15, 26, 27] and magnetron sputtering

[28–32] of stoichiometric targets. Finally, chemical vapor deposition of Nb_3Sn films was reported, e.g., in [33].

Tin Vapor Diffusion (TVD)

Among the previously listed Nb_3Sn film deposition techniques, the TVD process has several advantages that are worth being explained in further detail. First, the absence of solids other than the Nb substrate during the diffusion process is an important prerequisite for the deposition of phase-pure films.

Secondly, the deposition of tin from the gaseous phase can be performed in high-temperature furnaces allowing for substrate temperatures well above 1000°C, which facilitate the diffusion of tin into the Nb and the formation of large textured Nb_3Sn grains. Such high temperatures are difficult to achieve with the coevaporation and sputtering techniques for two reasons. The deposition temperature must be adjusted such that the rate of evaporation of volatile constituents like Sn is lower than the deposition rate. As a result, the critical temperature of the deposited films usually displayed characteristic dependences on the deposition temperature [29]. Corresponding substrate temperatures remained typically below 750°C. Furthermore, the required thermal insulation of the substrate holder from the other parts of the deposition chambers complicated the preparation from the technical standpoint.

Finally, the tin partial pressure can be adjusted in the TVD process by the temperature T_{Sn} of the tin source, independently of the substrate temperature T_{sub} [23]. This potential was exploited by the construction of a ultra-high vacuum furnace equipped with one heater for the Sn crucible and another, separate, heater for the Nb substrates [23, 34]. The TVD process is determined by the two temperatures T_{Sn} and T_{sub}, and in addition, by the duration of the tin diffusion which affects the resulting thickness of the Nb_3Sn layer. Figure 5.5 is a schematic representation of the temperature profiles which were adopted during typical TVD runs. An initial tempering at 200°C for 0.3 h (phase I in Fig. 5.5) was used to desorb the gases adherent to the Nb surface.

Starting at phase II, a temperature gradient of 200–300°C between the tin source and the Nb sample (e.g., a cavity resonator or a Nb-coated sapphire substrate) facilitated the formation of a thin Nb_3Sn film at the surface of the substrate [35]. During this temperature ramp, homogeneous film nucleation could be enforced by evaporation of $SnCl_2$ [36]. The substrate temperature was held for about 0.5 h around $T_{sub} \approx 500°C$, where the vapor pressure of $SnCl_2$ is as high as 20 kPa. The chemical decomposition of the tin halogenide is supposed to be related to residual hydrogen:

$$SnCl_2 + H_2 \Longrightarrow Sn + 2\,HCl\,. \tag{5.1}$$

The product HCl of this reaction could indeed be detected by means of mass spectrometry [37]. The vapor pressure of the tin source was kept low at

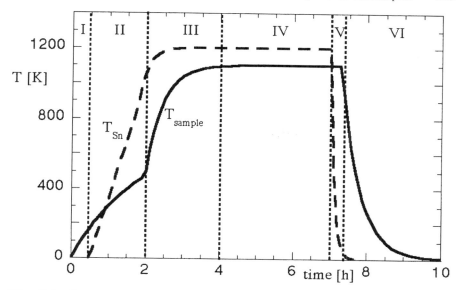

Fig. 5.5. Temperature–time diagram of typical TVD runs. The *solid* (*dashed*) curve denotes the temperature of the substrate (tin source). The phases I to VI are explained in the text. After [37].

$p_{Sn} \approx 0.7$ mPa in phase II. During the next interval (phase III), the heaters of the tin source and the substrate were adjusted to reach the final temperatures $T_{Sn} = 1200°C$ ($p_{Sn} \approx 0.6$ Pa) and $T_{sub} = 1000°C$. The finite temperature difference $T_{Sn} - T_{sub} > 0$ was chosen to compensate the enhanced vapor pressure of Sn at the Nb_3Sn surface compared to that above the Sn source [14], and also to create a gradient of the Sn concentration above the substrate and thus a unidirectional mass flow of Sn atoms into the Nb.

Phase IV represents the main part of the TVD process. Its duration was typically 3 h, which was sufficient to form a stoichiometric Nb_3Sn layer of (1–2) µm thickness in the case of bulk or sheet Nb substrates. Systematic studies of the thickness d_F of the Nb_3Sn films versus different durations of phase IV revealed the power-law dependence [19, 34, 38]

$$d_F(t) = d_{F,0}\, t^n, \tag{5.2}$$

with $n = 0.34$–0.42 when t is in hours. The prefactor $d_{F,0}$ depended on the choice of T_{Sn} and T_{sub}. For the parameters stated above, $d_{F,0} = (1.3\pm 0.1)$ µm was established. The power law in (5.2), which was also observed for, e.g., bronze-processed Nb_3Sn filaments [12], is in contrast to the $t^{1/2}$ dependence expected for bulk diffusion. It was attributed in [38] to the diffusion of tin from the gas-solid interface along the grain boundaries to the Nb_3Sn–Nb interface. Since the grain size (respectively: the density of grain boundaries) increases (decreases) during film growth, the effective diffusivity decreases with time, resulting in an attenuated film growth $n < 1/2$. The growth of TVD films was concluded to occur analogously to that of bronze-processed samples [39]. Similarly reduced exponents between $0.1 \leq n \leq 0.4$ were also reported for grain growth of ceramic materials by solid-state sintering, which is also a diffusion-mediated process [40]. It was observed in accordance with this interpretation that the average grain size corresponded approximately to the film thickness (Fig. 2.15 in Sect. 2.2).

In order to establish the stoichiometric Nb_3Sn phase, especially at the outermost layer of the substrate, a final tempering of 0.5 h was performed with the tin heater turned off (phase V). During this time, excess tin re-evaporated from the sample, leaving pure Nb_3Sn which is the only stable Nb-Sn phase at this temperature (Fig. 5.4). The sample cooled within 0.5 h from 1100°C to 500°C, and within 1–2 h to 200°C (phase VI).

Cross-sectional line scans of Nb_3Sn films prepared in the described way with Nb sheet material (thickness $d_S = 2$ mm) revealed within the measurement accuracy of 0.5 at% the desired tin concentration of 25 at% up to a depth of about 0.7 µm. The reduction from 90% to 10% of the 3:1 stoichiometry occured within 1 µm. The inductively measured critical temperature $T_c = 18.0$ K and, more importantly, the low resistivity $\rho(T_c) \geq 10\,\mu\Omega$cm indicated a high phase purity and atomic ordering of the TVD Nb_3Sn films.

Application of the TVD Process to the Deposition of Nb_3Sn onto Sapphire

The TVD process can be applied successfully to the deposition of phase-pure Nb_3Sn films on low-loss dielectric substrates like sapphire if two conditions are met. First, a Nb base layer ("precursor") must be provided. Second, the substrates must be chemically inert at the high conversion temperatures of 1000–1250°C (like, e.g., oxides). As reported recently in [25] and discussed in Sect. 2.2, both conditions could be satisfied with a two-step process and by using sapphire substrates up to 2″ in diameter.

In the first step, 0.3–3.0-µm-thick Nb films were magnetron sputtered onto the substrates. The Nb films were deposited at room temperature under 1–2 Pa partial pressure of argon, after the sputtering chamber had been evacuated to a residual pressure below 10^{-4} Pa. High deposition rates of 2–3 nm/s were achieved with plasma powers of 80–100 W. The T_c values of the Nb films varied between 7.8 and 9.3 K, indicating a variable oxygen content of up to 1.5 at% [41]. The Nb films were concluded from the results of other groups to be nanocrystalline [42].

The Nb precursor films were converted in the second step into stoichiometric Nb_3Sn by a TVD process exactly as described in the preceding paragraph. The homogeneous nucleation of the A15 phase was facilitated by the fine grain structure of the Nb precursor. In contrast to the TVD process performed with thick ($d_{Nb} \geq 2$ mm) Nb substrates, the thin-film technique ($d_{Nb} \leq 3$ µm) provides a finite Nb reservoir. Since the duration of the tin conversion was adjusted at the values illustrated in Fig. 5.5, the resulting thickness d_F of the Nb$_3$Sn films was determined mainly by d_{Nb}. It is therefore expected that the tin concentration at the film–substrate interface dropped steeply to zero for films with d_{Nb} below the typical Nb$_3$Sn thickness of about 1–2 µm. As for thick Nb substrates, the average grain size of the Nb$_3$Sn films scaled with d_F, reaching up to 3 µm in the thickest films. The pronounced phase purity and atomic ordering of the thin Nb$_3$Sn films could be concluded from $T_c \geq 18$ K and $\rho(T_c) \leq 10\,\mu\Omega\mathrm{cm}$ (c.f., Fig. 5.3 and the discussion of Fig. 2.17 in Sect. 2.2). These features turned out to be independent of the variable quality of the Nb precursors.

5.1.2 Preparation of YBa$_2$Cu$_3$O$_{7-x}$ Films

This section summarizes the major aspects related to the deposition of high-quality epitaxial YBa$_2$Cu$_3$O$_{7-x}$ films. The requirements for suitable substrates are discussed in Sect. 5.2.1. Other HTS compounds, especially Tl-2212, are also promising for certain microwave applications. However, the corresponding phase diagram is even more complicated than that of the 123 compound due to the larger number of constituents. Also, the formation of the Tl superconductor involves the volatile thallium oxide [43], and thus requires additional steps during the preparation process [44]. For the sake of clarity, the following description is therefore focused on the 123 compound, in a sequence similar to that of the previous section. The interested reader may find further information on the Tl process, e.g., in [45, 46].

Structure and Electronic Properties

The discovery of superconductivity in copper oxides at unprecedently high temperatures had initiated an intense research of their structure and electronic properties. Ten years later, a few dozens of HTS compounds had been synthesized. Their microscopic and electronic structures have been resolved in the frameworks of solid state physics and chemistry. Many details and surveys are documented in extended reviews and books (e.g., [47–50]). It is impossible, and therefore not attempted, to provide a general description of this topic. Only selected features which are of direct relevance for a comprehensive understanding of the transport properties of HTS (including those at microwave frequencies) are recalled.

The basic building block of the copper oxide superconductors is the perovskite ABX$_3$ [51], where A and B represent cations (in Y-123: A = Y, Ba and B = Cu). X stands for an anion, which is oxygen in the HTS compounds. The perovskite forms a three dimensional cubic arrangement of BX$_3$ octahedrons enclosing the cation A as depicted in Fig. 5.6. The various HTS unit cells can be considered to be derivatives of this perovskite structure [49]. A common feature is the occurrence of CuO$_2$ planes as shown in Fig. 5.7 for the YBa$_2$Cu$_3$O$_{7-x}$ compound. The layered arrangement is accompanied by

248   5. Technology of High-Temperature Superconducting Films and Devices

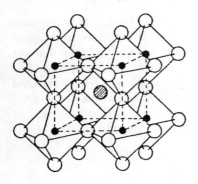

Fig. 5.6. Unit cell of the perovskite ABX_3 with A (*dashed circle*) and B (*filled circles*) cations, and X (*open circles*) an anion such as oxygen [51].

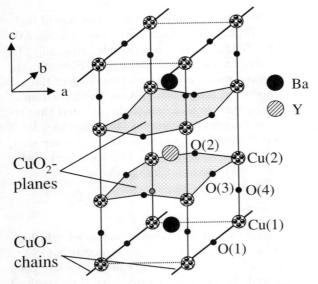

Fig. 5.7. Unit cell of $YBa_2Cu_3O_{7-x}$ [47]. The CuO_2 stacks along the c direction extend over the crystal (a,b) plane, the Cu-O chains extend along the b direction. The distinguishable electronic configurations of the Cu and O positions are consecutively numbered.

a reduced oxygen content of $7 - x$ instead of 9, which would follow for an undisturbed triplet of perovskite cells. The oxygen positions O(2) and O(3) indicated in Fig. 5.7 are subject to characteristic "breathing modes" of the lattice, which are assumed to be related to the strong pair interaction [52]. The two-dimensional structure of HTS causes a pronounced anisotropy of the electronic properties (Chap. 1).

The exact crystal structure of the Y-123 compound depends sensitively on the oxygen deficiency x. At $x = 1$, the unit cell displays a tetragonal symmetry [$a = b$, all O(1) positions are vacant], while $x = 0$ [$a < b$, all O(1)

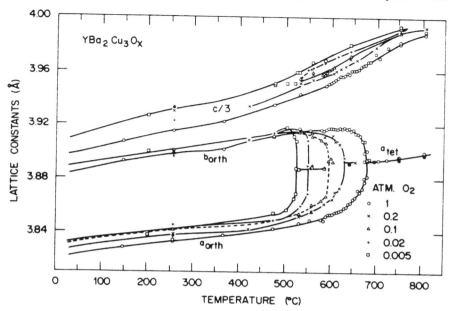

Fig. 5.8. Tetragonal-to-orthorhombic phase transition of $YBa_2Cu_3O_{7-x}$ versus temperature and oxygen partial pressure (*symbols*) [55].

positions are occupied] corresponds to an orthorhombic distortion with the lattice parameters $a = 0.382$ nm, $b = 0.389$ nm and $c = 1.168$ nm [53, 54]. Figure 5.8 displays this structural phase transition as a function of temperature and oxygen partial pressure ([55, 56], see also next paragraph). The b direction of the unit cell forms linear Cu-O chains. These chains provide a charge reservoir for the superconducting pair interaction which is mainly confined to the CuO_2 planes. In the framework of the two-band model, the CuO chains are assumed to present a separate superconducting subsystem which displays a small energy gap at oxygen deficiencies $x \leq 0.1$, and gapless superconductivity at larger x values (Sects. 1.3.3 and 2.3.4). The equivalent of the CuO chains in Y-123 are believed to be the Tl-O and Bi-O layers in the Tl and Bi compounds, respectively [57]. Spectroscopic investigations indicated unsaturated electronic spins at the O(4) positions ("apex oxygen") as well as at the positions of those Cu(1) atoms which face vacant O(1) positions [58, 59]. The two-band model accounts for these free spins by magnetic pair breaking.

The occurrence of superconductivity in the copper oxide superconductors is closely related to the number density of free charges, i.e., to the amount of doping [60]. The charge carriers are holes in most HTS compounds except for the NdCeCuO family which displays conduction by electrons [48]. The resulting electronic phase diagram of Y-123 is displayed in Fig. 5.9 in terms of the dependence of T_c on the oxygen deficiency x (referring to $YBa_2Cu_3O_{7-x}$) [47, 56, 60]. At an oxygen deficiency $x > 0.64$, Y-123 turns from a superconductor to an antiferromagnetic insulator (upper part of Fig. 5.9). It is this striking proximity of a magnetically ordered insulator and a high-temperature superconductor that led to the intense theoretical studies of magnetic pair interactions with a resulting d-wave symmetry of the order parameter (Sects. 1.3 and 2.3). At lower oxygen deficiencies ($x \leq 0.64$), the superconducting transition temperature increases, passes through a plateau at 60 K for $0.5 \leq x \leq 0.3$, rises to a maximum of about 92 K around $x = 0.12$ and decreases below 88 K as x vanishes (lower part of Fig. 5.9, [61, 62]).

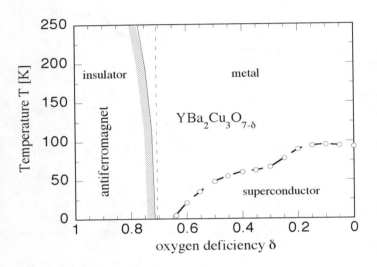

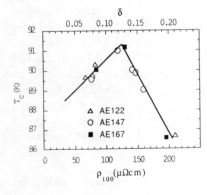

Fig. 5.9. *Top:* Electronic phase diagram of $YBa_2Cu_3O_{7-\delta}$ (from [67], after [56]). *Bottom:* expanded view of $T_c(\delta)$, in terms of the critical temperature versus resistivity and oxygen deficiency for three $YBa_2Cu_3O_{7-\delta}$ crystals in states of various oxygen concentration [61]. (With permission from Elsevier Science.)

The lower part of Fig. 5.9 displays the relationship between the critical temperature T_c, as determined from the midpoint of the diamagnetic transition, and the resistivity ρ (100 K) slightly above T_c for three differently oxygenated YBa$_2$Cu$_3$O$_{7-x}$ crystals [61]. This dependence illustrates the nonmonotonic behavior of T_c with the doping level. Optimum doping was observed around $x = 0.12$. Oxygenation states with $x < 0.12$ (respectively $x > 0.12$) are referred to as overdoped (underdoped). The transport properties in the normalconducting state (e.g., linear temperature dependence of the resistivity) and their dependences on the doping level were attributed to strong variations of the density of electronic states close to the Fermi level [47, 60, 63–65]. However, possible conclusions from the resistivity at T_c on the intrinsic electronic properties of HTS films are usually complicated by the strong impact of anisotropy and granularity, especially in view of the extremely short coherence lengths compared to the dimensions of the unit cell along and perpendicular to the CuO$_2$ planes [66].

The linear microwave response of HTS in the superconducting state was also found to vary markedly with the oxygen content (Sect. 2.3). Optimally doped and overdoped films displayed the lowest values of the surface resistance below T_c [68]. For operation near 77 K, optimally doped films performed best, while at lower temperatures overdoped Y-123 films displayed superior properties. The oxygenation and deoxygenation rates of the films were found to be sensitive to their background morphology [69]. The microstructure of the films, in turn, depended on the chosen deposition techniques and parameters.

Thermodynamic Phase Diagram: Metal Stoichiometry

The thermodynamic conditions for the formation of phase-pure YBa$_2$Cu$_3$O$_{7-x}$ can be divided into two groups. The first group concerns the correct stoichiometry of the metallic constituents while the second is related to the oxygen content and ordering.

Stoichiometric Y:Ba:Cu (1:2:3) powder is a common precursor material for the preparation of targets which are required for various prominent film deposition techniques [70] (see below). Properties of the targets such as phase purity, thermal or electrical conductivitiy affect the quality of the resulting films. Reflecting the progress in ceramic technology [40], HTS targets with different shapes and mass densities have become commercially available. Phase-pure YBCO powder is usually prepared by solid-state reaction of the metal oxides and of calcined carbonates [71, 72]. Alternate chemical routes have been applied where ultrafine powder can be obtained from the metal salts dissolved in proper acid solutions. The reactivity and phase purity of the precursor material depends sensitively on the reactor materials, processing temperatures and times, the ambient atmospheres and the grain size distribution.

Besides powder pressed into the desired shapes and subsequently sintered, the fabrication of polycrystalline layers, e.g., as shielding components (Sect. 5.2), also requires correct metal stoichiometry. Such HTS layers (typical thickness between 1 and 100 µm) are formed from µm-sized HTS grains deposited onto suitable substrates. In contrast, the deposition of epitaxial films (typical thickness between 0.1 and 1 µm) occurs on an atomic scale. The various thick-film techniques (e.g., spraying, dipping, printing, blading) are distinguished by the chemical and physical nature of the carrier (gas streams or chemical suspensions), the driving forces (kinetic, mechanical, electrical) and the required type of substrates (ceramics or metals). Solid-state connectivity between the individual as-deposited grains is usually obtained by diffusion processes at high temperatures. The resulting layers are polycrystalline with the grain boundaries acting as Josephson junctions [73]. Methods allowing for the fabrication of textured layers are therefore highly desirable [74]. Among the most advanced thick-film techniques are processes combining screen printing and melt texturing [75, 76], and electrophoretic deposition in ambient magnetic fields and sintering [77].

Regarding the phase purity required for the targets as well as for the thick layers, the possible phase transitions of the oxide material must be known and controlled in terms of the relative concentrations of the constituents, and of temperature and ambient atmosphere. The resulting phase diagram of the Y-123 compound is quasi-ternary since it is determined basically by Y_2O_3, BaO and CuO [78–82]. Figure 5.10 is one of the possible representations of the phase diagram in the proximity of the 1:2:3 stoichiometry. The hatched area indicates the liquidus surface of the 1:2:3 phase. Dotted lines show the intersection between the composition triangle $YBa_2Cu_3O_7$-$BaCuO_2$-CuO and the liquidus surfaces [81]. Not less than eleven phase transitions were identified in the temperature range between 850 and 1300°C which involve liquid phases [79]. In addition, the transition temperatures were found to decrease by about 30°C with the oxygen partial pressure increasing from ambient air to pure oxygen [71, 80, 83]. The various phase transitions are partially eutectic (i.e., different solidus phases form a homogeneous melt), peritectic (i.e., different solidus phases form a homogeneous melt together with a new solid), and congruent or incongruent melting (stoichiometry of the melt is equal to or differs from the stoichiometry of the solid phase). During cooling from the reaction temperature to ambient temperature, recrystallization from the liquid phases takes place. This process is likely to remain incomplete because of dynamic delays (i.e., when the cooling rate exceeds the characteristic time constants of the involved diffusion processes). Furthermore, since recrystallization starts at the surface of nuclei, the cores of these nuclei become isolated from the liquid phases during this process, with the result of nonstoichiometric segregations.

Oxygenation of the polycrystalline material sets in below about 700°C (see below). This process is facilitated by the existence of nonstoichiometric segregations, grain boundaries and fissures. The resulting oxygenation rate approaches therefore that of the basal CuO_2 planes which is much larger than that along the crystal c direction [84].

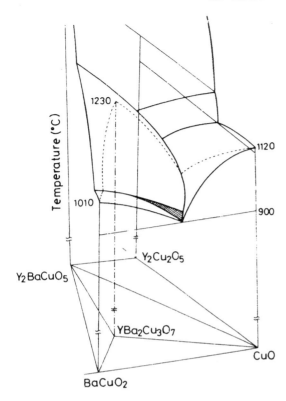

Fig. 5.10. Space model of the quasi-ternary phase diagram of $YBa_2Cu_3O_7$ with 1:2:3 metal stoichiometry [81].

Thermodynamic Phase Diagram: Oxygen Stoichiometry

Given that targets of the correct metal stoichiometry are available, or that the deposition of the individual metal constituents (e.g., in coevaporation processes) is adjusted to the 1:2:3 ratio, the stable formation of the tetragonal Y-123 compound depends still on the substrate temperature and the oxygen partial pressure [85, 86]. Figure 5.11 reproduces early results [85], which can still be considered typical for the currently applied deposition techniques.

At high temperatures and/or low oxygen partial pressures, the 123 phase decomposes into the impurity phases indicated in the figure. This region is separated by the so-called stability line towards reduced temperatures and/or elevated oxygen pressures. If atomic oxygen is supplied during the growth process, the region of stable film growth can be extended to lower oxygen partial pressures [87]. The film growth in the stability region is governed by thermodynamic [88] and kinetic [89, 90] processes, namely in terms of the chemical potentials and free energies of the deposited atoms and the substrate material. Experimental key parameters are the deposition rate p_D and the growth temperature T_S. At given T_S, the ratio of p_D to the evaporation rate of the deposited material determines the supersaturation ζ. Vice versa,

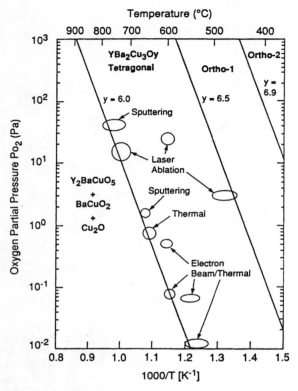

Fig. 5.11. Stability region of the growth of $YBa_2Cu_3O_{7-x}$ films in terms of temperature and oxygen partial pressure [85]. Indicated are typical parameter regions of prominent deposition techniques (see text). Oxygenation of the films takes place during cooling as indicated by the values of the oxygen deficiency $y = 7 - x$. (With permission from Elsevier Science.)

at a given deposition rate, T_S controls the atomic diffusivity [91]. The importance of both quantities for the growth of films was already mentioned in Sect. 5.1.1. Depending on the lattice matching between substrate and HTS unit cell (Sect. 5.2.1), and on the density of structural defects at the surface of the substrates, ζ and T_S determine the growth mode of the film. On defect-free surfaces, the three basic types known are two-dimensional ("layer by layer"), three-dimensional ("island"), and a combination of both [92]. With fixed T_S and decreasing supersaturation, the film growth changes from amorphous to two-dimensional, and further to island growth [91, 93]. Elevated temperatures and/or enhanced kinetic energy of the deposited particles (which can be varied, e.g., via the oxygen partial pressure) affect the growth mode in a similar way as enhanced supersaturation [85]. However, the finite lattice mismatch and the presence of substrate steps on an atomic scale, which cannot be prevented in real experiments, lead to the mixed growth mode ("spiral growth") becoming the prevailing one, except for very large supersaturation [94, 95]. The shape of, and the mutual distance between, the growth spirals correlates with the diffusivity. Altogether, the morphology of the films on a given substrate material is a fingerprint of the applied deposition parameters, such as the rate, growth temperature, oxygen pressure and particle energy.

It can be concluded from Fig. 5.11 that the formation of the superconducting orthorhombic phase requires an oxygenation of the as-deposited films. This transition happens during cooling of the films from the deposition temperature to temperatures below about 290°C [56a]. The change from tetragonal to orthorhombic symmetry nucleates at different sites simultaneously. It thus causes twinning, i.e., the formation of domains where the a and b directions are rotated by 90°. The final oxygen deficiency depends sensitively on the temperature profile and the rate of the cooling process, and on the oxygen partial pressure [47, 53, 56]. Different ordering phases were reported for the oxygen sublattice in YBCO [53, 56, 96]. At an oxygen deficiency $x = 1/3$, domains are nucleated in the YBCO films where the O(1) sites in every third CuO chain are completely vacant. Around the higher value $x \approx 1/2$, every second CuO-chain is oxygen deficient. Due to this ordering trend of the oxygen atoms, the YBCO films are expected to display inhomogeneous oxygenation levels on a microscopic scale. Due to their strong correlation with x (Sect. 2.3), the resulting transport properties will be inhomogeneous as well. Experimental indications of such a behavior were found, e.g., by Hall probe scanning [97].

According to Fig. 5.8, the oxygen content is related to the c-axis length of the unit cell, which can be deduced from x-ray diffractometry. However, the absolute value of c varies for different substrate materials and film morphologies [98]. The c-axis parameter correlates also with the critical temperature of the YBCO films (Fig. 5.9). Unfortunately, the relationship between T_c and x is ambiguous since it depends also on the cooling rate [99]. Relative variations of the oxygen deficiency at otherwise fixed conditions can be monitored reliably by measuring the c-axis parameter as well as T_c, but an absolute determination of the oxygen deficiency requires more sophisticated procedures.

Survey of Deposition Techniques for the Preparation of Epitaxial Films

The different techniques applied since 1987 to the deposition of epitaxial HTS films have been developed from thin-film processes which were previously well established (e.g., [70, 87, 100]). These methods can be distinguished in terms of the precursor material (compound targets, metals, metal/organic solutions), the deposited material (atoms, molecules, clusters) and the growth modes (see Fig. 5.11 and related discussion). Numerous reviews were devoted to this field [70, 87, 101–104]. In an early stage of HTS technology, and for the preparation of the Tl compounds, three-step processes have been applied [44, 70]. The low-temperature deposition ($T_S \approx 400°C$) of an amorphous film was followed by a recrystallization step around 900°C, and terminated by oxygenation during cooling. The high annealing temperatures induced chemical interdiffusion and decomposition, and the recrystallization of the already deposited films suffered from low atomic mobility. The growth of epitaxial films therefore relied on the development of two-step processes, in which the deposition and the recrystallization occurred simultaneously at temperatures around 700°C.

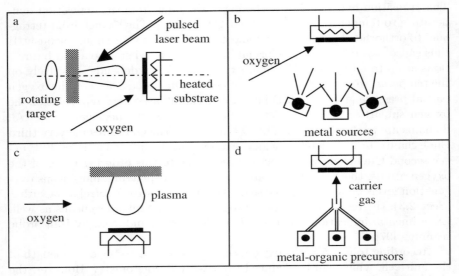

Fig. 5.12. Sketch of different film deposition techniques suitable for HTS material. (**a**) pulsed laser deposition (PLD), (**b**) physical vapor deposition (PVD), (**c**) sputtering, and (**d**) metal-organic chemical vapor deposition (MOCVD).

Figure 5.12 sketches four of the most prominent deposition techniques which were successfully applied to HTS films displaying comparably good microwave properties (Sect. 2.3): pulsed laser deposition (PLD, Fig. 5.12a), physical vapor deposition (PVD, Fig. 5.12b), sputtering (Fig. 5.12c) and metal-organic chemical vapor deposition (MOCVD, Fig. 5.12d). Some features of these methods are described in more detail below. A comparative evaluation of these techniques needs to take into account various physical aspects, e.g., the number of independent parameters, the reproducibility, the deposition rate and the geometry, and options for the deposition of multilayers. Economical aspects are also of concern, e.g., capital and operational costs, the production costs per film, the versatility (different substrates, large-area, double-sided), et cetera. The following paragraphs contain brief descriptions of the four techniques illustrated in Fig. 5.12. Because of the vigorous activity in this field it seems advisable to keep this part on a rather qualitative level.

The references were selected in relation to textbooks, original work on HTS, and recent developments in production of large-area and double-sided HTS films.

Pulsed Laser Deposition

Laser ablation is a novel, extremely versatile film deposition technique [105]. A high-energy pulsed laser beam is focused on a rotating stoichiometric target. Usually excimer lasers are used with wavelengths in the ultraviolet range. The very fast deposition of energy causes an explosive nonequilibrium evaporation of the target material. The evaporated particles, which form a narrow plume extending perpendicular to the target, can be collected on a substrate that has to be arranged properly relative to this plume. The process is equally well applicable to metals and insulators, and thus well suited for the in-situ deposition of multilayers.

Pulsed-laser ablation was applied to the deposition of epitaxial YBCO films shortly after the discovery of this HTS compound [106, 107]. Investigation of the surface impedance of such films revealed a very high quality [108] which, except for the low-temperature behavior, still belongs to the best Z_s data reported until now. A few microseconds after the laser pulse the deposition rates reach the order of 100 nm/min. The low duty-cycle, corresponding to repitition rates of a few Hz, enables this value to relax by about a factor of 10. According to the high energy of the ablated particles, the supersaturation and the diffusivity are high enough to enable two-dimensional film growth [109, 110]. Disadvantages such as the ablation of droplets, the deposition of inhomogeneously thick films and the fixed narrow focus of the laser plume can be avoided by proper modifications of the process (e.g., [111–113]).

Physical Vapor Deposition

The codeposition of vaporized metals is an old and extremely successful technique (Sect. 5.1.1). The metal vapor is typically generated by electron-beam or resistance heating. Both techniques were successfully applied to the deposition of high-quality epitaxial YBCO films (electron-beam: [87, 114], thermal coevaporation: [115]).

One major advantage of this technique is the precise control of the evaporation rates of the individual metals. Together with feedback circuits controlling the vapor composition and the substrate temperature, this process allows for the systematic optimization of the film morphology and the corresponding microwave properties (Sect. 2.3.3). Low surface resistance and linear microwave response up to high field amplitudes were observed in coevaproated films up to 700 nm in thickness [116].

Another advantage is the large distance between metal sources and substrate, which facilitates the deposition of large-area films. Homogeneous

YBCO films up to 9 inches in diameter could be prepared with this technique at relatively low substrate temperatures [117]. However, a low base pressure in the evaporation chamber is necessary to achieve sufficiently high deposition rates (e.g., below about 7 nm/min [114]). The required in-situ incorporation of oxygen into the growing HTS films (Fig. 5.11) has been accomplished by differential pumping, and through the use of activated oxygen.

Given a sufficient number of crucibles and/or electron guns, physical vapor deposition is also applicable to the in-situ deposition of multilayers.

Sputtering

The sputter technique is one of the well known glow-discharge processes [118]. High-energy cations (about 0.1–1 keV) are accelerated in a DC or RF power-induced plasma towards the target, which acts at the same time as cathode. Due to the bombardment, atoms and molecules are displaced from the target. The extension and the homogeneity of the plasma as well as the deposition rate can be enforced by a properly shaped magnetic field. The ablated material is deposited onto the substrates that are arranged slightly outside the main plasma plume. Care has to be taken to avoid resputtering of the deposited material off the growing film, especially if oxygen ions are contained in the sputtering gas. In contrast to DC, RF sputtering does not require metallic targets and is thus applicable to the deposition of multilayers.

Various modifications of the sputter process using stoichiometric compound targets were successfully applied to the deposition of high-quality epitaxial HTS films. The techniques comprise off-axis geometries such as the inverted cylindrical magnetron sputtering [70], or the hollow-cathode sputtering [119], as well as the planar on-axis high-pressure DC sputtering [120]. Resputtering is prevented in the latter process by providing a high oxygen partial pressure of around 200 Pa which thermalizes the deposited particles. While YBCO films prepared with this technique are very smooth and homogeneous and display high critical temperatures, the major drawback of the high pressure is a low deposition rate, of the order of 1 nm/min. The accordingly low diffusivity induces spiral growth of the YBCO films [109].

Proper scaling of the sputtering geometries allowed for the deposition of double-sided and large-area HTS films up to 3 inches in diameter [119, 121]. However, since the deposition parameters like plasma pressure and energy, distance between target and substrate, and growth temperature are strongly interdependent, empirical control algorithms are required to provide a sufficient degree of reproducibility.

Metal-Organic Chemical Vapor Deposition (MOCVD)

The organometallic chemical vapor deposition presents, in principle, a versatile alternative to the purely physical processes described before [122]. The elements to be deposited are chemically implemented into organic solutions. The specifically tailored nature of the solvents allows one to achieve a desired vapor pressure of each source at a prescribed temperature. A gas stream is used as a carrier to transport the vapor constituents to the heated substrate. The volatile reactants decompose there, thus leaving solely the metal atoms. MOCVD is well-known for its capability to cover large and/or curved surfaces.

Different research groups studied MOCVD for the deposition of HTS films [123–125]. However, since the evaporation rates depend sensitively on the latent heats of the solutions and thus on their precise composition, it turned out to be rather challenging to achieve a sufficiently reproducible synthesis of the Ba constituent. While some promising results were reported on the microwave properties of CVD-prepared YBCO films, this technique has not (yet) found wide application for depositing HTS films.

Concluding Remarks on the Preparation of Nb_3Sn and $YBa_2Cu_3O_{7-x}$ Films

In summarizing the phase diagrams and deposition techniques described for the metallic and the oxide high-temperature superconductors, some interesting analogies appear from an empirical point of view. First, both materials have in common structural phase transitions which are suspected to be related to the high transition temperatures. Second, there are common structural elements in the unit cells of the two compounds, namely the linear arrangements of Nb, or Cu and O. In both cases, these chains serve as charge reservoirs for the pair interaction and determine to a large extent the electronic band structure. Furthermore, doping of the chain elements was found to have a strong impact on the strongly correlated electronic properties of HTS. Though presently under investigation, it is tempting to anticipate a similar impact of doping of the Nb chains on the charge transport in Nb_3Sn. Finally, both superconductors display short coherence lengths, which make them very sensitive to the presence of nonstoichiometric segregations and grain boundaries. This aspect is much more strongly pronounced in the highly anisotropic layered copper oxide, where ξ_{ab} is comparable to the unit cell dimension, $\xi_{ab}/(ab)^{1/2} \approx 1$, than in the A15 compound where $\xi_0/a \approx 10$.

Physical vapor deposition turned out to be in both cases optimally suited to preparing phase-pure films. This coincidence reflects the importance of adjusting the composition of the materials as closely as possible to the correct stoichiometry. Sufficient supersaturation and high atomic diffusivity were found to be crucial to achieve homogeneous, dense and stoichiometric film structures. If required, the preparation of single-crystalline, even epitaxial, Nb_3Sn films should be achievable, similar to the remarkable development of epitaxial $YBa_2Cu_3O_{7-x}$ films.

5.1.3 Preparation of $YBa_2Cu_3O_{7-x}$ Grain Boundary Junctions

The availability of large-grained A15 (Sect. 5.1.1) or epitaxial HTS films (Sect. 5.1.2) is a prerequisite to achieve low surface impedance (Chaps. 1 and 2) and high critical fields (Chaps. 3 and 4), e.g., for exploitation in passive microwave devices (Chap. 6). However, regarding the potential for integrated circuits to contain active elements as well, the preparation of Josephson junctions is also desirable. This section therefore briefly sketches the Josephson technology of high-temperature superconductors. The discussion is focused on the copper oxides which, at the expense of a further development of the A15 materials, has attracted major interest since 1987 [126, 127].

HTS Josephson Technology

Compared to homogeneous films, the preparation of HTS Josephson junctions is more complicated, for two main reasons. First, the short coherence length demands material engineering on an atomic scale. Second, the complicated phase diagram of the multielement material makes it extremely difficult to deposit barrier layers without deteriorating the superconducting properties of the base electrodes at the interfaces on either side of the barrier.

In contrast, the Josephson technology of single-element superconductors like Nb has become well established [128–130]. Possible applications concern analog Josephson transmission lines or digital circuits for signal processing and computing. Different types of Josephson junctions can be distinguished from the electronic nature of the barrier material [131]. Insulating barriers mediate the transport of Cooper pairs and quasi-particles via direct tunneling. Normalconducting barriers couple the superconducting electrodes via a proximity effect. Geometric constrictions confine the charge transport ballistically. The situation is much more complicated in HTS junctions (see recent review [131]). The different observed transport mechanisms can be distinguished by the absence or presence of interfaces in the films. Depending on the type of interfaces (e.g., grain boundaries or multilayers), the junction parameters can be tailored, more or less. However, the resulting parameter spread is currently still inadequate for a reliable technology, which requires the reproducible and controllable fabrication of a large number of junctions on a chip, located at arbitrary positions. For HTS, if such a technology can be developed, it will likely be based on heteroepitaxial multilayers containing insulating barriers for the charge transport along or perpendicular to the CuO_2 planes (so-called "ramp-type" or "c-axis" junctions) [132]. The characteristic junction parameters can be engineered by controlling the barrier thickness, or by doping the barrier material to the desired charge transfer coupling.

While such a technology is presently still under development for HTS, Josephson circuits with a low scale of integration (for use in, e.g., SQUID devices) have been demonstrated. The preferred junction type was based on artificial grain boundaries, mostly fabricated from epitaxial films on bicrystal substrates [73]. However, besides its expensive fabrication, the position of the bicrystal boundary limits the possible geometrical arrangement of the junctions and the related circuits. A promising alternative has been the preparation of grain-boundary Josephson junctions in epitaxial films deposited over steep substrate steps [133].

Step-edge GBJs Prepared from Sputtered and Laser-Ablated Films

This paragraph sketches some of the key features of step-edge grain-boundary junctions (SEJ) in relation to the film growth. The described research [109] appears to present an almost negligible contribution to the worldwide efforts in this field. Still, the results provide a valuable illustration of the present challenges of the technology of HTS films and devices. They might also indicate possible approaches to achieve further improvements.

A consistent understanding of the microscopic structure and of the electronic properties of SEJs was attained by transport measurements of the critical current density J_{cJ} and the unit-areal normal resistance $R_J A$ [134], and by high-resolution transmission electron microscopy [135]. The preferred deposition process applied so far to the preparation of SEJs has been pulsed laser deposition. This technique was shown in Sect. 5.1.2 to provide high supersaturation and atomic diffusivity. For sufficiently steep and high steps (step angles above 60°, height comparable to the film thickness), two well-localized grain boundaries were found to nucleate at the lower and upper edges of the step, respectively. The different structural configurations of the grain boundaries resulted in distinguishable J_{cJ} values, with the lower value determining the performance of the devices. Further studies revealed that the current–voltage (IV) characteristics could be strongly affected by the morphology of the substrate steps [136, 137]. However, despite improving the shape and the flatness of the steps by electron-beam lithography and by the use of high-selective etch masks, on-chip variations of J_{cJ} of a factor 2–3 were measured in many cases, even with optimized steps. These results demonstrated the role of the film growth for the formation of homogeneous grain boundaries.

The transport properties of step-edge junctions prepared from differently deposited films were investigated, e.g., in [109a]. High-angle ($> 70°$) steps were produced in $10 \times 10 \, \mathrm{mm}^2$ LAO substrates by standard photolithography and argon ion beam etching. The process parameters of the individual fabrication steps had been optimized in a separate study [109b, 138]. Figure 5.13 is a SEM picture of the resulting step morphology which was found to be typical for steps up to 550 nm in height. Various identically prepared stepped substrates (step height 250 nm) were coated by different techniques with 200 nm thick YBCO films. High-pressure DC sputtered films (process 1, [121]) were deposited at the lowest ($T_S = 740°C$, process version 1a) and at the highest possible temperature ($T_S = 780°C$, 1b) which still allowed for the growth of high quality films. Off-axis laser ablation [112] at $T_S = 750°C$ was chosen as

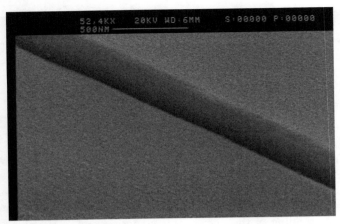

Fig. 5.13. Scanning-electron micrograph (SEM) picture of a 250-nm-high substrate step in LaAlO$_3$ [109].

an alternative technique (process 2). All processes yielded high-quality films as judged from $T_c = 89\text{--}92\,\text{K}$, $\Delta T_c = 0.35\text{--}0.70\,\text{K}$, $J_c(77\,\text{K}) > 2\,\text{MA/cm}^2$ and $R_s\,(87\,\text{GHz},\,77\,\text{K}) = 12\text{--}20\,\text{m}\Omega$.

More than 100 junctions (2 to 14 µm wide) were patterned from each type of film by UV lithography and argon ion beam etching. In contrast to the highly comparable transport properties of the homogeneous films, the IV characteristics of the step-edge junctions were distinctly different. Figure 5.14 shows the results which are typical of each deposition technique. The IV characteristics of junctions sputtered at the low substrate temperature (1a) deviated strongly from the "resistively shunted junction" (RSJ) behavior (Sect. 1.2), which has been found typical for high-quality HTS junctions [128, 131]. The critical temperature of the junctions was below 30 K. Some junctions of this type, which were measured before oxygen annealing, revealed charge transport across few narrow channels [139]. The IV curves of the junctions sputtered at high deposition temperatures (1b) had an opposite curvature as illustrated in the inset of Fig. 5.14. The data revealed an excess current, and a signature of the critical current of the second grain-boundary junction. The critical temperatures of these junctions were higher than for process 1a, ranging between 40 and 65 K. In contrast to the sputtered junctions, the laser-ablated grain boundaries (2) displayed RSJ-like IV characteristics, and showed much larger critical currents, and critical temperatures up to 88 K. Typical $I_{cJ}R_J$ products were 1.1–1.8 mV at 4.2 K and 0.1–0.22 mV at 77 K. The critical current could be suppressed by magnetic field. Microwave irradiation induced Shapiro steps in the IV curves. These features, which were not observed for the sputtered junctions, confirmed the expected high junction quality [134, 135]. Furthermore, comparison of 59 SEJs on four dif-

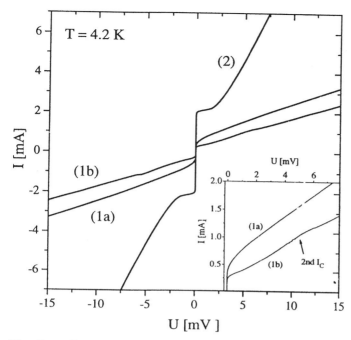

Fig. 5.14. Typical IV characteristics at 4.2 K of 4-µm-wide YBCO step-edge junctions on LAO prepared by low- and high-temperature sputtering (1a, 1b), and by pulsed-laser ablation (2). The *inset* illustrates the different curvatures of the IV curves at low bias voltages [109a].

ferent chips revealed a low parameter spread, which was comparable to the results reported for bicrystal and ramp-type junctions [131].

The microstructure of the different junction types was investigated by scanning (SEM) and transmission electron (TEM) and atomic force microscopy (AFM). AFM analysis revealed growth spirals as the predominant surface structure of the sputtered films. In contrast, two-dimensional film growth has been reported for the majority of laser-ablated films ([110], Sect. 5.1.2). The formation of trenches at the lower step-edge could be concluded from TEM investigations of the low-temperature sputtered junctions. While trenching was not observed in type 1b and 2 junctions, the microstructure of the grain boundaries in both types of sputtered films were significantly less well ordered than that of typical laser-ablated SEJs [135].

The transport properties and the microstructure of the sputtered and the laser-ablated junctions were found correlated with the atomic diffusivity D_S during the film deposition. Due to the low deposition temperature and the high oxygen partial pressure, process 1a yielded the lowest diffusivity. The formation of trenches and disordered grain boundaries was attributed to this low surface mobility. The increase of the deposition temperature from 740 to 780°C from process 1a to 1b resulted in a sufficiently high D_S value to prevent trenching. However, the diffusivity during pulsed laser ablation [110, 140], even in the off-axis geometry, was estimated to be several orders of magnitude higher than during the high-pressure sputtering. This result is in accordance with the two-dimensional film growth and the superior junction characteristics.

In conclusion, high surface mobility was proven to be crucial for the preparation of high-quality YBCO Josephson junctions with low parameter spread. While the analysis was based on step-edge junctions and on the specific comparison between sputtered and laser-ablated films, it seems plausible (and advisable) from first principles to extend this conclusion to other types of HTS junctions. Material engineering of the high-temperature superconductors on an atomic scale must necessarily account for the thermodynamic and kinetic processes occuring during all stages of the device fabrication.

5.2 Integration of Films into Devices

This section describes the main technical requirements that are essential to exploit the microwave response of high-temperature superconducting films in useful devices. Sections 5.2.1 and 5.2.2 deal with the choice of suitable substrate materials. The physical and chemical compatibility between the substrates and the HTS films are related to the microstructure and the electronic transport in the superconductor (Sect. 5.2.1). Furthermore, the substrates serve in many planar microwave devices as the storage volume for electromagnetic energy. It is therefore important to learn about the relevant dielectric loss mechanisms (Sect. 5.2.2). These aspects are especially relevant for HTS, whereas the availability of large-grained Nb_3Sn films on low-loss sapphire substrates does not require special attention. Section 5.2.3 summarizes the remaining empirical aspects which connect the material physics to the engineering of microwave devices. Issues treated in that section are microwave design and simulation tools, patterning, metallization and packaging. Supplementary information on these issues is contained in a few reviews and books [44, 49, 141–143].

5.2.1 Choice of Substrate Materials: Structural Compatibility

Basic Requirements for HTS Microwave Devices

Epitaxial growth of HTS films is the major requirement to achieve low surface impedance and high power handling. The meaning of "epitaxy" can be derived from the two greek words "epi" (at, on, above) and "taxis" (ordering, structure). Accordingly, the basic condition for epitaxial film growth is the nucleation of single-crystalline growth units at the surface of a "host crystal" (substrate), given that the matching of the lattice constants a_S and a_F of the substrate and of the film allows minimizing their total free energy.

It has been established that passive microwave applications based on epitaxial HTS films require single-crystalline substrates (up to about 3″ in diameter) with a small mismatch in lattice parameters and thermal expansion. The former condition is crucial to prevent dissipative defects resulting from the elastic strain in the film lattice, which scales like $|a_S - a_F|/a_F$ [144]. This aspect concerns especially the occurrence of microscopic cracks due to thermal strain, especially in large-area substrates [145]. Thermal strain can be induced by cooling the film from the deposition temperature to ambient, or from ambient to typical operating temperatures. Fractures in HTS films are deleterious to their transport properties at low and especially at high frequencies. Furthermore, deoxygenation and chemical decomposition are facilitated at internal and external surfaces of the copper oxides, with enhanced sensitivity to aging and corrosion as a probable consequence. The formation and the resulting pattern of microcracks were found to depend sensitively on structural and/or compositional inhomogeneities (stacking faults, misorientations, segregations, grain boundaries) of the substrate, the buffer layer and the HTS film [145, 146]. Optimized deposition techniques and proper adjustment of the film thickness are therefore required to avoid microcracking in HTS films.

Beside the structural requirements, application of HTS films to planar microwave devices also carries constraints on the permittivity $\text{Re}\{\varepsilon\} = \varepsilon_r$ and the dielectric losses ($\tan \delta = \text{Re}\{\varepsilon\}/\text{Im}\{\varepsilon\}$) of the substrates (Sect. 5.2.2). The reason is the simultaneous use of the substrate as a mechanical support of the film, and as a dielectric volume to store electromagnetic energy. The substrates should also be chemically inert at the conditions relevant to the preparation of the superconducting films. Other requirements concern, e.g., surface roughness, large-area availability, and costs.

Collection of Suitable Substrate Materials

Table 5.1 lists the structural, thermal, and electric properties of selected substrate materials. In addition to the materials relevant for YBCO films, unbuffered sapphire and Nb present possible substrate materials for high-quality Nb_3Sn films. The listed data were collected from various sources in order to provide a comprehensive overview. For original work and more detailed information, the reader is referred to the cited references.

Microwave applications of HTS films have thus far restricted the choice of substrate materials mostly to the low-loss dielectrics $LaAlO_3$ (LAO, despite its twinning) and MgO (despite the lattice mismatch and the hygroscopic surface). The high permittivity of LAO is uniquely suited for strong miniaturization. Its chemical inertness presents another important technological advantage. The structural quality and the transport properties of YBCO films on MgO were systematically optimized by physical vapor deposition

Table 5.1. Structural, thermal and electric properties of selected substrate materials suitable for the deposition of high-quality YBCO films [143, 147]. Al_2O_3 denotes single-crystalline sapphire. The relative lattice mismatch is estimated for YBCO from $|a_S - a_F|/a_F$ using the average lattice parameter of the CuO_2 planes. The thermal expansion coefficient [148] represents typical values around ambient temperature. The complex permittivity $\varepsilon_r(1 - i\tan\delta)$ was evaluated at 77 K and 10 GHz. The loss tangent of LAO is treated in more detail in Sect. 5.2.2. Listed for comparison are corresponding data for the superconductors $YBa_2Cu_3O_{7-x}$ and Nb_3Sn. [a] YBCO aligns with the [110] direction of the indicated materials, yielding the lowest mismatch for a single unit cell (in the case of $Y-ZrO_2$) and two unit cells (in the case of CeO_2).

Material	Lattice constant (nm)	Relative lattice mismatch	Thermal expansion (10^{-6}/K)	ε_r	$10^5 \tan\delta$	References
$NdGaO_3$	0.384–9	0.003	8–9	21–23	30–32	149, 150, 151
$SrTiO_3$	0.391	0.014	10–11	≈ 2200 (bulk)	≈ 100 (bulk)	149, 152, 153, 154, 155
$LaAlO_3$	0.379	0.017	9	23.6–25.3	0.1–1.7	149, 150, 151, 156, 157
$Y-ZrO_2$	$2^{1/2} \times 0.364$[a]	0.056	10	25–32	74	150, 158
MgO	0.421	0.092	13–14	9–10	0.2–1	149, 150, 155, 159, 160
Al_2O_3	0.348	0.097	5–9	9.3–11.6	≤ 0.1	145, 146, 149, 151, 161, 162
CeO_2 (buffer)	$2^{1/2} \times 0.541$[a]	0.001	11	14	–	145, 151b, 163, 164, 165
$YBa_2Cu_3O_7$	0.382–9	–	10–16	–	–	53, 54
Nb	0.330	–	7	–	–	3
Nb_3Sn	0.529	–	9.8	–	–	3

employing in-situ deposited homoepitaxial MgO base layers [87, 114]. Sapphire (as opposed to MgO) has attracted great interest because of the availability of large wafers ($\geq \varnothing 3''$), higher chemical stability, lower dielectric loss and high thermal conductivity (Sect. 3.2). Chemical reactions between Al in the sapphire and Ba in the YBCO film at the high deposition temperatures could be prevented by the deposition of thin (≤ 40 nm) layers of a chemically inert buffer like CeO_2 [163, 164, 166]. The formation of microcracks could be related to exceeding a critical thickness $d_{F,crit}$ of the YBCO film [167]. Recently, a phenomenological model based on the elastic properties of solids [168] provided a consistent estimation of $d_{F,crit}$ for substrates with different lattice mismatch and thermal expansion coefficient (e.g., $d_{F,crit} > 1$ μm for LAO and $d_{F,crit} \geq 100$ nm for CeO_2-buffered sapphire, CbS) [146]. In accordance with this model, the critical film thickness could be enhanced by introducing structural imperfections into the growing film (e.g., by a reduced deposition temperature). Up to 680 nm thick and up to 3″ diameter large double-sided YBCO films on CbS, with low R_s and good microwave power handling capability, have become available (Sect. 4.1 and [169–172]).

Because of the good lattice matching to YBCO, SrTiO$_3$ (STO), NdGaO$_3$ (NGO) and yttrium-stabilized ZrO$_2$ (YSZ) have also been used by some groups despite the high loss tangents. As in MgO, these materials are of special importance since they are available as bicrystals and hence enable the preparation of Josephson junctions. Finally, because of the possible wafer size and related costs, and the potential for hybrid superconductor-semiconductor integration, Si and GaAs were considered as possible substrate materials for YBCO films [115b, 173]. However, the large lattice mismatch ($|a_S - a_F|/a_F > 0.4$) and the different expansion coefficients limited the critical film thickness to values which seem to be too low for microwave applications. Also, the chemical reactivity required the use of buffer or even encapsulating layers, thus complicating the fabrication technology. Recent trends have been considering the possible use of cheap polycrystalline substrate materials or even buffered metal alloys. Beside YBCO, Tl-2212 films with good microwave performance were deposited onto LAO, NGO and CbS [44,45].

5.2.2 Choice of Substrate Materials: Complex Permittivity

The complex permittivity $\varepsilon(T, f)$ affects the geometry and the performance of microwave devices (see remarks in Sect. 2.1). Formally, ε can be expressed as

$$\varepsilon(T, f) = \varepsilon_r(T, f)[1 - \mathrm{i} \tan \delta(T, f)] . \tag{5.3}$$

In analogy to the complex conductivity σ of metals, the complex permittivity of dielectric materials gives rise to a dielectric surface impedance $Z_{s,\mathrm{diel}}$ [49]:

$$Z_{s,\mathrm{diel}} = \sqrt{\frac{\mu_0}{\varepsilon_0 \varepsilon}} \approx \frac{Z_0}{\sqrt{\varepsilon_r}} \left(1 + \frac{\mathrm{i}}{2} \tan \delta\right), \tag{5.4}$$

where the second expression employed (5.3) and the low-loss approximation $\tan \delta \ll 1$. In terms of the surface impedance, a formal analogy between σ and ε can be obtained by equating $\varepsilon = -\mathrm{i}\sigma/\omega\varepsilon_0$ [143]. In this picture ε_r presents a measure of the reactive response of the dielectric (energy storage). For instance, large ε_r values in dielectrics correspond to small penetration depths (or skin depths) in superconductors (or normal metals). The dielectric loss tangent corresponds to the conductivity ratio $y = |\sigma_1/\sigma_2|$ (see related discussion in Chap. 1). The absolute values of ε_r and $\tan \delta$ as well as their dependences on temperature and frequency are in general different and therefore treated separately in the following.

Relative Permittivity

According to the physical interpretation of ε_r, and along with the description of dielectric resonators [see (2.30) and (2.33a) in Sect. 2.1.3], the relative permittivity determines the achievable degree of circuit miniaturization. Alternatively, ε_r determines the propagation velocity of the waves guided in planar microwave devices. Materials with high ε_r, like LAO, might therefore be desirable for certain applications (while undesirable for other purposes, see [143]). Furthermore, the relative variation of ε_r with temperature $(\partial \varepsilon_r / \partial T)/\varepsilon_r$ affects the achievable frequency stability of, e.g., high-quality resonators for stable oscillators [see (2.27)]. The temperature dependence of ε_r of single-crystalline dielectrics was investigated in [174]. It basically reflects the change of polarizability with volume and is thus usually weak, in accordance with the results on numerous materials collected in [44]. A detailed analysis of the permittivity of LAO [156, 175] revealed a temperature-induced increase of ε_r from 23.6 at low temperatures to 24.0 at 300 K, at which the temperature coefficient $(\partial \varepsilon_r / \partial T)/\varepsilon_r$ was typically 9×10^{-5}/K.

Recent progress in fabricating high-quality films and patterned microwave circuits uncovered additional requirements for the relative permittivity. Lumped-element circuits provide the highest possible degree of miniaturization (Sect. 2.1.1) and are thus of primary interest for the design of novel devices operating at low signal power (Sect. 6.2). The performance of such devices is very difficult to predict and hence requires numerical simulations. Since the dielectric properties are major input parameters for such calculations, the successful design and experimental verification of the device performance depend sensitively on the isotropy and homogeneity of ε_r on a microscopic scale. These circumstances might cause problems in relation to the anisotropic permittivity of sapphire [44] or LAO [151, 157]. Similarly, the substrate thickness needs to be known (and even more: controlled) on a microscopic level. The relevant length scales are given by the smallest circuit elements and can easily reach the order of 10 μm.

Loss Tangent

The loss tangent of the substrate (respectively: the dielectric) limits the maximum attainable unloaded quality factor of planar (three-dimensional) resonant circuits and related devices like filters. Rewriting (2.16b), which was already discussed in Sect. 2.1.1,

$$\frac{1}{Q_0(T,f)} = \tan \delta(T,f) + \frac{R_s(T,f)}{\Gamma} + \frac{1}{Q_{\text{hous}}}, \tag{5.5}$$

reveals that the dielectric contribution dominates the total losses if the terms R_s/Γ, resulting from the superconductor, and $1/Q_{\text{hous}}$, resulting from the microwave enclosure, are small enough. This will be the case preferentially at low GHz frequencies (where $R_s/\Gamma \leq 3 \times 10^{-6}$) and with well-dimensioned housings ($1/Q_{\text{hous}} \leq 10^{-6}$, see Sect. 5.2.3) [176]. Moreover, recent investigations of variously grown LAO crystals showed strongly sample dependent absolute values and temperature dependences of $\tan \delta$ [149, 156, 175, 176a], as partially reproduced in Fig. 5.15.

The analysis of the temperature and frequency dependent loss tangent revealed two loss mechanisms [156]. At temperatures above about 150 K, most of the investigated samples displayed intrinsic dielectric losses, which agreed quantitatively with the theoretical analysis of Sparks, King, and Mills [177] (dashed lines in Fig. 5.15). According to their understanding, the loss tangent reflects the absorption of electromagnetic energy by phonons in the dielectric material. This interaction becomes dissipative due to finite relaxation rates of the phononic excitations. The resulting expression for $\tan \delta$ depends on the phonon density of states and on the frequencies of the relevant vibrational modes. In the absence of a precise knowledge of absolute values for these quantities, the measured value $\tan \delta$ (295 K, 4.1 GHz) $= 8 \times 10^{-6}$ was used to fit the temperature dependent data to the theory [156]. The authors noted a linear frequency dependence of $\tan \delta$ between 4 and 12 GHz, and a slightly steeper dependence $f^{1.15}$ in the extended range up to 1 THz. The temperature dependence of the loss tangent of LAO was also close to linear above 150 K. This behavior is in remarkable contrast with, e.g., the loss tangent of sapphire which displays a much stronger temperature dependence $T^{4.7}$ [161]. The different scaling behavior was attributed to the widely differing phonon structures of the two materials.

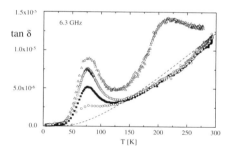 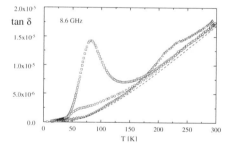

Fig. 5.15. Temperature dependent loss tangent of different LaAlO$_3$ crystals at 6.3 GHz (*left*) and 8.6 GHz (*right*) [156]. *Triangles* and *squares*, and *circles* denote Czochralski-grown and Verneuil-grown crystals. Further details are listed in [156]. The *dashed lines* represent calculated values of the intrinsic loss tangent due to dissipation of microwave energy by phonons.

Figure 5.15 demonstrates that the measured loss tangents exceeded the intrinsic contribution at temperatures below 150 K. Rather, at least one pronounced peak was observed for most samples, around 70 K, the shape and width of which varied with frequency. The peak height was strongly sample dependent, and more pronounced for Czochralski-grown than for Verneuil-grown crystals. At very low temperatures, the samples exhibited a residual loss tangent which decreased with increasing frequency.

The extremal behavior of $\tan\delta(T,f)$ was attributed to defect dipole relaxation processes [178]. Within the framework of this model, the dielectric losses can be parametrized by

$$[\tan\delta(T,\omega) - \tan\delta(0,\omega)] \propto \frac{\omega\tau}{1+(\omega\tau)^2}, \qquad (5.6a)$$

with $\omega = 2\pi f$. The relaxation time $\tau(T)$ is given by

$$\tau(T) = \tau_0 \exp\left(\frac{\Delta E}{k_B T}\right), \qquad (5.6b)$$

where ΔE denotes the activation energy for the underlying charge transport. As summarized in [156], τ_0 and ΔE provide characteristic signatures of the nature of the relaxation phenomena. Typical results, $\tau_0 \approx 0.3\,\mathrm{ps}$ and $\Delta E = 31\,\mathrm{meV}$, indicated the local motion of lattice or impurity (e.g., OH^-) ions between different equilibrium positions as the dominant mechanism for the observed losses. It was found that annealing in oxygen or hydrogen resulted in higher microwave dissipation [179]. The relative permittivity of LAO was not affected by the defect dipole relaxation. Altogether, the results indicated a promising potential of LAO as a high-permittivity low-loss substrate for use in passive HTS microwave devices. The Verneuil-type growth of LAO crystals presently limits the availability of wafers to diameters below about 30 mm. Alternatively, the identification and technological control of the defects dominating the loss tangent of Czochralski-grown crystals, though remaining challenging, seems straightforward (see triangles in Fig. 5.15 (right part) and data in [176a] as a proof of existence).

The above data were measured with as-grown dielectric samples, cut into cylindrical shape and polished. The samples formed part of a dielectric resonator enclosed in a Cu housing. A different method was reported in [176], where the loss tangent was deduced from the unloaded quality factor of disk resonators (Sect. 2.1.3). The resonators were fabricated from 0.5-mm- or 1-mm-thick LAO substrates covered from both sides with YBCO films. Quantitative analysis of the Q_0 data in the framework of the defect dipole relaxation model yielded, after subtraction of a residual loss tangent of 4×10^{-7}, the parameters $\tau_0 \approx 1.3\,\mathrm{ps}$ and $\Delta E = 22\,\mathrm{meV}$. These values differ from those reported for bare LAO samples [156] towards higher τ_0 values (about a factor of 4) and lower activation energies (by about 30%). Aside from systematic uncertainties related to the data analysis, this difference might indicate a possible alteration of the loss tangent due to the high-temperature deposition of the HTS films. As another conclusion, the Q_0 values of 2×10^5 required for the use of disk resonators in transmitter combiners of mobile communication systems (Sect. 6.2) could be achieved only at $T < 50\,\mathrm{K}$, due to the temperature dependent loss tangent of LAO. Epitaxial YBCO films on CbS substrates could be employed as an alternative. However, the disadvantages of the lower permittivity compared to LAO are larger dimensions of the resonator and weaker field confinement to the dielectric.

In summary, various microwave components and subsystems have been demonstrated on LAO, MgO and CbS, and are approaching the level of commercialization (Chap. 6). However, exploiting the full potential of HTS microwave devices might require the development of novel dielectrics, and a complete characterization and engineering of their temperature and frequency dependent microwave properties. Most desirable in this respect seem to be polarizable materials, the permittivity of which can be controlled by electric or magnetic fields without affecting the required low loss tangent.

5.2.3 Microwave Design

Suitable deposition methods for high-quality HTS films and proper substrate materials constitute the two major requirements for a real device. Still, in order to develop novel circuit layouts, or even to account for the properties of superconductors and dielectrics at high frequencies, reliable simulation techniques are needed. Furthermore, some guidelines concerning the fabrication and mounting of a superconducting device must be followed in order to prevent parasitic effects induced by, e.g., improper handling and/or unwanted coupling to a dissipative and noisy environment.

Simulation of Device Characteristics

Single microwave circuit elements usually consist of coplanar, microstrip or strip transmission lines (Sect. 2.1), all of which display specific advantages and disadvantages (single- or double-sided films, mono- or multilayer techniques, characteristic impedance, bandwidth, radiation losses, current enhancement at edges, sensitivity to substrate properties and so on). The optimum design will depend on the purpose of the device and has to be decided individually.

For a given type of transmission line, the surface impedance of the superconductor determines the complex propagation constant γ of the circuit. The parameter γ was evaluated at fixed temperature and frequency for stripline [180] and coplanar [181] geometries, using the complex surface impedance of the superconductor as a known input parameter. Alternatively, the temperature, frequency and power dependent surface impedance of HTS was parametrized in order to provide a more general approach [182, 183]. However, this phenomenological description of Z_s suffered from a strong dependence of the model parameters on the sample quality.

Many commercial computer-aided design codes for modelling planar microwave devices are available. However, the adventure of HTS and the demonstration of various devices like, e.g., multipole filters, uncovered significant weaknesses in several of these simulation techniques (e.g., [141, 184]). First, the large penetration depth of HTS compared to typical values of the film thickness d_F cannot be described by the standard default limits ($d_F \to \infty$ or $d_F = 0$). Second, the high permittivity ($\varepsilon_r > 18$) of some substrates adopted for HTS film deposition (see Table 5.1) was not accounted for in many commercial computer codes. Third, some codes for the design of microwave filters could not handle very small bandwidths ($\leq 1\%$) as these have become available with the high unloaded quality factors of HTS filters (Sects. 2.1 and 6.1). Finally, the high degree of miniaturization accessible with HTS leads to a high sensitivity of the device to parasitic effects. Such effects can result, for instance, from a residual inductive behavior of an interdigital capacitance, or from adjacent circuit elements which are coupled by fringe fields. The latter effect needs special consideration if coupled elements, e.g., for multipole filters, are concerned.

Patterning, Metallization and Mounting

A final and decisive step towards an operational microwave device is packaging, including patterning, metallization, bonding, cabling, mounting and cooling. Patterning of the HTS films is usually required to achieve a desired function. Fortunately, no deleterious effects could be detected for chemical and physical patterning procedures, neither in terms of temperature and power dependent surface impedance measurements [185], nor in terms of frequency intermodulation [186]. Chemical etching usually provides structures with geometrical edge definitions of the order of a few μm. Miniaturized circuit elements or highly integrated systems require, therefore, physical etching processes such as argon ion beam etching, e. g., through metallic masks.

Metallization of superconducting films might be required for several reasons. In many cases a galvanic coupling between the external microwave circuit (source, detector, processing network, etc.) and the device is needed. Since the microwaves are guided by the high-permittivity dielectric and the superconducting films, issues like impedance matching, thermal expansion and mechanical robustness need to be considered. Suitable microwave connections can be achieved by wire or ribbon bonding between standard connectors and metallized regions of the films. In other cases, homogeneous low-loss electrical contacts between HTS circuits and microwave housing might be necessary in order to prevent parasitic modes (e.g., stripline modes propagating in ill-defined air gaps), and to minimize losses between signal lines and ground planes. Low contact resistance and good mechanical adhesion of the metallic layers and the superconducting film can be provided by in-situ deposition of thin noble metal films like gold or silver. If the thickness of the normalconducting layer has to exceed the skin depth at a certain frequency, additional material can easily be deposited ex-situ. Annealing procedures and the deposition of cap layers, e.g., to protect YBCO films on CbS from moisture, might complete the technological steps that ensure proper functioning of the devices.

An appropriate dimensioning of the enclosure is essential for shifting parasitic resonances out of the frequency band of interest (e.g., [176a]). The packaging must also be mechanically robust to withstand the envisaged environmental conditions. This is of special concern in applications of HTS microwave systems in satellite systems [187]. Finally, to operate a superconductive system, sufficiently low operating temperatures have to be provided, sustained and controlled within prescribed time intervals and tolerances. Because of the many areas of potential applications, there is no general design solution. However, "efficient" cooling means to provide conditions for reliable physical operation and to keep costs, weight and maintenance intervals at an economically tolerable level. Refrigeration is therefore of major importance and thus will be treated separately in more detail in the following section.

5.3 Refrigeration

Refrigeration is a major concern for superconducting devices, irrespective of their type (e.g., filters, Josephson devices, or others). However, the technological driving forces for the application of superconductors conflict with those for the use of refrigerators (see also Fig. 6.9 in Sect. 6.2). The unique benefits of the former appear more and more clearly as their operating temperature is reduced (especially to values below about $T_c/2$). In contrast, the efficiency of coolers increases with temperature. The acceptance of cryogenic devices will therefore depend in a complicated way on a sensitive balance between total system performance and net economic efficiency. The following sections deal with those aspects that have to be considered for such a balance. After briefly recalling some basic principles of thermodynamics (Sect. 5.3.1), the cooling capability of liquefied gases is summarized in Sect. 5.3.2. Section 5.3.3 describes the advances and remaining challenges in the technology of closed-cycle cryocoolers.

5.3.1 General Considerations

Thermodynamic Principles

The transition to the superconducting state requires (trivially) the generation of cryogenic temperatures by extraction of heat from a cold stage. Much less trivial is the fact that the efficiency of extracting heat by expending mechanical work is smaller than unity, in contrast to the perfect conversion of mechanical work into heat (e.g., by friction). Heat and mechanical work are thus not equivalent. This discrepancy is the basic experience entering the first and second principles of thermodynamics [188]. There are two corresponding types of "thermodynamic" machines: one converts heat into mechanical work (like a steam engine) whereas the other applies mechanical work to extract heat (like a heat pump). The efficiency η of either type of machine is usually defined as the ratio

$$\eta = \frac{\text{extracted energy}}{\text{input energy}} \leq 1 \,. \tag{5.7}$$

Standard textbooks derive the efficiency of a reversible cyclic process by assuming an ideal gas. The cycle starts with an isothermal expansion at the warm temperature T_{hot}. Then follow an adiabatic expansion and an isothermal compression. The cycle is completed by an adiabatic compression. Reversibility is achieved if the system remains at any instant in equilibrium as, e.g., in slow processes when kinetic effects can be neglected. While reversible processes usually do not occur in reality, they define the maximum possible efficiency. This value, known as the "Carnot" efficiency η_{C}, is found for the heat machine to be

$$\eta_{\text{C}}^{\text{heat}} = \frac{\text{produced mechanical work}}{\text{input heat at } T_{\text{hot}}} = \frac{T_{\text{hot}} - T_{\text{cold}}}{T_{\text{hot}}} \,. \tag{5.8a}$$

It shall be noted that this efficiency is independent of the working gas (it be an ideal or a real gas), and depends solely on the temperatures of the hot (input) and cold (output) stages. A large temperature difference obviously leads to a high efficiency. A heat pump (or a refrigerator) operates in the reverse sequence. Accordingly, the resulting Carnot efficiency is high at small temperature differences between cold and warm stages:

$$\eta_{\text{C}}^{\text{cold}} = \frac{\text{extracted heat at } T_{\text{cold}}}{\text{input mechanical work}} = \frac{T_{\text{cold}}}{T_{\text{hot}} - T_{\text{cold}}} \,. \tag{5.8b}$$

The meaning of (5.8a) and (5.8b) can be further clarified by imagining a series connection of the two types of machines. The work produced by the first machine is used by the second for cooling. Such an arrangement should work without losses under ideal assumptions. We find indeed the total efficiency

$$\eta_{\text{C}}^{\text{cold}} \cdot \eta_{\text{C}}^{\text{hot}} = \frac{\text{extracted heat at } T_{\text{cold}}}{\text{input heat at } T_{\text{hot}}} = \frac{T_{\text{cold}}}{T_{\text{hot}}} \leq 1 \,. \tag{5.8c}$$

The amount of heat exchanged at the two temperatures corresponds to their ratio. This result is a direct consequence of the second principle of thermodynamics, that the entropy of thermodynamic systems always increases, or stays constant for reversible processes.

For low temperatures $T_{\text{cold}} \ll T_{\text{hot}}$, (5.8b) yields an efficiency which decreases almost linearly with temperature. For instance, for the two fixed temperatures 77 K (boiling point of nitrogen) and 4.2 K (boiling point of helium) one obtains at normal pressure (10^5 Pa) and at $T_{\text{hot}} = 300$ K maximum efficiencies $\eta_{\text{C}} = 35\%$ and 1.4%, respectively. Vice versa, the operating temperatures must stay above about 30 K in order to keep the Carnot efficiency above 10%.

These simple estimations illustrate the importance of high operating temperatures, i.e., values of T_{cold} not too far below T_{hot}. Even more serious, real refrigerators display technical efficiencies much below the Carnot value. Significant amounts of energy are lost during the conversion of electric energy to mechanical work. Further energy is dissipated in irreversible processes, e.g., due to turbulences of the streaming gas, friction in moving parts and so on. As a specific consequence of the cooling penalties, Nb_3Sn films will most likely be adopted for microwave devices only if the expense of refrigeration can be overcompensated by the performance of the device. Similar arguments might also apply to the oxide superconductors, which have to compete with conventional (until now: room temperature) systems.

One well-known cooling mechanism is the exploitation of the latent heat of liquefied gases [188] (see also Sect. 5.3.2). Heat is absorbed from the cold stage by inducing the first-order transition from liquid to the vapor phase. Alternatively, refrigeration can be obtained by expanding working gases (or gas mixtures) at constant entropy or constant enthalpy (Sect. 5.3.3). The latter process is known as the Joule-Thomson effect [189, 190]. The efficiency of gas expansion machines can be strongly improved by using heat exchangers. Recuperative systems exploit the steady backflow (DC) of cold gas towards the warm stage in order to enhance the cooling process. Regenerative systems use the large heat capacity of suitable materials (usually in the form of fine meshes or microscopic particles, which display a large specific surface) to extract heat in periodically driven (AC) systems. Closed-cycle coolers are usually based on AC operation. The cooler performance results from the flows of enthalpy and entropy and from the hydrodynamic work, and depends on the driving frequency. Strong refrigeration systems operate usually at low frequencies ($\approx$ Hz) and employ multiple cooling stages. More details about the physics and the technology of refrigerators can be found, e.g., in [189, 190].

Requirements for Realistic Refrigeration Systems

As indicated in the previous paragraph, there can hardly be commercial applications of high-temperature superconducting systems without appropriate cooling systems. The meaning of "appropriate" depends on the specific type of application. For HTS microwave systems, the basic concerns were summarized, e.g., in [191, 192] in a sequence which does not necessarily reflect the importance of each statement.

(1) The desired operating temperature T_{op} is a major input parameter. The cooling efficiency increases and, even more importantly, the required electrical input power, the size and the weight of the cryocooler decrease with increasing T_{op} [193]. On the other hand, the absolute levels of surface resistance $R_s(T)$ and penetration depth $\lambda(T)$ (Chaps. 1 and 2), the microwave field and power handling (Chaps. 3 and 4) improve with decreasing temperature, as does the temperature stability of these parameters. Furthermore, thermal noise becomes less pronounced at reduced temperatures. The resulting optimum temperature range for the application of HTS in microwave devices

276   5. Technology of High-Temperature Superconducting Films and Devices

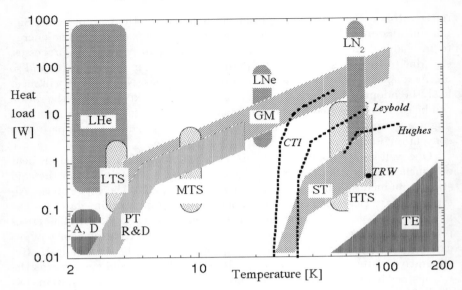

Fig. 5.16. Cooling power versus temperature for different cooling mechanisms. *Dark grey areas* indicate the regions accessible with the liquefied gases LHe, LNe, and LN_2. The *dotted areas* denote the regions for microwave applications of HTS ($T_c > 50$ K, e.g., YBCO), MTS (10 K $\leq T_c \leq 50$ K, e.g., Nb_3Sn) and LTS ($T_c \leq 10$ K, e.g., Nb). Abbreviations are: A = adiabatic demagnetization, D = dilution refrigeration, TE = thermoelectric refrigeration, GM = Gifford–McMahon, ST = Stirling, PT = pulse tube (R&D denotes the most advanced level). The figure is an update of [142], according to [197a].

is expected to be in the range between 50 and 80 K (see Figs. 5.16 and 6.9). Only for very high-performance applications might even lower temperatures be tolerable.

(2a) The net cooling power required for stable and reliable operation of a device needs to be known. Its value depends on the complexity of the system (single- or multi-element system), the internal power consumption (e.g., low-power operation in receivers or high-power in transmitters), the external heat loss due to mechanical support and electric leads, and the total thermal capacitance to be cooled. The latter is determined by the masses and thermal capacities of the materials used. A tentative interval of 100 mW to 10 W was assumed in Fig. 5.16 [194].

(2b) The total electrical input power relative to the net cooling power, sometimes called "technical efficiency" or "coefficient of performance" (COP), determines the choice and possible limitations of the refrigerator. The COP depends on the cooling capacity of the system (being higher for large systems) and on the operating temperature (passing through a maximum around 70 K). Considering an average value of $1\% \cdot \eta_C$ (respectively 5%) at 20 K

(80 K), the required plug-in powers amount to 1400 W per Watt (60 W per Watt) [190].

(2c) Weight and volume are of major relevance for the application of HTS microwave devices, especially in satellite-based systems [187]. The benefit of miniaturized and/or high-performance HTS circuits must not be compensated by bulky or heavy refrigerators. Such a counterbalance is likely to prevent HTS devices from being applied. At the beginning of 1997, typical data for commercial coolers were 200, 40, 15 and 1–2 kg per Watt cooling power at 4.2, 10, 20 and 80 K, respectively [193]. The cooling capability, on the other hand, decreases at reduced swept volumes (and thus cooler size) and/or increase of cooler frequency (which scales inversely with volume in some cooling cycles).

(3) Reliability has been a major issue for all types of refrigerating systems. A common statistical quantity for reliability is the "mean-time before failure" (MTBF) [191]. Other possible definitions can be based on the shelf life, the number of cool down cycles and so on. Liquefied gases handle extreme cooling situations and have very long MTBF, but require expensive and logistically complicated maintenance (Sect. 5.3.2). Closed-cycle cryocoolers were, until recently, developed mainly for two extreme areas: for military purposes (airborne environmental surveillance) or satellite-based systems (military and civilian). While for the former application relatively cheap systems with a low MTBF of a few 10^3 hours are available, refrigerators for the latter purpose have to be extremely reliable (MTBF$\approx 10^5$ h).

(4) While the costs of refrigeration systems seem to be of minor importance from an academic point of view, their impact on real-world applications of cryogenic systems cannot be overestimated. Until recently [194], a simple rule-of-thumb related the costs C of a cryocooler to its reliability: C [US\$] $\approx$ MTBF (h), i.e., $C \approx 10^3$ and 10^5 US\$ for the previously mentioned examples. The ongoing challenge is therefore, for obvious reasons, the development of sufficiently reliable (MTBF $\geq 10^4$ h) and at the same time cheap ($\leq 10^3$ US\$) cryocoolers. Feasibility studies have revealed various ways of simplifying the technical construction of the coolers and thus to reduce the related costs (e.g., by developing modular systems, reducing the number of accessories, simplifying the assembly etc.). However, there is a consensus between competitive parties that a breakthrough in this respect depends mainly on the reliable predictability of a large number of sales per year (about 10^3–10^4) [190b].

There are a few further issues to be addressed in relation to the suitability of coolers: thermal cooldown and relaxation times present an important criterion for a real device. The time scales depend on the device architecture and the packaging as well as on the net cooling power available at the operating temperature. Realistic values considered for civil applications are of the order of 10^1–10^2 minutes. The different types of cryocoolers differ also strongly in noise and fluctuations of the operating parameters. Stable and precise temperature control, low-level mechanical vibrations and low electromagnetic interferences are vital for applications of HTS to sensitive devices like SQUIDs, bolometric detectors, or mixers. Novel arrangements of cold heads and sample holders are being developed to minimize noise by means of mechanical, spatial, temporal and electronic compensation (e.g., [195, 196]).

Along with these numerous criteria, there are many different solutions possible. Furthermore, the rapid progress of HTS technology has been pushing the development of reliable miniaturized coolers [142, 197]. Various companies have begun to develop closed-cycle cryocoolers tailored to the specific needs of customers (dashed lines in Fig. 5.16). Different specifications complicate a direct comparison of the cooler performance, especially in the case of miniaturized systems which provide less cooling power.

5.3.2 Liquid Coolants

Liquefied gases such as liquid nitrogen have rarely been considered serious candidates for the refrigeration of HTS microwave systems because of the complex and frequent maintenance required. However, they have some remarkable features that might be relevant for specific applifications of HTS devices. Liquid coolants allow for immersion cooling or vapor cooling. Numerous revolutionary architectures of bath and gas flow cryostats have been developed since the pioneering work of H. K. Onnes (see Refs. [2, 3] in the Preface). Only a limited temperature range is accessible in bath cryostats and is determined by the thermodynamic phase diagrams of the gases (marked in Fig. 5.16 by crossed areas denoted by LHe, LNe, and LN_2).

A major advantage of immersion cooling is the very large cooling capacity for very large areas. As one extreme example, large-scale particle accelerators employing LTS cavities and magnets have been designed with cryostats having volumes of several 10^6 cm^3 and a refrigeration power of tens of kW at an efficiency of up to $20\% \cdot \eta_C$ even at $T = 2\,\mathrm{K}$ [198, 199]. A high reliablity with MTBF $= 5 \times 10^4$ h was reported, and was usually limited by the reliability of the helium compressor.

At the other extreme, gas flow cryostats have been constructed for high-precision measurements such as biomagnetic signal recording [200]. The volume of miniature portable cryostats ranges between 10^2 and 10^3 cm^3. The cryostats were completely made of nonmagnetic fiberglass materials with very low residual magnetic noise, allowing operation in unshielded environments at a level of 20 to 40 fT/(Hz)$^{1/2}$. The temperature could be stabilized within a few mK, and the low helium evaporation enabled continuous operation over some 10 to 100 h. Cryostats operating in upright and reverse positions have been built, as well as systems with separations of only a few mm between the cold sample and ambient temperature. Similar dewars have been constructed for liquid nitrogen (LN_2) [200, 201]. Due to the higher temperature, cold–warm distances below 1 mm were achieved. Since the latent heat of LN_2 is a factor of 40 higher than that of LHe, the need for refilling is correspondingly less frequent.

5.3.3 Closed-Cycle Cryocoolers

Gifford–McMahon and Stirling Coolers

Unless bulky systems need to be cooled, or extremely stable and quiet operating conditions are required, the use of closed-cycle cryocoolers (CCC) appears favorable. The most prominent types are based on the Gifford–McMahon and the Stirling thermodynamic cycles [189, 190] (GM and ST in Fig. 5.16). A certain volume of gas, e.g., some liters of helium, is periodically transferred between a compressing and an expanding stage with intermediate regenerative stages. A phase shift between the mass flow and the pressure cycle provides the net cooling effect [202]. The compressing stage in GM refrigerators is placed outside the cold stage, in contrast to the integral-type Stirling cooler. As a result, GM systems produce less vibrations, are less sensitive to gas contamination and had a higher reliability (MTBF $\geq 4 \times 10^4$ h) than ST systems (MTBF $\geq 5 \times 10^3$ h) [191]. Commercially available GM coolers typically provide 10 to 25 W refrigeration power at 20 K, but can hardly be miniaturized [193, 197]. Nevertheless, the helium compressor can be kept in a remote position, up to some 10 m away from the cold stage. Recent progress in the development of CCC resulted in cooling capacities of 0.5–1 W at 4.2 K with compressor powers of about 6–7 kW and technical efficiencies of $(1-3)\% \cdot \eta_C$. Similar performance was achieved with a two-stage GM cooler in combination with rare-earth regenerators, a three-stage GM cooler in combination with a Joule-Thomson expansion stage, and a three-stage GM cooler exploiting a specific thermodynamic cycle ("Boreas" cycle) [203]. These achievements have led to the recent commercialization of cryogen-free LTS magnets providing about 6 T at 5 K.

In contrast to the GM machines, Stirling cryocoolers are suited for applications where small and lightweight systems and small cooling capacities are required. Figure 5.17 presents a schematic view of a Stirling refrigerator. The displacer located at the cold stage provides the required phase shift between mass flow and pressure wave [202]. At a hypothetical 90° phase shift (realistic values are around 30°) between the volume variation at the cold end and the compressor cycle (performing the mechanical work W), the gas is compressed at the hot end and expanded at the cold end. As a result, the heat Q is absorbed at the cold head, and an amount Q_0 is produced at the hot stage. In contrast to the valve control of GM machines which dissipate energy, the reversible motion of the ST displacer results in a superior cooling efficiency. However, the cold moving displacer usually limits the maximum achievable lifetime of such systems. Various improvements have been reported recently, based on the clearance seal technology, the use of linear instead of rotary drive motors and the handling of gas contamination [191,

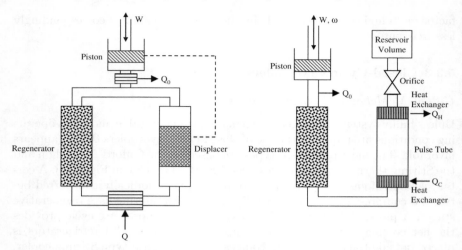

Fig. 5.17. Schematic of a Stirling cooler (*left*) and an orifice pulse tube refrigerator (*right*). Thermodynamic parameters like mechanical work W at frequency ω, and heat energies Q_0, Q, Q_H and Q_C produced or absorbed at the different stages are indicated [202].

193]. Modern Stirling coolers yield a minimum temperature of 30–55 K, a cooling capacity between 0.2 and 20 W at 80 K, and MTBF values between 10^3 and 5×10^4 h. The mechanical vibrations caused by the displacer motion can be depressed by anti-phase locking of two linear compressors on a common axis. However, magnetic interference of this cooler type might be a drawback for highly sensitive detectors [196].

Pulse Tube Coolers

A very promising modification of the Stirling cooler is the pulse tube cooler (PT in Fig. 5.16, schematic in Fig. 5.17), in which the mechanical displacer is eliminated from the expander. The function of the displacer is performed by an oscillating pressure wave (the "pulse"). The PT cooler has no moving parts in the cold stage and thus has great potential for high reliability and miniaturization. The concept of the pulse tube refrigerator has been known since 1963 [204], and has been continuously improved since then. Enhanced cooling capability has been obtained by modifications of the pulse tube [205], e.g., by implementation of an orifice [206], of multiple inlets [207], and of sophisticated valving. The required thermodynamic phase shift is usually obtained by mass flow through an orifice to a reservoir [202]. This process is irreversible, and to achieve an efficiency of pulse tube coolers, which is approximately $T_{\rm cold}/T_{\rm hot}$ [190], comparable to that of Stirling coolers has been challenging.

There is a variety of parameters which need to be adjusted for optimum operation, such as the average pressure, the amplitudes and shapes of the pressure waves, the gas volume and the frequencies [205]. Lower volumes, which are desirable for miniaturized systems, correlate in general with higher frequencies and lower efficiency. The involved thermodynamic processes can become rather complex. They are partially understood in terms of theory [202], partially numerical methods have been applied to evaluate optimum parameter sets [197]. Improvements to PT coolers are presently under intense development, for example, various drive types, combination of multiple stages for higher refrigeration power, miniaturization and noise cancellation have been tested [208]. The presently achieved performance of single-stage PT coolers provides cooling capacities of several W at 80 K and minimum temperatures of 30–60 K. Two-stage laboratory versions achieved a minimum temperature of about 2.5 K and a cooling capacity of 0.17 W (respectively: 0.5 W) at 4.2 K and with a compressor power of 1.7 kW (6.3 kW) [209]. The technical efficiency at 4.2 K reached $1\% \cdot \eta_C$. A modified version of such a PT cooler was proven powerful enough to liquefy helium, initially at a rate of 130 ml/h.

Cryogenic Systems and Perspectives

Efforts to integrate complete cryogenic HTS circuits started hesitatingly ([192]; Sect. 6.2), but recently have been proceeding more rapidly. An initial push came from the High-Temperature Superconductivity Space Experiment (HTSSE) conducted by the Naval Research Laboratory in Washington, D.C. [210, 211]. The final system package including about 10 HTS microwave subsystems (contributing a mass of about 3 kg), cold bus, ambient temperature frame, cryocooler and related electronics, and multilayer insulation weighed about 30 kg. Under optimized thermal packaging, the total heat load amounted to 350 mW at about 80 K. Demonstration of on-orbit operation of this system is proceeding. Progress in cryogenic systems was recently announced by various companies, in terms of rack- or mast-mountable HTS microwave filter packages or complete cryogenic nuclear magnetic resonance probes (Sect. 6.2).

In summary, GM, ST and PT refrigerators provide sufficient refrigeration power for the cooling of HTS microwave systems (Fig. 5.16). Improved reliability, miniaturized designs and excellent noise performance have been demonstrated. Comparing the present state of the art in CCC technology with that of a few years ago [142], rapid progress can still be observed. The number of proprietary claims and patent applications in this area is increasing. Besides being a prerequisite for applications of HTS devices, the availability of compact and cheap CCCs is also expected to push the development of cold semiconductor electronics, hybrid semi-superconducting technology, as well as cryogenic physics in general.

6. Passive High-Temperature Superconducting Microwave Devices

The best motivation for applications is the persuading fascination of having discovered something new – and useful.

Just as for surface impedance measurements of HTS, there have been innumerable publications on passive HTS microwave components. It is therefore impossible, and also impractical, to consider all of them. But it seems advisable to illustrate the basic concepts of a few elementary types of devices which represent key components of operational systems. (A similar procedure was applied, e.g., in [1, 2].) The interested reader may find advanced information on the design and engineering of superconducting microwave devices in recent books (see, e.g., references in the preface).

While microwave filters are merely one example among a variety of passive and active microwave devices, they have been attracting major commercial interest. For this reason, as well as for illustration purposes, Sect. 6.1 contains a brief introduction to filter theory. The superb performance of HTS microwave filters, as well as of other types of passive devices, is promising for applications in communication and radar systems, in medical diagnostics, and for environmental surveillance and prospecting. Section 6.2 summarizes the potential benefits and prospects of such applications. Finally, Sect. 6.3 is devoted to a brief description of microwave Josephson devices. While Josephson technology constitutes a field of research and development of its own [3], it is its overlap with microwave technology that shall be discussed in this section. The variety offered by the combination of the two technologies might present a key for future fully integrated superconducting device applications.

6.1 Basic Features of Filters

Microwave filters are composed of a number of individual resonators that are coupled mutually and externally. Therefore, the unique features of superconducting resonators, that is their being either extremely small or displaying unprecedentedly high quality factors, map into the specific properties of superconducting filters. This shall be illustrated in the following in three steps. Section 6.1.1 describes the "evolution" from a single resonator towards a two-pole filter. While being purely phenomenological, this description aims to introduce some of the basic parameters of filters. Section 6.1.2 provides additional background on the various filter types, and on the design of filters with prescribed frequency response. The correspondence between the properties of filters and those of the constituting resonators is treated in Sect. 6.1.3.

6.1.1 From Resonators to Filters

Single Resonators

The characteristic parameters of superconducting resonators were treated in Chap. 2 in relation to measurement systems for fundamental investigations. In this chapter, resonators are considered key elements of microwave devices, and a distinctly different point of view arises. For the sake of consistency, some features are briefly repeated in the following.

The unloaded quality factor Q_0 of a resonator is defined as the ratio of the circulating power $P_{c,r}$ to the dissipated power $P_{d,r}$:

$$Q_0 = \frac{P_{c,r}}{P_{d,r}} = \gamma_{\text{cond}} \frac{R_s}{Z_0} + \gamma_{\text{diel}} \tan \delta . \tag{6.1}$$

The additional index "r" refers to the properties of a single resonator in order to distinguish it from those of a complete filter (which will be identified by the index "f"). The circulating power is proportional to the stored energy [see (2.10) in Sect. 2.1.1], and to the square of the microwave field amplitude H_s [see (2.14)]. Dissipation of microwave energy can be due to the finite surface resistance $R_s(T, f, H_s)$ of the superconductor, and due to the loss tangent $\tan \delta(T, f)$ of the dielectric (Sect. 5.2). In order to couple energy into and out of the resonator, the unloaded quality factor is reduced to the loaded value Q_L:

$$\frac{1}{Q_L} = \frac{1}{Q_0} + \frac{2}{Q_e} . \tag{6.2}$$

Equation (6.2) assumes symmetric coupling at the input and the output of the two-fold-coupled resonator. The power transmitted through the resonator, normalized to the power available at the input, defines the transmission coefficient $|S_{21}|^2 \equiv t^2$ [see (2.18) and (2.19)]:

$$t^2(u) = \frac{t_0^2}{1 + 4Q_L^2 u^2} , \tag{6.3a}$$

where t_0^2 is the transmission at resonance, $f = f_0$ (or $u = 0$):

$$t_0^2 = \frac{4Q_0^2/Q_e^2}{(1 + 2Q_0/Q_e)^2} , \tag{6.3b}$$

and u is the normalized frequency deviation relative to the resonance:

$$u = \frac{f - f_0}{f_0} . \tag{6.3c}$$

Equation (6.3) is valid only close to the resonant frequency, i.e., for small values $u \ll 1$. High-power transmission requires strong coupling, and thus $Q_e \ll Q_0$. This condition is in contrast to fundamental measurements where $Q_e \geq Q_0$ is usually preferrable (Sect. 2.1).

In conventional filter theory, the transducer loss ratio L is an important figure of merit (see, e.g., [4]). It is related to the transmission coefficient, and usually expressed in decibels:

$$L_1(u) = -20 \log[t(u)] \approx 10 \log \left[\left(1 + \frac{Q_e}{2Q_0}\right)^2 + u^2 Q_e^2 \right] . \tag{6.4a}$$

The second part of (6.4a) resulted from inserting (6.3) for $t(u)$. The index "1" indicates the number of resonators considered. At resonance, the loss ratio can be approximated by

$$L_1(0) \approx 20 \log \left(1 + \frac{Q_e}{2Q_0}\right) \approx 4.343 \frac{Q_e}{Q_0} . \tag{6.4b}$$

The prefactor 4.343 is exactly $10/\ln 10$, and the approximation in (6.4b) refers to the assumption $Q_e/2Q_0 \ll 1$ (expansion of the logarithm around the argument of value 1).

It shall be noted that L, in general, should not be confused with the insertion loss IL. The latter is defined as [4]

$$\text{IL} = \frac{\text{power absorbed in a load without circuit}}{\text{power absorbed in the load with circuit}} . \tag{6.5}$$

According to this definition, the value of IL depends on the internal resistance R_{Sc} of the source as well as on the load resistance R_L. However, in the case $R_{Sc} = R_L$ the insertion loss and the transducer loss ratio are equal. We will tacitly assume this condition to hold in the following.

Figure 6.1 displays the frequency dependence of L_1 for a single strongly coupled resonator with $Q_e = 0.1 Q_0$ for two values of Q_0. The inset provides a magnified view near resonance. The attenuation at center frequency ($u = 0$) is equal for the two cases since $L_1(0)$ is determined by the ratio Q_e/Q_0 [see (6.4b)]. In contrast, the 3-dB bandwidth [i.e., the difference of frequencies $u_\pm$ at which $L(u_\pm) - L(0) = 3\,\text{dB}$] of the high-$Q$ resonator is much smaller, and the attenuation outside this passband (usually called the stopband) is much larger than for the low-Q resonator.

Although this result is very well known and appears hardly worth mentioning, it introduces some of the major benefits related to employing superconducting resonators instead of conventional (normalconducting) ones. Figures of merit of filters such as the bandwidth and the skirt [the steepness of $L(u)$ at the edges of the passband] are found to benefit from the high Q value of the resonators. This, in turn, is due to the extremely low surface resistance R_s of superconductors (given a sufficiently low loss-tangent of the involved dielectric). In a similar way [2], the low R_s improves the radiation efficiency of antennas and the insertion loss of transmission lines. Analogously,

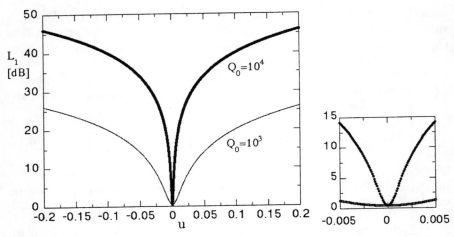

Fig. 6.1. Frequency dependent transducer loss ratio $L_1(u)$ calculated from (6.4) for $Q_e = 0.1 Q_0$ and $Q_0 = 10^3$ and 10^4, respectively. The *inset* magnifies $L_1(u)$ at $u \approx 0$.

the surface reactance affects the center frequency of filters and its stability, the energy stored in the near field of radiation elements, as well as the kinetic inductance, the propagation velocity and the dispersion of transmission lines.

Identical Pairs of Resonators

An important step towards making a filter is the coupling of two (identical) resonators such that they can exchange energy. The coefficient k of mutual coupling is defined as usual ([5], see also Sect. 6.3). The resulting loss coefficient $L_2(u)$ was calculated explicitly as [6, 7]

$$L_2(u) = 10 \log \left[\frac{\beta^2}{4} \left(\frac{1}{q} + q \right)^2 + 2u^2 Q_e^2 \left(\frac{1}{q^2} - 1 \right) + 4u^4 \frac{Q_e^4}{q^2 \beta^2} \right], \quad (6.6a)$$

where the following definitions allowed for a compact notation of L_2:

$$\beta \equiv 1 + \frac{Q_e}{Q_0}, \quad (6.6b)$$

$$q \equiv \frac{k}{1/Q_e + 1/Q_0}. \quad (6.6c)$$

The midband attenuation (at $u = 0$) is accordingly given by

$$L_2(0) = 20 \log \left[\frac{\beta q}{2} \left(\frac{1}{q^2} + 1 \right) \right]. \quad (6.6d)$$

One major difference from the result for a single resonator is the introduction of the new parameter k (or, equivalently, the auxiliary variable q), which strongly modifies the frequency response. Obviously, the cases $q \ll 1$, $q = 1$, and $q \gg 1$ result in different $L_2(u)$ behavior [see (6.6a)]. This is illustrated in Fig. 6.2 for $Q_0 = 10^3$ and with $Q_e = 0.1 Q_0$ as before.

If very strongly coupled ($q \gg 1$), the two resonators display two distinct minima of $L_2(u)$. The frequencies $u_{\min}$ of these minima merge as q approaches unity, as can be verified by inspection of (6.6):

$$u_{\min} = \pm \frac{q^2 \beta^2}{4 Q_e^2} \left(1 - \frac{1}{q^2}\right) \quad \text{for } q \geq 1 . \tag{6.7}$$

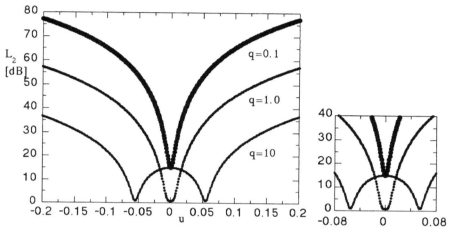

Fig. 6.2. Frequency dependent transducer loss ratio $L_2(u)$, calculated from (6.6) for $Q_e = 0.1 Q_0$, $Q_0 = 10^3$ and $q = 0.1$, 1.0, and 10, respectively. The *inset* magnifies $L_2(u)$ at $u \approx 0$.

The in-band variation of $L_2(u)$, $\Delta L_2 = L_{\max} - L_{\min}$ is usually called "ripple", and referred to below as L_r. For the symmetrically coupled twin resonator, (6.6) yields the specific result:

$$\Delta L_2 = -20 \log \left(\frac{2q}{1+q^2}\right) , \tag{6.8}$$

which depends solely on the "internal" coupling strength q (6.6c).

Exactly at $q = 1$ (critical coupling), the first argument of the logarithm in (6.6a) reaches a minimum, and the u^2 term vanishes. As a result, the two resonances constitute a frequency band of flat attenuation at the level $L_2 = 2 L_1(0)$ [see (6.4b)], and with zero ripple, $\Delta L_2 = 0$ [see (6.8)]. The fractional bandwidth w of the passband can be defined as before: Given a criterion for the maximum tolerable in-band attenuation L_{crit} (e.g., $L_{\text{crit}} = 3\,\text{dB}$), w is fixed by $w = u_+ - u_-$, where the frequencies $u_\pm$ are determined such that $\Delta L_\pm = L(u_\pm) - L(0) = L_{\text{crit}}$.

For very weakly coupled resonators ($q \ll 1$), only one resonance appears, but with enhanced attenuation at center frequency. Altogether, the filter characteristic at given Q_0 values of the resonators is completely determined by the values of Q_e and q.

Figure 6.3 compares the insertion loss of the single resonator from Fig. 6.1 with that of the twin resonator (Fig. 6.2) for $q = 1$ at the two different Q values $Q_0 = 10^3$ and 10^4. Comparing $N = 1$ and $N = 2$ at a given Q_0 value shows that the coupled resonators provide a flatter passband and a steeper stopband rejection. This trend reflects the typical effect of increasing the number N of resonant elements (usually called "poles") in a certain filter design. Comparing $Q_0 = 10^3$ and 10^4 at $N = 2$ shows that enhanced Q factors also lead to sharpened skirts of the filter, though at the expense of a reduced bandwidth. This disadvantage can be compensated, e.g., by further increasing N [8], thus requiring powerful filter design procedures.

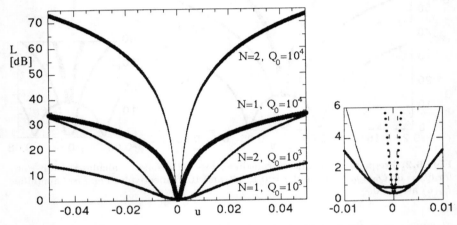

Fig. 6.3. Frequency dependent transducer loss ratio $L(u)$ of the single resonator ($N = 1$, *dots*, Fig. 6.1) and the twin resonator ($N = 2$, *bars*, Fig. 6.2) with $Q_e = 0.1 Q_0$ and $q = 1$ for the two Q values $Q_0 = 10^3$ (*small*) and $Q_0 = 10^4$ (*large symbols*). The *inset* magnifies $L(u)$ at $u \approx 0$.

6.1.2 Different Types of Filter Characteristics

The interdependences between fractional bandwidth, steepness of the skirts, and the occurrence of ripples in the passband constrain the filter design. In principle, filters could be made by increasing the number of resonators $N = 1, 2, \ldots$ as indicated in Sect. 6.1.1. However, the identical properties of the resonators and the symmetry of the coupling coefficients lead to a specific frequency characteristic which might not be appropriate for the desired function. Filter design "reverses", therefore, the logical sequence: the desired bandwidth, the in-band loss ratio and the stopband rejection are prescribed, and the number of required elements as well as the appropriate external and mutual couplings are results of the design procedure [5, 8]. The filter characteristics can be tailored to very different applications (e.g., bandpass, low-pass, high-pass, allpass, impedance matching circuits or group delay adapters). It is worth mentioning that the group delay $\tau_{\mathrm{gr}} = \mathrm{d}\Phi_{21}/\mathrm{d}\omega$ (Sect. 2.1.2) of a filter corresponds to the group velocity $v_{\mathrm{gr}} = \mathrm{d}\omega/\mathrm{d}k$ of the propagating waves (k = wave number = $\mathrm{d}\Phi_{21}/\ell$, where ℓ is the electrical length of the filter). Designing a filter is, in this sense, equivalent to constructing the appropriate dispersion relation $\omega(k)$.

The following examples focus on three well-known types of frequency response that are relevant for many applications. Before going into details it is helpful to introduce some general definitions. First, the passband attenuation $L(u)$ can formally be written as

$$L(u) = 10 \log \left(\alpha + |K(u)|^2 \right) , \tag{6.9}$$

where $K(u)$ is called the characteristic function of the considered filter type [9]. One common constraint on the choice of $K(u)$ is $K_0 \equiv K(u=0) = 0$. We see from (6.4a) and (6.6a) for the cases of a "one-pole" filter ($N = 1$) and the symmetric "two-pole" filter ($N = 2$) that $|K(u)|^2$ is an even polynomial of order $2N$. The parameter α in (6.9) denotes the attenuation at center frequency. In the case of infinite Q_0 values α equals 1, and the midband attenuation $L_0 \equiv L(0)$ vanishes. We will consider this case several times in the following.

In many real cases the ripple L_{r} amounts just to a fraction of a decibel. Although it affects the frequency response of the filter markedly, it may be difficult to measure. A much more appropriate quantity in this respect is the reflection coefficient $\rho^2(u) = 1 - t^2(u)$ (we assume dissipation loss to be negligible). The reflection loss $R \equiv -10 \log(\rho^2)$ in dB is the analog of the transmission loss $L = -10 \log(t^2)$, to which it is related by

$$L = -10 \log \left(1 - 10^{-R/10} \right) = -10 \log \left(1 - \rho^2 \right) . \tag{6.10}$$

Tschebyscheff (or Equal-Ripple) Filters

Figure 6.4 compares one possible type of frequency response of a bandpass filter, which consists of five poles, with an $N = 2$ filter. As before, N can be found from $L(u)$ by the number of zeroes. The ripple was adjusted in Fig. 6.4 to $L_{\mathrm{r}} = 1\,\mathrm{dB}$ (corresponding to $R = 6.9\,\mathrm{dB}$), and the fractional bandwidth was fixed at $w = 2\%$ for the two cases. Therefore, both frequency characteristics intersect at $u_\pm = \pm w/2$ at the level $L_{\mathrm{r}}(\pm w/2) = 1\,\mathrm{dB}$. The figure, which magnifies the region around the passband, reveals clearly that the steepness of the filter skirts improves drastically as the number of filter elements is increased.

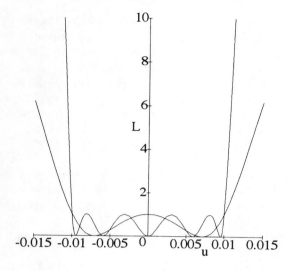

Fig. 6.4. Bandpass characteristics $L(u)$ of a 5-pole and a 2-pole Tschebyscheff filter, both having a fractional bandwidth of 2% and a ripple of 1 dB.

The frequency response shown in Fig. 6.4 is characterized by N zeroes of L in the passband. The characteristic function $K(u)$ of such a filter (for N assumed odd) could be [9]:

$$K(u) = Cu(u^2 - u_{0,1}^2)(u^2 - u_{0,2}^2)\cdots(u^2 - u_{0,(N-1)/2}^2), \quad (6.11)$$

where C is a constant and $u_{0,k}$ denotes the zeroes of $K(u)/u$. Another choice of $K(u)$ is a Tschebyscheff polynomial of order N [10]. The corresponding attenuation $L^T(u)$ is [8, 9]:

$$L^T(u) = 10\log\left\{1 + \varepsilon\cos^2\left[N\cos^{-1}\left(\frac{2u}{w}\right)\right]\right\} \quad (6.12a)$$

in the passband $2u \leq w$. The stopband attenuation ($2u \geq w$) involves the hyperbolic cosine:

$$L^T(u) = 10\log\left\{1 + \varepsilon\cosh^2\left[N\cosh^{-1}\left(\frac{2u}{w}\right)\right]\right\}. \quad (6.12b)$$

The exponents "-1" at the cosine and at the hyperbolic cosine stand for the inverse function. In both cases is w the equal-ripple band edge, and ε follows from

$$\varepsilon = 10^{L_r/10} - 1 = \frac{\rho^2}{1-\rho^2}. \quad (6.12c)$$

At small frequencies, away from the band edge ($2u \ll w$), (6.12a) can be expanded with respect to $2u/w$. The approximative result for N being odd is

$$L^T(u) \approx 10\log\left[1 + \varepsilon\sin^2\left(\frac{2Nu}{w}\right)\right]. \quad (6.13)$$

For N even, the sine must be replaced by cosine. Equation (6.13) thus yields an obvious rule on how to extract the parity and the number of poles from the frequency characteristic. Furthermore, it shows that the ripples of the Tschebyscheff response are equal for every pole, in contrast to the polynomial nature of the frequency characteristic resulting from equal elements (Sect. 6.1.1). The minimum in-band attenuation given by (6.12a) and (6.13) equals zero. This condition can be approached in a real filter only if $\beta = 1$ [see (6.6) for infinite Q_0] and/or if it is very strongly coupled such that $Q_e/Q_0 \ll 1$.

While Tschebyscheff-type bandpass characteristics can be formally synthesized on the basis of (6.12), nothing has yet been said about the external and the mutual coupling coefficients ($k_{j,j+1}$, $j = 1, N-1$) required for a filter with given N, w, f_0 and L_r. Various methods to design such filters are known. Before completing the analysis of the Tschebyscheff filter, it is therefore advisable to introduce one prominent example.

Low-Pass Prototype Filters

The following paragraphs shall introduce the relevant parameters needed in Sect. 6.1.3 to relate the frequency response of filters to the properties of the constituent resonators. These correspondences, in turn, will be the basis to discuss the progress in the development of HTS microwave filters in Sect. 6.2.

The design of the various types of filters is strongly facilitated by reducing the problem to the design of an "equivalent" low-pass prototype filter as sketched in Fig. 6.5 [5, 8]. Such a filter may consist of a chain of series inductors with inductance $L'_{2k} = g_{2k}$ and shunt capacitors with capacitance $C'_{2k-1} = g_{2k-1}$ ($k = 1 \ldots N$). The generator is represented by a resistance $R'_0 = g_0$. The load is assumed to be the conductance $G'_{n+1} = g_{n+1}$ or the resistance $R'_{n+1} = g_{n+1}$, depending on whether the Nth circuit element is inductive or capacitive. The primed values denote normalized quantities, while unprimed variables stand for absolute values. The normalization refers to a unit-value source resistance $g_0 = 1$, and to a unit-valued width of the passband: $\omega'_1 = 11$. In the case of a bandpass filter, the corresponding condition would mean $u_1 = w/2$. The g_k are referred to as the element values of the low-pass prototype filter. An equivalent prototype would result from exchanging in Fig. 6.5 the sequence inductor, shunt capacitor, and so on by the dual sequence shunt capacitor, inductor, and so on. Analysis of the filter elements g_k is sufficient to construct a large variety of filters with prescribed frequency response. The power, as well as the limitations, of this design technique were extensively illustrated in [5, 7, 8].

As one example, the element values g_k of the low-pass prototype can be determined such that the Tschebyscheff response according to (6.12) is achieved. The expressions are recursive and involve various auxiliary definitions [11, 12]. They will therefore not be repeated here. An extended listing

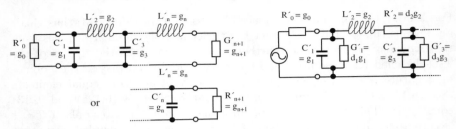

Fig. 6.5. *Left*: Schematic of a low-pass prototype defining the filter parameters g_k, $k = 0\ldots n+1$. *Right*: Modification of the circuit layout for the consideration of dissipative elements. (From: G. L. Matthaei, L. Young, and E. M. T. Jones: *Microwave Filters, Impedance-Matching Networks, and Coupling Structures* (Artech House, Dedham, Massachusetts, 1980), Chap. 2.)

of the element values for N-pole filters with different ripple and $N = 1$ up to 15 can be found, e.g., in [8, 9].

Lossy filter elements can be considered in the prototype filter by introducing finite quality factors for the inductors and the capacitors, respectively,

$$Q_k = \frac{\omega_1 L_k}{R_k} \quad \text{or} \quad Q_k = \frac{\omega_1 C_k}{G_k}. \tag{6.14a}$$

R_k and G_k represent the parasitic resistance and conductance of the kth element as sketched in the right-hand part of Fig. 6.5. In the case of a bandpass filter (BPF) designed from a low-pass prototype, the corresponding Q value of the k_{th} reactive element of the prototype is

$$Q_k = w Q_{0,k}, \tag{6.14b}$$

where w is the fractional bandwidth as before and $Q_{0,k}$ is the midband unloaded quality factor of the k_{th} resonator of the BPF. Each reactive element of the prototype is thus accompanied by a dissipative element $d_k g_k$, where the dissipation factor d_k is given by

$$d_k = \frac{\omega_1'}{Q_k}. \tag{6.14c}$$

The increase of the transducer loss ratio, $\Delta L(\omega_1')$, due to the finite Q of the circuit elements was estimated by Cohn for a doubly resistively terminated filter in the limit of low dissipation ($\sum_k d_k g_k \ll 1$) at midband [13]:

$$\Delta L_0 \approx 4.343 \sum_{k=1}^{N} d_k g_k. \tag{6.15a}$$

An alternative expression was derived in [14] under the assumption of uniform and low dissipation ($d_k = d \ll 1$ for $k = 1\ldots N$):

$$\Delta L(\omega') \approx 8.686\, d\, \frac{d\Phi_{21}(\omega')}{d\omega'}. \tag{6.15b}$$

The latter equation relates the dissipation losses to the group delay at center frequency.

$$\tau'_{\text{gr}}(0) = \left.\frac{\mathrm{d}\Phi_{21}(\omega')}{\mathrm{d}\omega'}\right|_{\omega'\to 0} = \frac{1}{2}\sum_{k=1}^{N} g_k \,. \tag{6.16a}$$

It will later prove convenient to introduce the following notation for the sum of the g-values:

$$S_g \equiv \sum_{k=1}^{N} g_k = N\bar{g} \,, \tag{6.16b}$$

where $\bar{g}$ denotes the average over all element values.

Once the formalism of the low-pass prototype circuit is at hand, it is straightforward to "translate" between a bandpass characteristic and the corresponding low-pass prototype. The mapping rule for the reduced frequency is, in the limit of narrow bandwidth,

$$\frac{\omega'}{\omega'_1} = \frac{2u}{w} \,, \tag{6.17}$$

where $u = (\omega - \omega_0)/\omega_0$, and w is the fractional bandwidth. The exact conversion of a BPF circuit into an equivalent low-pass prototype depends in detail on the nature and the arrangement of the resonators and of the coupling elements between them. For the case of series-type resonators, the resulting expressions for the input ($Q_{e,0}$) and output coupling ($Q_{e,N}$) are [5]:

$$Q_{e0,N} = \frac{\omega'_1}{w} g_j g_{j+1}, \; j = 0, N \,, \tag{6.18a}$$

and for the mutual coupling coefficients $k_{j,j+1}$ [$j = 1 \ldots (N-1)$]

$$k_{j,j+1} = \frac{w}{\omega'_1} \frac{1}{\sqrt{g_j g_{j+1}}} \,. \tag{6.18b}$$

For the two examples considered in Fig. 6.4, the g_k can be found from the extended tabulation in [8]. The filter design is completed after having fixed the corresponding Q_e and k_j. Table 6.1 summarizes the prototype element values and the values of the coupling coefficients resulting from (6.18). It can be seen that the group delay [see (6.16a)] of the 5-pole filter is higher by a factor of 3.7 than for the 2-pole filter. Enhanced sensitivity to dissipation losses at the band edges and stronger requirements on the power-handling capability (Sect. 6.1.3) are the price to be paid for the steepened filter skirts.

Other Convenient Filter Types

The Tschebyscheff characteristic considered so far is favorable if filters with steep skirts are desired. However, there might be applications where the passband ripple has to be avoided. Maximally flat response of an N-pole filter is achieved if $L(\omega')$ has as many zero derivatives at $\omega'_1 = 0$ as possible. This condition can be met by choosing the Nth power of ω' for the characteristic function $K(\omega')$ as proposed in [15]. We have already found an example

Table 6.1. Element values of the Tschebyscheff filters considered in Fig. 6.4, with $w = 2\%$, $L_r = 1\,\text{dB}$, and $N = 2$ and $N = 5$. The external and mutual couplings resulting from (6.18) are also listed, as well as the sum S_g and the average value $\bar{g}$ [see (6.16b)].

$N = 2$		$N = 5$	
$g_0 = 1.0000$		$g_0 = 1.0000$	
$g_1 = 1.8219$		$g_1 = 2.1349$	
$g_2 = 0.6850$	$k_{1,2} = 8.95 \times 10^{-3}$	$g_2 = 1.0911$	$k_{1,2} = 6.55 \times 10^{-3}$
$g_3 = 2.6599$		$g_3 = 3.0009$	$k_{2,3} = 5.53 \times 10^{-3}$
		$g_4 = 1.0911$	$k_{3,4} = 5.53 \times 10^{-3}$
		$g_5 = 2.1349$	$k_{4,5} = 6.55 \times 10^{-3}$
		$g_6 = 1.0000$	
$Q_{e,0} = 182$	$S_g = 2.5069$	$Q_{e,0} = 213$	$S_g = 9.4529$
$Q_{e,N} = 182$	$\bar{g} = 1.2535$	$Q_{e,N} = 213$	$\bar{g} = 1.8906$

for $N = 2$ in the case of the twin resonators in Sect. 6.1.1 at critical coupling $q = 1$. The general expression for the passband attenuation $L(\omega')$ of a Butterworth (or maximally flat) filter is

$$L^{\text{B}}(\omega') = 10 \log \left[1 + \varepsilon \left(\frac{\omega'}{\omega_1'} \right)^{2N} \right], \qquad (6.19)$$

where ε is defined by (6.12c). Figure 6.6 compares the low-pass characteristics of a Tschebyscheff filter with a Butterworth filter, for the case $N = 7$. The maximum tolerable loss in the passband was set, for illustration, to $5\,\text{dB}$ ($R = 3.3\,\text{dB}$). The figure reveals peculiar differences between the two filter types. The steep skirts of the Tschebyscheff response result in (more or less) strongly pronounced ripple. In contrast, the flat passband of the Butterworth type has, at a given number of poles, the disadvantage of shallower stopband rejection. For a specific application, a compromise between flatness of the passband and steepness of the skirts must therefore usually be searched for.

In contrast to the Tschebyscheff filter, the prototype element values g_k of the Butterworth filter can be calculated in a straightforward way [12],

$$g_0 = g_{N+1} = 1,$$

and

$$g_k = 2 \sin \left[2^{-N} \left(2^k - 1 \right) \pi \right], \qquad k = 1 \ldots N. \qquad (6.20)$$

The transformation from the low-pass prototype to a bandpass characteristic follows analogously, as described before.

We now turn to a third type of filter response. The design of a Tschebyscheff filter with steep skirts usually requires an appreciable number of poles N, $N \gg 2$. One obvious disadvantage is the size of the resulting filter package. A possible solution to improve the skirts without increasing N is the

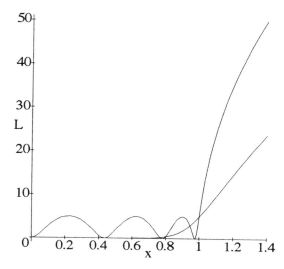

Fig. 6.6. Frequency characteristic $L(x \equiv \omega')$ in dB of a 7-pole low-pass prototype Tschebyscheff filter [from (6.12)] with a ripple of 5 dB in comparison with a 7-pole Butterworth filter [from (6.19)].

introduction of poles in the stopband, at which $L(\omega'_\infty > 1)$ diverges. Hence, the characteristic function $K(\omega')$ becomes a rational function like [9],

$$K^C(\omega') = C \frac{\omega'(\omega'^2 - \omega'^2_{0,1})(\omega'^2 - \omega'^2_{0,2}) \ldots (\omega'^2 - \omega'^2_{0,(N-1)/2})}{(\omega'^2 - \omega'^2_{\infty,1})(\omega'^2 - \omega'^2_{\infty,2}) \ldots (\omega'^2 - \omega'^2_{\infty,(N-1)/2})} \quad (6.21)$$

for N odd, where $\omega'_{0,k}$ denote the zeroes and $\omega'_{\infty,k}$ the poles of K^C. Cauer stated that the parameters of K^C, and thus of the passband attenuation [see (6.9)], can be derived from Jacobian elliptic functions [16]. This filter type is therefore called a Cauer (or elliptic) filter. Figure 6.7 compares the response of such a filter (left) with Tschebyscheff (middle) and Butterworth (right) versions for $N = 5$. Indicated are the ripple L_r and the stopband attenuation L_S at bandwidth w_S. At equal ripple, the filter skirts improve (i.e., w_S decreases) from the right to the left.

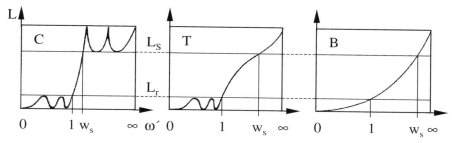

Fig. 6.7. Comparison of the frequency response $L(\omega')$ of 5-pole filters (after [9]). Left, Cauer (or elliptic); middle, Tschebyscheff (or equal ripple); right, Butterworth (or maximally flat).

An elliptic filter can be identified from its frequency response by an equal number of zeroes in the passband as there are poles in the stopband. The technical realization of such a filter involves, besides coupling of the nearest-neighbors, well adjusted cross-coupling between resonators farther apart. Proper dimensioning of all coefficients for external, mutual and cross-couplings is usually a challenging task. However, it can be facilitated by dropping the condition for a truly elliptic response, and by adjusting exactly two poles at the edges of the passband instead. Such a filter type is referred to as quasi-elliptic [17].

6.1.3 Relations Between the Properties of Filters and the Constituting Resonators

We can now formulate the conditions which must be met by the resonators (and thus by the material they are made of) in order to achieve a presribed filter response. We will no longer distinguish between different types of frequency response, but generalize the discussion to the low-pass prototype filter elements. Further details can be found, e.g., in [18–20].

Quality Factor

Equation (6.15) provides the basis for discussing the impact of finite unloaded Q factors of the resonators on the frequency characteristic of the complete filter. If we assume equal dissipation in all resonators, we can rewrite (6.15) using d_k from (6.14c) and $\bar{g}$ from (6.16b):

$$\Delta L_0 \approx 4.343 \frac{N}{w\,Q_0} \bar{g}, \tag{6.22a}$$

where $w = (\omega_+ - \omega_-)/\omega_0$ is the fractional bandwidth for a given L criterion [e.g., $L(\pm w) = 3\,\text{dB}$]. Curve fitting of the parameters of Tschebyscheff filters with $N \geq 3$ [21] revealed $\bar{g} \approx 2(N-2)/N$. The asymptotic value of $\bar{g}$ for large N is therefore 2, in accordance with the data in Table 6.1. The midband attenuation is found, in this limit, to scale as $\Delta L_0 \propto N$. The attenuation $\Delta L(u)$ at the band edges $u_\pm$ exceeds ΔL_0 in proportion to the group delay [see (6.15b)]:

$$\Delta L(u_\pm) = \Delta L_0 \frac{\tau_\pm}{\tau_0}. \tag{6.22b}$$

The unloaded quality factor Q_0, which fixes ΔL_0, is determined by the total power balance via

$$\frac{1}{Q_0} = \gamma_{\text{cond}} \frac{R_s}{Z_0} + \gamma_{\text{diel}} \tan\delta + \frac{1}{Q_{\text{para}}}. \tag{6.23}$$

The latter equation contains, in comparison with (6.1), a very important supplement in terms of the parasitic losses that occur, e.g., in the filter housing (Sect. 5.2.3). The first additive term in (6.23) comprises the HTS surface resistance $R_s(T)$, weighted by the geometric factor γ_{cond}. Except for lumped-element designs (small γ_{cond}), this term is, at $f_0 = 1\,\text{GHz}$ and $T = 50\text{--}70\,\text{K}$, of order 10^{-7} to 10^{-6} (Sect. 2.3). The second term denotes the loss tangent of the dielectric ($\gamma_{\text{diel}} \approx 1$ for many devices), within which the electromagnetic energy is stored. In planar geometries, where the dielectric is the substrate, typical values of $\tan\delta$ are, at 70 K, of the order of 10^{-5} for LAO and 10^{-6} for CbS (Sect. 5.2.2), indicating the profitable use of sapphire. The third term quantifies the ratio of the power dissipated in the housing to the power $P_{c,f} = \omega_0 W$ circulating in the filter. This quantity depends strongly on the design of the HTS circuit and housing. For microstrip circuits mounted into gold-plated housings, values around 10^{-4} are often observed, with Q_{para} thus limiting the unloaded Q factor: $Q_0 \leq Q_{\text{para}}$. Optimized filter design and proper dimensioning of the housing have recently enabled reduction of parasitic losses to $Q_{\text{para}} \geq 10^5$ [22] in lumped-element devices, and above 10^6 in disk resonator filters [21, 23].

Equation (6.22a) reveals a three-fold benefit of the high Q_0 values of superconducting filters. First, for a given design (N, w, S_g), higher Q values reduce the midband attenuation of the filter. Such an effect is highly desirable, e.g., for high-power transmission in communication systems (Sect. 6.2.2). Second, enhanced quality factors allow, at constant ΔL_0, for a reduced bandwidth w without affecting other filter properties, which is attractive for narrow-band preselection and noise reduction filters (Sect. 6.2.2). Finally, as Q_0 is increased, improved filter layouts (larger N and/or $\bar{g}$) become accessible at prescribed passband attenuation and bandwidth. This case is beneficial for multipole filters and filters with large group delay [1b, 2].

Similar arguments hold for the (higher) attenuation at the edges of the passband [see (6.22b)], which is proportional to ΔL_0. The relative increase $\tau_\pm/\tau_0$ of the group delay from the center frequency ($u = 0$) to the band edges ($u = u\pm$) depends on the ripple L_r, and hence on the reflection coefficient R. It was found in [21] to scale with N ($N \geq 3$) and R like

$$\Delta L(u_\pm) \approx \Delta L_0 \frac{4.5(N+4)}{R}. \qquad (6.24)$$

Low midband attenuation (e.g., through high Q_0 values) is therefore exceptionally important in multipole filters ($N \gg 2$). Alternatively, the filter layout has to be changed from Tschebyscheff-type to elliptic or at least quasi-elliptic (at the expense of a more complicated design).

Another potential benefit of high Q_0 values concerns the ratio of the midband attenuation L_0 to the stopband rejection L_S (Fig. 6.7). The bandwidth w_S, at which the required L_S value is achieved, is related to the fractional width w of the passband. An approximative expression was derived for the case of an equal-element matched filter ($g_k = 1$ for all $k = 1 \ldots N$) [7, 13]:

$$\frac{w_S}{w} \approx 10^{(L_S+6.02)/20N}, \qquad (6.25)$$

where $10\log(4) \approx 6.02$. Using (6.22a) with $\bar{g} = 1$ relates L_0 and L_S:

$$L_0 \approx \frac{4.343N}{w_{\mathrm{S}}Q_0} 10^{(L_{\mathrm{S}}+6.02)/20N} . \tag{6.26}$$

Inspection of (6.26) shows that high Q_0 values allow for a reduction of $w_{\mathrm{S}} \propto 1/Q_0$ at otherwise constant parameters. This result quantifies the steepening of the filter skirts which was already mentioned in Sect. 6.1.1. Alternatively, N can be reduced at prescribed L_{S} if high-Q resonators are employed, resulting in simpler design and smaller filter size.

Circulating Power

Many filter applications involve strong microwave signals, e.g., in transmitters of communication systems [18]. It is therefore important to relate the filter parameters to the power $P_{\mathrm{c,f}}$ circulating in it, and further to the field and power dependent surface impedance of the resonators [19, 20]. The power $P_{\mathrm{d,f}}$ dissipated in the filter at center frequency is [c.f., definition of $L(u)$ in (6.4b), and (6.22a)]:

$$P_{\mathrm{d,f}}(0) = P_{\mathrm{inc}} \frac{\Delta L_0 (\text{in dB})}{4.343} , \tag{6.27a}$$

where P_{inc} denotes the incident power, which equals the transmitted power in the considered case of matched filters and weak dissipation. The average power dissipated in a single resonator is approximately one Nth of the total value (if $N \gg 2$). It is in fact about 20% lower in the outermost resonators, which are strongly coupled to the external circuit [18, 24]. Combining (6.27a) with (6.22a), we thus arrive at

$$P_{\mathrm{d,r}}(0) \approx P_{\mathrm{inc}} \frac{1}{wQ_0} \overline{g} . \tag{6.27b}$$

The circulating power is $P_{\mathrm{c,r}}(0) = Q_0 P_{\mathrm{d,r}}(0)$, and thus in terms of the normalized group delay at midband [see (6.16a)]:

$$P_{\mathrm{c,r}}(0) \approx P_{\mathrm{inc}} \frac{2}{Nw} \tau'_{\mathrm{gr}}(0) . \tag{6.28}$$

High circulating powers obviously occur not only at high-power operation, but also if large group delays are involved as, e.g., in narrow-band filters.

The power handling of a filter is determined by the largest value occurring in any one of the filter elements. Due to the transfer of energy through N coupled resonators, the maximum power loading usually occurs in the second resonator after the input [18, 24, 25]. Further analysis shows that the group delay reaches its maximum neither at midband (where it is in fact minimum) nor at the band edges $u_\pm$. It rather occurs at frequencies $\pm u_{\max}$ slightly outside the passband, where the largest amount of energy remains stored in the filter (being neither reflected nor transmitted) [21]. The corresponding scaling of the circulating power is therefore

$$P_{\mathrm{c,r}}^{\max} = P_{\mathrm{c,r}}(0) \frac{\tau_{\max}}{\tau_0} , \tag{6.29}$$

where $\tau_{\max} = \tau(\pm u_{\max})$. The curve fitting of the Tschebyscheff parameters [21] yielded empirical scaling rules for $\tau_{\max}(N, L_r)$. Taking these into account and using (6.28) with $\tau'_{gr}(0) \approx (N-2)$, leads to the approximation ($N \geq 3$):

$$P_{c,r}^{\max} \approx P_{inc} \frac{2.4(N-2)(N+4)}{Nw\sqrt{R}}. \qquad (6.30)$$

While $P_{c,r}(0)$ approaches $2P_{inc}/w$ for large N, the maximum value of the circulating power increases progressively as more and more filter elements are added.

Equation (6.30) is the desired expression that relates the filter design to the power handling capability of the most strongly excited (second) HTS resonator. The microwave field amplitude $H_{s,\max}$ excited in this resonator is given by [see (2.14) in Sect. 2.1.1]:

$$H_{s,\max} = \gamma_H \sqrt{P_{c,r}^{\max}}. \qquad (6.31)$$

The coefficient γ_H is determined by the volume of the filter element and by the microwave field distribution. Small circuits (e.g., lumped-elements or microstriplines) with field enhancement at their contours are deleterious to power handling. In contrast, two- or even three-dimensional resonators are appropriate for high-power filtering. Using a conservative value of $\gamma_H \leq 40\,\text{A/m/W}^{1/2}$ (or $50\,\mu\text{T/W}^{1/2}$), the $H_{s,crit}$-data of Sect. 4.1.2 translate into $P_{c,r}$ values up to $1.4\,\text{MW} \times [1-(T/T_c)^2]^2$. Such values, unless reduced by defects or improper handling, would suffice for most presently known specifications of high-power filters, even at 77 K (Sect. 6.2.2).

6.2 Passive Devices Related to Resonant Elements

An extended analysis of microwave device applications using epitaxial HTS films is available in various reviews [2, 26] and books [27, 28], that are devoted to the science and engineering of HTS. In contrast to these detailed treatments, the following sections try to summarize the potential benefits that the reported efforts have in common. The focus will be on passive devices based on resonators (filters, antennas and oscillators), although appreciable progress was also reported, e.g., on wide-band nondispersive LTS and HTS delay lines for spectral analysis [26, 29, 30]. Active HTS elements such as three-terminal devices [31, 32], parametric amplifiers or oscillators [3] are still under development and will not be considered here.

Section 6.2.1 distinguishes the categories to which the various HTS microwave device applications can be attributed. Section 6.2.2 sketches major technological challenges that remain to be solved on the way towards commercialization. Selected examples illustrate the ongoing novel approaches toward achieving high-performance HTS microwave devices.

6.2.1 Survey of Passive Microwave Device Applications

Potential Figures of Merit (FOM)

The development of the material sciences of HTS has been accompanied by a series of reviews indicating the potential benefit of superconducting microwave devices (e.g., [1b, 2, 33–35]). According to the analysis contained there, the unique features of high-temperature superconductors span four areas where device and system applications are most promising:

(1) Miniaturization ("smaller than existing"). Microwave devices can be miniaturized by means of superconducting circuit elements for three reasons. First, the high permittivity $\varepsilon_r = 9$ (sapphire) to 25 (LaAlO$_3$) of the substrates suitable for HTS films [Table 5.1] reduces the guided wavelength by a factor of $\varepsilon_r^{1/2} \approx 3$–5. Typical dimensions of the circuit elements can be reduced in planar devices by the same factor or even by its square, depending on whether the miniaturization affects one or two dimensions. If, furthermore, a three dimensional component (like a waveguide filter) can be substituted by a planar version (like a microstrip filter), the gain in volume and weight will be even of the order of $\varepsilon_r^{3/2}$. Second, the latter reasoning applies only if the miniaturization does not result in reduced performance, i. e, reduced Q factors. This condition can be met because of the extremely low surface resistance of superconductors which, according to (2.8) and (2.9), balances the reduction in size:

$$Q_0 = \frac{\omega \mu_0}{R_s} \frac{V}{A} \frac{\overline{H}_V^2}{\overline{H}_A^2}, \qquad (6.32)$$

where V and A denote the volume and the surface area, respectively, over which the microwave fields extend. Neglecting parasitic effects such as cross-coupling between adjacent circuit elements, or technological difficulties related to the photolithography of microscopic objects, (6.32) indicates a potential for miniaturization that corresponds to the reduction of R_s of HTS films below that of the best normal conductor. This improvement factor is about 10^2 at 77 K and 2 GHz for YBCO (Sect. 2.3), and increases with decreasing temperature and frequency. Third, the high permittivity and the high quality factors of HTS resonators result in a confinement of fringe fields, in real space (dielectric polarization) and in frequency (narrow bandwidth). This confinement facilitates the decoupling of closely spaced groups of miniaturized elements and enhances the versatility of circuit layouts.

(2) Performance ("better than existing"). Equation (6.32) can also be interpreted in a reverse manner. Given the geometric dimensions of, e.g., a single resonator, its Q factor increases as the surface resistance decreases. This case was discussed in detail in Sect. 6.1.3 in relation to microwave filters. Potential benefits are steep skirts, small fractional bandwidths and improved stopband rejection of bandpass filters, large group delays of allpass sections and so

on. However, market opportunities for superconducting devices will depend on the net system advantage. Unfortunately, the performance of a complete subassembly or system is not determined by the performance of the superconducting component alone. Rather, the necessity to provide cooling has to be taken into consideration (Sect. 5.3). The issue of improved performance has therefore been related so far in most cases to the miniaturization of HTS circuits, including the use of miniaturized coolers.

(3) Unique features ("without analog"). Superconductors display properties which are not known in other conductors. The superconducting transition temperature leads to a steep drop of $R_s(T)$ which can be exploited, e.g., for switches or bolometric detectors. Also, the microwave field penetration depth is independent of frequency up to the THz range (Chap. 1). This feature is fundamental for the construction of nondispersive filters or for kinetic mixers and photodetectors. The latter type of application is additionally supported by the short relaxation times of high-temperature superconductors (Sect. 3.1). Finally, the quantization of magnetic flux and the Josephson effect are key features of extremely sensitive and/or strongly nonlinear microwave devices having no conventional analog. Devices of this type are considered in more detail in Sect. 6.3.

(4) Monolithic integration ("all on one"). Summarizing the above potential benefits reveals the challenging opportunity for monolithic integration of various functional circuits (analog and digital, active and passive) on a single chip. Furthermore, hybrid technology combining the unique features of superconductors with the advanced possibilities of semiconductor devices appears attractive for future generations of signal processing [27, 36]. Since this area requires technologies that are not sufficiently developed yet (for technological, economic and political reasons), related device applications will not be considered further.

Technologically and Economically Attractive Areas of HTS Device Applications

The potential benefits of superconducting circuit elements concern a variety of devices, subassemblies or systems which provide exciting features. Not only smaller, better or novel components have become feasible. Rather, the creation of multi-functional circuit boards based on the development of an interdisciplinary technology presents one of the most challenging aspects of the upcoming century. However, only a fraction of this potential has been considered attractive for the near- or medium-term.

The worldwide emerging industry devoted to the commercialization of superconductivity has conducted annual summits since 1992. The aim of these ISIS meetings (International Superconductivity Industry Summit) has been to promote international cooperation and information exchange between industry, government and academic institutions in order to stimulate rapid product development. During the 4th ISIS meeting held in 1995 in Washington, DC, a global status report was prepared that provided a review of the superconductivity industry and its potential [37]. According to this study, the following areas of superconductive electronics (here: passive device applications) have been identified as being most promising.

Communication systems including cellular phones, satellite communications and advanced navigation aids are likely to be improved by the use of superconductors. Scientific research instruments will also benefit from superconducting electronics in, e.g., detectors for nuclear magnetic resonance (NMR) systems and magnetic resonance imaging (MRI). In addition to civilian applications, modern radar, electronic guidance and control, and advanced navigation systems are of interest. The envisaged ground-based and air-borne applications have in common the need for enhanced performance [signal-to-noise ratio (SNR) of front-ends, low phase noise, adaptivity, bandwidth, dynamic range]. In addition, low volume and weight are important for multi-functional applications, and especially for space-based systems. Economic driving forces in this field are the reduction of costs for the payloads of satellites (presently about 10^4–10^5 US\$/kg) and the opportunity for enhanced functionality.

Near-term applications of HTS devices in mobile communication systems depend on the specific technical and political boundary conditions (e.g., [2, 38–40]). One issue is the improvement of coverage, i.e., the reduction of the number of base stations required to serve a given cell size. This aspect is relevant in rural areas where the quality of the signal transmission is usually limited by noise, and hence can be improved by increasing the SNR of the receivers (e.g., [41, 42]). In contrast, interference limits the performance of mobile communication in urban environments. Here, the development of multi-channel narrow-band front ends is desirable for exploiting the allocated frequency bands, and for increasing the number of calls per channel [2]. Further miniaturization and enhanced lifetimes of the battery-powered hand sets might be another economically important outcome. On the transmit side, low insertion loss and narrow bandwidth help to improve the electric power efficiency. Future systems are finally expected to strongly benefit from adaptively operating base stations, which provide higher functionality, larger capacity and reduced error rates.

Satellite-based communication systems will mainly profit from the reduced volume and weight of, e.g., input multiplexers, by substituting waveguide assemblies by planar microwave circuits [18, 41]. Improved microwave power management through the use of high-Q high-power output multiplexers might contribute to lighter systems on the transmit site, given the availability of sufficiently reliable and efficient coolers.

The performance of magnetic resonance systems for chemical and medical applications can be improved, and their spreading extended, by the implementation of HTS receive coils [43]. A major challenge is the achievement of high quality factors in strong ambient magnetic fields without deterioration of the required extreme field homogeneity (see below). Finally, improved radar and navigation systems can be achieved, e.g., by the implementation of low-phase noise oscillators (see below) and the use of rapidly switchable filter banks [41].

Overview of Passive HTS Microwave Devices

Table 6.2 lists representative applications of HTS films in passive microwave devices in relation to the topics 1 to 3 at the beginning of this section. Some of the listed items (marked by "#") are briefly described below. Further information on this subject can be found in the cited references. Devices marked by an asterisk are considered later in Sect. 6.2.2.

(1a) *Lumped-element circuits.* The geometric dimensions D of distributed circuit elements are determined by the guided wavelength, $\lambda_g \approx D$. Quasi-miniaturized devices can be obtained by designing dual-mode resonators ("two in one") which display two degenerate resonant frequencies [44]. In contrast, lumped-element circuits, forming a network of interdigital and patch capacitors, and narrow-strip or loop inductors, allow for a much higher degree of miniaturization, $\lambda_g \ll D$. Reduced size is achieved at the expense of higher dissipative losses [large γ_{cond} in (6.23)] and reduced power handling capability [large γ_H, (6.31)]. Lumped-element devices present therefore a unique opportunity to benefit from HTS in front-ends where normalconducting structures would suffer from untolerably high insertion loss. Besides the advantage of extremely compact low-loss circuits at frequencies $f \leq 10$ GHz ($\lambda_g \geq 6$ mm for $\varepsilon_r = 25$), lumped-element circuits offer the possibility of building synthetic transmission lines with prescribed dispersion and scattering parameters [28, 41, 45]. Because of the possibility to place loosely coupled adjacent cells close to each other, very narrow-band filters can be designed which would require large areas if built from distributed elements [46]. Lumped-element circuits have therefore been preferred in the U.S. to build HTS filters for preselection filters and input multiplexers with up to 20 poles per channel in order to exploit the exotic frequency allocation scheme [38].

Lumped-element receiver filters for mobile phone systems, fabricated from Y-123 and Tl-2212 films, are now commercially available. The advantage of Tl-2212 over Y-123 is the higher operating temperature of 77 K compared to 60 K [47]. Typical filter specifications are a center frequency $f_0 \approx 2$ GHz, bandwidth 15 MHz ($w = 7.5 \times 10^{-3}$), stopband attenuation $L_S > 55$ dB at $w_S/w = 1.3$ (± 2.13 MHz off the band edges), passband ripple $L_r < 0.1$ dB per channel and return loss $R > 14$ dB. Inspection of these figures in terms of the analysis presented in Sect. 6.1.3 shows that the applied design must be (quasi-)elliptic, and that the single resonators display high unloaded Q-factors ($\geq 10^4$), which are characteristic of superconductors. The commercial packages consist of a multichannel superconducting filter and a cooled preamplifier stage. Noise figures, improved by at least 1 dB, could be concluded from field trials.

Table 6.2. Summary of HTS microwave device applications (without Josephson devices, Sect. 6.3). The devices marked by "#" are discussed below, those marked by an asterisk in Sect. 6.2.2.

Category	Type of device	References
(1) Miniaturization	(1a) Lumped-element filters #	[28, 41]
	(1b) Antenna arrays	[28, 35, 48, 49]
	(1c) Stable oscillators *	[1a, 27]
(2) Performance	(2a) Antennas #	[28, 35, 48, 49]
	(2b) Narrow-band filters *	[2, 41]
	(2c) High-power filters *	[2, 23, 25, 42]
(3) Unique features	(3a) Switches, limiters, modulators #	[41, 50]
	(3b) Kinetic mixers and detectors	[1b, 51]
	(3c) Wide-band delay lines	[26, 28]

As another example, a quasi-elliptic 8-pole channel filter for input multiplexers of satellite transponders was developed with double-sided Y-123 films on ⌀3″ CbS [52]. The desired filter specifications ($f_0 = 4.0$ GHz, $w \approx 1\%$, $R \geq 23$ dB, $L_S = 30$ dB and 40 dB at $w_S = 1.25\%$ and 1.5% could be achieved by means of improved film quality, filter design and trimming methods. If the same filter was based on a design without cross-coupling, a minimum number of about 19 poles (instead of 8) would have been needed according to (6.25).

(2a) Antennas. The application of HTS to single antennas and antenna arrays was reviewed in [28, 35, 48, 49]. It was pointed out that it is essential to evaluate the benefit of superconducting components in terms of a complete antenna system, which consists of the radiating elements as well as of the matching network. In the case of arrays, the distribution, decoupling and beam-forming networks also have to be considered. A major benefit of superconducting antennas is the possible miniaturization of the radiating and network circuit elements without degrading the radiation efficiency. Furthermore, the narrow spacing of adjacent elements enables the development of interlaced multi-beam multifrequency arrays, e.g., for adaptive antenna systems.

A very important example of the beneficial use of HTS antennas are nuclear magnetic resonance systems [37] which allow the investigation of complex multiple-nuclei systems (e.g., [53]). The nuclear spins are oriented by a static magnetic field B_0, and then forced to precess by an AC field directed perpendicular to B_0. The frequency of precession, the Larmor frequency ω_L, is related to B_0 and to the nuclear magnetic moment n_m by

$$\omega_L = \gamma_{gyr} \frac{B_0}{\mu_0} \quad \text{or} \quad f_L \text{ (MHz)} = n_m \frac{e\, B_0 \,[\text{in (T)}]}{M_p}, \tag{6.33}$$

with γ_{gyr} the gyromagnetic ratio, $n_{\mathrm{m}} = \mu_{\mathrm{nucl}}/\mu_{\mathrm{K}}$ the nuclear magnetic moment relative to the nuclear magneton $\mu_{\mathrm{K}}/\mu_0 = he/4\pi M_{\mathrm{p}} = 5.051 \times 10^{-27}$ J/T, $e = 1.602 \times 10^{-19}$ C the elementary charge and $M_{\mathrm{p}} = 1.673 \times 10^{-27}$ kg the rest mass of the proton. For relevant nuclei such as ^{1}H, ^{13}C, ^{14}N and ^{31}P, the n_{m} values are 2.792, 0.702, 0.404 and 1.131, leading to $\omega_{\mathrm{L}}/B_0 = 6.159$–42.564 MHz/T. The time dependent precession of a certain nucleus provides information about its chemical environment and thus about the structural and dynamic properties of the material under investigation.

Based on the availability of superconducting magnets with strong ($B_0 \approx$ T) and homogeneous fields ($\Delta B_0/B_0 \leq 10^{-5}$) over extended volumes ($\leq 10^6$ cm^3), nuclear magnetic resonance has gained major scientific and economic relevance in chemistry and pharmaceutics for high-resolution investigation of synthesized molecules (NMR), and in medicine for nonradiative magnetic resonance imaging (MRI) [54]. The figure-of-merit (FOM) in all three cases is the SNR of the receive coil. Increased SNR permits reduced sampling times $\propto$ SNR^{-2}, smaller sample volumes, and/or reduced field strengths B_0. These benefits are reflected in system performance and costs, and patient comforts [43]. While a detailed analysis needs to consider many input parameters [55–58], the SNR can be estimated from

$$\mathrm{SNR} \approx \frac{F(V_{\mathrm{spl}}, V_{\mathrm{rec}}) B_1 \omega_{\mathrm{L}}^2}{\{\Delta f[\alpha_{\mathrm{rec}}(V_{\mathrm{rec}}) T_{\mathrm{rec}} \omega_{\mathrm{L}} L_{\mathrm{rec}}/Q_{\mathrm{rec}} + \alpha_{\mathrm{spl}}(V_{\mathrm{spl}}) T_{\mathrm{spl}} R_{\mathrm{spl}}]\}^{1/2}}, \quad (6.34)$$

where the subscripts "rec" and "spl" stand for the receive coil and the sample, respectively. The signal [numerator in (6.34)] scales with the filling factor F, which depends on absolute values and geometric arrangements of the two volumes V_{spl} and V_{rec}. It also scales with the magnetic field amplitude B_1 per unit current flowing in the coil. Both factors should be as large as possible in an optimized coil design. For the low frequencies of interest ($f_{\mathrm{L}} \leq 400$ MHz), the antennas consist of lumped-element coils resonated by appropriately dimensioned capacitors.

The two noise contributions [denominator in (6.34)] from the sample and from the coil are weighted by the parameters α_{rec} and α_{spl}. The noise is usually thermal, and thus scales with the temperatures T_{rec} and T_{spl}. The coil resistance is $\omega_{\mathrm{L}} L_{\mathrm{rec}}/Q_{\mathrm{rec}}$ with L_{rec} the coil inductance and Q_{rec} its quality factor. The sample resistance R_{spl} often scales like ω_{L}^2 [55, 56]. It can be seen from (6.34) that the use of HTS instead of normalconducting receiver coils is beneficial if R_{spl} is low, i.e. at low magnetic fields ($f_{\mathrm{L}} \leq 10$ MHz), or in the case of small α_{spl} (fields of view ≤ 5 cm). Possible applications are NMR spectroscopy or microscopy, both of which benefit primarily from reduced sampling times, and low-field MRI ($B_0 \leq 100$ mT) which enables local rather than whole-body imaging scans. Novel applications would be in open systems, e.g., for emergency rooms, orthopedics or interventional MRI [43]. All of these factors have been tested with planar Tl-2212 and Y-123 coils [59–61], and significant SNR improvements were achieved: a factor of 5 at 25 K in a 400 MHz NMR spectrometer [60], ≤ 8 in a 64 mT MRI system at 77 K [61, 62], and up to 10 in a NMR microscope at 10 K [63]. Recently, a Y-123 double-spiral antenna, mounted on a nonmagnetic single-stage pulse-tube refrigerator, was successfully operated in a commercial 0.2-T C-magnet system at cm-distances from the tissue [64].

Such improvements, which exceed the performance achievable with cryogenic normalconducting receiver coils, have led to an early commercialization of HTS probes for magnetic resonance stations. In addition to the design and development of the bare HTS detector various other technological challenges were met: The dewars had to be optimized in order to not disturb the magnetic fields B_0 and B_1, to not introduce additional noise sources, and to not reduce the filling factor too strongly. The diamagnetic properties of superconductors cause field distortions, which were minimized in the layout of high-resolution NMR spectrometers. The high Q values of 10^3 to 10^4 of the receiver coils reduced the bandwidth below 3 kHz at $f_L = 3$ MHz, although typical MRI systems require some 10 kHz. Therefore, superconductive matching and/or tuning circuits had to be designed at a tolerable increase of the insertion loss [59]. A current challenge is optimizing the nonlinear microwave response of the HTS coils in ambient magnetic fields, especially if they are also used as the signal input, i.e., at high power (Sect. 4.3).

(3a) *Switched devices.* The S–N transition between the superconducting and the normal state is the key property for the construction of switches, modulators, programmable filter banks or delay lines, and digital phase shifters [50]. Triggered by a control parameter, switching between an "on" (S) state with low insertion loss and an "off" (N) state with high isolation can be obtained. With proper adjustment of the amplitude, the sequence and the duration of the control signals, switching elements also provide a basis for making modulators and samplers. Low-frequency currents [65], microsecond or nanosecond microwave pulses [66, 67] or femtosecond laser pulses [68, 69] were identified as possible control parameters (see also [27]). As an alternative to the kinetically induced S–N transition, the controlled variation of the transmission parameters of a device segment by means of thermal [70] or photoconductive [71] switches, or ferroelectric capacitors [50, 72] (see below) were also reported.

The quality factor K_{sw} of a switch is the attainable transmission-to-isolation ratio. It is given by the resistance values R_N and R_S of the switching element in the N and S states [50]:

$$K_{sw} = \frac{R_N(f, T_c, d_F)}{R_S(f, T, d_F)} \ . \tag{6.35}$$

K_{sw} depends on the signal frequency f, the operating temperature T and the thickness d_F of the superconducting film, and possibly on the device geometry. To achieve high R_N values, thin narrow strips are usually used. Besides high K_{sw} values, high switching speeds are desirable for the envisaged applications (e.g., radar systems). While the S–N transition is kinetic in nature and thus very fast, the reverse transition (N–S) is governed by the thermal properties of the device and is hence much slower. Switching times below 2 μs and 80 ps were reported for S–N transitions in current and laser-controlled devices, respectively. In contrast, the N–S transition occurred in the ns range. The thermal switches were even slower (about 50 and up to 500 ms for S–N and N–S) and highly power consumptive (about 0.1 W).

Four time regimes can be distinguished during the dynamic response ([73], see also Chap. 3): heating of the electronic subsystem of the HTS (time scale $\tau_e \approx 1$ ps), pair breaking ($\tau_p \leq 40$ ps), phonon escape from the film to the substrate ($\tau_{esc} \approx 5$ ns), and thermal relaxation of the complete package ($\tau_{th} \leq$ μs). With the duration of the control pulses adjusted between τ_e and τ_p, thermal effects can be completely avoided. If, furthermore, small HTS switching volumes and operating temperatures close to T_c are considered, ultimate speeds of the order of 100 GHz appear feasible at an average power consumption of around 10 μW.

6.2.2 Remaining Challenges

The progress in the development of commercial passive HTS microwave components has been considerably faster than that of other technologies [74]. It is therefore not suprising that further improvements are needed in several areas. The two following examples present the prerequisites for a matured HTS microwave technology. The section concludes with two further examples that illustrate promising achievements in the development of novel HTS devices for communication systems.

HTS Compatible Tuning Elements

Since narrow-bandwidth filters are needed for many applications of HTS, tuning them becomes an important issue. Considering, for example, fractional bandwidths of filters below 1% around 1 GHz means to tune the center frequencies on a scale of 10 MHz, with high precision and reproducibility. Moreover, the Q factors of the filter elements and the mutual coupling coefficients should remain unaffected by the tuning process. Beside tunable filters, steerable systems like adaptive antenna arrays require controlled shifting of the phases of the different input and output channels. In all cases, a continuous adjustment (in contrast to step-by-step trimming procedures) of the phase or of the resonant frequency is highly desirable.

Tuning can be obtained in general by varying the specific inductance (geometric and kinetic) and/or the capacitance of the circuit elements. Beside the phase, the characteristic impedance of the line is also affected, which leads to an inherent trade-off between tunability and impedance matching, unless the circuit inductance can be adjusted separately. The major FOM of tuners is the fractional tuning bandwidth w_{tun} times the quality factor Q_0 of the tunable circuit element (see below and [50]). In order to achieve high w_{tun} values, the tuner has to be coupled strongly to, e.g., the resonators of a filter. In turn, the strong coupling makes the device highly susceptible to parasitic losses that tend to reduce the attainable Q values.

Up to now, ferroelectric, ferrimagnetic, dielectric and superconductive tuning elements have successfully been demonstrated with HTS filters and resonators. Figure 6.8 provides a schematic distinction between the different approaches in terms of the parameter $w_{\text{tun}}Q_0$. Each type of tuner has specific advantages, as well as disadvantages, which usually restrict its applicability to specific situations. Furthermore, the various tuning mechanisms turn out to limit the attainable Q factors roughly to 10^3, 10^4, and $\geq 10^5$, as indicated in the figure.

Ferroelectric tuning was possible by using incipient ferroelectrics deposited as thin-film capacitors onto transmission line devices [72, 75]. Applying an electric field gradient affects the permittivity ε_r of the capacitor and therefore the phase (or resonant frequency) of the HTS circuit. The quality factor K_{tun} of a ferroelectric tuner can be defined as [50]:

$$K_{\text{tun}} = \frac{(\nu - 1)^2}{\nu} Q_{\text{diel}}^2 , \tag{6.36a}$$

where $1/Q_{\text{diel}} = (\tan \delta_1 \tan \delta_2)^{1/2}$ is the geometric average of the dielectric quality factors corresponding to the loss tangents at the edges of the tuning band. The parameter $\nu = \varepsilon_{\max}/\varepsilon_{\min}$ in (6.36a) quantifies the total voltage-induced variation of the permittivity. Assuming a variation of the frequency $f \propto \varepsilon_r^{-1/2}$, ν can be related to the tuning range w_{tun} via

$$\nu \equiv \frac{\varepsilon_{\max}}{\varepsilon_{\min}} = \left(\frac{1 + w_{\text{tun}}/2}{1 - w_{\text{tun}}/2}\right)^2 \approx 1 + 2w_{\text{tun}} . \tag{6.36b}$$

Combining both equations shows $K_{\text{tun}} \approx (2w_{\text{tun}}Q_{\text{diel}})^2$.

Frequency shifts of up to 300 MHz at a bias voltage of 50 V and $f_0 = 11$ GHz ($w_{\text{tun}} \approx 3\%$) were reported in [72]. The polarizability of the ferroelectric films was found to increase with increasing thickness at almost constant attenuation [76]. However, the investigated ferroelectric materials (e.g., $SrTiO_3$) display a large loss tangent ($\tan \delta \leq 10^{-2}$, see Table 5.1 in Sect. 5.2). K_{tun} was correspondingly limited at values to around 10^4 (the parameter $w_{\text{tun}}Q_0$ shown in Fig. 6.8 corresponds to half the square-root of K_{tun}). The implementation of ferroelectric tuning elements in HTS filters has mainly been restricted up to now to military applications. Wider application still requires the development of suitable materials with high permittivity and polarizability, low voltage-independent loss tangent and compatibility with HTS.

Digital tuning elements were reported for bandstop filters containing fast photoconductive switches (e.g., [71]). Switching times of 8 μs were observed for the transition from the dark state (reject, filter is "on") to the illuminated

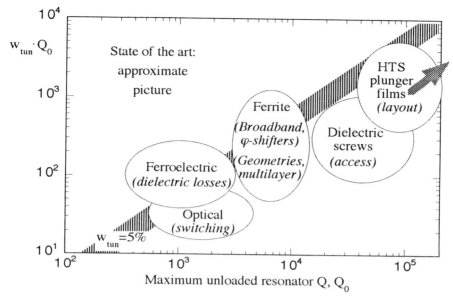

Fig. 6.8. Schematic (and approximate) distinction between different approaches towards HTS compatible tuning elements. The *hatched diagonal* indicates the fractional tuning range of 5%.

state (pass, filter is "off"). The reverse switching needed 340 µs. Typical resonator Q were of the order of a few 10^3 at 77 K. The tuning range assumed in Fig. 6.8 reflects the 3-dB filter bandwidth of 10–100 MHz.

Magnetic tuning based on the complex permeability of ferrimagnetic materials is well known in conventional microwave technology [77]. The FOM of ferrimagnetic phase shifters is, in accordance with the definition of K_{tun}, given by the differential phase shift in degrees per dB insertion loss. Recently, this tuning mechanism was successfully demonstrated also at cryogenic temperatures with Y-123 transmission line structures deposited onto a low-loss ferrite [78]. Differential phase shifts of 200°–500° between the two extreme polarizations of the ferrite and an insertion loss below 0.3 dB were measured between 9 and 12 GHz for a nonreciprocal arrangement using a Yttrium-Iron-Garnet (YIG) torroid. The hysteresis of the ferrite limited the useful frequency range to values above 5 GHz. Typical Q values of meander line resonators and published data on the equivalent tuning range of resonant HTS-ferrite devices [78] are indicated in Fig. 6.8. Optimized performance was stated to involve proper choice of the ferrite material, improved circuit designs, and compatibility between the magnetic materials and the HTS films. Progress towards a more compact circuit layout, and the realization of a 3-element HTS circulator have also been reported. While the ferrimagnetic approach has proven suitable for broadband applications such as phased-array antennas, the versatility of possible circuit layouts may be limited by the necessity of providing magnetic control fields parallel to the HTS device, and circular-polarized microwave fields perpendicular to the control field for phase-shifting. Compact designs require, furthermore, multilayer technology.

In conventional waveguide filters, dielectric (usually low-loss sapphire) screws are employed for mechanically adjusting the resonant frequencies and coupling coefficients [7]. This method has also been adopted for compact HTS filters with high Q values (Q_0 up to 10^5) and reasonable tuning ranges [79, 80]. However, the basic disadvantage of dielectric trimming is the limited access to the srews during operation, complicating in-situ tuning.

Finally, HTS films have been employed as superconducting shields in high-Q resonators for stable oscillators [23a, 81]. In-situ piezo-driven displacement of the films provided sensitive tuning without marked reductions of the quality factor. One disadvantage of this technique is the close proximity between the tuning film and the resonator, needed for a sufficiently large tuning range. Furthermore, HTS plunger films appear to be promising mainly for planar devices rather than lumped-element or transmission line circuits.

In summary, reliable tuners for HTS microwave devices need further improvements in the quality factor K_{tun}, tuning speed, power dissipation (in the tuned circuit), power consumption (by the tuning mechanism), stability of operation and potential for monolithic integration.

Optimum Operating Temperature

Figure 6.9 illustrates the payoff between high-performance superconductive devices and closed-cycle coolers with high efficiency (c.f., Sect. 5.3). The unloaded quality factor Q_0 or the critical magnetic field amplitude H_{crit} (Chap. 4) for linear microwave response can be considered to be typical FOM of the former. The resulting temperature dependence is schematically represented by $1 - (T/T_c)^2$ in the figure. In contrast, the FOM of coolers (their technical efficiency) increases with increasing temperature. The temperature dependence of the Carnot efficiency η_C [see (5.8b) in Sect. 5.3] was adopted in Fig. 6.9 for simplicity.

The figure illustrates that the competitiveness of superconducting devices compared to conventional (room temperature) technology improves as the microwave performance of the HTS films can be improved at a given temperature, or as the critical temperature can be increased. Especially for the potential applications of HTS microwave devices in satellite-based communication systems, the cooler efficiency dictates the minimum tolerable operating temperature. Desired operating temperatures fall between 70 and 80 K. As a consequence, the linear and nonlinear microwave response of the HTS films must be optimized to the intrinsic limits of the material. Possible ways to achieve this condition can be:

1. The use of Tl-2212 films instead of Y-123 films due to their higher T_c values.

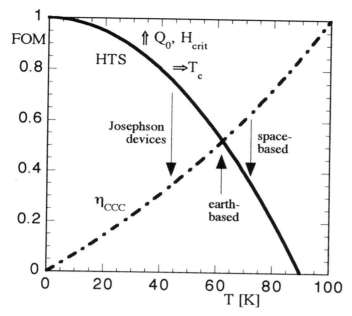

Fig. 6.9. Schematic temperature dependences of the FOM of superconducting devices (*solid line*) and of closed-cycle cryocoolers (*dash-dotted*). *Single-* and *double arrows* indicate different applications and ways for further improvement, as explained in the text.

2. Optimization of the oxygen stoichiometry of Y-123 films for maximum T_c (Sect. 2.3).
3. Optimization of the film thickness in order to reduce the effective surface resistance towards the bulk value (Sect. 1.1). This is mainly relevant if power dissipation in the film limits the performance of the device, e.g., at temperatures close to T_c.
4. If the quality factor is limited by dielectric losses, sapphire instead of LaAlO$_3$ must be used for high-temperature operation.
5. High H_{crit} values require the optimization of the HTS film technology in terms of very low defect densities. Low R_s and high T_c values support an extended linear region due to the prevention of microwave heating and vortex penetration (Sect. 3.2).
6. The microwave packaging must be optimized with respect to power dissipation, e.g., by preventing parasitic modes or by using superconducting housings.
7. The cryogenic packaging must be optimized with respect to low heat capacity and heat loads, and at the same time yield stable and sufficient cooling of the HTS device.

The conditions for microwave applications in ground-based communication systems appear to be much less stringent, since the envisaged operating temperatures are around 60 K.

The converse argument is that if a specific application demands some unique property of superconductors, like a certain value of the critical current of Josephson junctions, then the maximum tolerable operating temperature is dictated by the device (e.g., $T \leq 40$ K for HTS RSFQ circuits). The properties of the cooler (efficiency, noise etc.) might then be adapted to the required circuit specification.

Novel Filter Layouts

Many of the discussed near-term applications of passive HTS microwave devices concern the replacement of existing components by smaller and/or better superconducting ones. The market potential is accordingly lowered by the existence of well-approved conventional versions, and by the hesitating acceptance of system engineers. It is further limited by the number of operating systems, as well as by the continuing progress of the competitive alternatives. It appears, therefore, challenging to develop novel circuits that display unprecedented functionality, and that favorably fit to the demands of upcoming systems.

One possible approach is the development of high-Q multipole and, at the same time, miniaturized preselection filters with elliptic frequency response for use in communication systems [82–84]. Figure 6.10 is a schematic representation of an 8-pole quasi-elliptic filter composed of half-wavelength square-loop resonators occupying a total area of 72 mm×52 mm on RT duroid ($\varepsilon_r = 10.8$, $\lambda_g/2D \approx 2.8$) [83]. The characteristic feature of the layout is the symmetry of the single resonators, which results in different coupling properties along the two orthogonal directions. This symmetry provides high flexibility in arranging the filter elements. Proper dimensioning of the mutual orientation of the resonators and the spacings between them allows one to achieve a desired coupling scheme. The right-hand part of Fig. 6.10 compares the simulated frequency response with data initially measured with a normalconducting prototype. The center frequency is around 970 MHz, and the fractional bandwidth amounts to 5%. The noteworthy feature characterizing this filter layout is the achievement of 40 dB and 60 dB stopband rejection at a narrow fractional bandwidth of 7.5% and 8.5%, respectively. The insertion loss of about 3 dB is enhanced by dissipation in the normalconductor, as reflected by the moderate Q values. A modified HTS filter version has been developed for field trials at 1.8 GHz in GSM-1800 ("Global System Mobile") base stations, and for later implementation in UMTS ("Universal Mobile Telecommunication System") networks [82a]. The HTS filter (39 mm×22.5 mm) is even smaller than the Cu version due to the higher ε_r of MgO. The reduced insertion loss of around 1 dB reflects the high Q values around 5×10^4 of the single resonators.

6.2 Passive Devices Related to Resonant Elements

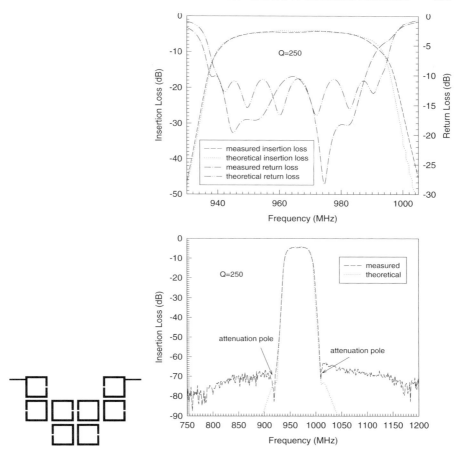

Fig. 6.10. *Left*: Layout of an 8-pole quasi-elliptic filter composed of half-wavelength square-loop resonators [83]. *Right*: Filter response on different frequency scales. (© 1999 IEEE).

As discussed in Sect. 6.1, high quality factors are crucial to achieve steep filter skirts, and thus to exploit the unique potential of superconducting resonators. At the low frequencies of interest ($f_0 = 1$–4 GHz), the attainable Q values are limited mainly by losses in the housing and, at temperatures above 50 K, by the loss tangent of LAO. Optimized Q factors require therefore a reduction of fringe fields at the edges of the resonator lines which could couple to the housing. Figure 6.11 shows the layout of a 9-pole lumped-element resonator on LAO with dimensions 39 mm×22.5 mm ($\varepsilon_r = 24$, $\lambda_g/D \geq 3.5$), designed for the resonant frequency $f_0 = 1.78$ GHz [22]. The symmetric layout of the resonators again provides the above mentioned advantages. It also minimizes the fringe fields by cancelling the field contributions of parallel

Fig. 6.11. Layout of a 9-pole quasi-elliptic HTS filter at 1.78 GHz with high Q values [22]. Main coupling (*solid lines*) and cross-couplings (*dashed lines*) are indicated in the *right-hand part*.

lines. The orthogonal arrangement of the resonators enables close spacing. The coupling strength between the resonators can be adjusted by, e.g., the use of dielectric screws [44].

The envisaged specifications of the filter are $f_0 = 1.77\,\text{GHz}$, $w = 0.85\%$, $N = 9$, $L_S \geq 30\,\text{dB}$ at 0.8 MHz outside the passband, $R \geq 18\,\text{dB}$ and $L_0 = 0.3\,\text{dB}$. To achieve such a low insertion loss requires unloaded Q values of around 5×10^4 [see (6.26)] which, at this small size, is without analog in conventional technology. Furthermore, the stopband rejection of 30 dB at a fractional bandwidth of $w_S = 1.1\,w$ would require more than 15 poles if built in a Tschebyscheff design [see (6.25)], illustrating the significant benefit of the quasi-elliptic layout.

Despite its strong miniaturization, a single HTS resonator was found to display unloaded Q factors above 9×10^4, 5×10^4 and 1.8×10^4, $T = 30\,\text{K}$, 60 K and 77 K [85]. The Q_0 values around 60–70 K suffice to meet the desired specifications, bringing even sufficiently high operating temperatures into reach. The filter housing could be concluded from the high Q_0 factors to contribute to a residual value well above 10^5. The wide signal lines and the field cancellation turned out to be advantageous also for a good power handling of the miniaturized resonators. A degradation of the quality factor by 10% was observed at circulating powers exceeding 50 W. Based on this promising result, the development of a complete operational HTS filter is in progress.

High-Q Resonators for High-Power Filters and Low-Phase Noise Oscillators

While strong miniaturization and extreme selectivity (steep skirts) are most desirable for receive filters, low insertion loss and high power handling capability are the major issues of transmitter combiner filters, e.g., for output multiplexers in communication systems. The potential system advantages concern an improved electric power efficiency and reduced size, both of which, however, demand high-power operation. For instance, the application of high-power filters in the transmit site of the US personal communication system (PCS) requires specifications like $f_0 = 2\,\text{GHz}$, $w = 0.75\%$ and the handling of about 3 kW incident power [42]. These figures correspond to a circulating power of 500–800 kW. Furthermore, the demand for low insertion loss ($L_0 \leq 0.2$ dB) translates into very high Q factors of the order of 10^5 [23].

The construction of such filters seriously limits the design and the quality of the employed superconductor and of the dielectric materials. In accordance with the discussion in Sects. 2.1.3 and 6.1.3, two or three dimensional resonators are generally required for this type of application. Representative of the latter class, HTS-shielded dielectric resonators have been employed, operated in the TE_{011} mode and composed of low-loss sapphire at an aspect ratio of $\varnothing/H \approx 2$ [1b, 18, 27, 86, 87]. Dielectric materials with a sufficiently low loss tangent but higher permittivity compared to sapphire would be desirable for further miniaturization. For a given frequency and temperature, the optimum design of a dielectric resonator is then determined by a balance between R_s and $\tan\delta$, and by the resulting temperature stability of resonant frequency and Q factor (Sect. 2.1.3 and discussion of ε_r and $\tan\delta$ in Sect. 5.2.2).

Early success with Tl-2212 films provided Q values of up to 3×10^6 and linear response up to circulating powers above 10 kW at 5.6 GHz and 77 K [86]. Later work demonstrated $Q_0 = 2 \times 10^5$ and $P_{\text{circ}} = 400$ kW in a YBCO-shielded LAO resonator at 5.5 GHz and 40 K [87]. Four-channel 4-pole output multiplexers ($f_0 = 4$ GHz, $w = 1\%$) based on HTS-shielded dielectric resonators were demonstrated with P_{circ} (77 K) ≥ 11 kW [18]. A two-pole filter ($f_0 = 6.4$ GHz, $w = 0.4\%$, $L_0 \approx 0.01$ dB, $R \geq 25$ dB) and subsequently a quasi-elliptic four-pole filter ($f_0 = 3.72$ GHz, $w = 0.86\%$, $L_0 \approx 0.04$ dB) were developed on the basis of YBCO-shielded dual-mode LAO hemispheres. Typical quality factors at $T = 77$ K were $Q_0 = 0.87 \times 10^5$ and remained at this high level up to an incident power of 180 W [79, 88].

An approach towards planar high-power filters was based on coupled microstrip resonators [89]. As summarized in [1b, 25], the design considered the minimization of the field-to-power conversion coefficient γ_H. The resonators were made of thick, wide Y-123 microstriplines in order to further improve the power handling capability. A 5-pole side-coupled filter ($f_0 = 2$ GHz, $w = 1\%$) was fabricated from $\varnothing 2''$ double-sided YBCO films on LAO. The filter response remained constant up to input powers of 10 W at 58 K and 36 W at 40 K. According to (6.30) in Sect. 6.1.3 and the related discussion, these figures correspond to circulating powers between 3 and 10 kW.

As introduced in [90] and described in some detail in Sect. 2.1.3, the disk resonator concept adopting edge-current free rotational symmetric modes presents a promising compromise between planar layout and good power handling capability [19]. This concept is currently being pursued by various European, American and Japanese companies, using Y-123 or Tl-2212 films (e.g., [23, 42, 44, 80, 91]). Figure 6.12 displays recent results on the quality factor and the microwave power handling of a $\varnothing 2''$ Y-123 microstrip disk resonator on LAO as measured in the TM_{010} mode in pulsed and CW

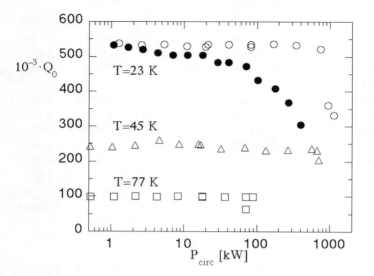

Fig. 6.12. Unloaded quality factor versus criculating power, measured with a microstrip disk resonator at 1.6 GHz in a test housing mounted on a GM cooler [23a, 92]. The *open* (*filled*) symbols were obtained in pulsed (slow sweep) measurements.

operation (Sect. 2.1.2) [23b, 92]. The Q_0 values approached 10^6 at low temperatures, limited by losses in the housing. Still, at 77 K the loss tangent of LAO allowed for a Q factor of 10^5 [i.e., $\tan \delta$ (1.6 GHz, 77 K)$\leq 10^{-5}$]. The response at 1.6 GHz remained linear up to $P_{\rm circ} \approx 1$ MW and 0.1 MW at $T \leq 50$ K and 77 K, respectively. Comparable power handling (30 kW at 40 K and around 10 kW at 77 K) was reported for a tunable filter version, which remained unaffected by the presence of a plunger film for frequency tuning over a range of more than 60 MHz at 2 GHz [23b]. The results demonstrated the excellent power performance of disk resonators, which exceeds that of lower-dimensional circuits and resembles that of the best HTS-shielded dielectric resonators.

Microwave heating deteriorated the power handling of the disk resonator to levels of about 0.1 MW at 23 K, indicating the remaining potential to improve the cryogenic layout of the filter package. At 23 K and at 45 K, a steep reduction of Q_0 was observed at comparable power levels $P_{\rm circ} = 0.8$ MW. This value corresponds to a magnetic field amplitude of around 25 mT ($\mu_0 \gamma_{\rm H} \approx 30\,\mu{\rm T/W}^{1/2}$). According to the results described in Sect. 4.1, the observed field breakdown is likely to be due to local defects in the superconducting films [93].

Quantitatively similar results, namely $Q_0(T) \leq 10^6$ and $P_{\rm circ}(T) \leq 1$ MW, were reported for a stripline modification of the disk resonator [42]. The edges of the two stacked double-sided ⌀3″ LAO wafers were metallized, thus forming essentially a fully enclosed cavity. As a result, the field-to-power

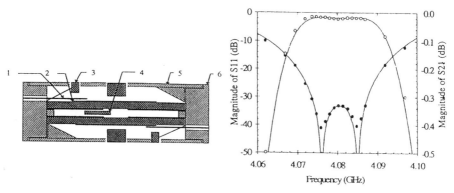

Fig. 6.13. *Left*: Layout of a 4-GHz two-pole disk resonator filter ($w = 0.4\%$ at $L_0 = 0.025\,\text{dB}$) [80] (1, coupling pin; 2, disk resonator; 3, trimming screws; 4, trimming strip; 5, lid; 6, housing). *Right*: Scattering parameters $|S_{11}|^2$ (*filled*) and $|S_{21}|^2$ (*open symbols*) in dB measured at $T = 77\,\text{K}$, and calculated for a two-pole Tschebyscheff filter with $Q_0 = 5 \times 10^4$ (*solid lines*).

conversion coefficient γ_H was lower than in the "open" microstrip design of the disk resonator. The quality factor was again limited below 10^5 by the loss tangent of the substrate at $T \geq 50\,\text{K}$. The temperature dependence of the power handling indicated the important role of microwave heating, in accordance with the previously described results.

A two-pole prototype version of an elliptic four-pole disk resonator filter at $f_0 = 4\,\text{GHz}$ was presented in [80]. Figure 6.13 sketches the layout employing two double-sided ∅1″ YBCO films on 1-mm-thick LAO. Dielectric screws and strips were used to trim resonant frequencies and coupling coefficients. The scattering parameters measured at 77 K (right-hand part of Fig. 6.13) revealed an insertion loss $L_0 < 0.025\,\text{dB}$ within a fractional bandwidth of $w \approx 0.4\%$, and a return loss $R > 30\,\text{dB}$. The frequency response corresponded to Q_0 values of 5×10^4, which were dominated by losses in the dielectric. The midband attenuation remained in CW operation below $L_0 = 0.1\,\text{dB}$ up to an incident (respectively: circulating) power of 100 W (about 30 kW). The disk resonator concept has been extended to the development of a three-pole Tschebyscheff filter in the stripline approach [94], and to a quasi-elliptic four-pole filter by coupling two pairs of microstrip resonator stacks as shown in the left part of Fig. 6.13 [95].

Another area of HTS microwave applications is miniaturized low phase-noise oscillators, e.g., for use in digital communication or radar systems. High quality factors ($\geq 10^6$) and high frequency stability of the resonators are crucial for successful operation [26, 27, 96]. Most approaches have therefore been based on HTS-shielded dielectric resonators composed of lowest-loss sapphire, which allows for maximum Q at minimum volume. Thermal drifts of the resonant frequency $\partial(\ln f_0)/\partial T$ are mainly caused by the temperature dependent permittivity of the dielectric, $\partial \varepsilon_r/\partial T > 0$. This unwanted effect can be compensated for with sapphire-rutile composites, since the temperature coefficient of rutile $\partial \ln \varepsilon_r/\partial T < 0$ is negative [81, 97]. Alternatively and/or additionally, piezoelectrically driven HTS tuning circuits [81] have been applied to achieve a stable frequency. Values of $\partial(\ln f_0)/\partial T \approx 10^{-11}/\text{K}$ were anticipated for 10 GHz and 77 K [97]. Stable operation at 24 GHz with phase noise of -130 dBc/Hz at 1 kHz off the carrier is expected for improved noise figures of the feedback amplifiers, even with the resonator mounted on a Stirling cooler [81].

6.3 Passive Devices Based on Josephson Junctions

Designs of superconductive microwave devices become more versatile if, in addition to components fabricated from homogeneous films (Sect. 6.2), the integration of Josephson junctions is considered. The exciting field of superconducting electronics has been reviewed, e.g., in [98, 99].

The unique features of Josephson junctions [100], which give rise to important functions such as wide-band detection and emission of electromagnetic fields, or analog and digital signal processing, are outlined in Sect. 6.3.1. Section 6.3.2 reviews possible applications of Josephson devices in metrology, magnetometry, signal detection and processing, and computing. The physics and the technology of Josephson junctions form a fascinating discipline in themselves, and have been the focus of many research groups and innumerable publications. Therefore, the present description is not intended to be complete, but to provide introductory impressions of this field. In order to illustrate the richness of the physical aspects involved even in simple junction devices, Sect. 6.3.3 presents some features of single-junction SQUIDs at microwave frequencies which appear interesting, e.g., for analog signal processing in communication systems. The description here is different from the usual approach which focuses on the properties of the SQUID while considering the microwave resonator as one part among others of the readout circuit. Here, the passive microwave device plays the major role with the SQUID giving it specific nonlinear flux-sensitive properties.

6.3.1 Unique Features of Josephson Junctions and SQUIDs

Some of the physical properties of Josephson junctions (JJs) were already considered in previous parts of this book. Section 1.2 introduced the characteristic voltage V_c and the Josephson penetration depth λ_J as characteristic junction parameters. The microwave response of JJs was discussed on the basis of the Josephson equations [see (1.34)], and interpreted with the RSJ model [101, 102] and the resulting lumped-element circuit [103]. The main theme of that section was the impact of Josephson junctions on the linear surface impedance of HTS. The short coherence length of A15 and HTS materials was found to lead in many cases to granular effects. In addition, Sects. 3.1.4 and 3.4.1 dealt with the nonlinear response of Josephson junctions, causing enhanced magnetic field dependence of the surface impedance of granular films.

In contrast to dealing with unavoidable grain-boundary junctions in extended film areas, this section focuses on the properties of Josephson junctions, which are desired for specific operations at high frequencies or at fast clock rates. These unique features become accessible by integrating engineered JJs into properly designed microwave circuits (including those discussed in Sect. 6.2). To aid the reader, some of the basic relationships needed to describe the Josephson behavior are summarized in the following.

Basic Features of Josephson Junctions

Key parameters of JJs are the critical Josephson current I_{cJ} and the characteristic voltage $V_c = I_{cJ} R_J$ [3, 104]. According to the analysis in Sects. 1.2.2 and 3.1.4, the Josephson equation

$$I(t, \Phi) = I_{cJ} F\left[\psi(t, \Phi)\right] + \frac{V(t)}{R_J} \tag{6.37a}$$

and the voltage drop across the junction

$$V(t) = \frac{\Phi_0}{2\pi} \frac{\partial \psi}{\partial t}, \tag{6.37b}$$

reveal the inductive nature of Josephson junctions, parametrized by the inductance $L_J \propto I_c^{-1}$. L_J as well as the surface impedance Z_{sJ} of Josephson junctions depend explicitly on the relation $F(\psi)$ between the Josephson current and the time and magnetic flux dependent phase difference of the Cooper pair wave function across the junction. For HTS grain-boundary Josephson junctions engineered from substrate steps and 24° bicrystal interfaces, a sinusoidal relationship $F(\psi) = \sin(\psi)$ could be experimentally verified [105–107]. The situation is less clear for large-angle (e.g., 45°) bicrystal boundaries [107] and for c-axis oriented microbridges [108]. Investigations of the surface impedance of such junctions could provide further information about their dynamic transport properties. For simplicity, we will assume $F(\psi) = \sin(\psi)$ in the following. The Josephson inductance can then be described at low bias currents $I < I_{cJ}$ as:

$$L_J = L_{J0} \frac{1}{[1 - (I/I_{cJ})^2]^{1/2}} \tag{6.38a}$$

with the zero-current value

$$L_{J0} = \frac{\Phi_0}{2\pi} \frac{1}{I_{cJ}}. \tag{6.38b}$$

The dynamic response of JJs in the adiabatic limit, which is of special interest in microwave applications, can be described well by the RSJ model. This limit implies that the frequencies of interest be small compared to the gap frequency, $\omega \ll \Delta/\hbar$, and to the inverse relaxation time τ_J of the junction-equivalent L–R circuit [103], $\omega \ll 1/\tau_\mathrm{J}$. Using (6.38b), τ_J can be rewritten as

$$\tau_\mathrm{J} = \frac{L_{\mathrm{J}0}}{R_\mathrm{J}} = \frac{\Phi_0}{2\pi} \frac{1}{I_{\mathrm{cJ}} R_\mathrm{J}} \ . \tag{6.39}$$

The nonlinear surface impedance $Z_{\mathrm{sJ}}(I,T)$ of Josephson junctions is determined by the Josephson inductance $L_\mathrm{J}(I,T)$ and the dimensionless frequency $\Omega(T) = \omega \tau_\mathrm{J}(T)$. For isotropic SIS junctions (Superconductor-Insulator-Superconductor), the characteristic voltage $I_{\mathrm{cJ}} R_\mathrm{J}$ is related to the energy gap Δ of the superconducting electrodes [109]:

$$V_\mathrm{c}(T) = I_{\mathrm{cJ}} R_\mathrm{J} = \frac{\pi}{2e} \Delta(T) \tanh\left(\frac{\Delta(T)}{2 k_\mathrm{B} T}\right) . \tag{6.40}$$

Equation (6.40) requires minor modifications if strong-coupling superconductors are considered [110, 111]. However, large energy gaps are beneficial for short relaxation times and thus for high-frequency applications of Josephson junctions: τ_J is of the order of 400 fs and 40 fs for an energy gap of $\Delta = 3$ meV and 30 meV, respectively, typical of conventional superconductors and oxide superconductors. However, the possible two-band nature or the d-wave pairing symmetry of the HTS materials (Chap. 1) shed some doubts on the accessibility of this potential benefit.

As ω approaches the gap frequency $\omega_\Delta \equiv \Delta/\hbar$, the Josephson inductance L_J increases to $1.57 L_{\mathrm{J}0}$, reflecting the frequency dependence of the critical current $I_{\mathrm{cJ}}(\omega)$ [112]. Even above ω_Δ, the decrease of $I_{\mathrm{cJ}}(\omega)$ is shallow and extends up to about $10\omega_\Delta$, leaving significant potential for submillimeter wave applications of Josephson devices [3, 112]. In terms of circuit analysis, Josephson junctions are thus nondispersive over an extremely wide bandwidth. However, the same is not true for the superconducting transmission lines required to arrange a number of junctions in a reasonable network. Power dissipation and dispersion of the superconductor itself (Chap. 2) might therefore reduce the technologically accessible frequency range.

The prefactor $\Phi_0/2\pi = h/4\pi e$ in (6.37)–(6.39) is characteristic of the physics of Josephson junctions. Referring to (6.37b), its physical importance becomes obvious. Josephson junctions biased at $V > 0$ in the voltage state (i.e., at $I > I_{\mathrm{cJ}}$) excite high-frequency currents at the "internal" frequency $f_\mathrm{V} = \omega_\mathrm{V}/2\pi$:

$$f_\mathrm{V} = \frac{|V|}{\Phi_0} = 483.59\,\mathrm{GHz} \times V\,(\mathrm{mV}) \ . \tag{6.41}$$

Vice versa, microwave irradiation of Josephson junctions at a frequency f_ext leads to mixing of the external and internal frequencies due to the highly nonlinear Josephson reactance [113]. If the Nth harmonic $N f_\mathrm{ext}$ coincides with the Josephson oscillation f_V, the mixing product at intermediate frequency occurs at DC. As a result, the time-averaged IV characteristics (IVC) of Josephson junctions display current steps (the Shapiro steps, [114]) at voltages $\pm N f_{ext} \Phi_0$. This so-called AC Josephson effect (in contrast to the occurrence of a DC supercurrent at $V = 0$) is the key to high-frequency applications such as Josephson voltage standards, mixers and voltage controlled oscillators (see below). Shapiro steps in the current–voltage characteristics of HTS JJs which are already above the limit accessible with LTS–SIS junctions have been observed up to several THz [51].

Besides the inductive nature of Josephson junctions, the high-frequency displacement of charges across the barrier experiences a geometry dependent junction capacitance C_J. The combination of C_J and L_J leads to resonant behavior, and thus defines another intrinsic time constant, given by the plasma frequency ω_p [115]:

$$\tau_p = \frac{1}{\omega_p} = \sqrt{L_{J0} C_J} \ . \tag{6.42}$$

The time τ_p can be considered to be an approximate measure of the highest resolution that can be achieved in time domain measurements. However, for small values of C_J (e.g., in the case of leaky barriers as in most types of HTS junctions), the Josephson dynamics are determined by the larger value $\tau_J > \tau_p$.

Comparing (6.39) and (6.42) reveals that the ratio τ_p^2/τ_J defines a third characteristic time scale of Josephson junctions: the charging time of the Josephson capacitance via its normal resistance, $\tau_p^2/\tau_J = R_J C_J$. The dimensionless Stewart-McCumber parameter $\beta_C = (\tau_p/\tau_J)^2 \propto V_c R_J C_J$ quantifies the effect of capacitive shunting on the time-averaged IVC. For the case $\beta_C < 1$ (respectively: $\beta_C > 1$), the junctions are called overdamped (underdamped), and the resulting IVC is nonhysteretic (hysteretic). Detailed analysis of the dynamics of Josephson junctions and circuits under various operating modes [3, 112] identifies a dominant role of the macroscopic parameters $R_J C_J$ and τ_p for time-domain applications such as the generation, detection and processing of pulses and switching transients. Control of the nature and the thickness of the tunnel barriers is therefore a crucial prerequisite of Josephson technology.

The response of Josephson junctions to time-varying magnetic fields is different for small and long junctions, where the geometrical width is smaller, respectively larger, than the Josephson penetration depth λ_J (Sect. 1.2 and, e.g., [103]). Penetration and propagation of vortices in long junctions depend on the geometry of the applied fields and currents [104]. The dynamics reveal unique features that can be exploited for an additional class of high-frequency applications, prominent representatives of which are the flux-flow oscillator [116] and the flux-flow transistor (see review [31]). However, the electrodynamic properties of long Josephson junctions are complicated, and numerical simulation is generally required to solve for specific boundary conditions. Characteristic parameters of long JJs are the Swihart velocity of the electromagnetic waves propagating in a barrier of effective permittivity ε_r, and the dielectric thickness $d_\mathrm{J}/\varepsilon_\mathrm{r}$ of the barrier [117]. Some approaches have been successful to fabricate long HTS junctions for flux-flow devices [118]. Existing data yielded a Swihart velocity of $(0.01$–$0.04)\,c$ and an electrical thickness of 0.1–$0.4\,\mathrm{nm}$ [118, 119], which resemble data obtained with conventional SIS Josephson junctions [3].

Josephson Junctions Enclosed in Superconducting Loops

In order to enhance the sensitivity of Josephson devices to magnetic flux, one or more junctions are usually enclosed in superconducting loops. The most prominent class of loop-shunted junctions is the SQUID. It forms the key element of sensitive magnetometers as well as of digital components like analog-to-digital converters or shift registers [120–122].

The flux dependence of SQUIDs and related devices is determined by the external flux Φ_x threading the loop of inductance L_S, and by the self-flux $-L_\mathrm{S}I$ generated by the current I flowing in it. It proves convenient (see preceding section) to express the magnetic flux in dimensionless units, namely in terms of the variable $\varphi \equiv 2\pi\Phi/\Phi_0$. Then Φ_x translates into φ_x, and $-L_\mathrm{S}I$ becomes $-\beta_\mathrm{L}\iota$, where $\iota = I/I_\mathrm{cJ}$ and

$$\beta_\mathrm{L} = \frac{2\pi}{\Phi_0} L_\mathrm{S} I_\mathrm{cJ} = \frac{L_\mathrm{S}}{L_\mathrm{J0}} \ . \tag{6.43}$$

Different from the capacitance parameter β_C, the parameter β_L relates the inductance of the Josephson junction to that of the loop circuit. It is of paramount importance for any SQUID circuit since it quantifies its "shielding efficiency". The loop accepts exactly one flux quantum under critical conditions $(I = I_\mathrm{cJ})$ if $\beta_\mathrm{L} = 2\pi$. If circuits need to be designed which prevent (respectively: favor) the trapping of flux quanta, the quantity $\beta_\mathrm{L}/2\pi$ has to be adjusted to values smaller (larger) than unity. Since the junction critical current I_cJ is usually determined such that the Josephson coupling energy $\Phi_0 I_\mathrm{cJ}/2\pi$ exceeds the thermal energy $k_\mathrm{B}T$ by a prescribed factor $1/\Gamma$ (with Γ a noise parameter), the design of such circuits requires proper layout of the loop inductances. Small L_S values are required especially in the case of high operating temperatures, as they have become accessible with the oxide superconductors. Regarding the appreciable kinetic contribution to L_S resulting from the large penetration depth in HTS (Chap. 2), multilayer structures must be used, which present additional technological challenges [51].

As an instructive example of simple microwave Josephson devices, let us consider a SQUID consisting of a single Josephson junction, which is inductively coupled to a transmission line (TL) as sketched in Fig. 6.14. The TL section has the inductance L, and the SQUID loop inductance is L_S. The whole device can be replaced in the small-signal limit $|I|/I_\mathrm{cJ} < 1$ by an effective flux dependent inductance $L_\mathrm{eff}(\varphi)$, which can be evaluated from the corresponding equivalent circuit (displayed in the right-hand part of Fig. 6.14):

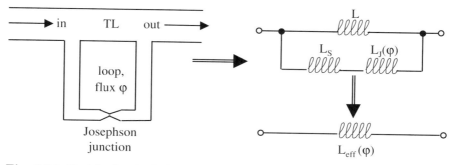

Fig. 6.14. Sketch of a single-junction SQUID coupled to a transmission line (*left*) and equivalent circuits (*right*).

$$L_{\text{eff}} = \frac{L(L_S + L_J)}{L + (L_S + L_J)} = L \left(1 + \frac{k^2 \beta_L \cos(\varphi)}{1 + \beta_L \cos(\varphi)} \right)^{-1}, \quad (6.44\text{a})$$

where the second equality used (6.37a) and (6.38) for $L_J[I(\varphi)]$ and (6.43) for the normalized SQUID inductance β_L. The coupling coefficient k in (6.44a) describes, in accordance with its usual definition (Sect. 6.1), the exchange of magnetic energy between the transmission line and the SQUID:

$$k^2 = \frac{\text{energy exchanged between TL and SQUID}}{\text{energy stored in the TL}} = \frac{M^2}{L L_S} \leq 1 \quad (6.44\text{b})$$

with M the mutual inductance. In case of direct galvanic coupling between SQUID loop and transmission line, $M = L$ and accordingly $k^2 = L/L_S$. Equation (6.44) reveals the important result that, as long as $(1 + k^2) < 1/\beta_L$ holds, a continuous variation of the total flux $\varphi_{\text{tot}} = \varphi_x + \beta_L \cdot \iota$ threading the SQUID loop leads to a periodic variation of L_{eff} with period 2π and amplitude

$$\left| \frac{\Delta L_{\text{eff}}}{L} \right| = \frac{2k^2 \beta_L}{1 - [\beta_L(1 + k^2)]^2}. \quad (6.45)$$

The 2π periodicity reflects the quantization of magnetic flux in the loop. In the realistic case of weak coupling ($k^2 \ll 1$) and for $\beta_L < 1$ (known as the dispersive regime), (6.45) yields for the relative variation of the effective inductance the value $2k^2 \beta_L / (1 - \beta_L^2)$. Large effects require therefore an adjustment of β_L close to, but still below unity.

The presence of the loop inductance L_S does not only affect the flux dependence of the device, but also modifies its dynamic features [3]. The investigation of the switching behavior of dissipative (or hysteretic) SQUIDs having $\beta_L > 1$ revealed a reduction of the delay time compared to that of the bare junction [112].

Further consequences of (6.44) and (6.45) as well as the implications of $\beta_L > 1$ are treated in more detail in Sect. 6.3.3.

Remarks on the Preparation of Josephson Junctions

The fabrication of devices that could exploit the benefits of large bandwidth, fast switching times, and the potential for monolithic integration with passive circuits relies on the controlled fabrication of engineered Josephson junctions. For conventional superconductors like Nb and Pb, various routes to fabricate Josephson junctions with low parameter spread and high yield were developed [104]. Regarding especially the fabrication of homogeneous Nb-SIS junctions, the discovery of a suitable barrier material (AlO$_x$, [123]) was crucial for the technological progress in this field, which has led to a reproducible yield of 2–3-µm-wide junctions with a spread of critical current densities below about 3% and a scale-of-integration above 10^4 junctions per chip [51].

An analogous approach to A15 tunnel junctions has been complicated by the high preparation temperatures (Sect. 5.1.1), the roughness of the polycrystalline films and the lack of appropriately wetting insulators [124]. SIS junctions of these materials are therefore not available. The discovery of HTS raised hopes of extending Josephson electronics to even higher frequencies. Since 1987, the "traditional" fabrication techniques have also been applied to HTS, as described in several reviews [98, 99, 120, 121, 125]. However, in contrast to the LTS junctions, HTS JJs usually exhibit V_c values much below the theoretical limit [see (6.40)], and large on-chip and even larger chip-to-chip parameter spreads (Sect. 5.1.3). Moreover, V_c often scaled with the critical current density, universally for different junction types and over many orders of magnitudes like

$$V_c \propto J_c^q \qquad (6.46)$$

with $q \approx 0.5$. Also, an antiphase correlation between critical-current and resistance fluctuations was observed [125a]. These unconventional properties were interpreted in terms of an intrinsically shunted junction model [126]. The junction barrier is assumed to be inherently insulating, but to contain a high density of localized states, providing a resistive shunt across the junction. They furthermore restrict the tunneling of Cooper pairs (mainly via direct tunneling) and of quasi-particles (via direct and resonant tunneling) to different conduction channels. The engineering of HTS JJs means, therefore, to control the properties of the effective barrier on an atomic scale of the order of the coherence length. As a consequence, applying HTS to devices containing about 10^2 junctions or more with prescribed properties is prevented by the large parameter spread [51]. Anticipating success in a reproducible fabrication technology of HTS JJs, the design and development of analog and digital devices is nevertheless rapidly proceeding.

6.3.2 Survey of Josephson Junction Microwave Devices

Classification of Josephson Device Applications

Shortly after the theoretical prediction of Josephson tunneling and its experimental verification [51, 100, 127], the junctions were identified as being key elements in numerous applications. Table 6.3 summarizes the resulting potential in terms of the required scale of integration. The abbreviations SSI, MSI, and LSI stand for small-, medium-, and large-scale integration, defined by the number of junctions per chip. The tolerable spread of junction parameters usually decreases towards higher integration.

The first row of Table 6.3 lists applications between DC and GHz frequencies. Devices devoted to the generation, detection and processing of millimeter and sub-millimeter wave and/or broadband signals are in the second row. Instrumentations based explicitly on oxide high-temperature superconductor (mainly Y-123) Josephson junctions (respectively: SIS junctions from conventional superconductors, mainly Nb/AlO$_x$/Nb) are indicated by the letter "H" ("L"). Digital circuits are printed in bold letters, while devices requiring phase synchronization are printed in italics. The table does not account for the different specifications of the listed devices in terms of, e.g., sensitivity, operating temperature, and so on. Details and reviews on this topic are given in the references to this chapter.

The low yield and the large parameter spread of HTS junction circuits are reflected in the lack of devices containing more than about 50 JJs. Furthermore, the present impossibility to prepare SIS junctions from HTS films leads to devices for mixing, pulse forming and waveform analyzing as another domain of LTS junctions. In contrast, the major benefit of using HTS instead of LTS junctions in SSI and MSI devices is the lowered cryogenic burden due to higher operating temperatures. Possible applications concern non-destructive evaluation (NDE, [128, 129]), geological prospecting [130] and magneto-cardiography (MCG, [51]). The high field sensitivity of about 1–10 fT/(Hz)$^{1/2}$ required for magneto-encephalography (MEG) is still slightly below the routinely achieved performance of HTS systems, and therefore in commercial applications predominantly LTS systems are used [131].

Josephson devices for metrology include flux comparators which can be applied, e.g., for extremely sensitive voltage measurements (pV-meter, sensitivity up to 10^{-15} V/(Hz)$^{1/2}$ at 4.2 K [132]) or susceptometry (sensitivity up to $\Delta\chi/\chi \leq 10^{-7}$, [133]). Recent developments aimed to implement miniaturized SQUID sensors in scanning tip microscopes for imaging physical quantities that can be converted into magnetic flux (susceptibility, conductivity, etc.) [134–136]. Such instruments have already become commercially available.

The layered crystal structure of HTS and the related anisotropy have brought about a novel phenomenon which had no counterpart in conventional superconductors, namely, intrinsic Josephson coupling between adja-

Table 6.3. Potential of Josephson junctions for device applications. SSI, MSI, and LSI stand for small-, medium-, and large-scale integration. The types of superconductors ("L" for LTS and "H" for HTS) are indicated only if the alternative version is lacking. If none is specified, implementations can be in both types. Digital devices are printed in boldface. Italics denote devices which require phase synchronization of the junctions. Further explanations are given in the text.

Frequency regime	SSI (10^0–5×10^1 JJs/chip)	MSI (5×10^1–10^3 JJs/chip)	LSI (10^3–5×10^4 JJs/chip)
DC to GHz	Metrology NDE Geological prospection Microscopy	MCG MEG (L)	**DAC (L)**
GHz to THz	Hilbert transform spectroscopy RSJ mixer (H) SIS mixer (L)	Phase shifter (H) *Oscillator (H)* **Shift register (H)** Pulse former (L) Wave form analyzer (L)	*Voltage standard (L)* *Oscillator (L)* **ADC, DSP (L)** **Computing (L)**

cent CuO_2 planes [137]. This feature has been verified not only in single-crystals of the Bi and Tl compounds (2212 and 2223), but also in mesas prepared from thin Bi and Tl films as well as in oxygen deficient Y-123 films. The intrinsic Josephson effect provides an ultimate packing density (LSI) which might find applications for on-chip high-frequency sources [138] (not yet attributable to the areas listed in Table 6.3).

Selected Josephson Junction Microwave Devices

This paragraph contains explanatory remarks on some of the potential applications considered in Table 6.3. The focus of this description is on the benefit of analog Josephson devices for high-frequency applications. Compared to the passive microwave devices treated, e.g., in Sect. 6.2, HTS Josephson components are still less well developed. However, important steps towards useful high-frequency components, and first demonstrations of prototype circuits have been reported. Most of these devices have been fabricated from grain-boundary Josephson junctions on bicrystal (MgO, NGO, YSZ) or stepped substrates (LAO) with dielectric properties suitable for microwave applications (Sect. 5.2).

Josephson devices operating in the frequency domain or in the time domain (bottom row of Table 6.3) comprise detectors, mixers and voltage controllable oscillators or ultrafast switches for detection, pulse forming, sampling, counting and analog-to-digital conversion (ADC) [3, 104, 112, 122, 139].

Furthermore, Josephson junctions are the building blocks of logic circuits, the most prominent example being the "rapid single flux quantum" (RSFQ) logics. RSFQ circuits access ultrafast ($> 100\,\text{GHz}$) computing and digital signal processing (DSP) [140, 141]. The challenge is the design of circuits tailored to a specific function, and which put minimum constraints on the tolerable parameter spread of the constituents. Since RSFQ circuits consist of 10^2–10^4 junctions, HTS are not yet applicable to this field.

Signal operations such as detecting, mixing, switching etc. could in principle also be performed with film devices without junctions (c.f., Sect. 6.2). Whether homogeneous structures or Josephson elements are better suited depends on the required sensitivity, dynamic range and noise performance. Josephson devices allow improved sensitivity and dynamic range because of the lower critical current densities (10^4–10^5 rather than 10^6–$10^7\,\text{A/cm}^2$) and response times, $\tau_J \geq 40\,\text{fs}$ compared to $\tau_\Delta \geq 10\,\text{ps}$ (Sects. 3.1.1 and 6.3.1).

(1) *Signal detection and spectroscopy using Josephson junctions.* RSJ-like current–voltage characteristics can be obtained reproducibly with HTS Josephson junctions. Accordingly, many different kinds of microwave detection and mixing experiments have been reported, while only few of them considered the figures of merit in relation to real applications.

The FOMs of detectors are high responsivity and sensitivity, and a large receiving bandwidth. The responsivity of a detector to a high-frequency signal is defined as the ratio of the output response (e.g., a current or a voltage) to the power of the detected signal. The maximum sensitivity is determined by the signal value which equals the noise floor. The noise-equivalent power (NEP) of a detector is given by the ratio of the maximum sensitivity to the responsivity at a postdetection bandwidth $\Delta f = 1\,\text{Hz}$. For instance, the intrinsic current responsivity of a symmetric SIS junction is one electron per photon, e/hf_S, where f_S is the frequency of the detected signal [142]. The corresponding minimum NEP is given by $hf_S(I_0/e)^{1/2}$, with I_0 the dark current of the junction. Since direct detectors are not sensitive to the phase of the signals, there is no quantum limit to the noise, in contrast to phase-sensitive devices like frequency mixers (see below). An ideal junction would thus display NEP$\to 0$. The best SIS detector exhibited a noise-equivalent power of $10^{-15}\,\text{W/Hz}^{1/2}$ [3, 142a].

Shapiro steps constitute the basis for the detection of intense monochromatic microwave signals in RSJ-like detectors. It can be shown that the performance (i.e. the NEP) improves with increasing characteristic voltage V_c [3, 104]. However, very weak signals can also be detected with RSJ-like Josephson junctions [3, 143]. The *IVC* is modified under microwave irradiation by an amount $\Delta I(V)$ which, in the small-signal limit, is much smaller than the equilibrium value $I_0(V)$. The response was found to be proportional to the Hilbert transformation of the spectral density $S_I(f)$ characterizing the detected radiation. It is therefore possible to deconvolute $S_I(f)$ from the experimental data [143b]:

$$S_\mathrm{I}(f) = \frac{1}{\pi} \int_{-\infty}^{\infty} \frac{H(f_\mathrm{V})}{f_\mathrm{V} - f} \mathrm{d} f_\mathrm{V} \,, \tag{6.47a}$$

where f_V is given by (6.41). The normalized response function $H(f_\mathrm{V})$ is

$$H(f_\mathrm{V}) = \frac{8\Phi_0}{\pi} \frac{\Delta I(V) I(V) V}{(I_\mathrm{cJ} R_\mathrm{J})^2} \,. \tag{6.47b}$$

The application of Hilbert transform spectroscopy can be advantageous for high-speed and/or broadband measurements in the mm and sub-millimeter wave range. A frequency resolution of better than 1 GHz and an energy resolution of 10^{-13} J/s were reported for a resistively shunted SIS junction with $R_\mathrm{n} = 0.7\,\Omega$ at 95 GHz [144]. A spectrum analyzer with 2-GHz resolution and a measurement time of a few ms was developed for plasma diagnostics in the mm-wave range [145]. Recently, Hilbert transform spectroscopy has been demonstrated with Y-123 grain boundary junctions on NGO bicrystals at 77 K in the frequency range 60 to 800 GHz [143a,b]. The absence of resonances at voltages $V_N = N f_\mathrm{ext} \Phi_0$ for $N > 1$ proved the response to be in the small-signal limit. The spectral resolution was limited to 2.9 GHz at 693 GHz by the linewidth of the junction radiation which, in turn, was determined by thermal fluctuations. The spectrometer was successfully applied to investigate the harmonic content of commercial mm-wave oscillators.

(2) *Josephson mixers*. Sensitive mixers have important applications in radioastronomy and environmental spectroscopy [142]. Compared to Schottky diodes, a major advantage of mixers using SIS tunnel junctions with suppressed Josephson current is the low power requirement of the local oscillator. The noise temperature is quantum-statistically limited to $f h/2 k_\mathrm{B} = 24\,K \times f$ (THz), which has been approached in the best mixer junctions. A complete receiver front-end, consisting of a planar antenna, a Nb SIS mixer, a flux-flow oscillator as local oscillator (LO) and a matching network, integrated together on a 4 mm×4 mm×0.2 mm quartz chip, were reported in [146]. The oscillator delivered $\geq 0.1\,\mu\mathrm{W}$ at 450 GHz and exhibited a linewidth of 750 kHz at 280 GHz. However, the frequency range accessible with LTS-SIS mixers is limited at about 1 THz by the energy gap of the Nb electrodes. Unfortunately, the promise of higher gap frequency in HTS junctions is questioned by the unconventional pairing mechanism, and by the technological problems in the fabrication of thin-film tunnel junctions. Alternative approaches to wide-band detection and mixing are based on measuring the temperature rise of electrons, which are loosely coupled to the phonon system of thin conducting films (hot electron devices) [142].

While optimum performance can be obtained with externally shunted SIS mixers, HTS JJs are inherently resistive. The corresponding maximum conversion gain K_JJ of RSJ mixers was theoretically evaluated in [147] as a function of junction properties and matching circuit:

$$K_\mathrm{JJ} = \delta^2 \frac{4 R_\mathrm{d} R_\mathrm{IF}}{(R_\mathrm{d} + R_\mathrm{IF})^2} \frac{R_\mathrm{d}}{R_\mathrm{J}} \,. \tag{6.48}$$

Here, δ is a dimensionless parameter determined by the sensitivity of the critical Josephson current to the power of the local oscillator signal, $\partial I_{\rm c}/\partial P_{\rm LO}$. It was found to range between 0.4 and 1.2 for different junctions [148, 149]. $R_{\rm d}(V) = [\partial I(V)/\partial V]^{-1}$ denotes the dynamic resistance of the Josephson junction at the bias voltage V, and $R_{\rm IF}$ is the resistance at the intermediate frequency. As a result of (6.48), high values of $R_{\rm d}$ and $I_{\rm c}$ are better for achieving high conversion gain and efficient matching. The theoretical minimum noise temperature of RSJ mixers was reported in [51] to be about $10\, f\, h/k_{\rm B}$.

HTS Josephson mixers have been investigated by numerous groups ([150] and references therein). A conversion gain $K_{\rm JJ} = -11\,{\rm dB}$ was reported at a LO power of $-20\,{\rm dBm}$, and at $f_{\rm LO} = 11\,{\rm GHz}$ and $f_{\rm IF} = 1\,{\rm GHz}$. The minimum noise figure was $10.4\,{\rm dB}$ (corresponding to a noise temperature of $2900\,{\rm K}$) at $17\,{\rm K}$ at a reduced gain of $K_{\rm JJ} = -16\,{\rm dB}$. Mismatches between the low-resistance junction and the microwave circuit limited the performance. Series arrays of junctions have been investigated in order to increase the junction input resistance in detector or mixer applications. Voltage responsivities of $1100\,{\rm V/W}$ at frequencies up to $150\,{\rm GHz}$ and two-tone mixing up to $670\,{\rm GHz}$ were observed in a series array of up to ten Y-123 GBJs on bicrystal MgO and stepped LAO substrates [151]. The dynamic effects indicated a response time of about $500\,{\rm fs}$, corresponding to $V_{\rm c}\,(40\,{\rm K}) \approx 3\,{\rm mV}$ [152].

(3) *Analog signal processing.* Another class of Josephson microwave devices is based on the controllability of the Josephson inductance $L_{\rm J}(B) = \Phi_0/2\pi I_{\rm c}(B)$ by external magnetic fields, e.g., for integrated superconducting tunable filters and phase shifters. The principle of operation can be understood from (6.44) and (6.45), by considering a transmission line with each line segment containing a separate SQUID [153]. This concept is known as the "distributed Josephson inductance" (DJI), which is a special version of a Josephson transmission line. The flux dependent inductance leads to a flux dependent phase constant of the transmission line, which scales like $L_{\rm eff}^{1/2}$. As concluded from (6.45), $k^2 \approx 1$ and $\beta_{\rm L} \leq 1$ optimize the achievable phase shift. In this case, large $V_{\rm c}$ values are required in order to keep the dissipation losses in the Josephson junctions low. Furthermore, at microwave currents exceeding $I_{\rm cJ}$, large k values limit the power handling capability, especially in resonant structures. SQUID-based phase shifters are therefore mainly suited for low-power devices like lumped-element filters (Sect. 6.2). Another advantage of such devices is the availability of small inductances providing an enhanced fractional modulation per SQUID. Improved power handling could be attained by decreasing the coupling strength between SQUIDs and the microwave device. To compensate for the reduced inductance variation, an increased length of the transmission line and thus of the number of SQUIDs is required.

HTS microstrip DJI phase shifters were demonstrated, e.g., for resonant structures with 40 [154] and for wide-band nonresonant transmission lines with more than 1000 JJs [155]. The 2-μm wide Y-123 step-edge junctions (SEJ) on 0.5 mm thick LAO substrates [V_c(77 K) ≈ 140 μV] provided, at 10 GHz, phase shifts of about 60^0 below 40 K and at microwave power levels below −60 dBm (resonant) and up to −20 dBm (nonresonant). The loaded Q and the insertion loss stayed above 200 and below 2 dB for the two device concepts. A major drawback of the devices was the need for many operational SQUIDs with comparable properties ($\beta_L < 1$) in order to achieve a purely dispersive response.

Another approach concerned a lumped-element bandstop filter with a center frequency of 5.6 GHz, coupled to a single Y-123 SEJ SQUID on LAO. The SQUID provided a frequency shift of 4 MHz at low temperatures, while the 3 dB-bandwidth (corresponding to $Q_L \approx 100$) remained essentially unchanged [156]. As an alternative to SQUID-coupled transmission lines, linear Josephson arrays of up to 900 HTS SNS junctions were operated between 16 and 20 GHz at temperatures around 60 K [157]. The maximum phase shift was 25° at an incremental dissipation loss of 0.5 dB. The best 300-JJ array yielded 90°/dB and an absolute shift of 18°. These figures are well below the performance achieved with the ferrimagnetic phase shifter mentioned in Sect. 6.2.3. However, given optimized designs of microwave circuit and junction network, the Josephson devices might be advantageous for modulation frequencies in the high MHz-range which are preferable e.g., for phased-array antennas.

It is well known that nonlinear transmission lines composed of variable capacitances (e.g., [158]) can provide parametric amplification or pulse sharpening. Roughly, the pulse forming comes from the fact that different signal values $I(t)$ experience different propagation constants $\propto [L\,C(I)]^{1/2}$. Similar effects can be derived for variable-inductance devices, e.g., employing Josephson junctions [139, 159–161]. The sharpest possible time edges of the signal pulses are then determined by the L–R relaxation time τ_J. Experimental verifications of this approach using HTS Josephson junctions were reported for the aforementioned DJI phase shifter [155] and for coplanar transmission lines [162]. In the latter work, a series array of 60 Y-123 nanobridge junctions on LAO [average V_c(77 K) = 200 μV] provided a sharpening of a 20-ps rise- and fall-time input pulse to a falling edge of 12 ps of the output pulse. This value was close to the expectation, though limited by the measurement resolution. Additionally, in a shunt arrangement employing about 100 Tl-2212 step-edge junctions on LAO between inner conductor and groundplanes [typical V_c(77 K) = 1 mV], a rise time of the output voltage of 1 ps for a 10 GHz sinusoidal input was concluded indirectly from spectral analysis with a Josephson detector. The current amplitude of the processed pulses had to be adjusted below the critical current of the junctions (about 50 μA and 20 mA for the two devices) to prevent excessive power dissipation. The results are promising with respect to the principle of operation, while applications of the bare effect (in contrast to multi-functional circuits) are presently not in sight.

(4) *Voltage standard, voltage-controlled oscillators, and waveform synthesizers.* The key feature of Josephson junctions as an active microwave element (i.e., delivering power at a specified frequency) is the AC Josephson effect. Appropriate biasing of a JJ network by a homogeneous microwave field will lock all junctions to the same frequency and phase. As a result of the constructive interference, DC voltages can be synthesized with high precision. This concept is exploited in Josephson voltage standards, which have become the primary standards in industrial countries [51]. The required phase coherence puts serious constraints on the design and the fabrication of the Josephson arrays. Based on the successful development of the LTS-SIS technology, voltage accuracies as high as 10^{-10} have been achieved with LSI arrays (more than 10^4 JJs/chip) at a drive frequency of around 70 GHz [163]. Different groups have been developing LTS oscillator arrays for compact 1-V voltage standards at a reduced (and thus less expensive) source frequency of 10 GHz [164, 165]. Present efforts aim to develop programmable voltage sources (digital synthesizers) and high-precision digital-to-analog converters (DAC) for applications, e.g., in industrial metrology [166].

The analog of the voltage standard (power transfer from microwave to DC) is the voltage controlled oscillator (VCO, power transfer from DC to millimeter wave). Such devices are inherently well suited for tunable oscillators, especially above 200 GHz where compact sources delivering power in the range of 10 to 100 μW are lacking. A Josephson VCO consists of a network of phase-locked junctions biased at the voltage V, which emits high-frequency radiation at f_V. Due to the low output power and the very low impedance of a single Josephson junction, and with respect to spectral purity and linewidth, LSI-arrays are required for realistic oscillators. Although there are different locking mechanisms applicable, which differ in array architecture and biasing scheme [167], complete phase locking can be achieved only below a maximum tolerable spread $\Delta I_{cJ}/I_{cJ} \leq 10\%$. The design of Josephson VCOs requires, therefore, careful control and adjustment of all relevant parameters, leaving hardly any degree of freedom. The junction critical current determines the maximum deliverable power of the array, and its immunity to noise. The junction width must be optimized for homogeneous current injection to prevent self-field effects. Usually, the resulting current density J_{cJ} should be as high as possible (about $10\,\text{kA/cm}^2$) with minimum possible spread. The characteristic voltage V_c limits the oscillator frequency to the lower bound $f_{\min} \approx V_c/\Phi_0$. The two parameters I_{cJ} and V_c fix the normal resistance R_J which, at a given number M of junctions in the array, affects the architecture (series versus parallel arrays or combinations) and the matching network. The array size M is of crucial importance for the total available power of the array and for the linewidth. Parasitic reactances L and C limit the maximum operating frequency to values $2\pi f_{\max} \approx (LC)^{-1/2}$. The parameters β_C and

β_L thus need to be adjusted to be around unity to achieve stable operation and maximum accessible tuning range. Given the high I_cJ values needed for high-power oscillators, very small inductances of the order of $L \leq 1$ pH are required.

LTS Josephson junction arrays composed of 500 resistively shunted Nb-SIS junctions delivered about 47 µW power at 400 GHz to a 68-Ω load resistor [168]. Later designs involved 1968 junctions delivering 0.85 mW (-0.7 dBm) at 240 GHz to a 56-Ω load [169]. The detected power stayed above 0.1 mW (-10 dBm) in the frequency range 100–300 GHz. The estimated linewidth at 394 GHz was between 0.1 and 1.0 MHz.

The research on HTS junction arrays has been concerned so far mainly with the development of coupling schemes that maximize the tolerable parameter spread [170, 171]. Phase locking in a four-junction double-series array shunted by a superconducting loop was reported in [172]. The loop design required the use of Y-123 bicrystal GBJs on STO substrates to provide $\beta_\mathrm{C} \approx 1$. In order to adjust β_L to the required value $2/\beta_\mathrm{C}$, the critical current of the junctions had to be suppressed by a magnetic field. The experimental conditions for coherent Josephson emission were fulfilled at temperatures below 50 K. Coherent emission of five Y-123 bicrystal junctions on YSZ were reported in [173]. The emitted power was 5–7 nW at 470 GHz, as derived from the Shapiro steps in the current-voltage characteristic of a 25-Ω detector junction. Similar results (4 nW at 150 GHz) were obtained at a 9.5-Ω detector junction in an array of five ramp-type junctions on LAO, which were incorporated into weakly coupled microstrip resonators [151]. The resonances forced the junctions to radiate coherently, though at the expense of tunability. Coherent microwave emission at 100–300 GHz has since then been observed in arrays of 10^1–10^2 junctions [151, 174]. The linewidths usually agreed with the RSJ model including thermal fluctuations. The effective noise temperature of different types of junctions in the mm and submillimeter wave regime was found to be close to the physical temperature. All available data on the locking of HTS junctions in arrays suffered from the large spread of I_cJ. Low-temperature scanning electron microscopy yielded statistical variations of I_cJ in series arrays of 10–30 Y-123 GBJs on STO bicrystals of typically 20% [175]. An even larger parameter spread was found for biepitaxial and step-edge junctions.

6.3.3 Single-Junction SQUIDs Coupled to Microwave Devices

SQUID magnetometers employing a single Josephson junction were developed shortly after Josephson's prediction of superconducting tunneling in weak links [176]. They are usually referred to as RF SQUIDs since the drive frequencies are between 20 and 100 MHz. However, operation at up to 10 GHz was reported, justifying the more general nomenclature "microwave" SQUIDs. The principles of operation, dynamic features and applications of such devices have been extensively treated, e.g., in [177–181]. Various applications in magnetometry (MCG, NDE, geophysical prospecting), metrology (dilatometry, thermometry, galvanometry, susceptometry etc.) and digital signal processing are uniquely based on the high field, energy and time resolutions of SQUID devices. The potential advantages of employing oxide superconductors (enhanced operating temperatures and reduced separations between sensor and object) have led to rapid development of HTS magnetometers and gradiometers that are comparable or even competitive with LTS devices [179].

As with LTS materials, the perfection achieved in fabricating HTS Josephson junctions for small-scale integrated devices made it feasible to focus on the optimization of DC SQUIDs. This type is based on a direct-current readout of the voltage drop across two parallel junctions enclosed in a superconducting loop. However, microwave SQUIDs have several advantages over their DC counterparts. First, only a single junction is required for the basic magnetometer cell. Second, recent theoretical analysis revealed superior energy resolution of microwave SQUIDs at high noise parameters Γ (i.e., at elevated operating temperatures), and at high loop inductances, facilitating the problematic impedance matching between loop and flux antenna [181]. Third, the crossover frequency between white noise and $1/f$ noise in microwave SQUIDs can be as low as 0.5 Hz, which is lower than in typical DC SQUIDs, even at a comparable level of the white noise [182]. Finally, different regimes of operation (dispersive or hysteretic) can be accessed by trimming the junction critical current to values $I_{cJ} < (\text{or} >) \Phi_0/2\pi L_S$.

Microwave Response of Single-Junction SQUIDs

There is some interest in the scattering parameters of microwave circuits coupled to single-junction SQUIDs. The unique nonlinear properties of loop-shunted junctions can be exploited for amplifiers, attenuators, phase shifters, mixers or signal-converters [177], which make possible monolithic integration with other HTS microwave devices. Moreover, ultimate performance of microwave SQUID magnetometers can be achieved only by optimizing the microwave tank circuit [182–185]. The following analysis introduces the basic features of the microwave response of SQUIDs, with some allusions to the treatment of bare resonators (Sects. 2.1 and 6.1).

The total flux φ_{tot} in a SQUID equals the external flux φ_x, reduced by the shielding current flowing in the loop [c.f., discussion of (6.43)]:

$$\varphi_{\text{tot}} = \varphi_x - \left[\frac{L_S}{L_{J0}} \sin(\varphi_{\text{tot}}) + \frac{L_S}{R_J} \frac{\partial \varphi_{\text{tot}}}{\partial t} \right] . \tag{6.49a}$$

The Josephson equation (6.37a) was used for this expression, giving an implicit equation for φ_{tot}. The external flux φ_x is usually the sum of a DC flux and a component at frequency $\omega = 2\pi f$:

$$\varphi_x = \frac{2\pi}{\Phi_0} \left[\Phi_{\text{DC}} + \Phi_{\text{HF}} \cos(\omega t) \right] . \tag{6.49b}$$

The following discussion focuses on the adiabatic limit $\Omega_\tau = \omega\tau \ll 1$ (with τ the dynamic response time $L_{\text{S}}/R_{\text{n}}$), where the time derivative in (6.49a) can be neglected.

As long as $\beta_{\text{L}} < 1$, the response $\varphi_{\text{tot}}(\varphi_x)$ of the SQUID (6.49) is single-valued. In contrast, as β_{L} exceeds unity, the flux balance exhibits positions of infinite slopes $\partial\varphi_{\text{tot}}/\partial\varphi_x$ where the response becomes unstable and switches by $\pm$ one flux quantum. The SQUID becomes hysteretic since switching back and forth between adjacent flux states occurs at different φ_x values. The switching time depends on the relaxation time τ [122]. The first flux transition occurs when the external flux exceeds the critical value

$$\varphi_{\text{crit}} = \cos^{-1}\left(-\frac{1}{\beta_{\text{L}}}\right) + \beta_{\text{L}} \sin\left[\cos^{-1}\left(-\frac{1}{\beta_{\text{L}}}\right)\right] - [\Phi_{\text{DC}}]_{\text{mod}(N)} , \tag{6.50a}$$

where the exponent "-1" indicates the inverse function arccos. The last term in (6.50a) represents the flux quantization in the loop. It is $[\Phi_{\text{DC}}]_{\text{mod}(N)} = 0$ for $0 \leq |\Phi_{\text{DC}}/\Phi_0 - N| < 1/2$ and $1/2$ for $1/2 \leq |\Phi_{\text{DC}}/\Phi_0 - N| < 1$, N being a positive integer. In the following, the extremal values $\Phi_{\text{DC}} = 0$ and $\Phi_{\text{DC}} = \Phi_0/2$ are referred to as the "0" state and the "1/2" state, respectively. With the external flux increasing above φ_{crit}, the SQUID response becomes 2π-periodic. It assumes two extremal shapes as the DC flux is swept between the "0" and the "1/2" states. This modulation is usually exploited in magnetometry. For large $\beta_{\text{L}} \gg 1$, (6.50a) can be linearized. The condition for the critical flux then reduces to

$$\varphi_{\text{crit}} = \frac{\pi}{2} + \beta_{\text{L}} - [\Phi_{\text{DC}}]_{\text{mod}(N)} . \tag{6.50b}$$

A pioneering and detailed analysis of strongly hysteretic microwave SQUIDs in the adiabatic limit was performed in [176]. The following discussion is a generalization of this approach, enabling a direct comparison between theory and experimental results [184].

The microwave response of the whole SQUID circuit is in general not only determined by the reduced SQUID inductance β_{L}, but also by τ and by the coupling coefficient $k = M^{1/2}/(L_{\text{T}} L_{\text{S}})$, where L_{T} is the inductance of the tank resonator. In addition, the loaded quality factor Q_{L} of the resonator is an important parameter, as can be immediately concluded from a simple consideration of the involved energies [c.f., (6.44b)]:

$$\kappa \equiv k^2 Q_{\text{L}} = \frac{E_{\text{exc}}}{W} \frac{\omega W}{P_{\text{diss}}} = \frac{\omega E_{\text{exc}}}{P_{\text{diss}}} . \tag{6.51}$$

Here, E_{exc} is the magnetic energy exchanged ("shared") between the SQUID loop and the resonator. In order to achieve a high sensitivity to the ambient magnetic energy, this quantity should be small. W is the magnetic energy stored in the resonator, and P_{diss} is the total power loss in the resonator due to dissipation and coupling to the external circuit. Dissipation and dispersion in the SQUID loop can be resolved best in the tank circuit if the resonator displays high Q_L values (large W or low P_{diss}). However, the coupling coefficient κ has to be adjusted properly in order to find an optimum balance between the voltage responsivity of the SQUID to the sensed DC flux (see below) and the resulting noise performance [177].

The scattering parameters of a resonator (whether or not coupled to a SQUID) depend unambiguously on the resonant frequency f_0 and the loaded quality factor Q_L (Sect. 2.1). The response of a microwave SQUID is therefore completely determined if its reactance and resistance are known, in addition to those of the bare resonator. These parameters can be evaluated numerically and, in the linearized approximation for $\beta_L \gg 1$, also analytically. They are derived from the Fourier components E' and E'' of the electromotive force across the SQUID, which are out-of-phase or in-phase, respectively, with the excitation signal [176]. The effective microwave inductance and resistance of the SQUID are then given in normalized units by $E'\Phi_0/\Phi_{\text{HF}}$ and $E''\Phi_0/\Phi_{\text{rf}}$. Figure 6.15 summarizes numerical results for several values of β_L and Ω_τ. For $\beta_L \gg 1$ and $\Omega_\tau \ll 1$, the impedance reduces to the behavior originally modeled in [176]. Solid and dashed lines indicate the modulation by external DC flux between the "0" and the "1/2" states. In contrast, for $\beta_L \approx 1$ or Ω_τ approaching 1, microwave SQUIDs display complicated dynamic features which, e.g., alter the performance as magnetometers [177] (shown in Fig. 6.15 for the "0" state). The numerical results were confirmed by measurements with Y-123 step-edge junctions on LAO with $\beta_L \geq 3$, $L_S \approx 100\,\text{pH}$ and $\omega/2\pi = 3\,\text{GHz}$ at 77 K [184].

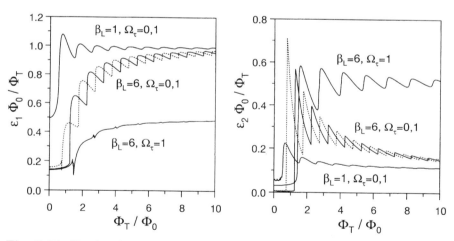

Fig. 6.15. Simulated microwave inductance $E'\Phi_0/\Phi_{\text{HF}}$ ($\varepsilon_1\Phi_0/\Phi_T$, *left*) and resistance $E''\Phi_0/\Phi_{\text{HF}}$ ($\varepsilon_2\Phi_0/\Phi_T$, *right*) of a single-junction SQUID versus the reduced microwave flux amplitude Φ_{HF}/Φ_0 for different values of β_L and Ω_τ [184a]. The *solid* (*dashed*) lines denote the "0" ("1/2") state.

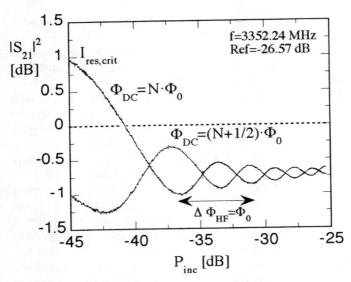

Fig. 6.16. Typical power sweep of $\Delta|S_{21}|^2$ obtained with a Y-123 step-edge junction SQUID on LAO for the "0" and the "1/2" states at 77 K and 3 GHz [186]. The *dashed line* denotes the reference value of $|S_{21}|^2$.

Figure 6.16 displays typical results for the dependence of the microwave transmission coefficient $|S_{21}|^2$ on the field amplitude for the two extremal states "0" and "1/2". The measurement was performed with a network analyzer at constant frequency f ("power sweep"), where f was adjusted to the resonant frequency at low excitation signals. Due to the large value $\beta_L \approx 3$, the flux-induced variation of f_0 was of the order of 10^{-4} and thus affected $|S_{21}|^2$ only a little. As a result of the above analysis, the reduced SQUID inductance β_L can be derived from Fig. 6.16 in relation to (6.50b):

$$\varphi_{\mathrm{crit}} = \frac{I_{\mathrm{res,crit}}}{\Delta I_{\mathrm{res}}}, \tag{6.52}$$

where $I_{\mathrm{res}} \propto P_2^{1/2} \propto |S_{21}|$ is the microwave current amplitude in the microwave SQUID, and P_2 is the power transmitted through the coupling port. $I_{\mathrm{res,crit}}$ corresponds to the P_2 value at which the first flux transition occurs. The period ΔI_{res} reflects the variation of the microwave flux fed into the SQUID loop by one Φ_0. Inspection of the data in Fig. 6.16 revealed $\beta_L \approx 3.5$ ($L_S = 120\,\mathrm{pH}$).

The coupling parameter k is related to the shift Δf of the resonant frequency due to increasing the microwave flux from a low value $\varphi_{\mathrm{HF}} < \varphi_{\mathrm{crit}}$ to a high level $\varphi_{\mathrm{HF}} \gg \varphi_{\mathrm{crit}}$ [184a]:

$$k = \sqrt{2A\frac{\Delta f}{f_0}\left(1 - A\frac{\Delta f}{2f_0}\right)}. \tag{6.53}$$

The parameter $A(\beta_L, \Omega_\tau)$ accounts empirically for the effect of finite relaxation times. In the adiabatic limit, $A \approx \beta_L/(1+\beta_L)$ was deduced from the computations. Once k is known, κ follows from the Q_L measurement which can be performed in the usual manner (Sect. 2.1). The time constant τ could be related analogously to measurable quantities. It was estimated from the total variation of the Q factor at low and high drive amplitudes, $\varphi_{HF} < \varphi_{crit}$ and $\varphi_{HF} \gg \varphi_{crit}$, respectively:

$$\omega\tau \approx \frac{1+\beta_L}{\beta_L}\Delta\left(\frac{1}{Q_L}\right). \tag{6.54}$$

Inspection of the experimental results displayed in Fig. 6.16 yielded dynamic response times between 5 and 30 ps and typical R_J values of about $5\,\Omega$ for the investigated step-edge junctions [184]. The R_J value was in good agreement with data deduced from time-averaged IV curves.

Remarks on Magnetometer Operation

One important figure of merit of microwave SQUIDs, especially when applied to magnetometry, is the peak-to-peak variation of the scattering parameter $|S_{m1}|^2$ induced by a change of the external DC flux. The index m equals 2 if the tank circuit is a two-port, and $m = 1$ in the reflected mode. Typical magnetometer readout electronics employ microwave diodes, yielding a voltage $V \propto |S_{m1}|$. The resulting signal modulation $V(\Phi)$ is called the transfer function of the SQUID. Further FOMs are the spectral density S_Φ of the flux noise, the energy resolution $E_N = S_\Phi/(2L_S)$, and the magnetic field resolution $B_N = S_\Phi^{1/2}/A_{\text{eff}}$. $A_{\text{eff}} = (\partial B/\partial \Phi)^{-1}$ quantifies the efficiency with which the sensed magnetic fields can be collected by the SQUID loop. Altogether, large $\partial V/\partial \Phi$ and A_{eff}, and low S_Φ and L_S are desired for optimum performance.

The optimization of the FOMs requires knowledge about the interdependence between the properties of the magnetometer and of the tank resonator. The voltage induced across the resonant inductor L_T is determined by the flux in the loop and the mutual inductive coupling between the SQUID and the tank-circuit (Fig. 6.14) [177]:

$$V_{\text{res}} = \omega L_T \frac{\Phi}{M}. \tag{6.55}$$

The index "res" identifies the voltage as a quantity internal to the resonator, which has to be distinguished from the one which is measurable in the external readout circuit. The second factor, Φ/M, denotes the current I_{res} flowing in the resonator. Adopting the definition of the coupling constant k, and assuming $k^2 Q = 1$, (6.55) yields for the transfer function

$$\frac{\partial V_{\text{Res}}}{\partial \Phi} = \omega\sqrt{Q_L \frac{L_T}{L_S}}. \tag{6.56}$$

In the case of a two-port resonator (consideration of a singly coupled resonator leads to an analogous conclusion), the resonator current I_{res} gives rise to a transmitted power

$$P_2 = \frac{1}{Q_2}\left(\frac{I_{\text{res}}}{\gamma_I}\right)^2, \tag{6.57}$$

where γ_I (in units of $\text{A}/\text{W}^{1/2}$) is the coefficient describing the conversion of current into power. Its value can be estimated for lumped-element devices (neglecting edge-current enhancement) as $\gamma_I \approx (2/\omega L_T)^{1/2}$ (Sect. 2.1.1). In (6.57), Q_2 quantifies the coupling between the resonator and the external circuit. In real circuits, the transmitted power is usually amplified by a factor η, and detected with an effective responsivity R (in units of $\text{V}/\text{W}^{1/2}$). Combining (6.56) and (6.57) thus leads to the voltage modulation measurable in the external circuit:

$$\frac{\partial V_{\text{ext}}}{\partial \Phi} = R\sqrt{\eta}\sqrt{\frac{Q_L}{Q_2}}\frac{1}{\gamma_I}\frac{1}{\sqrt{L_T L_S}}. \tag{6.58}$$

High-sensitivity detection (large R and η) is crucial for efficient signal readout. Large sensor outputs also require strongly coupled (low Q_2 values) high-Q_0 resonators. In this respect, a hysteretic microwave SQUID magnetometer resembles a microwave filter that displays flux-modulated insertion loss and bandwidth. Of further benefit is a low microwave current concentration (low γ_I values), i.e., the "filter" should have reasonable power handling. Accordingly, single-junction SQUIDs coupled to dielectric-shielded resonators were reported for LTS [187] as well as for HTS [188] devices. Finally, (6.58) constrains the inductance values of the tank-circuit and SQUID loop. The resonator inductance L_T should be as low as possible, and the driving frequency $\propto L_T^{-1/2}$ accordingly high, while staying below about $1/\tau$. Low L_S values are desired for efficient "flux-to-voltage" conversion. However, there is a payoff between voltage modulation and impedance matching to the large pickup areas, which are needed for efficient "field-to-flux" conversion. Multiloop sensors, flux focusers (so-called washer-type SQUIDS) and spiral flux transformers, either monolithically integrated or flip-chip mounted to the SQUID sensors, present possible solutions, leading to typical values of $L_S \approx 100\,\text{pH}$.

As important as (and related to) large voltage modulation is low noise. The total flux noise of microwave SQUIDs is expected to decrease with increasing tank frequency [189]. At 0.1 to 1 GHz, typical voltage modulations are about 10–100 $\mu\text{V}/\Phi_0$. The resulting noise spectral density is often limited by the noise of the preamplifier [177, 182]. The achievable energy resolution is also limited by noise, and by the adjustment of L_S, to typical values of about $10^3\,h$ ($h = $ Planck's constant).

Continuously improving performance of HTS microwave washer-type SQUIDs fabricated from step-edge junctions on LAO and operating between 0.6 and 0.9 GHz was reported in [185, 190]. As indicated in Fig. 6.17, circular or square-shaped coplanar resonators surrounded the washer ($\leq \varnothing 15\,\mathrm{mm}$), to which the SQUID was flip-chip coupled. A coplanar resonator was favored despite the high γ_I values because of the avoidance of a ground plane, which caused additional flux noise. At 77 K, the resonators displayed Q_0 values up to 2×10^4. Loop inductances of $L_\mathrm{S} = 150$ and 260 pH yielded $\partial V/\partial \Phi = 150\text{--}200\,\mathrm{\mu V}/\Phi_0$, $S_\Phi^{1/2} = 8.5\text{--}10\,\mu\Phi_0/\mathrm{Hz}^{1/2}$ and $B_\mathrm{N} = 15\text{--}30\,\mathrm{fT}/\mathrm{Hz}^{1/2}$, above 100 Hz. The effective area of the microwave SQUIDs was $2\text{--}3\,\mathrm{nT}/\Phi_0$.

Fig. 6.17. Schematic of coplanar tank resonators of a flip-chip coupled microwave HTS SQUID, enclosing a washer-type flux focuser [190].

Nonlinear Signal Processing

Besides magnetometry, potential microwave applications of HTS SQUIDs are systems where nonlinear signal operations have to be performed. Due to the weak superconductive coupling, Josephson elements reveal inherently nonlinear response, especially at low signal levels. The sinusoidal relationship between current and flux results in a unique oscillatory frequency response, characterized by a series of Bessel functions with flux-dependent arguments [191]. Additional nonlinearity is introduced if the SQUID operates in the hysteretic mode, i.e., if it performs step-like quantum transitions. The possible use of single-junction SQUIDs for nonlinear signal processing is briefly illustrated below with three examples: two-tone frequency intermodulation, quantization, and demodulation of harmonic microwave signals.

(1) *Two-tone frequency intermodulation.* The microwave response of hysteretic YBCO microwave SQUIDs to two time-harmonic signals was investigated in [192]. Starting for simplicity with the dispersive regime ($\beta_\mathrm{L} \leq 1$), the flux balance (6.49) can be expressed in an explicit form, which is for the "0" state [191]

$$\varphi_\mathrm{tot} = \varphi_\mathrm{x} + 2 \sum_{m=1}^{\infty} \frac{(-1)^m}{m} J_m(m\beta_\mathrm{L}) \sin(m\varphi_\mathrm{x}) \tag{6.59a}$$

with J_m the mth order Bessel function of the first kind. For a harmonic two-tone excitation

$$\varphi_x(t) = \varphi_\mathrm{HF} \left[\sin(2\pi f_1 t) + \sin(2\pi f_2 t) \right] \tag{6.59b}$$

the amplitudes of the total flux at the carrier frequencies f_1 and f_2 are identical, $\varphi_{1,0} = \varphi_{2,0}$. The output results from the superposition of a linear and a nonlinear response:

$$\varphi_{1,0} = \varphi_{\mathrm{HF}} + 4 \sum_{m=1}^{\infty} \frac{(-1)^m}{m} J_m(m\beta_{\mathrm{L}}) J_0(m\varphi_{\mathrm{HF}}) J_1(m\varphi_{\mathrm{HF}}) . \quad (6.60)$$

The signals at the first intermodulation frequency, $f_{1+} = 2f_2 - f_1$ and $f_{1-} = 2f_1 - f_2$ (Sect. 3.3), are also symmetric with respect to f_0, but determined solely by nonlinear effects:

$$\varphi_{1\pm} = -4 \sum_{m=1}^{\infty} \frac{(-1)^m}{m} J_m(m\beta_{\mathrm{L}}) J_1(m\varphi_{\mathrm{HF}}) J_2(m\varphi_{\mathrm{HF}}) . \quad (6.61)$$

The dependences of φ_{1-} and φ_{2-} (at $f_{2-} = 3f_1 - 2f_2$) on the response $\varphi_{1,0}$ are depicted in Fig. 6.18 for the "0" state. Since $|f_{n\pm} - f_{n+1,\pm}|/f_0 \ll 1$, the flux amplitudes are proportional to the measurable voltage amplitudes, $\varphi_{n\pm} \propto f_0 V_{n\pm}$ (c.f., Fig. 6.19 and below).

To obtain results also for the hysteretic mode ($\beta_{\mathrm{L}} \geq 1$), the SQUID response was simulated in the time-domain, and then Fourier-transformed to the frequency domain. Both methods, which are valid only in the adiabatic limit $\omega_0 \tau \ll 1$, yielded identical results at $\beta_{\mathrm{L}} = 1$. The results for φ_{1-} and φ_{2-} at $\beta_{\mathrm{L}} = 3$ are also shown in Fig. 6.18. An oscillatory variation is observed for all cases. With β_{L} exceeding unity, the average level of φ_{n-} increases, the number and the sharpness of the extrema decrease, and an antiphase behavior forms. This behavior can be considered the dynamic analog of the DC flux-modulated power transmission of microwave SQUIDs at single-tone excitation (Fig. 6.16).

The two-tone response of hysteretic microwave SQUIDs was verified by measurements with the hysteretic SQUID of Fig. 6.16 ($\beta_{\mathrm{L}} \approx 3$) at 3 GHz. The loaded quality factor of the tank resonator was $Q_{\mathrm{L}} \approx 3000$, corresponding to a 3-dB bandwidth of $\Delta f_{1/2} = 1\,\mathrm{MHz}$. Two generators provided signals at f_1 and $f_2 = f_1 + \Delta f$, centered at the resonant frequency $f_0 = (f_1 + f_2)/2 = 3.3\,\mathrm{GHz}$ and separated by $\Delta f = 100\,\mathrm{kHz}$. The power levels P_{inc} of both signals were varied between -60 and $-30\,\mathrm{dBm}$. This power range was sufficient to cover microwave current amplitudes I_{res} from below $I_{\mathrm{res,crit}}$ up to a level inducing three quantum transitions per cycle. The value of $I_{\mathrm{res,crit}}$ and the period ΔI_{res} between subsequent hysteresis cycles were determined from a preceding single-tone experiment (Fig. 6.16). The measured frequency spectrum was symmetric with respect to f_0, as expected from the symmetric power feed [see (6.59b)]. The signal amplitude P_0 at f_1 and f_2 mainly reflected the linear response of the SQUID: P_0 (dB) $\propto P_{\mathrm{inc}}$ (dB) $(1 + \delta_{\mathrm{nl}})$, where $\delta_{\mathrm{nl}} \ll 1$ presents a small nonlinear contribution [second term in (6.60)]. Varying an external DC flux between the "0" and "1/2" states induced a variation of P_0 by $\Delta P_0 = 1.5\,\mathrm{dB}$. Furthermore, nth order intermodulation signals $P_{n\pm}$ were detected at the frequencies $f_{n-} = (n+1) \times f_1 - n f_2$ and $f_{n+} = (n+1)f_2 - n f_1$

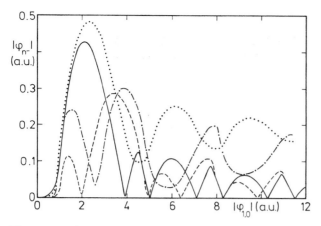

Fig. 6.18. Simulated dependence of the normalized intermodulation flux amplitudes at f_{1-} ($\beta_L = 1$, *solid line*; $\beta_L = 3$, *dotted*) and f_{2-} ($\beta_L = 1$, *dashed*; $\beta_L = 3$, *dash-dotted*) on the flux transmitted at f_0 for the "0" state [192].

with $n = 1$, 2 and 3. The power levels of these signals were unaffected by the bandpass characteristic of the tank circuit since $n\,\Delta f \ll \Delta f_{1/2}$. In contrast to the small DC flux-induced variation ΔP_0, the first (respectively: second) order intermodulation varied by about 6 dB (10 dB). It is important to note that the signals $P_{n\pm}$ resulted solely from frequency conversion in the SQUID with no linear contribution from P_{inc} [see (6.61)]. This feature might be exploited to improve the sensitivity of microwave SQUID magnetometers, e.g., in terms of heterodyne detectors.

Figure 6.19 illustrates the dependence of the measured IMD voltage $V_{n-} \propto (P_{n-})^{1/2}$ on the voltage V_0 at the carrier frequency for the two extremal DC-flux states on a linear scale. An oscillatory behavior can be resolved with "antiphase" relations between both intermodulation voltages at a given DC flux. For a given order n, anticorrelations are also observed between the voltages obtained for the "0" and "1/2" states, with an additional oscillation of V_{n-} in the "1/2" state at low V_0 values. These results agree qualitatively with the theoretical expectation (Fig. 6.18). The remaining differences are attributed to dynamic effects, since the time constant τ of the SQUID loop was close to $1/\omega_0$ (see preceding paragraph). From the ideal behavior in Fig. 6.18, a maximum conversion gain of $\varphi_{1-}/\varphi_{1,0} \approx 0.5$ can be extracted. This ratio corresponds to a power conversion of -6 dB. Regarding the low source power level (which was about -45 dBm in the present experiment), these figures seem attractive for the development of mixers or modulators based on hysteretic SQUIDs rather than on single junctions (Sect. 6.3.2). However, projections on the achievable noise figure of such a device remain to be analyzed.

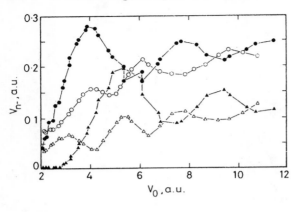

Fig. 6.19. Measured dependence of the intermodulation voltages at f_{1-} (*circles*) and f_{2-} (*triangles*) on the flux transmitted at f_0 for the "0" state (*filled symbols*) and the "1/2" state (*open symbols*) [192].

(2) *Quantization of microwave signals.* The following example illustrates the implications of the sinusoidal nonlinearity of Josephson devices for microwave signal processing [193]. It is well known that the hysteretic and periodic response of microwave SQUIDs at $\beta_L > 1$ provides a means to quantize high-frequency signals [122, 176]. The detailed analysis is complicated and therefore not repeated here. Briefly, microwave signals with amplitudes exceeding the critical value φ_{crit} [see (6.50)] induce voltage spikes at every instant when a flux quantum jumps into or out of the SQUID loop. The number of pulses is therefore a direct measure of the signal strength. Furthermore, the polarity of the voltage spikes distinguishes between increasing and decreasing input levels. The pulse height increases in proportion to the frequency of the signal. However, the hysteretic behavior introduces dissipation and thus noise. Furthermore, the frequency of the digitized output (i.e., the temporal sequence of the pulses) is always an integer multiple of the signal frequency.

The latter drawback can be prevented in dispersive SQUIDs ($\beta_L < 1$) if biased at a DC flux $\Phi_{\mathrm{DC}} = \Phi_0/4$. Evaluation of (6.59a) yields for the signal $\varphi_x = \pi/2 + \varphi_{\mathrm{HF}}\sin(\omega t)$:

$$\varphi_{\mathrm{tot}}(t) = \varphi_x(t) - 2J_1(\beta_L)\left[J_0(\varphi_{\mathrm{HF}}) + 2\sum_{n=1}^{\infty} J_{2n}(\varphi_{\mathrm{HF}})\cos(2n\omega t)\right]. \quad (6.62)$$

The Fourier representation of the sine term in (6.59a) was limited to the lowest order $m = 1$ of the sum, which is justified in the case of sufficiently small β_L values. The interesting feature of (6.62) is the generation of an output at DC which varies with the amplitude φ_{HF} of the input signal like the zero-order Bessel function J_0. Since $J_0(x)$ is a quasi-periodic transcendent function of x, counting the zeros $x_{0N} = x_{01} + (N-1)\pi$ ($N = 1, 2, \ldots$) at large N yields a digital representation of the analog input signal.

(3) *Demodulation of phase-encoded microwave signals.* The previous examples considered roughly the response of microwave SQUIDs to harmonic excitations with varying frequency or amplitude. The concluding example, which

acknowledges the importance of transmitting digitally encoded information in communication systems [194], concerns a phase-modulated voltage signal of the form:

$$V(t) = V_0 + V_{HF} \cos(\omega t + \Theta) . \qquad (6.63a)$$

The phase Θ, which carries the binary information, jumps between the digitized values 0 and $\pi/2$. Demodulation of phase-encoded microwave signals requires the comparison of the detected signal with a reference signal of known phase. This can be delivered by the Josephson junction itself if it is biased at constant voltage $V_0 > 0$ (Sect. 6.3.1). The total flux induced in the SQUID loop by the voltage (6.63a) is

$$\varphi_x(t) = \int V(t)dt = \varphi_{DC} + \varphi_0 t + \varphi_{HF} \sin(\omega t + \Theta) , \qquad (6.63b)$$

where a DC flux bias was added, and all fluxes are expressed in normalized units as before. The latter result is to be substituted into (6.59a). For simplicity, we will not consider the complete formula, but restrict attention to the terms $\sin(m\,\varphi_x)$. Taking into account (6.63b), the Fourier expansion of the lowest term ($m = 1$) yields

$$\sin(\varphi_x) = \sum_{n=-\infty}^{\infty} J_n(\varphi_{HF}) \sin\left[(\varphi_0 + n\omega)t + n\Theta + \varphi_{DC}\right] . \qquad (6.64)$$

The junction induces a DC signal, which depends on the phase Θ as desired, if the voltage bias is adjusted to the condition for Shapiro steps: $\varphi_0 = \pm n\,\omega$ or $f_V = \pm n\,f$ [see (6.41)]. According to (6.59a), the total flux conserves this phase-dependent DC signal.

The key element of a Josephson demodulator circuit is based on the flux-dependent inductance $L_{eff}(\varphi_{tot})$ considered in Sect. 6.3.1. The presence of a phase-encoded microwave signal will thus not only be converted to DC, but will also modify the resulting value of L_{eff}. Given an appropriate readout circuit, the SQUID operates effectively as a phase decoder. One figure of merit of such a device is the fractional variation $\ell \equiv 1 - L_{eff}(\Theta = \pi/2)/L_{eff}(\Theta = 0)$. This parameter was investigated numerically in terms of the variables n, φ_{DC}, β_L and k [193]. The largest inductance modulation was found for $n = -1$ and $\varphi_{DC} = 3\pi/2$ ($\Phi_{DC} = 3\Phi_0/4$). Within the range of stable operation, $\beta_L(1 + k^2) < 1$, large values of β_L and k were favorable. Maximum ℓ values around 50% were concluded for $\beta_L, k \approx 0.6$–0.7.

Appendix

The following listings provide a survey of the constants, notations, definitions and abbreviations which are used throughout this book, unless otherwise stated. All formulas are evaluated in standard (SI) units. Vector are indicated by bold italic letters.

A.1 Fundamental Constants

Name	Variable	Value	Magnitude	Unit
Boltzman constant	k_B	1.3806	10^{-23}	J/K
Electronic unit charge	e	1.6022	10^{-19}	As
Euler constant	γ	1.7811	1	1
Flux quantum	$\Phi_0 = h/2e$	2.068	10^{-15}	Vs
Planck's constant	h	6.6262	10^{-34}	Js
Reduced Planck constant	$\hbar = h/2\pi$	1.0546	10^{-34}	Js
Vacuum permeability	μ_0	4π	10^{-7}	Vs/Am
Velocity of light	c	2.9979	10^8	m/s
Vacuum permittivity	$\varepsilon_0 = (c^2\mu_0)^{-1}$	8.8543	10^{-12}	As/Vm
Wave impedance of free space	$Z_0 = (\mu_0/\varepsilon_0)^{1/2}$ $\approx 120\pi$	376.73	1	V/A

A.2 Frequently used Abbreviations, Symbols and Definitions

AC	Alternating current
BCS	Bardeen, Cooper, Schrieffer
β	Microwave power coupling coefficient
D	Demagnetization factor
δ	Reduced energy gap $\Delta(T=0\,\mathrm{K})/k_\mathrm{B}T_\mathrm{c}$
DC	Direct current
DOS	density of energy states
DUT/ FUT	Device under test/ superconducting film under test
$\varnothing$	Diameter
EM	Electromagnetic
$f(E)$	Fermi distribution function of energy
FF	Flux flow
FOM	Figure of merit
$\varphi_\pm$	Scaling functions of the surface impedance
$G[\Omega]$	Geometry factor
GL	Ginzburg–Landau
GM	Gifford–McMahon
HF	High-frequency
HTS	High-temperature superconductor
JJ	Josephson junction
LTS	Low-temperature superconductor
NC	Normal conducting
OP	Order parameter
PT	Pulse tube
Q_0, Q_L	Unloaded and loaded quality factors
r	Differential loss tangent
S	Scattering parameter
SC	Superconducting
SQUID	Superconducting quantum interference device
ST	Stirling
TE	Transverse electric (EM mode identification)
TFM	Two-fluid model
TM	Transverse magnetic (EM mode identification)
TVD	Tin vapor diffusion
WC	Weak coupling
$x_\mathrm{n}, x_\mathrm{s}$	Fractional number density of unpaired and paired charge carriers
y	Conductivity ratio, $y = \mathrm{Re}\{\sigma\}/\mathrm{Im}\{\sigma\}$
Ψ	Quantum mechanical wave function

A.3 Basic Physical Quantities

$\boldsymbol{A}$ (Vs/m)	Magnetic vector potential (real space)
$\boldsymbol{a}$ (Vs/m)	Fourier transform of magnetic vector potential (momentum space)
a (m)	Grain size
α	Anisotropy factor ($\alpha = \lambda_c/\lambda_{ab}$)
$\boldsymbol{B}$ (T) or (Vs/m^2)	Magnetic induction
C (J/K)	Heat capacity
C_V (J/Km^3)	Heat capacity per unit volume
χ (J/m^3)	Pinning constant
$\boldsymbol{D}$ (As/m^2)	Displacement current
Δ (J)	Energy gap
d (m)	Thickness
δ (m)	Skin depth
ε_r	Permittivity
$\boldsymbol{E}$ (V/m)	Electric field strength
F (J), f (J/m^3)	Helmholtz free energy, Helmholtz free energy density
f (Hz)	Frequency
G (J), g (J/m^3)	Gibbs free energy, Gibbs free energy density
Γ (Hz)	Relaxation rate, scattering rate
$\boldsymbol{H}$ (A/m)	Magnetic field strength, or surface current
η (Js/m^3)	Flux-flow viscosity
I (A)	Current
i	Imaginary unit (i$^2 = -1$)
ι	Reduced current (I/I_c)
$\boldsymbol{J}$ (A/m^2)	Electric current density
$\boldsymbol{J}_\mathrm{th}$ (W/m^2)	Heat current density
K (m^{-2})	Kernel of the current–field relation
$\boldsymbol{k}$ (m^{-1})	Wave vector
κ	Ginzburg–Landau parameter
L (Vs/A)	Inductance
l (m)	Geometric length
ℓ (m)	Electronic mean free path
λ (m)	Penetration depth of magnetic field
λ_coupl	Electron–phonon coupling strength
N (As/m^3)	Charge density
n (m^{-3})	Number density of electric unit charges
m (kg)	Mass
μ_r	Magnetic permeability
P (VA) or (W)	Power
p (kg m/s)	Momentum

$R_s\,(\Omega)$ — Surface resistance
$R_{th}\,(\text{Km}^2/\text{W})$ — Thermal resistance
$\rho\,(\Omega\text{m})$ — Resistivity
$\boldsymbol{q}\,(\text{m}^{-1})$ — Wave vector (Fourier transform of momentum space)
$\sigma\,(1/\Omega\text{m})$ — Electrical conductivity
$\sigma_{th}\,(\text{W/Km})$ — Thermal conductivity
$T\,(\text{K})$ — Temperature
$\tan\delta$ — Dielectric loss tangent, $\tan\delta = \text{Re}\{\varepsilon\}/\text{Im}\{\varepsilon\}$
$\tau\,(\text{s})$ — Relaxation time
$V\,(\text{V})$ — Voltage
$W\,(\text{J})$ — Stored energy
$W_1, W'\,(\text{J/m})$ — Energy per unit length (of flux lines)
$\Omega\,(\text{Hz})$ — Characteristic frequency
$\omega\,(\text{Hz})$ — Circular frequency ($\omega = 2\pi f$)
$X_s\,(\Omega)$ — Surface reactance
$\xi\,(\text{m})$ — Coherence length
$Z_s\,(\Omega)$ — Surface impedance (complex)

A.4 HTS Compounds, Substrates and Deposition Techniques

ReBCO — Rare-earth $Ba_2Cu_3O_{7-x}$
TBCCO, Tl-2212 — $Tl_2Ba_2CaCu_2O_y$
TBCCO, Tl-2223 — $Tl_2Ba_2Ca_2Cu_3O_y$
YBCO, Y-123 — $YBa_2Cu_3O_{7-x}$

CbS — CeO_2-buffered sapphire
LAO — $LaAlO_3$
NGO — $NdGaO_3$
STO — $SrTiO_3$

C — Metal/organic chemical vapor deposition
EC — Electron-beam coevaporation
L — Pulsed laser ablation
S — High oxygen pressure DC sputtering
TC — Thermal coevaporation

References

Preface

1. J. G. Bednorz and K. A. Müller: Z. Phys. B **64**, 189 (1986).
2. H. K. Onnes: Proc. Royal Acad. Amsterdam **11**, 168 (1908).
3. H. K. Onnes: Commun. Phys. Lab. Univ. Leiden Nr. 120b (1911).
4. T. van Duzer and C. W. Turner: *Principles of Superconductive Devices* (Elsevier, New York, 1981).
5. S. T. Ruggiero and D. A. Rudman (Eds.): *Superconducting Devices* (Academic, London, 1990).
6. J. Hinken: *Superconductor Electronics* (Springer, Berlin, Heidelberg, 1989).
7. Z.-Y. Shen: *High-Temperature Superconducting Microwave Circuits* (Artech House, Boston, 1994).
8. M. J. Lancaster: *Passive Microwave Device Applications of High-Temperature Superconductors* (Cambridge University Press, Cambridge, 1997).
9. N. Newman and W. G. Lyons: J. Supercond. **6**, 119 (1993); and N. Newman: In *High-Temperature Superconductors*, ed. by J. J. Pouch, S. A. Alterovitz, R. R. Romanofsky (TransTech., Switzerland, 1994).
10. M. Hein: In *Studies of High-Temperature Superconductors*, ed. by A. Narlikar (Nova Sciences, New York, 1996), Vol. 18, pp. 141–216.
11. H. J. Chaloupka: In *Applications of Superconductivity*, NATO-ASI Series, ed. by H. Weinstock (Kluwer, Dordrecht, 1999).
12. For recent reviews see: a. J. Mannhart and H. Hilgenkamp: Supercon. Sci. Technol. **10**, 880 (1997).
    b. Proc. Euroconference: *Polarons: Condensation, Pairing, Magnetism*, 9–17 June 1998, Erice, Sicily, Italy, ed. by A. Bussmann-Holder, K. A. Müller, V. Z. Kresin, J. Superconductivity **12** (1), 1999.
13. G. W. Mitschang: IEEE Trans. Appl. Supercond. **5**, 69 (1995).
14. H. Piel: In *Applied Superconductivity 1995*, Inst. Phys. Conf. Ser. No 148, ed. by D. Dew-Hughes (IOP, Bristol, 1995), p. 1.
15. For an overview see, e.g., references to Chap. 6, Sect. 6.2.
16. M. Weger and I. B. Goldberg: In *Solid State Physics*, ed. by H. Ehrenreich, F. Seitz and D. Turnbull (Academic, New York, 1973), Vol. 28, p. 1.
17. M. Tinkham: Introduction to Superconductivity (McGraw-Hill, New York, 1996).
18. R. D. Parks (Ed.): Superconductivity (Marcel Dekker, New York, 1969); Vols. I and II.
19. F. London and H. London: Proc. Roy. Soc. A **149**, 71 (1935).
20. a. R. Beyers and T. M. Shaw: Solid State Physics **42**, 135 (1989).
    b. D. M. Ginsberg (Ed.) *Physical Properties of High-Temperature Superconductors II* (World Scientific, Singapore, 1990).

Chapter 1

1. A. M. Portis: *Electromagnetic Fields, Sources and Media* (Wiley, New York, 1974).
2. C. J. Gorter and H. Casimir: Phys. Z. **35**, 963 (1934).
3. F. London and H. London: Proc. Roy. Soc. A **149**, 71 (1935).
4. M. J. Lancaster: *Passive Microwave Device Applications of High-Temperature Superconductors* (Cambridge University Press, Cambridge, 1997).
5. A. M. Portis: *Electrodynamics of High-Temperature Superconductors*, Lecture Notes in Physics, Vol. 48 (World Scientific, Singapore, 1993).
6. D. R. Tilley and J. Tilley: *Superfluidity and Superconductivity*, Graduate Student Series in Physics (Adam Hilger, Bristol, Boston, 1986).
7. G. E. H. Reuter and E. H. Sondheimer: Proc. Roy. Soc. A **195**, 336 (1948).
8. A. B. Pippard: Rep. Prog. Phys. **23**, 176 (1960).
9. C. Kittel: *Introduction to Solid State Physics* (Wiley, New York, 1968).
10. T. P. Orlando, E. J. McNiff, Jr., S. Foner, M. R. Beasley: Phys. Rev. B **19**, 4545 (1979).
11. M. Tinkham: *Introduction to Superconductivity* (McGraw-Hill, New York, 1975).
12. J. Halbritter: J. Supercond. **5**, 171 (1992).
13. P. G. deGennes: *Superconductivity of Metals and Alloys* (W. A. Benjamin, New York, Amsterdam, 1966).
14. J. R. Waldram: Adv. Phys. **13**, 1 (1964).
15. D. S. Linden, T. P. Orlando, W. G. Lyons: IEEE Trans. Appl. Supercond. **4**, 136 (1994).
16. D. C. Mattis, J. Bardeen: Phys. Rev. **111**, 412 (1958).
17. D. M. Ginsberg and L. C. Hebel: In *Superconductivity*, ed. by R. D. Parks (Marcel Dekker, New York, 1969), Vol. 1, p. 193.
18. J. R. Waldram: *Superconductivity of Metals and Cuprates* (IOP, Bristol, 1996).
19. D. A. Bonn, R. Liang, T. M. Riseman, D. J. Baar, D. C. Morgan, K. Zhang, P. Dosanjh, T. L. Duty, A. MacFarlane, G. D. Morris, J. H. Brewer, W. N. Hardy: Phys. Rev. B **47**, 11 314 (1993).
20. P. Kneisel, H. Küpfer, W. Schwarz, O. Stolz, J. Halbritter: IEEE Trans. Magn. **13**, 496 (1977).
21. G. Arnolds, H. Heinrichs, R. Mayer, N. Minatti, H. Piel, W. Weingarten: IEEE Trans. Nuclear Science **26**, 3775 (1979).
22. S. Ramo, J. R. Whinnery, T. VanDuzer: *Fields and Waves in Communication Electronics* (Wiley, New York, 1984).
23. S. Orbach-Werbig: Dissertation, University of Wuppertal, Report WUB-DIS 94-9 (1994).
24. N. Klein, H. Chaloupka, G. Müller, S. Orbach, H. Piel, B. Roas, L. Schultz, U. Klein, M. Peiniger: J. Appl. Phys. **67**, 6940 (1990).
25. S. Hensen: Dissertation, University of Wuppertal (1999).
26. M. A. Hein, Supercond. Sci. Technol. **10**, 867 (1997).
27. G. Müller, N. Klein, A. Brust, H. Chaloupka, M. Hein, S. Orbach, H. Piel, D. Reschke: J. Supercond. **3**, 235 (1990).
28. D. N. Basov, R. Liang, D. A. Bonn, W. N. Hardy, B. Dabrowski, M. Quijada, D. B. Tanner, J. P. Price, D. M. Ginsberg, T. Timusk: Phys. Rev. Lett. **74**, 598 (1995).
29. U. Dähne, N. Klein, H. Schulz, N. Tellmann, K. Urban: *Applied Superconductivity*, Inst. Phys. Conf. Ser. No 148, ed by. D. Dew-Hughes (IOP, Bristol, 1995), p. 1095.

30. Ya. G. Ponomarev, B. A. Aminov, N. B. Brandt, M. A. Hein, C. S. Khi, V. Z. Kresin, G. Müller, H. Piel, K. Rosner, S. V. Thesnokov, E. B. Tsokur, D. Wehler, K. Winzer, T. Wolf, A. V. Yarygin, K. T. Yusupov: Phys. Rev. B **52**, 1352 (1995).
31. D. A. Bonn, S. Kamal, Kuan Zhang, Ruixing Liang, D. J. Baar, E. Klein, W. N. Hardy: Phys. Rev. B **50**, 4051 (1994).
32. Zhi-Yuan Shen: *High-Temperature Superconducting Microwave Circuits* (Artech House, Boston, London, 1994).
33. a. P. Chaudari, J. Mannhart, D. Dimos, C. C. Tsuei, J. Chi, M. M. Oprysko, M. Scheuermann: Phys. Rev. Lett. **60**, 1653 (1988).
    b. J. Mannhart, P. Chaudari, D. Dimos, C. C. Tsuei, T. R. McGuire: Phys. Rev. Lett. **61**, 2476 (1988).
34. T. W. Lee and C. M. Falco: Appl. Phys. Lett. **38**, 567 (1981).
35. A. Barone and G. Paterno: *Physics and Applications of the Josephson Effect* (Wiley, New York, 1982).
36. R. Gross: In *Interfaces in Superconducting Systems*, ed. by S. L. Shinde and D. Rudman (Springer, New York, 1992).
37. T. L. Hylton, A. Kapitulnik, M. R. Beasley, J. P. Carini, L. Drabeck, G. Grüner: Appl. Phys. Lett. **53**, 1343 (1988).
38. V. Ambegaokar and A. Baratoff: Phys. Rev. Lett. **10**, 486 (1963).
39. a. D. E. McCumber, J. Appl. Phys. **39**, 2503, 3113 (1968).
    b. W. C. Stewart: Appl. Phys. Lett. **12**, 277 (1968).
40. T. L. Hylton and M. R. Beasley: Phys. Rev. B **39**, 9042 (1989).
41. a. N. Newman and W. G. Lyons: J. Supercond. **6**, 119 (1993).
    b. N. Newman: In *High-Temperature Superconductors*, ed. by J. J. Pouch, S. A. Alterovitz, R. R. Romanofsky (TransTech., Switzerland, 1994)
    c. M. Hein: In *Studies of High-Temperature Superconductors*, ed. by A. Narlikar (Nova Sciences, New York, 1996), Vol. 18, p. 141.
42. A. M. Portis and D. W. Cooke: Supercond. Sci. Technol. **5**, S395 (1992).
43. A. M. Portis, D. W. Cooke, E. R. Gray, P. N. Arendt, C. L. Bohn, J. R. Delayen, C. T. Roche, M. Hein, N. Klein, G. Müller, S. Orbach, H. Piel: Appl. Phys. Lett. **58**, 307 (1993).
44. J. C. Gallop, A. L. Cowie, L. F. Cohen: *Applied Superconductivity 1997*, Inst. Phys. Conf. Ser. No 158, ed. by H. Rogalla and D. H. A. Blank (IOP, Bristol, 1997), Vol. 1, p. 65.
45. M. A. Hein, J. Strupp, H. Piel, A. M. Portis, R. Gross: J. Appl. Phys. **75**, 4581 (1994).
46. O. G. Vendik, A. B. Kozyrev, A. Y. Popov: Revue Phys. Appl. **25**, 255 (1990).
47. J. Halbritter: J. Appl. Phys. **68**, 6315 (1990) and **71**, 339 (1992).
48. J. Bardeen, L. N. Cooper, J. R. Schrieffer: Phys. Rev. **108**, 1175 (1957).
49. M. Prohammer and J. P.Carbotte: Phys. Rev. B **43**, 5370 (1991).
50. R. D. Parks (Ed.): *Superconductivity* (Marcel Dekker, New York, 1969), Vol. 1 and 2.
51. N. N. Bogoliubov: Nuovo Cimento **7**, 794 (1958).
52. J. G. Valatin: Nuovo Cimento **7**, 843 (1958).
53. V. Z. Kresin, G. Deutscher, S. A. Wolf: J. Supercond. **1**, 327 (1988).
54. T. P. Sheahen: Phys. Rev. **149**, 368 (1966).
55. J. Halbritter: Z. Phys. **266**, 209 (1974).
56. J. P. Turneaure, J. Halbritter, H. A. Schwettman: J. Supercond. **4**, 341 (1991).
57. A. A. Abrikosov, L. P. Gor'kov, I. M. Khalatnikov: Zh. Eksperim. Teor. Fiz. **35**, 265 (1958) [English translation: Sov. Phys. JETP **8**, 182 (1959)].
58. a. L. C. Hebel and C. P. Slichter: Phys. Rev. **107**, 901 (1957).
    b. L. C. Hebel: Phys. Rev. **116**, 79 (1959).

59. J. Halbritter: Externer Bericht 3/69-2 (Kernforschungszentrum Karlsruhe, 1969); Externer Bericht 3/70-6 (Kernforschungszentrum Karlsruhe, 1970).
60. J. P. Turneaure: Ph.D. Dissertation, Stanford University (1967).
61. J. Hinken: *Superconductor Electronics* (Springer, Berlin, Heidelberg, 1989).
62. A. A. Abrikosov and L. P. Gor'kov: Zh. Eksperim. I Teor. Fiz. **39**, 1781 (1960) [English transl.: Soviet Phys. JETP **12**, 1243 (1961)].
63. S. Skalski, O. Betbeder-Matibet, P. R. Weiss: Phys. Rev. **136**, A1500 (1964).
64. K. Maki: In *Superconductivity*, ed. by R. D. Parks (Marcel Dekker, New York, 1969), Vol. 2, p. 1035.
65. M. A. Biondi, M. P. Garfunkel, W. A. Thomson: Phys. Rev. **136**, A1471 (1964).
66. J. R. Clem: Ann. Phys. (NY) **40**, 269 (1966).
67. R. Blaschke, U. Klein, G. Müller: Verhandlungen DPG (VI) **17**, 988 (1982).
68. R. Blaschke, A. Philipp, J. Halbritter: Proc. Int. Conf. on Superconductivity in D- and F-Band Metals, Karlsruhe, ed. by W. Buckel and W. Weber (1978), p. 593.
69. K. I Wysokinski and T. Domanski: Phys. Rev. B **45**, 5005 (1992).
70. D.W. Cooke, E. R. Gray, H. H. S. Javadi, R. J. Houlton, B. Rusnak, E. A. Meyer, P. N. Arendt, N. Klein, G. Müller, S. Orbach, H. Piel, L. Drabeck, G. Grüner, J. Y. Josefowicz, D. B. Rensch, F. Krajenbrink: Solid State Commun. **73**, 297 (1990).
71. D. J. Scalapino: In *Superconductivity*, ed. by R. D. Parks (Marcel Dekker, New York, 1969), Vol. 1, p. 449.
72. a. G. E. Eliashberg: Zh. Eksperim. I Teor. Fiz. **38**, 966 (1960) [English transl.: Soviet Phys. JETP **11**, 696 (1960)].
  b. J. P. Carbotte: Rev. Mod. Phys. **62**, 1027 (1990).
73. W. McMillan: Phys. Rev. **167**, 331 (1968).
74. R. Dynes: Solid State Commun. **10**, 615 (1972).
75. V. Z. Kresin: Phys. Lett. A **122**, 434 (1987).
76. P. B. Allen and R. C. Dynes: Phys. Rev. B **12**, 905 (1975).
77. S. B. Nam: Phys. Rev. **156**, 470 (1967) and 487 (1967).
78. F. Marsiglio, J. P. Carbotte, J. Blezius: Phys. Rev. B **41**, 6457 (1990).
79. R. Blaschke and R. Blocksdorf: Z. Phys. B **49**, 99 (1982).
80. G. M. Eliashberg, G. V. Klimovitch, A. V. Rylyakov: J. Supercond. **4**, 393 (1991).
81. S. M. Anlage and D. H. Wu: J. Supercond. **5**, 395 (1992).
82. K. Okuda, S. Kawamata, S. Noguchi, N. Itoh, K. Kadowaki: J. Phys. Soc. Japan **60**, 3226 (1991).
83. N. Phillips, R. Fisher, J. Gordon: *Progress in Low-Temperature Physics*, ed. by D. Brewer (North-Holland, Amsterdam, 1992), Vol. 13, p. 267.
84. a. V. Z. Kresin and S. A. Wolf: Phys. Rev. B **51**, 1229 (1995).
  b. V. Z. Kresin, H. Morawitz, S. A. Wolf: *Mechanisms of Conventional and High-T_c Superconductivity* (Oxford University Press, New York, 1993).
  c. V. Z. Kresin and S. A. Wolf: Phys. Rev. B **46**, 6458 (1992).
  d. V. Kresin: Solid State Commun. **63**, 725 (1987).
85. S. D. Adrian, M. E. Reeves, S. A. Wolf, V. Z. Kresin: Phys. Rev. B **51**, 6800 (1995).
86. A. A. Golubov, M. R. Trunin, A. A. Zhukov, O. V. Dolgov, S. V. Shulga: JETP Lett. **62**, 497 (1995).
87. V. Z. Kresin and S. A. Wolf: Physica C **198**, 328 (1992).
88. A. Andreone, V. Z. Kresin, S. A. Wolf: J. Supercond. **5**, 339 (1992).

89. a. A. A. Golubov and I. I. Mazin: Phys. Rev. B **55**, 15146 (1997).
  b. A. A. Golubov, M. R. Trunin, A. A. Zhukov, O. V. Dolgov, S. V. Shulga: J. Phys. I France **6**, 2275 (1996).
  c. A. A. Golubov, O. V. Dolgov, E. G. Maksimov, I. I. Mazin, S. V. Shulga: Physica C **235–240**, 2383 (1994).
90. S. D. Adrian, S. A. Wolf, O. Dolgov, S. Shulga, V. Z. Kresin: Phys. Rev. B **56**, 7878 (1997).
91. Y. N. Ovchinnikov and V. Z. Kresin: Phys. Rev. **54**, 1251 (1996).
92. V. Z. Kresin, S. A.Wolf, Y. N. Ovchinnikov: J. Supercond. **9**, 431 (1996).
93. D. J. Scalapino, E. Loh, J. Hirsch: Phys. Rev. B **34**, 8190 (1986).
94. D. J. Scalapino: Phys. Rep. **250**, 329 (1995).
95. D. Xu, S. K. Yip, J. A. Sauls: Phys. Rev. B **51**, 16233 (1995).
96. M. Prohammer and J. P. Carbotte: Phys. Rev. B **43**, 5370 (1991).
97. a. P. J. Hirschfeld, W. O. Puttika, D. J. Scalapino: Phys. Rev. B **50**, 10250 (1994).
  b. P. J. Hirschfeld, W. O. Puttika, D. J. Scalapino: Phys. Rev. Lett. **71**, 3705 (1993).
98. P. J. Hirschfeld and N. Goldenfeld: Phys. Rev. B **48**, 4219 (1993).
99. P. A. Lee: Phys. Rev. Lett. **71**, 1887 (1993).
100. S. Hensen, G. Müller, C. T. Rieck, K. Scharnberg: Phys. Rev. B **56**, 6237 (1997).
101. a. C. T. Rieck, W. A. Little, J. Ruvalds, A. Virosztek: Phys. Rev. B **51**, 3772 (1995).
  b. A. Virosztek and J. Ruvalds: Phys. Rev. B **42**, 4064 (1990).
102. J. Mao, D. H.Wu, J. L. Peng, R. L. Greene, S. M. Anlage: Phys. Rev. B **51**, 3316 (1995).
103. J. Mao, S. M. Anlage, J. L. Peng, R. L. Greene: IEEE Trans. Appl. Supercond. **5**, 1997 (1995).
104. V. Kresin, S. Wolf, Y. Ovchinnikov, A. Bill, S. Adrian, O. Dolgov, S. Shulga: J. Low-Temp. Physics **106**, 159 (1997).
105. J. R. Schrieffer: J. Low-Temp. Physics **99**, 397 (1995); Bull. APS **41**, 74 (1996).
106. a. J. Mannhart, H. Hilgenkamp, B. Mayer, C. Gerber, J. Kirtley, K. A. Moler, M. Sigrist: Phys. Rev. Lett. **77**, 2782 (1996).
  b. J. Mannhart and H. Hilgenkamp: Supercond. Sci. Technol. **10** (1997).
107. Proc. Euroconference *Polarons: Condensation, Pairing, Magnetism*, 9–17 June 1998, Erice, Sicily, Italy, ed. by A. Bussmann-Holder, K. A. Müller, V. Z. Kresin: J. Superconductivity **12** (1), 1999.
108. a. D. J. VanHarlingen: Rev. Mod. Phys. **67**, 515 (1995).
  b. C. C. Tsuei, J. R. Kirtley, C. C. Chi, L. S. Yu-Jahnes, A. Gupta, T. Shaw, J. Z. Sun, M. B. Ketchen: Phys. Rev. Lett. **73**, 593 (1994).
109. a. L. N. Bulaevskii, V. V. Kuzii, A. A. Sobyanin: Sov. Phys. JEPT Lett. **25**, 290 (1977).
  b. B. I. Spivak and S. A. Kivelson: Phys. Rev. B **43**, 3740 (1991).
110. V. Z. Kresin and S. A. Wolf: J. Supercond. **7**, 865 (1994).
111. K. E. Kihlstrom, R. H. Hammond, J. Talvacchio, T. H. Geballe, A. K. Green, V. Rehn: J. Appl. Phys. **53**, 8907 (1982).
112. M. R. Beasley: Physica C **185–189**, 227 (1991).
113. Z. X. Shen, D. S. Dessau, B. O. Wells, D. M. King, W. E. Spicer, A. J. Arko, D. Marshall, L. W. Lombardo, A. Kapitulnik, P. Dickinson, S. Doniach, J. DiCarlo, A. G. Loeser, C. H. Park: Phys. Rev. Lett. **70**, 1553 (1993).
114. J. Kane, Chen Qun, K. W. Ng, H. J. Tao: Phys. Rev. Lett. **72**, 128 (1994).
115. B. A. Aminov, B. Aschermann, M. A. Hein, F. Hill, M. Lorenz, G. Müller, H. Piel: Phys. Rev. B **52**, 13631 (1995).

116. B. A. Aminov, M. A. Hein, G. Müller, H. Piel, D. Wehler, Ya. G. Ponomarev, K. Rosner, K. Winzer: J. Supercond. **7**, 361 (1994); and Ya. G. Ponomarev, unpublished.
117. M. R. Beasley: IEEE Trans. Appl. Supercond. **5**, 141 (1995).
118. S. Y. Lin, P. Chaudari, N. Khare: In *Advances in Superconductivity VII*, ed. by K. Yamaguchi and T. Morishita (Springer, Tokyo, 1995), p. 39.
119. D. A. Wollmann, D. J. VanHarlingen, W. C. Lee, D. M. Ginsberg, A. J. Leggett: Phys. Rev. Lett. **71**, 2134 (1993).
120. N. Klein, N. Tellmann, H. Schulz, K. Urban, S. A. Wolf and V. Z. Kresin: Phys. Rev. Lett. **71**, 3355 (1993).

Chapter 2

1. M. J. Lancaster: *Passive Microwave Device Applications of High-Temperature Superconductors* (Cambridge University Press, Cambridge, 1997).
2. A. M. Portis: *Electrodynamics of High-Temperature Superconductors*, Lecture Notes in Physics, Vol. 48 (World Scientific, Singapore, 1993).
3. G. Müller: Proc. 4th Workshop on RF Superconductivity, ed. by Y. Kojima, KEK Report 89-21, KEK, Tsukuba, Japan (1990), p. 267.
4. N. Klein: Proc. 5th Workshop on RF Superconductivity, ed. by D. Proch, DESY M-92-01, Deutsches Elektronen Synchrotron, Hamburg, Germany (1992), p. 285.
5. A. P. Jenkins, K. S. Kale, D. Dew-Hughes: In *Studies of High-Temperature Superconductors*, ed. by A. Narlikar (Nova Sciences, New York, 1996), Vol. 17, p. 179.
6. Z.-Y. Shen: *High-Temperature Superconducting Microwave Circuits* (Artech House, Boston, London, 1994).
7. a. S. Ramo, J. R. Whinnery, T. VanDuzer: *Fields and Waves in Communication Electronics*, 2nd ed. (Wiley, New York, 1984).
   b. R. E. Collins: *Foundations for Microwave Engineering*, (McGraw-Hill, New York, 1966).
   c. A. J. Baden-Fuller: *Microwaves: An Introduction to Microwave Theory and Techniques* (Pergamon, Oxford, 1990).
8. a. M. Bartsch et al.: Computer Physics Commun. **72**, 22 (1992).
   b. Computer code "em", Sonnet Software, release 2.3, Liverpool, N.Y. 13090 (1991).
9. a. J. C. Slater: Rev. Mod. Phys. **18**, 441 (1946).
   b. J. D. Jackson: *Classical Electrodynamics* (deGruyter, Berlin, New York, 1962).
10. J. Halbritter: J. Supercond. **10**, 91 (1997) and previous papers cited therein.
11. K. Lange and K. H. Löcherer (Eds): *Taschenbuch der Hochfrequenztechnik, Meinke; Gundlach* (Springer, Berlin, 1992), Chap. C.
12. a. N. Newman, W. G. Lyons: J. Supercond. **6**, 119 (1993)
    b. N. Newman: In *High-Temperature Superconductors*, ed. by J. J. Pouch, S. A. Alterovitz, R. R. Romanofsky (TransTech., Switzerland, 1994)
    c. M. Hein: In *Studies of High-Temperature Superconductors*, ed. by A. Narlikar (Nova Sciences, New York, 1996), Vol. 18, p. 141.
13. W. H. Chang: J. Appl. Phys. **50**, 8129 (1980).
14. O. G. Vendik, I. B. Vendik, T. B. Samoilova: IEEE Microwave Theory Tech. **45**, 173 (1997).
15. D. M. Sheen, S. M. Ali, D. E. Oates, R. S.Withers, J. A. Kong: IEEE Trans. Appl. Supercond. **1**, 108 (1991).

16. A. Porch, M. J. Lancaster, R. Humphreys: IEEE Trans. Microwave Theory Tech. **43**, 306 (1995).
17. a. U. Klein: Dissertation, University of Wuppertal, Report WUB-DI 81-2 (1981).
    b. J. P. Turneaure and I. Weissmann: J. Appl. Phys. **39**, 4417 (1968).
18. a. C. Reece, T. Powers, P. Kushnik: CEBAF TN-91-031, CEBAF-SRF-05-03-EXA (Newport News, USA, 1991).
    b. T. Hays, H. Padamsee, R. W. Röth: Proc. 7th Workshop on RF Superconductivity, ed. by B. Bonin, Gif-sur-Yvette, France (1995), Vol. II, p. 437.
    c. Ke-Jun Kang, Yi-Xiang Wei, B. Dwersteg: Proc. 5th Workshop on RF Superconductivity, ed. by D. Proch, DESY M-92-01, Deutsches Elektronen Synchrotron, Hamburg, Germany (1992), p. 527.
19. a. D. Reschke: Dissertation, University of Wuppertal, Report WUB-DIS 95-5 (1995).
    b. D. W. Reschke and R. W. Röth: Proc. 6th Workshop on RF Superconductivity, ed. by J. Graber, CEBAF, Newport News, USA, 1118 (1993).
20. H. Piel: Proc. Workshop on RF Superconductivity, ed. by M. Kuntze, Kernforschungszentrum Karlsruhe, KfK 3019.85 (1980).
21. G. L. Ragan: *Microwave Transmission Circuits* (McGraw-Hill, New York, 1948).
22. S. Orbach-Werbig: Dissertation, University of Wuppertal, Report WUB-DIS 94-9 (1994).
23. a. S. Hensen, S. Orbach-Werbig, G. Müller, H. Piel, N. G. Chew, J. A. Edwards, R. G. Humphreys: In *Applied Superconductivity*, ed. by H. C. Freyhardt (DGM, Oberursel, 1993), p. 1053.
    b. S. Hensen, G. Müller, C. T. Rieck, K. Scharnberg: Phys. Rev. B **56**, 6237 (1997).
    c. S. Hensen: Dissertation, University of Wuppertal (1999).
24. D. Kajfez, P. Guillon (Eds.): *Dielectric Resonators* (Artech House, Dedham, 1986).
25. Y. Kobayashi and S. Tanaka: IEEE Trans. Microwave Theory Tech. **28**, 1077 (1980).
26. a. W. Diete, B. Aschermann, H. Chaloupka, M. Jeck, T. Kamppeter, S. Kolesov, G. Müller, H. Piel, H. Schlick: In *Applied Superconductivity*, Inst. Phys. Conf. Ser. No 148, ed. by D. Dew-Hughes (IOP, Bristol, 1995), p. 1107.
    b. T. Kaiser, W. Diete, M. Getta, M. A. Hein, G. Müller, M. Perpeet, H. Piel: Particle Accel. **60**, 171 (1998).
    c. T. Kaiser: Dissertation, University of Wuppertal, Report WUB-DIS 98-13 (1998).
27. G. Kirchhoff: *Gesammelte Abhandlungen* (Barth, Leipzig, 1882–1891).
28. Z.-Y. Shen, C. Wilker, P. Pang, D. W. Face, C. F. Carter III, C. M. Harrington: IEEE Trans. Appl. Supercond. **7**, 1283 (1997).
29. G. C. Liang, D. Zhang, C. F. Shih, M. E. Johannson, R. S. Withers, A. C. Anderson, D. E. Oates: IEEE Trans. Appl. Supercond. **5**, 2652 (1995).
30. H. Chaloupka, M. Jeck, B. Gurzinski, S. Kolesov: Electronics letters **32**, 1735 (1996).
31. G. Müller, B. Aschermann, H. Chaloupka, W. Diete, M. Getta, B. Gurzinski, M. Hein, M. Jeck, T. Kaiser, S. Kolesov, H. Piel, H. Schlick, R. Theisejans: IEEE Trans. Appl. Supercond. **7**, 1287 (1997).
32. S. Kolesov, H. Chaloupka, A. Baumfalk, T. Kaiser: J. Supercond. **10**, 179 (1997).
33. C. E. Reece: Proc. 3rd Workshop on RF Superconductivity, ed. by K. W. Shepard, Argonne National Laboratory, Report ANL-PHY-88-1 (1988), p. 545.

34. D. E. Oates, P. P. Nguyen, G. Dresselhaus, M. S. Dresselhaus, C. W. Lam, S. M. Ali: J. Supercond. **5**, 361 (1992).
35. a. S. M. Anlage, D. E. Steinhauer, C. P. Vlahacos, B. J. Feenstra, A. S. Thanawalla, W. Hu, S. K. Dutta, F. C. Wellstood: IEEE Trans. Appl. Supercond. **9**, June 1999.
    b. D. E. Steinhauer, C. P. Vlahacos, S. K. Dutta, F. C. Wellstood, S. M. Anlage: Appl. Phys. Lett. **71**, 1736 (1997).
    c. C. P. Vlahacos, R. C. Black, S. M. Anlage, A. Amar, F. C. Wellstood: Appl. Phys. Lett. **69**, 3272 (1996).
36. a. M. Golosovsky, A. Lann, D. Davidov: Ultramicroscopy **71**, 133 (1998).
    b. M. Golosovsky, A. Galkin, D. Davidov: IEEE Trans. Microw. Theory Tech. **44**, 1390 (1996).
    c. M. Golosovsky and D. Davidov: Appl. Phys. Lett. **68**, 1579 (1996).
37. a. T. Wei, X. D. Xiang, P. G. Schultz: Appl. Phys. Lett. **68**, 3506 (1996).
    b. I. Takeuchi, C. Gao, F. Duewer, T. Wei, X. D. Xiang, S. P. Pai, T. Venkatesan: IEEE Trans. Appl. Supercond. **9**, June 1999.
38. G. Hampel, P. Kolodner, P. L. Gammel, P. A. Polakos, E. deObaldia, P. M. Maniewich, A. Anderson, R. Slattery, D. Zhang, G. C. Liang, C. F. Shih: Appl. Phys. Lett. **69**, 571 (1996).
39. a. R. C. Black, F. C. Wellstood, E. Dantsker, A. H. Miklich, D. Koelle, F. Ludwig, J. Clarke: IEEE Trans. Appl. Supercond. **5**, 2137 (1995).
    b. R. C. Black, F. C. Wellstood, E. Dantsker, A. H. Miklich, D. T. Nemeth, D. Koelle, F. Ludwig, J. Clarke: Appl. Phys. Lett. **66**, 99 (1994).
40. J. S. Martens, V. M. Hietala, D. S. Ginley, T. W. Zipperian, G. K. G. Hohenwarter: Appl. Phys. Lett. **58**, 2543 (1991).
41. T. E. Harrington, J. Wosik, S. A. Long: IEEE Trans. Appl. Supercond. **7**, 1861 (1997).
42. a. R. Schwab, R. Spörl, J. Burbach, P. Severloh, R. Heidinger, J. Halbritter: In *Applied Superconductivity 1997*, Inst. Phys. Conf. Ser. No 158, ed. by H. Rogalla, D. H. A. Blank, (IOP, Bristol, 1997), Vol. 1, p. 61.
    b. R. Schwab and R. Heidinger: Proc. MIOP'97, Sindelfingen, 22.-24. April 1997, Germany, p. 106.
    c. R. Heidinger, F. Königer, G. Link: Conference Digest *15th Internat. Conf. on Infrared and Millimeter Waves*, Proc. SPIE 1514, 184 (1990).
43. Y. Lemaitre, L. M. Mercandalli, B. Dessertenne, D. Mansart, B. Marcilhac, J. C. Mage: Physica C **235–240**, 643 (1994).
44. a. L. Hao, J. C. Gallop, F. Abbas: In *Applied Superconductivity 1997*, Inst. Phys. Conf. Ser. No 158, ed. by H. Rogalla, D. Blank (IOP, Bristol, 1997), Vol. 1, p. 275.
    b. L. Hao, J. C. Gallop: IEEE Trans. Appl. Supercond. **9**, June 1999.
45. a. C. A. Krulle, T. Doderer, D. Quenter, R. P. Huebener, R. Pöpel, J. Niemeyer: Appl. Phys. Lett. **59**, 3042 (1991).
    b. R. P. Huebener: In *Superconducting Quantum Electronics*, ed. by V. Kose (Springer, Berlin, 1989), p. 205.
    c. R. P. Huebener: Rep. Prog. Phys. **47**, 175 (1984).
46. R. B. Hammond: Supercond. Technol. Inc., Santa Barbara, private communication (1996).
47. a. J. C. Culbertson, H. S. Newman, C. Wilker: J. Appl. Phys. **84**, 2768 (1998).
    b. H. S. Newman, J. C. Culbertson, Microwave and Optical Technol. Lett. **6**, 725 (1993).
    c. J. C. Culbertson, H. S. Newman, U. Strom, J. M. Pond, D. B. Chrisey, J. S. Horwitz, S. A. Wolf: J. Appl. Phys. **70**, 4995 (1991).
48. S. Cho and D. Choi: Physica C **271**, 319 (1996).

49. T. Kaiser, M. A. Hein, G. Müller, M. Perpeet: Appl. Phys. Lett. **73**, 3447 (1998).
50. a. A. Forkl, T. Dragon, H. Kronmüller, H.-U. Habermeier, G. Mertens: Appl. Phys. Lett. **57**, 1067 (1990).
    b. C. Jooss, R. Warthmann, A. Forkl, H.-U. Habermeier, B. Leibold, H. Kronmüller: Physica C **266**, 235 (1996).
51. a. P. Leiderer, J. Boneberg, P. Brüll, V. Bujok, S. Herminghaus: Phys. Rev. Lett. **71**, 2646 (1993).
    b. J. Eisenmenger, S. Kambach, S. Saleh, A. Tihi, P. Leiderer, M. Wallenhorst, H. Dötsch: J. Low-Temp. Phys. **105**, 1123 (1996).
52. M. A. Hein, C. Bauer, W. Diete, S. Hensen, T. Kaiser, G. Müller, H. Piel: J. Supercond. **10**, 109 (1997).
53. A. Oral, S. J. Bending, R. G. Humphreys, M. Henini: Supercond. Sci. Technol. **10**, 17 (1997).
54. B. Lehndorff, H. G. Kürschner, H. Piel: In *Applied Superconductivity*, Inst. Phys. Conf. Ser. No 148, ed. by D. Dew-Hughes (IOP, Bristol, 1995), p. 691.
55. J. Blitz: *Electrical and Magnetic Methods of Nondestructive Testing* (Adam Hilger, London, 1991).
56. K. Wegendt: Cryogenics **35**, 155 (1995).
57. A. T. Fiory, A. F. Hebard, P. M. Mankiewich, R. E. Howard: Appl. Phys. Lett. **52**, 2165 (1988).
58. A. Bock, R. Kürsten, M. Bruhl, N. Dieckmann, U. Merkt: Phys. Rev. B **54**, 4300 (1996).
59. C. Thomsen: In *Light Scattering in Solids VI*, ed. by M. Cardona and G. Güntherodt, Topics in Applied Physics, Vol 68 (Springer, Heidelberg, 1991).
60. C. A. Bryant and J. B. Gunn: Rev. Sci. Instrum. **36**, 1614 (1965).
61. E. A. Ash and G. Nicholls: Nature **237**, 510 (1972).
62. H. Piel and R. Romjin: CERN/EF/RF 80-3 (1980).
63. T. Kaiser, C. Bauer, W. Diete, M. A. Hein, J. Kallscheuer, G. Müller, H. Piel: In *Applied Superconductivity 1997*, Inst. Phys. Conf. Ser. No 158, ed. by H. Rogalla, D. H. A. Blank, (IOP, Bristol, 1997), Vol. 1, p. 45.
64. T. P. Orlando, E. J. McNiff, Jr., S. Foner, M. R. Beasley: Phys. Rev. B **19**, 4545 (1979).
65. F. Marsiglio, J. P. Carbotte, J. Blezius: Phys. Rev. B **41**, 6457 (1990).
66. A. Andreone, C. Cantoni, A. Cassinese, A. DiChiara, R. Vaglio: Phys. Rev. B **56**, 7874 (1997).
67. K. E. Kihlstrom, R. H. Hammond, J. Talvacchio, T. H. Geballe, A. K. Green, V. Rehn: J. Appl. Phys. **53**, 8907 (1982).
68. G. Müller, P. Kneisel, D. Mansen, H. Piel, J. Pouryamout, R. W. Röth: Proc. 5th Europ. Particle Accelerator Conf., ed. by S. Myers et al. (IOP, Bristol, Philadelphia, 1996), Vol. 3, p. 2085.
69. P. Kneisel, O. Stolz, J. Halbritter: IEEE Trans. Magn. **15**, 21 (1979).
70. V. Z. Kresin and S. A. Wolf: Physica C **198**, 328 (1992).
71. R. Mailfert, B. W. Batterman, J. J. Hanak: Phys. Lett. A **24**, 315 (1967).
72. R. Flükiger, H. Küpfer, J. L. Jorda, J. Muller: IEEE Trans. Magn. **23**, 980 (1987).
73. R. A. Hein: In *The Science and Technology of Superconductivity*, ed. by W. D. Gregory, W. W. Mathews, E. A. Edelsack (Plenum, New York, London, 1985).
74. R. G. Ross, Jr. (Ed.): *Cryocoolers 8*, (Plenum, New York, 1995).
75. R. Radebaugh: Plenary Talk at the Applied Superconductivity Conference, 13.-18. Sept. 1998, Palm Desert, California, IEEE Trans. Appl. Supercond. **9**, June 1999.

76. P. Kneisel, H. Küpfer, W. Schwarz, O. Stolz, J. Halbritter: IEEE Trans. Magn. **13**, 496 (1977).
77. a. G. Arnolds, R. Blaschke, H. Piel, D. Proch: IEEE Trans. Magn. **15**, 27 and 613 (1979).
    b. G. Arnolds and D. Proch: IEEE Trans. Magn. **13**, 500 (1977).
    c. G. Arnolds, H. Heinrichs, R. Mayer, N. Minatti, H. Piel, W. Weingarten: IEEE Trans. Nuclear Science **26**, 3775 (1979).
78. a. M. Peiniger, M. Hein, N. Klein, G. Müller, H. Piel, P. Thüns: Proc. 3rd Workshop on RF Superconductivity, ed. by K. W. Shepard, Argonne Natinal Laboratory, Report ANL-PHY-88-1 (1988), p. 503.
    b. M. Peiniger: Dissertation, University of Wuppertal, Report WUB-DIS 89-1 (1989).
    c. D. Dasbach, G. Müller, M. Peiniger, H. Piel, R. W. Röth: IEEE Trans. Magn. **25**, 1862 (1989).
    d. M. Peiniger and H. Piel: Trans. Nuclear Science **32**, 3610 (1989).
79. B. Hillenbrand, H. Martens, H. Pfister, K. Schnitzke, Y. Uzel: IEEE Trans. Magn. **13**, 491 (1977).
80. H. H. Farell, H. Gilmer, M. Suenaga: J. Appl. Phys. **45**, 4025 (1974).
81. M. Suenaga: In S. Foner, B. B. Schwartz: *Superconductor Materials Science, Metallurgy, Fabrications and Applications* (Plenum, New York, 1981), p. 201.
82. a. L. H. Allen, M. R. Beasley, R. H. Hammond, J. P. Turneaure: IEEE Trans. Magn. **23**, 1405 (1987).
    b. L. H. Allen, W. J. Anklam, M. R. Beasley, R. H. Hammond, J. P. Turneaure: IEEE Trans. Magn. **21**, 525 (1985).
    c. L.H. Allen, M. R. Beasley, R. H. Hammond, J. P. Turneaure: IEEE Trans. Magn. **19**, 1003 (1983).
83. A. Andreone, A. Cassinese, A. Di Chiara, M. Ivarone, F. Palomba, A. Ruosi, R. Vaglio: IEEE Trans. Appl. Supercond. **7**, 1772 (1997).
84. a. D. A. Rudman, F. Hellman, R. H. Hammond, M. R. Beasley: J. Appl. Phys. **55**, 3544 (1984).
    b. D. A. Rudman and M. R. Beasley: Phys. Rev. B **30**, 2590 (1984).
    c. D. A. Rudman, R. E. Howard, D. F. Moore, R. B. Zubeck, M. R. Beasley: IEEE Trans. Magn. **15**, 582 (1979).
85. T. W. Lee and C. M. Falco: Appl. Phys. Lett. **38**, 567 (1981).
86. a. C. T. Wu, R. T. Kampwirth, J. W. Hafstrom: J. Vac. Sci. Technol. **14**, 134 (1977).
    b. R. T. Kampwirth, J. W. Hafstrom, C. T. Wu: IEEE Trans. Magn. **13**, 315 (1977).
87. M. Perpeet, M. A. Hein, G. Müller, H. Piel, J. Pouryamout, W. Diete: J. Appl. Phys. **82**, 5021 (1997).
88. a. M. Perpeet, W. Diete, M. A. Hein, S. Hensen, T. Kaiser, G. Müller, H. Piel, J. Pouryamout: In *Applied Superconductivity 1997*, Inst. Phys. Conf. Ser. No 158, ed. by H. Rogalla, D. H. A. Blank (IOP, Bristol, 1997), Vol. 1, p. 101.
    b. M. Perpeet, W. Diete, M. A. Hein, S. Hensen, T. Kaiser, G. Müller, H. Piel, J. Pouryamout: Proc. 8th Workshop on RF Superconductivity, 6–10 October 1997, ed. by V. Palmieri and A. Lombardi Abano Terme, Padova, Italy, LNL-INFN(REP) 133/98 (1998), Vol. IV, p. 934.
89. J. Halbritter: J. Appl. Phys. **68**, 6315 (1990); Z. Phys. B **71**, 411 (1988).
90. J. H. Claassen, M. E. Reeves, J. R. Soulen, Jr.: Rev. Sci. Instrum. **62**, 996 (1991).
91. A. Andreone, A. Cassinese, M. Iavarone, R. Vaglio, I. I. Kulik, V. Palmieri: Phys. Rev. B **52**, 4473 (1995).
92. W. DeSorbo: Phys. Rev. **132**, 107 (1963).

93. M. Perpeet: University of Wuppertal, Dissertation (1999).
94. L. R. Testardi and L. F. Mattheis: Phys. Rev. Lett. **41**, 1612 (1978).
95. J. P. Turneaure, J. Halbritter, H. A. Schwettman: J. Supercond. **4**, 341 (1991).
96. B. Mühlschlegel: Z. Phys. **155**, 313 (1959).
97. J. Halbritter: Z. Phys. **238**, 466 (1970).
98. M. Tinkham: *Introduction to Superconductivity* (McGraw-Hill, New York, 1975).
99. F. Marsiglio, J. P. Carbotte, R. Akis, D. Achkir, M. Poirier: Phys. Rev. B **50**, 7203 (1994); and private communication (June 1998).
100. C. Kittel: *Introduction to Solid State Physics* (John Wiley & Sons, New York, 1986).
101. R. Bormann, D.Y. Yu, R. H. Hammond, T. H. Geballe, S. Foner, E. J. McNiff, Jr.: IEEE Trans. Magn. **21**, 1140 (1985).
102. M. Perpeet, A. Cassinese, M. A. Hein, T. Kaiser, G. Müller, H. Piel, J. Pouryamout: IEEE Trans. Appl. Supercond. **9**, June 1999.
103. M. Hein, S. Kraut, E. Mahner, G. Müller, D. Opie, H. Piel, L. Ponto, D. Wehler, M. Becks, U. Klein, M. Peiniger: J. Supercond. **3**, 323 (1990).
104. R. M. Scanlan, W. A. Fietz, E. F. Koch: J. Appl. Phys. **46**, 2244 (1975).
105. P. W. Anderson: J. Phys. Chem. Solids **11**, 26 (1959).
106. V. Hoffstein and R. W. Cohen: Phys. Lett. A **29**, 603 (1969).
107. E. Schachinger and M. Prohammer: Physica C **156**, 701 (1988).
108. M. K. Wu, J. R. Ashburn, C. J. Torng, P. H. Hor, R. L. Meng, L. Gao, Z. J. Huang, Y. Q. Wang, C. W. Chu: Phys. Rev. Lett. **58**, 908 (1987).
109. P. H. Hor, R. L. Meng, Y. Q. Wang, L. Gao, Z. J. Huang, J. Bechtold, K. Forster, C. W. Chu: Phys. Rev. Lett. **58**, 1891 (1987).
110. K. Yamafuji and T. Morishita (Eds.): *Advances in Superconductivity VII* (Springer, Tokyo, 1995) Vol. I, Sect. 3.
111. G. Liang: Proc. 6th Workshop on RF Superconductivity, ed. by R. M. Sundelin, Continuous Electron Beam Accelerator Facility, CEBAF, Newport News, Virginia, USA (1993), p. 307.
112. T. Sugimoto, M. Yoshida, K. Yamaguchi, K. Sugawara, Y. Shiohara, S. Tanaka: Appl. Phys. Lett. **57**, 928 (1990).
113. L. Ranno, R. M. Defourneau, J. P. Enard, J. Perriere, D. Martinez-Garcia: J. Alloys & Compounds **195**, 251 (1993).
114. P. Wagner, F. Hillmer, U. Frey, T. Kluge, H. Adrian, T. Steinborn, A. Elschner, L. Ranno, M. Siegel, I. Heyvaert, Y. Bruynseraede: In *Applied Superconductivity*, ed. by H. C. Freyhardt (DGM, Oberursel, 1993), p. 579.
115. J. N. Eckstein, I. Bozovic, G. F. Virshup: In *Superconducting Superlattices and Multilayers*, Proc. SPIE 2157 (1994).
116. R. B Hammond, G. V. Negrete, L. C. Bourne, D. D. Strother, A. H.Cardona, M. M. Eddy: Appl. Phys. Lett. **57**, 825 (1990).
117. a. W. L. Holstein, L. A. Parisi, C. Wilker, R. B. Flippen: Appl. Phys. Lett. **60**, 2014 (1992).
    b. D. W. Face, F. M. Pellicone, R. J. Small, L. Bao, M. S. Warrington, C. Wilker: IEEE Trans. Appl. Supercond. **9**, June 1999.
118. S. M. Morley, A. P. Jenkins, L.Y. Su, M. J. Adams, D. Dew-Hughes, C. R. M. Grovenor: IEEE Trans. Appl. Supercond. **3**, 1753 (1993).
119. A. Piehler, J. P. Ströbel, N Reschauer, R. Löw, R. Schönberger, K. F. Renk, M. Kraus, J. Daniel, G. Saemann-Ischenko: In *Applied Superconductivity*, ed. by H. C. Freyhardt (DGM, Oberursel, 1993), p. 583.
120. Z. G. Ivanov, H. Olin, E. Ohlsson, E. A. Stepantsov, A. Tzalenchuk, T. Claeson: In *Applied Superconductivity*, ed. by H. C. Freyhardt (DGM, Oberursel, 1993), p. 591.

121. M. Manzel, H. Bruchlos, M. Kuhn, M. Klinger, J. H. Hinken: In *Applied Superconductivity*, ed. by H. C. Freyhardt (DGM, Oberursel, 1993), p. 995.
122. W. L. Holstein, L. A. Parisi, Z.-Y. Shen, C. Wilker, M. S. Brenner, J. S. Martens: J. Supercond. **6**, 191 (1993).
123. M. Manzel, S. Huber, H. Bruchlos, M. Köhler, W. Diete, S. Hensen, G. Müller, J. Keppler, C. Neumann, M. Klauda: In *Applied Superconductivity*, Inst. Phys. Conf. Ser. No 148, ed. by D. Dew-Hughes (IOP, Bristol, 1995), p. 1155.
124. M. Hagen, M. Hein, N. Klein, A. Michalke, G. Müller, H. Piel, R. W. Röth, F. M. Mueller, H. Sheinberg, J. L. Smith: J. Magn. Magn. Mat. **68**, L1 (1987).
125. a. D. Opie, H. Schone, M. Hein, G. Müller, H. Piel, D. Wehler, V. Folen, S. Wolf: IEEE Trans. Magn. **27**, 2944 (1991).
    b. D. B. Opie, H. E. Schone, M. Hein, G. Müller, H. Piel, H. P. Schneider, V. Folen, A. Frank, W. M. Golding, S. Wolf: Proc. 45th Annual Symp. on Frequency Control, Los Angeles, 1991.
126. P. Chaudari, J. Mannhart, D. Dimos, C. C. Tsuei, C. C. Chi, M. M. Oprysko, M. Scheuermann: Phys. Rev. Lett. **60**, 1653 (1988).
127. a. D. Dimos, P. Chaudhari, J. Mannhart, F. K. LeGoues: Phys. Rev. Lett. **61**, 219 (1988).
    b. D. Dimos, P. Chaudhari, J. Mannhart: Phys. Rev. B **41**, 4038 (1990).
128. T. L. Hylton, A. Kapitulnik, M. R. Beasley, J. P. Carini, L. Drabeck, G. Gruner: Appl. Phys. Lett. **53**, 1343 (1988).
129. O. Ishii, T. Konaka, M. Sato, Y. Koshimoto: Jap. J. Appl. Phys. **29**, L1075 and L2177 (1990).
130. N. McN. Alford, T. W. Button, M. J. Adams, S. Hedges, B. Nicholson, W. A. Phillips: Nature **349**, 680 (1991).
131. J. Wosik, L. M. Xie, R. Chau, A. Samaan, J. C. Wolfe, V. Sevamanickam, K. Salama: Phys. Rev. B **47**, 8968 (1993).
132. P. P. Nguyen, D. E. Oates, G. Dresselhaus, M. S. Dresselhaus: Phys. Rev. B **48**, 6400 (1993).
133. a. M. Hein, F. Hill, G. Müller, H. Piel, H. P. Schneider, M. Strupp: IEEE Trans. Appl. Supercond. **3**, 1745 (1993).
    b. M. Hein, S. Kraut, E. Mahner, G. Müller, D. Opie, L. Ponto, D. Wehler, M. Becks, U. Klein, M. Peiniger: J. Supercond. **3**, 323 (1990).
    c. M. Hein, G. Müller, H. Piel, L. Ponto, M. Becks, U. Klein, M. Peiniger: J. Appl. Phys. **66**, 5940 (1989).
    d. M. Hein: Dissertation, University of Wuppertal, Report WUB-DIS 92-2 (1992).
134. J. R. Clem: Physica C **153–155**, 50 (1988).
135. R. Gross, P. Chaudari, D. Dimos, A. Gupta, G. Koren: Phys. Rev. Lett. **64**, 228 (1990).
136. V. Ambegaokar and B. I. Halperin: Phys. Rev. Lett. **22**, 1364 (1969).
137. U. Welp, W. K. Kwok, G. W. Crabtree, K. G. Vandervoort, J. Z. Liu: Phys. Rev. Lett. **62**, 1908 (1989).
138. a. D. A. Bonn, Kuan Zhang, S. Kamal, Ruixing Liang, P. Dosanjh, W. N. Hardy, C. Kallin, A. J. Berlinsky: Reply: Phys. Rev. Lett. **72**, 1391 (1994).
    b. D. A. Bonn, R. Liang, T. M. Risemann, D. J. Baar, D. C. Morgan, K. Zhang, P. Dosanjh, T. L. Duty, A. MacFarlane, G. D. Morris, J. H. Brewer, W. N. Hardy: Phys. Rev. B **47**, 11 314 (1993).
139. A. Porch, J. R. Powell, M. J. Lancaster, J. A. Edwards, R. G. Humphreys: IEEE Trans. Appl. Supercond. **3**, 1987 (1995).
140. U. Dähne, N. Klein, H. Schulz, N. Tellmann, K. Urban: In *Applied Superconductivity*, Inst. Phys. Conf. Ser. No 148, ed. by D. Dew-Hughes (IOP, Bristol, 1995), p. 1095.

141. A. A. Golubov, M. R. Trunin, D. Wehler, J. Dreibholz, G. Müller, H. Piel: Physica C **213**, 139 (1993).
142. a. N. Klein, N. Tellmann, S. A. Wolf, V. Z. Kresin: J. Supercond. **7**, 459 (1994).
    b. N. Klein, U. Poppe, N. Tellmann, H. Schulz, W. Evers, U. Dähne, K. Urban: IEEE Trans. Appl. Supercond. **3**, 1102 (1993).
    c. N. Klein, N. Tellmann, H. Schulz, K. Urban, S. A. Wolf, V. Z. Kresin: Phys. Rev. Lett. **71**, 3355 (1993).
143. a. S. M. Anlage and D. H. Wu: J. Supercond. **5**, 395 (1992).
    b. S. M. Anlage, B. W. Langley, G. Deutcher, J. Halbritter, M. R. Beasley: Phys. Rev. B **44**, 9764 (1991).
144. V. A. Gasparov, M. R. Mkrtchyan, M. A. Obolensky, A.V. Bondarenko: Physica C **231**, 197 (1994).
145. M. A. Hein, A. Cassinese, S. Hensen, T. Kaiser, G. Müller, M. Perpeet: J. Supercond. **12**, 129 (1999).
146. S. Orbach, S. Hensen, G. Müller, H. Piel, M. Lippert, G. Saemann-Ischenko, S. A. Wolf: J. Alloys & Compounds **195**, 555 (1993).
147. W. Diete, M. Getta, M. Hein, T. Kaiser, G. Müller, H. Piel, H. Schlick: IEEE Trans. Appl. Supercond. **7**, 1236 (1997).
148. D. A. Bonn, S. Kamal, Kuan Zhang, Ruixing Liang, D. J. Baar, E. Klein, W. N. Hardy: Phys. Rev. B **50**, 4051 (1994).
149. J. Mao, D. H. Wu, J. L. Peng, R. L. Greene, S. M. Anlage: Phys. Rev. B **51**, 3316 (1995).
150. R. C. Taber, P. Merchant, R. Hiskes, S. A. DiCarolis, M. Narbutovskih: J. Supercond. **5**, 371 (1992).
151. N. Klein, G. Müller, H. Piel, J. Schurr: IEEE Trans. Magn. **25**, 1362 (1989).
152. P. Kneisel, R. W. Röth, H. G. Kürschner: Proc. 7th Workshop on RF Superconductivity, ed. by B. Bonin, Gif-sur-Yvette, France (1995), Vol. II, p. 449.
153. J. R. Delayen: Proc. 3rd Workshop on RF Superconductivity, ed. by K. W. Shepard, Argonne National Laboratory, Report ANL-PHY-88-1 (1988), p. 469.
154. B. Avenhaus: PhD thesis, University of Birmingham (1996).
155. J. P. Turneaure, J. Halbritter, H.A. Schwettmann: J. Supercond. **4**, 341 (1991).
156. A. Gladun, N. Cherpak, A. Gippius, S. Hensen, M. Lenkens, G. Müller, S. Orbach, H. Piel: Cryogenics **32**, 1071 (1992).
157. T. Jacobs, S. Sridhar, C. T. Rieck, K.Scharnberg, T.Wolf, J. Halbritter: J. Phys. Chem. Solids **56**, 1945 (1995).
158. A. Porch, M. J. Lancaster, R. G. Humphreys, N. G. Chew: IEEE Trans. Appl. Supercond. **3**, 1719 (1993).
159. M. Prohammer and J. P. Carbotte: Phys. Rev. B **43**, 5370 (1991).
160. S. Orbach-Werbig, A. Golubov, S. Hensen, G. Müller, H. Piel: Physica C **235–240**, 1823 (1994).
161. A. P. Mourachkine and A. R. F. Barel: In *Studies of High-Temperature Superconductors*, ed. by A. Narlikar (Nova Sciences, New York, 1996), Vol. 17, p. 221.
162. H. Piel, M. Hein, N. Klein, U. Klein, A. Michalke, G. Müller, L. Ponto: Physica C **153–155**, 1604 (1988).
163. N. Klein, G. Müller, H. Piel, B. Roas, L. Schultz, U. Klein, M. Peiniger: Appl. Phys. Lett. **54**, 757 (1989).
164. K. Zhang, D. A. Bonn, R. Liang, D. J. Baar, W. N. Hardy: Appl. Phys. Lett. **62**, 3019 (1993).
165. F. Gao, J. F. Whitaker, C. Uher, S. Y. Hou, J. M. Phillips: IEEE Trans. Appl. Supercond. **5**, 1970 (1995).

166. a. M. Nuss, K. W. Goosen, P. M. Mankiewich, M. L. O'Malley: Appl. Phys. Lett. **58**, 2561 (1991).
    b. M. C. Nuss, K. W. Goosen, P. M. Mankiewich, M. L. O'Malley, J. L. Marshall, R. E. Howard: IEEE Trans. Magn. **27**, 863 (1991).
167. a. N. G. Chew, J. A. Edwards, R. G. Humphreys, J. S. Satchell, S. W. Goodyear, B. Dew, N. J. Exon, S. Hensen, M. Lenkens, G. Müller, S. Orbach-Werbig: IEEE Trans. Appl. Supercond. **5**, 1167 (1995).
    b. N. G. Chew, S. W. Goodyear, J. A. Edwards, J. S. Satchell, S. E. Blenkinsop, R. G. Humphreys: Appl. Phys. Lett. **57**, 2016 (1990).
168. a. D. A. Bonn, S. Kamal, R. Liang, W. N. Hardy, C. C. Homes, D. N. Basov, T. Timusk: Czech. J. Phys. **46**, 3195 (1996).
    b. W. N. Hardy, S. Kamal, D. A. Bonn: NATO-ASI Series (Plenum, New York, 1998).
169. H. Srikanth, B. A. Willemsen, T. Jacobs, S. Sridhar, A. Erb, E. Walker, R. Flükiger: Phys. Rev. B **55**, R14 733 (1997).
170. M. R. Trunin: J. Supercond. **11**, 381 (1998).
171. A. A. Golubov, M. R. Trunin, A. A. Zhukov, O. V. Dolgov, S. V. Shulga: JETP Lett. **62**, 496 (1995).
172. S. D. Adrian, M. E. Reeves, S. A. Wolf, V. Z. Kresin: Phys. Rev. B **51**, 6800 (1995).
173. S. Orbach-Werbig, A. A. Golubov, S. Hensen, G. Müller, H. Piel: Physica C **235–240**, 1823 (1994).
174. a. A. A. Golubov, M. R.Trunin, A. A. Zhukov, O. V. Dolgov, S. V. Shulga: J. Phys. I France **6**, 2275 (1996).
    b. A. A. Golubov, O. V. Dolgov, E. G. Maksimov, I. I. Mazin, S. V. Shulga: Physica C **235–240**, 2383 (1994).
175. A. Virosztek and J. Ruvalds: Phys. Rev. B **42**, 4064 (1990).
176. A. A. Golubov, O. V. Dolgov, S. V. Shulga, M. R. Trunin, A. A. Zhukov: unpublished (1997).
177. Ya. G. Ponomarev, C. S. Khi, K. K. Uk, M. V. Sudakova, S. N. Tchesnokov, M. V. Lomonosov, M. A. Lorenz, M. A. Hein, G. Müller, H. Piel, B. A. Aminov, A. Krapf, W. Kraak: Physica C **315**, 85 (1999).
178. R. Kleiner, A. S. Katz, A. G. Sun, R. Summer, D. A. Gajewski, S. H. Han, S. I. Woods, E. Dantsker, B. Chen, K. Char, M. B. Maple, R. C. Dynes, J. Clarke: Phys. Rev. Lett. **76**, 2161 (1996); and M. Mößle and R. Kleiner, Phys. Rev. B **59**, 4486 (1999).
179. C. O'Donovan and J. P. Carbotte: Phys. Rev. B **52**, 16 208 (1995).
180. R. Combescot and X. Leyronas: Phys. Rev. Lett. **75**, 3732 (1995).
181. R. Modre, I. Schürrer, E. Schachinger: Phys. Rev. B **57**, 5496 (1998).

Chapter 3

1. J. Halbritter: J. Supercond. **10**, 91 (1997).
2. A. M. Portis: *Electrodynamics of High-Temperature Superconductors*, Lecture Notes in Physics, Vol. 48 (World Scientific, Singapore, 1993).
3. a. M. Golosovsky, M. Tsindlekht, D. Davidov: Supercond. Sci. Technol. **9**, 1 (1996).
   b. M. Golosovsky, M. Tsindlekht, H. Chayet, D. Davidov: Phys. Rev. B **50**, 470 (1994).
4. a. J. R. Powell: PhD thesis, University of Birmingham (1996).
   b. I. S. Ghosh, L. F. Cohen, V. Fry, T. Tate, A. D. Caplin, J. C. Gallop, S.Sievers, R. Somekh, S. Hensen, M. Lenkens: IEEE Trans. Appl. Supercond. **5**, 1756 (1995).

5. D. H. Wu, J. C. Booth, S. M. Anlage: Phys. Rev. Lett. **75**, 525 (1995).
6. a. N. Belk, D. E. Oates, D. A. Feld, G. Dresselhaus, M. S. Dresselhaus: Phys. Rev. B **56**, 11966 (1997); Phys. Rev. B **53**, 3459 (1996).
   b. P. P. Nguyen, D. E. Oates, G. Dresselhaus, M. S. Dresselhaus, A. C. Anderson: Phys. Rev. B **51**, 6686 (1995).
   c. S. Revenaz D. E. Oates, D. Labbé-Lavigne, G. Dresselhaus, M. S. Dresselhaus: Phys. Rev. B **50**, 1178 (1994).
7. J. Owilaei and S. Sridhar: Phys. Rev. Lett. **69**, 3366 (1992).
8. V. L. Ginzburg and L. D. Landau: Zh. Eksperim. i Teor. Fiz. **20**, 1064 (1950) [English transl.: *Men of Physics: L. D. Landau*, Vol. 1, ed. by D. TerHaar (Pergamon, New York, 1965), p. 138].
9. H. Padamsee, D. Proch, P. Kneisel, J. Mioduszewski: IEEE Trans. Magn. **17**, 947 (1981).
10. M. A. Hein, W. Diete, S. Hensen, T. Kaiser, M. Lenkens, G. Müller, H. Piel, H. Schlick: IEEE Trans. Appl. Supercond. **7**, 1264 (1997).
11. H. Padamsee: IEEE Trans. Magn. **19** 1322 (1983).
12. G. Müller: Proc. 3rd Workshop on RF Superconductivity, ed. by K. W. Shepard, Argonne Nat. Lab., report ANL-PHY-88-1 (1988), p. 331.
13. L. Kramer: Phys. Rev. **170**, 475 (1968).
14. C. P. Bean and J. P. Livingston: Phys. Rev. Lett. **12**, 14 (1964).
15. J. Halbritter: J. Supercond. **8**, 691 (1995).
16. A. Barone and G. Paterno: *Physics and Applications of the Josephson Effect* (Wiley, New York, 1982).
17. M. Nahum, S. Verghese, P. L. Richards, K. Char: Appl. Phys. Lett. **59**, 2034 (1991).
18. T. B. Samailova: Supercond. Sci. Technol. **8**, 259 (1995).
19. R. Tidecks: *Current-induced nonequilibrium phenomena in quasi-one dimensional superconductors* (Springer, Berlin, Heidelberg, NewYork, 1990).
20. I. O. Kulik: Zh. Eksp. Teor. Fiz. **57**, 600 (1969).
21. P. Bura: Appl. Phys. Lett. **8**, 155 (1966).
22. a. E. Keskin, K. Numssen, J. Halbritter: IEEE Trans. Appl. Supercond. **9**, June (1999).
    b. J. Halbritter: J. Appl. Phys. **68**, 6315 (1990); **71**, 339 (1992).
23. a. M. Hein, H. Piel, M. Strupp, M. R. Trunin, A. M. Portis: J. Magn. Magn. Mat. **104–107**, 529 (1992).
    b. A. M. Portis, D. W. Cooke, E. R. Gray, P. N. Arendt, C. L. Bohn, J. R. Delayen, C. T. Roche, M. Hein, N. Klein, G. Müller, S. Orbach, H. Piel: Appl. Phys. Lett. **58**, 307 (1991).
24. M. A. Golosovsky, H. J. Snortland, M. R. Beasley: Phys. Rev. B **51**, 6462 (1995).
25. A. Andreone, C. Aruta, A. Cassinese, A. DiChiara, F. Palomba, R. Vaglio: Extended Abstracts, Internat. Supercond. Electronics Conf. ISEC'97, ed. by H. Koch, Berlin, Germany (1997).
26. L. P. Gorkov: Zh. Eksp. Teor. Fiz. **36**, 1918 (1959); **37**, 1407 (1959).
27. D. R. Tilley and J. Tilley: *Superfluidity and Superconductivity* (Adam Hilger, Bristol, Boston, 1986).
28. L. D. Landau and E. M. Lifschitz: *Statistische Physik*, Teil 1, Lehrbuch der theoretischen Physik, (Akadamie, Berlin, 1979).
29. M. Tinkham: *Introduction to Superconductivity*, (McGraw-Hill, New York, 1975).
30. D. Xu, S. K. Yip, J. A. Sauls: Phys. Rev. B **51**, 16233 (1995).
31. C. Caroli, P. G. deGennes, J. Matricon: Phys. Lett. **9**, 307 (1964).
32. J. R. Clem: J. Low-Temp. Phys. **18**, 427 (1975).

33. R. D. Sherman: Phys. Rev. B **8**, 173 (1973).
34. T. vanDuzer and C. W. Turner: *Principles of Superconductive Devices and Circuits* (Elsevier North-Holland, New York, 1981).
35. C.-W. Lam, D. M. Sheen, S. M. Ali, D. E. Oates: IEEE Trans. Appl. Supercond. **2**, 58 (1992).
36. M. Cyrot: Rep. Prog. Phys. **36**, 103 (1973).
37. T. Dahm and D. J. Scalapino: J. Appl. Phys. **81**, 2002 (1997); Appl. Phys. Lett. **69**, 4248 (1996).
38. A. L. Fetter and P. C. Hohenberg: In *Superconductivity*, ed. by R. D. Parks (Marcel Dekker, New York, 1969), Vol. 2, p. 817.
39. V. L. Ginzburg: Zh. Eksperim. Teor. Fiz. **34**, 113 (1958) [English transl.: Soviet Phys. JETP **7**, 78 (1958)].
40. P. Kneisel, R. W. Röth, H. G. Kürschner: Proc. 7th Workshop RF superconductivity, ed. by B. Bonin, Gif-sur-Yvette, France (1995), Vol. II, p. 449.
41. P. G. de Gennes: Solid State Commun. **3**, 127 (1965).
42. J. Matricon and D. Saint-James: Phys. Lett. **24A**, 241 (1967).
43. H. J. Fink and A. G. Presson: Phys. Lett. **25A**, 378 (1967).
44. H. J. Fink, D. S. McLachlan, B. Rothberg-Bibby: In *Progress in Low Temperature Physics*, ed. by D. F. Brewer (North-Holland, 1978).
45. A. J. Dolgert, S. J. DiBartolo, A. T. Dorsey: Phys. Rev. B **53**, 5650 (1996); and Erratum: Phys. Rev. B **56**, 2883 (1997).
46. V. P. Galaiko: Zh. Eksperim. Teor. Fiz. **50**, 717 (1966) [English transl.: Soviet Phys. JETP **23**, 475 (1966)].
47. A. A. Abrikosov: Zh. Eksperim. Teor. Fiz. **32**, 1442 (1957) [English transl.: Soviet Phys. JETP **5**, 1174 (1957)].
48. E. H. Brandt: Rep. Prog. Phys. **58**, 1465–1594 (1995).
49. C. Jooss, A. Forkl, H. Kronmüller: Physica C **268**, 87 (1996).
50. C. R. Hu: Phys. Rev. B **6**, 1756 (1972).
51. J. L. Harden and V. Arp: Cryogenics **3**, 105 (1963).
52. a. J. I. Gittleman and B. Rosenblum: J. Appl. Phys. **39**, 2617 (1968); Phys. Rev. Lett. **16**, 7345 (1966); Proc. IEEE **52**, 1138 (1964).
    b. M. Cardona, J. I. Gittleman, B. Rosenblum: Phys. Lett. **17**, 92 (1965).
    c. M. Cardona, G. Fischer, B. Rosenblum: Phys. Rev. Lett. **12**, 101 (1964).
53. W. H. Hartwig and C. Passow: In *Applied Superconductivity*, Vol. II, ed. by V. L. Newhouse (Academic, New York, 1975).
54. a. H. Suhl: Phys. Rev. Lett. **14**, 226 (1965).
    b. M. J. Stephen and H. Suhl: Phys. Rev. Lett. **13**, 797 (1964).
55. a. A. R. Strnad, C. F. Hempstead, Y. B. Kim: Phys. Rev. Lett. **13**, 794 (1964).
56. J. Bardeen and M. J. Stephen: Phys. Rev. **140**, A1197 (1965).
57. Y. Shapira and L. J. Neuringer: Phys. Rev. **154**, 375 (1965).
58. A. M. Campbell: J. Phys. C **2**, 1492 (1969); **C4**, 3186 (1971).
59. a. M. W. Coffey and J. R. Clem: Phys. Rev. B **45**, 9872, 10527 (1992).
    b. Phys. Rev. Lett. **67**, 386 (1991).
60. E. H. Brandt: Phys. Rev. Lett. **67**, 2219 (1991).
61. C. P. Bean: Rev. Mod. Phys. **36**, 31 (1964); Phys. Rev. Lett. **8**, 250 (1962).
62. S. Sridhar: Appl. Phys. Lett. **65**, 1054 (1994).
63. J. McDonald, J. R. Clem, D. E. Oates: Phys. Rev. B **55**, 11823 (1997).
64. F. Auracher and T. van Duzer: J. Appl. Phys. **44**, 848 (1973).
65. a. M. Perpeet, M. A. Hein, H. Piel, S. Beuven, M. Siegel, E. Sodtke: In *Applied Superconductivity*, Inst. Phys. Conf. Ser. No 148, ed. by D. Dew-Hughes (IOP, Bristol, 1995), p. 1255.
    b. M. A. Hein, S. Beuven, M. Gottschlich, M. Perpeet, H. Piel, M. Siegel: J. Supercond. **9**, 241 (1996).

c. M. A. Hein, M. Strupp, H. Piel, A. M. Portis, R. Gross: J. Appl. Phys. **75**, 4581 (1994).
66. a. Y. M. Habib, D. E. Oates, G. Dresselhaus, M. S. Dresselhaus, L. R. Vale, R. H. Ono: Appl. Phys. Lett. **73**, 2200 (1998); D. E. Oates, Y. M. Habib, C. J. Lehner, L. R. Vale, R. H. Ono, G. Dresselhaus, M. S. Dresselhaus: IEEE Trans. Appl. Supercond. **9**, June 1999.
    b. Y. M. Habib, D. E. Oates, G. Dresselhaus, M. S. Dresselhaus: IEEE Trans. Appl. Supercond. **7**, 2553 (1997).
    c. D. E. Oates, P. P. Nguyen, Y. Habib, G. Dresselhaus, M. S. Dresselhaus, G. Koren, E. Polturak: Appl. Phys. Lett. **68**, 705 (1996).
67. a. L. M. Xie, J. Wosik, J. C. Wolfe: Phys. Rev. **B54**, 15494 (1996).
    b. J. Wosik, L. M. Xie, M. F. Davis, N. Tralshawala, P. Gierlowski, J. H. Miller, Jr.: In *High-Temperature Microwave Superconductors and Applications*, ed. by J. Hodge: Proc. SPIE **2559**, 76 (1995).
68. N. Ogawa, M. Sigrist, K. Ueda: J. Phys. Soc. Japan **65**, 545 (1996).
69. A. A. Abrikosov: Soviet Physics JETP **19**, 988 (1964).
70. L. Civale, T. K. Worthington, A. Gupta: Phys. Rev. B **48**, 7576 (1993).
71. R. A. Klemm and J. R. Clem: Phys. Rev. B **21**, 1868 (1980).
72. M. W. Denhoff and Suso Gygax: Phys. Rev. B **25**, 4479 (1982).
73. W. Lawrence and S. Doniach: Proc. 12th Internat. Conf. on Low-Temperature Physics, ed. by Eizo Kanda (Academic, Kyoto, 1971), p. 361.
74. A. Sudbo, E. H. Brandt, D. A. Huse: Phys. Rev. Lett. **71**, 1451 (1993).
75. J. A. Osborn: Phys. Rev. **67**, 351 (1945).
76. R. A. Klemm: J. Low-Temp. Phys. **39**, 589 (1981).
77. R. A. Klemm: Phys. Rev. B **47**, 14630 (1993).
78. G. E. Volovik: JETP Lett. **58**, 469 (1993).
79. P. I. Soininen, C. Kallin, A. J. Berlinsky: Phys. Rev B **50**, 13883 (1994).
80. Y. Ren, J.-H. Xu, C. S. Ting: Phys. Rev. Lett. **74**, 3680 (1995).
81. R. M. Easson, P. Hlawiczka, J. M. Ross: Phys. Lett. **20**, 465 (1966).
82. J. Halbritter: KFZ Karlsruhe, Externer Bericht 3/69-6 (1969).
83. A. V. Gurevich and R. G. Mints: Rev. Mod. Phys. **59**, 941 (1987).
84. a. M. A. Hein, C. Bauer, W. Diete, S. Hensen, T. Kaiser, G. Müller, H. Piel: J. Supercond. **10**, 109 (1997).
    b. C. Bauer: University of Wuppertal, External report WUD 97-11 (1997).
85. R. R. Mansour, B. Jolley, S. Ye, F. S. Thomson, V. Dokas: IEEE Trans. Microwave Theory Tech. **44**, 1322 (1996).
86. M. Rabinowitz: J. Appl. Phys. **42**, 88 (1971); Appl. Phys. Lett. **19**, 73 (1971).
87. H. Padamsee: IEEE Trans. Magn. **19**, 1322 (1983).
88. J. Tückmantel: CERN/EF/RF 84-6 (1984).
89. a. H.-G. Kürschner: University of Wuppertal, External report WUD 92-9 (1992).
    b. G. Müller: Proc. 3rd Workshop on RF Superconductivity, ed. by K. W. Shepard, Argonne Nat. Lab., Argonne, report No. ANL-PHY-88-1 (1988), p. 331.
    c. H. Elias: University of Wuppertal, External report WUD 85-12 (1985).
90. a. A. N. Reznik, A. A. Zharov, M. D. Chernobrovtseva: IEEE Trans. Appl. Supercond. **5**, 2579 (1995).
    b. A. N. Reznik, A. I. Smirnov, M. D. Chernobrovtseva: Supercond. Phys. Chem. Technol. **6**, 186 (1993).
91. a. J. Wosik, L. M. Xie, K. Nesteruk, D. Li, J. H. Miller, Jr., S. A. Long: J. Supercond. **10**, 97 (1997).
    b. J. Wosik, L.-M. Xie, D. Li, J. H. Miller, Jr., S. A. Long: Czech. J. Phys. **46**, 1133 (1996).

92. H. S. Carslaw and J. C. Jaeger: *Conduction of Heat in Solids* (Clarendon, Oxford, 1973).
93. E. F. Nogotov: *Applications of Numerical Heat Transfer* (McGraw-Hill, New York, 1978).
94. W. H. Press, B. P. Flannery, S. A. Teukolsky, W. T. Vetterling: *Numerical Recipes in C* (Cambridge University Press, Cambridge, 1988).
95. J. D. Jackson: *Classical Electrodynamics*, (Wiley, New York, 1975).
96. See review by C. Uher: J. Supercond. **3**, 337 (1990). For the reported simulations, Ref. 113 of this review was used: V. Florentiev, A. Inyushkin, A. Taldenkov, O. Melnikov, A. Bykov: In *Progress in High-Temperature Superconductivity*, ed. by R. Nicolsky (World Scientific, Singapore, 1990), p. 462.
97. S. J. Hagen, Z. Z. Wang, N. P. Ong: Phys. Rev. **B40**, 9389 (1989).
98. M. Nahum, S. Verghese, P. L. Richards, K. Char: Appl. Phys. Lett. **59**, 2034 (1991).
99. S. Zeuner, M. Lengfellner, J. Betz, K. F. Renk, W. Prettl: Appl. Phys. Lett., **61**, 973 (1992).
100. A. Bock: Phys. Rev. B, **51**, 15506 (1995).
101. P. C. Michael, I. U. Trefny, B. Yarar: J. Appl. Phys. **72**, 107 (1992).
102. Landolt-Börnstein: *Numerical data and functional relationships in science and technology: New Series*, ed. by K. H. Hellwege (Springer, Berlin, New York, 1981), group III: Crystal and Solid State Physics, Vol. 16a.
103. K. Stierstadt: *Physik der Materie* (VCH, Weinheim, 1989), Chap. 16.
104. W. A. Little: Can. J. Phys. **37**, 334 (1959).
105. F. W. de Wette, A. D. Kulkarny, J. Prade, U. Schröder, W. Kress: Phys. Rev. B **42**, 6707 (1990).
106. W. C. Lee, K. Sun, L. L. Miller, D. C. Johnston, R. A. Klemm, S. Kim, R. A. Fisher, N. E. Phillips: Phys. Rev. B **43**, 463 (1991).
107. Specific Heat – Nonmetallic Solids: In *Thermophysical Properties of Matter: The TPRC data series; a comprehensive compilation of data*, Thermophysical Properties Research Center (TPRC), Purdue University, ed. by Y. S. Touloukian and E. H. Buyco (IFI/Plenum, New York, 1970).
108. H. A. Schwettmann: IEEE Trans. Nucl. Science **22**, 1118 (1975).
109. S. Isagawa and K. Isagawa: Cryogenics **20**, 677 (1980).
110. B. Fernandes and R. Parodi: Cryogenics **24**, 433 (1984).
111. H. Tautz: *Wärmeleitung und Temperaturausgleich: Die mathematische Behandlung in-stationärer Wärmeleitungsprobleme mit Hilfe von Laplace-Transformationen* (Chemie, Weinheim, 1971).
112. M. A. Hein: In *Studies of High-Temperature Superconductors*, ed. by A. Narlikar, (Nova Sciences, New York, 1996), Vol. 18, p. 141.
113. H. Safa: Proc. 7th Workshop on RF superconductivity, ed. by B. Bonin, Gif-Sur-Yvette, France, (1995), Vol. II, p. 413.
114. X. Cao and D. Proch: Proc. 5th Workshop on RF Superconductivity, ed. by D. Proch, DESY M-92-01 (Deutsches Elektronen Synchrotron, Hamburg, Germany, 1992), Vol. 2, p. 721.
115. J. Wosik, L. M. Xie, D. McFall, T. Hogan, S. A. Long: IEEE Trans. Appl. Supercond. **9**, June 1999.
116. R. B. Flippen: Phys. Lett. **17**, 193 (1965).
117. J. Halbritter: Proc. 1972 Appl. Supercond. Conf., Annapolis, Maryland (1972), p. 662
118. E. Schuon und H. Wolf: *Nachrichten-Meßtechnik*: In *Nachrichtentechnik, Bd. 9*, ed. by H. Marko (Springer, Berlin, Heidelberg, New York, 1981).
119. T. Dahm and D. J. Scalapino: to be published in IEEE Trans. Appl. Supercond. (1999).

120. B. A. Willemsen, T. Dahm, B. H. King, D. J. Scalapino: IEEE Trans. Appl. Supercond. **9**, June 1999.
121. Z.-Y. Shen: *High-Temperature Superconducting Microwave Circuits* (Artech House, Boston, London, 1994).
122. O. G. Vendik and I. B. Vendik: IEEE Microwave Theory Tech. **45**, 173 (1997).
123. O. G. Vendik and I. B. Vendik: IEEE Microwave Theory Tech. **46**, 851 (1998).
124. T. C. L. G. Sollner, J. P. Sage, D. E. Oates: Appl. Phys. Lett. **68**, 1003 (1996).
125. P. Seidel, M. A. Hein, H. J. Chaloupka: unpublished (1995).
126. a. H. J. Chaloupka: In *Applications of Superconductivity*, NATO-ASI Series, ed. by H. Weinstock (Kluwer, Dordrecht, 1999).
    b. J. Mitterer, M. Jeck, H. J. Chaloupka: unpublished (1994).
127. H. Chaloupka, M. Jeck, S. Kolesov, O. Vendik: Proc. 22nd Europ. Microwave Conf., Helsinki (1992), p. 189.
128. D. P. Choudhury, B. A. Willemsen, J. S. Derov, S. Sridhar: IEEE Trans. Appl. Supercond. **7**, 1260 (1997).
129. a. D. E. Oates, Y. Habib, C. Lehner, J. Herd, R. Ono, L. Vale: unpublished (1997).
    b. J. S. Herd, D. E. Oates, J. Halbritter: IEEE Trans. Appl. Supercond. **7**, 1299 (1997).
130. L. F. Cohen, A. Cowie, J. C. Gallop, I. S. Ghosh, I. N.Goncharov: J. Supercond. **10**, 85 (1997).
131. J. H. Claassen: In *Applied Superconductivity 1997*, Inst. Phys. Conf. Ser. No 158, ed. by H. Rogalla and D. H. Blank (IOP, Bristol, 1997), Vol. 1, p. 57.
132. M. A. Hein, C. Bauer, W. Diete, S. Hensen, T. Kaiser, V. Z. Kresin, G. Müller: J. Supercond. **10**, 485 (1997).
133. A. Ya. Kirichenko, M. B. Kosmyna, A. B. Levin, N. T. Cherpak: JETP Lett. **50**, 290 (1989).
134. A. P. Kharel, A. V. Velichko, J. R. Powell, A. Porch, M. J. Lancaster, R. G. Humphreys: Phys. Rev. B **58**, 11189 (1998).
135. P. Russer: J. Appl. Phys. **43**, 2008 (1972).
136. T. L. Hylton, A. Kapitulnik, M. R. Beasley, J. P. Carini, L. Drabeck, G. Gruner: IEEE Trans. Magn. **25**, 810 (1989); Appl. Phys. Lett. **53**, 1343 (1988).
137. C. Attanasio, L. Maritato, R. Vaglio: Phys. Rev. B **43**, 6128 (1991).
138. A. M. Portis and D. W. Cooke: Supercond. Sci. Technol. **5**, S395 (1992).
139. D. E. Oates, P. P. Nguyen, G. Dresselhaus, C. W. Lam, S. M. Ali: J. Supercond. **5**, 361 (1992).
140. P. P. Nguyen, D. E. Oates, G. Dresselhaus, M. S. Dresselhaus: Phys. Rev. B **48**, 6400 (1993).
141. L. Elfassy and J. Waldram, Interdisciplinary Research Centre in Superconductivity at the University of Cambridge: private communication (1995).
142. a. J. S. Herd, D. E: Oates, H. Xin, S. J. Berkowitz: IEEE Trans. Appl. Supercond. **9**, June 1999.
    b. Y. M. Habib, C. J. Lehner, D. E. Oates, L. R. Vale, R. H. Ono, G. Dresselhaus, M. S. Dresselhaus: Phys. Rev. B **57**, 13833 (1998).
143. A. F. G. Wyatt, V. M. Dmitriev, W. S. Moore, F. W. Sheard: Phys. Rev. Lett. **16**, 1166 (1966).
144. A. H. Dayem and J. J. Wiegand: Phys. Rev. **155**, 419 (1967).
145. L. T. Aslamazov and A. I. Larkin: Sov. Phys. JETP **47**, 1136 (1978).
146. J. C. Gallop, A. L. Cowie, L. F. Cohen: In *Applied Superconductivity 1997*, Inst. Phys. Conf. Ser. No 158, ed. by H. Rogalla and D. Blank (IOP, Bristol, 1997), Vol. 1, p. 65.
147. A. V. Velichko, preprint cond-mat 9903123 (1999).
148. S. Sridhar and J. E. Mercereau: Phys. Rev. B **34**, 203 (1986).

149. Y. N. Ovchinnikov: Zh. Eksp. Teor. Fiz. **59**, 128 (1971) [English transl.: Sov. Phys. JETP 32, **72** (1971)].
150. Y. N. Ovchinnikov and V. Z. Kresin: Phys. Rev. B **54**, 1251 (1996).
151. M. Hein, A. Cassinese, S. Hensen, T. Kaiser, G. Müller, M. Perpeet: J. Supercond. **12**, 129 (1999).
152. V. Z. Kresin: Lawrence Berkeley Laboratory, private communication (1998).
153. R. Shaw, B. Rosenblum, F. Bridges: IEEE Trans. Magn. **13**, 811 (1977).
154. S. Ghamati, H. Suhl, W. Vernon, G. Webb: IEEE Trans. Magn. **21**, 831 (1985).
155. R. Hecht: RCA Review **25**, 453 (1964).
156. Dong-Ho Wu and S. Sridhar: Phys. Rev. Lett. **65**, 2074 (1990).
157. R. Liang, P. Dosanjh, D. A. Bonn, W. N. Hardy, A. J. Berlinsky: Phys. Rev. B **50**, 4212 (1994).
158. C. Heinzel, C. Neumann, T. Ritzi, P. Ziemann: In *High T_c Superconductor Thin Films*, ed. by L. Correra (Elsevier, North-Holland, 1992).
159. L. Burlachkov, Y. Yeshurun, M. Konczykowski, F. Holtzberg: Phys. Rev. B **45**, 8193 (1992).

Chapter 4

1. A. M. Portis: *Electrodynamics of High-Temperature Superconductors*, Lecture Notes in Physics, Vol. 48 (World Scientific, Singapore, 1993).
2. Z.-Y. Shen: *High-Temperature Superconducting Microwave Circuits* (Artech House, Boston, London, 1994).
3. M. J. Lancaster: *Passive Microwave Device Applications of High-Temperature Superconductors* (Cambridge University Press, Cambridge, 1997).
4. N. Newman and W. G. Lyons: J. Supercond. **6**, 119 (1993).
5. M. Hein: In *Studies of High-Temperature Superconductors*, ed. by A. Narlikar (Nova Sciences, New York, 1996), Vol. 18, pp. 141–216.
6. C. Wilker, Z.-Y. Shen, P. Pang. W. L. Holstein, D. W. Face: IEEE Trans. Appl. Supercond. **5**, 1665 (1995).
7. M. Hein, S. Hensen, G. Müller, S. Orbach, H. Piel, M. Strupp, N. G. Chew, J. A. Edwards, S. W. Goodyear, J. S. Satchell, R. G. Humphreys: In *High-Temperature Superconductor Thin Films*, ed. by L. Correra (Elsevier, Amsterdam, 1992), p. 95.
8. a. R. Wördenweber, R. Einfeld, R. Kutzner, A. G. Zaitsev, M. A. Hein, T. Kaiser, G. Müller: IEEE Trans. Appl. Supercond. **9**, June 1999.
   b. W. Diete, M. Getta, M. Hein, T. Kaiser, G. Müller, H. Piel, H. Schlick: IEEE Trans. Appl. Supercond. **7**, 1236 (1997).
   c. M. A. Hein, C. Bauer, W. Diete, S. Hensen, T. Kaiser, G. Müller, H. Piel: J. Supercond. **10**, 109 (1997).
9. a. C. Zuccaro: Dissertation, RWTH Aachen, 1998.
   b. N. Klein, N. Tellmann, U. Dähne, A. Scholen, H. Schulz, G. Höfer, H. Kratz: IEEE Trans. Appl. Supercond. **5**, 2663 (1995).
10. L. F. Cohen, A. Cowie, J. C. Gallop, I. S. Ghosh, I. N. Goncharov: J. Supercond. **10**, 85 (1997).
11. J. Wosik, L. M. Xie, K. Nesteruk, D. Li, J. H. Miller, Jr., S. A. Long: J. Supercond. **10**, 97 (1997).
12. J. Mao, S. M. Anlage, J. L. Peng, R. L. Greene: IEEE Trans. Appl. Supercond. **5**, 1997 (1995).
13. A. V. Velichko, N. T. Cherpak, E. V. Izhyk, A. Ya. Kirichenko, A. V. Moroz: Supercond. Sci. Technol. **11**, 716 (1998).

14. a. A. Porch, M. J. Lancaster, R. G. Humphreys: IEEE Trans. Microwave Theory Tech. **43**, 306 (1995).
    b. B. Avenhaus, A. Porch, M. J. Lancaster, S. Hensen, M. Lenkens, S. Orbach-Werbig, G. Müller, U. Dähne, N. Tellmann, N. Klein, C. Dubourdieu, J. P. Senateur, O. Thomas, H. Karl, B. Stritzker: IEEE Trans. Appl. Supercond. **5**, 1737 (1995).
15. W. Rauch, E. Gornik, G. Sölkner, A. A. Valenzuela, F. Fox, H. Behmer: J. Appl. Phys. **73**, 1866 (1993).
16. U. Salz, S. Hofschen, R. Schneider: IEEE Trans. Appl. Supercond. **3**, 2816 (1993).
17. H. Chaloupka, M. Jeck, S. Kolesov, O. Vendik: Proc. 22nd Europ. Microwave Conf., Helsinki (1992), p. 189.
18. T. Yoshitake, H. Tsuge, T. Inui: IEEE Trans. Appl. Supercond. **5**, 2571 (1995).
19. a. B. A. Willemsen, K. E. Kihlstrom, T. Dahm, D. J. Scalapino, B. Gowe, D. A. Bonn, W. N. Hardy: Phys. Rev. B **58**, 6650 (1998).
    b. B. A. Willemsen, J. S. Derov, J. H. Silva, S. Sridhar, IEEE Trans. Appl. Supercond. **5**, 1753 (1995).
20. a. A. Cassinese, A. Andreone, C. Aruta, M. Iavarone, F. Palomba, R. Vaglio: Particle Accel. **60**, 161 (1998).
    b. A. Andreone, C. Aruta, A. Cassinese, A. DiChiara, F. Palomba, R. Vaglio: Extended Abstracts, Internat. Supercond. Electronics Conf., ed. by H. Koch, Berlin, Germany (1997), p. 239.
    c. A. Andreone, C. Cantoni, A. Cassinese, A. DiChiara: In *Applied Superconductivity*, Inst. Phys. Conf. Ser. No 148, ed. by D. Dew-Hughes (IOP, Bristol, 1995), p. 1115.
21. O. Vendik and E. Kollberg: Microwave & RF **7**, 118 (1993).
22. a. Y. M. Habib, C. J. Lehner, D. E. Oates, L. R. Vale, R. H. Ono, G. Dresselhaus, M. S. Dresselhaus: Phys. Rev. B **57**, 13833 (1998).
    b. D. E. Oates, P. P. Nguyen, G. Dresselhaus, M. S. Dresselhaus, G. Koren, E. Polturak: J. Supercond. **8**, 725 (1995).
23. T. Kaiser: Dissertation, University of Wuppertal, Report WUB-DIS 98-13 (1998).
24. T. Bollmeier, W. Biegel, B. Schey, B. Stritzker, W. Diete, T. Kaiser, G. Müller: J. Alloys and Compounds **251**, 176 (1997).
25. M. Lenkens: Dissertation, University of Wuppertal, Report WUB-DIS 96-5 (1996).
26. A. Porch, M. J. Lancaster, R. G. Humphreys, N. G. Chew: J. Alloys & Compounds **195**, 563 (1993).
27. M. Lorenz, H. Hochmuth, D. Natusch, H. Börner, K. Kreher, W. Schmitz: Appl. Phys. Lett. **68**, 3332 (1996).
28. M. Lorenz, H. Hochmuth, J. Frey, H. Börner, J. Lenzner, G. Lippold, T. Kaiser, M. A. Hein, G. Müller: In *Applied Superconductivity 1997*, Inst. Phys. Conf. Ser. No. 158, ed. by H. Rogalla and D. H. A. Blank (IOP, Bristol, 1997), Vol. 1, p. 283.
29. W. Diete, H. Chaloupka, M. A. Hein, M. Jeck, T. Kaiser, G. Müller: unpublished (1996).
30. C. Bauer: University of Wuppertal, External report WUD 97-11 (1997).
31. A. Ghis, S. Pfister, J. C. Villegier, M. Nail, J. P. Maneval: IEEE Trans. Appl. Supercond. **3**, 2136 (1993).
32. M. A. Heusinger, A. D. Semenov, R. S. Nebosis, Y. P. Gousev, K. F. Renk: IEEE Trans. Appl. Supercond. **5**, 2595 (1995).

33. F. Hegmann, D. Jacobs-Perkins, S. H. Moffat, C. C. Wang, R. A. Hughes, M. Currie, P. M. Fauchet, T. Y. Hsiang, J. S. Preston, R. Sobolewski: Appl. Phys. Lett. **67**, 285 (1995).
34. T. Kaiser, B. Aminov, A. Baumfalk, A. Cassinese, H. Chaloupka, M. Hein, S. Kolesov, H. Medelius, G. Müller, Perpeet, H. Piel, E. Wikborg: J. Supercond. **12**, 343 (1999).
35. J. McDonald, J. R. Clem, D. E. Oates: Phys. Rev. B **55**, 11823 (1997).
36. C. W. Lam, D. M. Sheen, S. M. Ali, D. E. Oates: IEEE Trans. Appl. Supercond. **2**, 58 (1992).
37. P. P. Nguyen, D. E. Oates, G. Dresselhaus, M. S. Dresselhaus, A. C. Anderson: Phys. Rev. B **51**, 6686 (1995).
38. J. S. Herd, J. Halbritter, K. G. Herd: IEEE Trans. Appl. Supercond. **5**, 1991 (1995).
39. J. Halbritter: J. Appl. Phys. **68**, 6315 (1990).
40. T. Kaiser, C. Bauer, W. Diete, M. A. Hein, J. Kallscheuer, G. Müller, H. Piel: In *Applied Superconductivity 1997*, Inst. Phys. Conf. Ser. No 158, ed. by H. Rogalla and D. H. A. Blank (IOP, Bristol, 1997), Vol. 1, p. 45.
41. R. W. Ralston, D. E. Oates, A. C. Anderson, W. G. Lyons: Extended Abstracts, Internat. Conf. Supercond. Electronics, Boulder, Colorado (1993), p. 48.
42. A. Porch, B. Avenhaus, F. Wellhöfer, P. Woodall: In *Applied Superconductivity*, Inst. Phys. Conf. Ser. No 148, ed. by D. Dew-Hughes (IOP, Bristol, 1995), p. 1039.
43. M. A. Hein: Supercond. Sci. Technol. **10**, 867 (1997).
44. G. Müller, B. Aschermann, H. Chaloupka, W. Diete, M. Getta, B. Gurzinski, M. Hein, M. Jeck, T. Kaiser, S. Kolesov, H. Piel, H. Schlick, R. Theisejans: IEEE Trans. Appl. Supercond. **7**, 1287 (1997).
45. a. R. Schwab, R. Heidinger, J. Geerk, F. Ratzel, M. Lorenz, H. Hochmuth: Proc. 23rd Internat. Conf. on Infrared and Millimeter Waves, Colchester, UK, 7–11 September 1998, ISBN 0-9533839-0-3, p. 390.
    b. R. Heidinger and R. Schwab, Forschungszentrum Karlsruhe: private communication (1997).
    c. R. Heidinger, F. Königer, G. Link: Conference Digest *15th Internt. Conf. on Infrared and Millimeter Waves*, Proc. SPIE **1514**, 184 (1990).
46. a. J. Geerk: Forschungszentrum Karlsruhe, private communication (1997).
    b. H. Kittel: Dissertation, Forschungszentrum Karlsruhe, FZKA-Report 5645 (1995).
47. M. Lorenz, H. Hochmuth, D. Natusch, G. Lippold, V. L. Svetchnikov, T. Kaiser, M. A. Hein, R. Schwab, R. Heidinger: IEEE Trans. Appl. Supercond. **9**, June 1999.
48. G. C. Liang, D. Zhang, C. F. Shih, M. E. Johansson, R. S. Withers, A. C. Anderson, D. E. Oates: IEEE Trans. Appl. Supercond. **5**, 2652 (1995).
49. S. Kolesov, H. Chaloupka, A. Baumfalk, T. Kaiser: J. Supercond. **10**, 179 (1997).
50. J. Kallscheuer: University of Wuppertal, External Report WUD 97-35 (1997).
51. J. C. Culbertson, H. S. Newman, C. Wilker: J. Appl. Phys. **84**, 2768 (1998).
52. T. Kaiser, M. A. Hein, G. Müller, M. Perpeet: Appl. Phys. Lett. **73**, 3447 (1998).
53. H. S. Newman and J. C. Culbertson: Microwave and Optical Technol. Lett. **6**, 725 (1993).
54. M. Perpeet: Dissertation, University of Wuppertal, (1999).
55. A. Andreone, A. Cassinese, A. DiChiara, M. Iavarone, F. Palomba, A. Ruosi, R. Vaglio: IEEE Trans. Appl. Supercond. **7**, 1772 (1997).

56. a. M. Perpeet, A. Cassinese, M. A. Hein, T. Kaiser, G. Müller, H. Piel, J. Pouryamout: IEEE Trans. Appl. Supercond. **9**, June (1999).
   b. M. Perpeet, W. Diete, M. A. Hein, S. Hensen, T. Kaiser, G. Müller, H. Piel, J. Pouryamout: In *Proc. 8th Workshop on RF Supercond.*, ed. by V. Palmieri and A. Lombardi, Abano Terme (Padova), Italy, LNL-INFN(REP) 133/98 (1998), Vol. IV, p. 934.
57. H. Piel: In Proc. Workshop on RF Superconductivity, ed. by M. Kuntze, Kernforschungszentrum Karlsruhe, KfK 3019.85 (1980).
58. G. D. Cody and R. W. Cohen: Rev. Mod. Phys. **36**, 121 (1964).
59. G. Müller, P. Kneisel, D. Mansen, H. Piel, J. Pouryamout, R. W. Röth: In Proc. 5th. Europ. Part. Acc. Conf. EPAC, ed. by S. Myers, A. Pacheco, R. Pascual, C. Petit-Jean-Genaz, J. Poole (IOP, Bristol, 1996), Vol. 3, p. 2085.
60. a. G. Müller: Habilitation thesis, University of Wuppertal (1990).
    b. M. Peiniger: Dissertation, University of Wuppertal, Report WUB-DIS 89-1 (1989).
61. I. E. Campisi: IEEE Trans. Magn. **MAG-21**, 134 (1985).
62. T. Hays, H. Padamsee, R. W. Röth: In *Proc. 7th Workshop on RF Superconductivity*, ed. by B. Bonin, Gif-sur-Yvette, France (1995), Vol. II, p. 437.
63. a. M. Golosovsky, M. Tsindlekht, D. Davidov: Supercond. Sci. Technol. **9**, 1 (1996).
    b. M. Golosovsky, M. Tsindlekht, H. Chayet, D. Davidov: Phys. Rev. B, 470 (1994).
64. S. Hensen: Dissertation, University of Wuppertal, (1999).
65. S. Hensen, G. Müller, C. T. Rieck, K. Scharnberg: Phys. Rev. B **56**, 6237 (1997).
66. R. G. Humphreys, J. S. Satchell, N. G. Chew, J. A. Edwards, S. W. Goodyear, S. E. Belkinsop, O. D. Dosser, A. G. Cullis: Supercond. Sci. Technol. **3**, 38 (1990).
67. a. J. R. Powell, A. Porch, R. G. Humphreys, C. E. Gough, M. J. Lancaster: In *Applied Superconductivity 1997*, Inst. Phys. Conf. Ser. No 158, ed. by H. Rogalla and D. H. A. Blank (IOP, Bristol, 1997), Vol. 1, p. 125.
    b. J. Powell: PhD thesis, University of Birmingham (1996).
68. C. Dubordieu, J. P. Sénateur, F. Weiss, O. Thomas, S. Hensen, S. Orbach-Werbig, G. Müller: In *Applied Superconductivity 1995*, Inst. Phys. Conf. Ser. No 148, ed. by D. Dew-Hughes (IOP, Bristol, 1995), p. 827.
69. J.Halbritter: In *Proc. 4th EuroCeramics – High-T_c Superconductors, Part II*, ed. by A. Barone, D. Fiorani, A. Tampieri (Gruppo Editoriale Faenza Editrice, Italy, 1995), Vol. 7, p. 267.
70. The magneto-optic imaging was performed by Christian Jooss and Ralph Warthmann at the *MPI für Metallforschung*, Stuttgart, Germany (1996).
71. C. Jooss, R. Warthmann, A. Forkl, H. U. Habermeier, B. Leibold, H. Kronmüller: Physica C **266**, 235 (1996).
72. E. H. Brandt: Phys. Rev. B **52**, 15442 (1995).
73. M. Benkraouda and J. R. Clem: Phys. Rev. B **53**, 5716 (1996).
74. A. Oral, S. J. Bending, R. G. Humphreys, M. Henini: Supercond. Sci. Technol. **10**, 17 (1997).
75. C. P. Bean: Rev. Mod. Phys. **36**, 31 (1964).
76. M. A. Hein, W. Diete, S. Hensen, T. Kaiser, M. Lenkens, G. Müller, H. Piel, H. Schlick: IEEE Trans. Appl. Supercond. **7**, 1264 (1997).
77. a. M. A. Hein, C. Bauer, W. Diete, S. Hensen, T. Kaiser, V. Z. Kresin, G. Müller: J. Supercond. **10**, 485 (1997).
    b. M. A. Hein, A. Cassinese, S. Hensen, T. Kaiser, G. Müller, M. Perpeet: J. Supercond. **12**, 129 (1999).

78. H. Kinder, P. Berberich, B. Utz, W. Prusseit: IEEE Trans. Appl. Supercond. **5**, 1575 (1995).
79. L. F. Cohen, A. Cowie, J. C. Gallop, I. S. Ghosh, I. N. Goncharov: J. Supercond. **10**, 85 (1997).
80. J. Halbritter: J. Supercond. **10**, 91 (1997).
81. J. Wosik, L. M. Xie, K. Nesteruk, D. Li, J. H. Miller, Jr., S. A. Long: J. Supercond. **10**, 97 (1997).
82. J. C. Gallop, A. L. Cowie, L. F. Cohen: In *Applied Superconductivity 1997*, Inst. Phys. Conf. Ser. No 158, ed. by H. Rogalla and D. H. A. Blank (IOP, Bristol, 1997), Vol. 1, p. 65.
83. S. Sridhar and J. E. Mercereau: Phys. Rev. B **34**, 203 (1986).
84. D. P. Choudhury, B. A. Willemsen, J. S. Derov, S. Sridhar: IEEE Trans. Appl. Supercond. **7**, 1260 (1997).
85. Y. N. Ovchinnikov and V. Z. Kresin: Phys. Rev. B **54**, 1251 (1996).
86. A. P. Kharel, A. V. Velichko, J. R. Powell, A. Porch, M. J. Lancaster, R. G. Humphreys: Phys. Rev. B **58**, 11189 (1998).
87. P. P. Nguyen, D. E. Oates, G. Dresselhaus, M. S. Dresselhaus: Phys. Rev. B **48**, 6400 (1993).
88. See, e.g., Proc. 8th Workshop on RF Superconductivity, ed. by V. Palmieri Abano Terme (Padova), Italy (1997).

Chapter 5

1. B. Aune: In Proc. 7th Workshop on RF Superconductivity, ed. by B. Bonin, Gif-sur-Yvette, France (1995), Vol. I, p. 325.
2. H. Piel: CERN Accelerator School, *Advanced Acclerator Physics*, 16–27 September 1985, Oxford, England, University of Wuppertal, External Report WUB 86-14 (1986).
3. J. Muller: Rep. Prog. Phys. **43**, 641 (1980).
4. J. R. Gavaler: Appl. Phys. Lett. **23**, 480 (1973).
5. a. B. T. Matthias, J. K. Hulm, E. J. Kunzler: Physics Today **34**, 34 (1981).
   b. B. T. Matthias, T. H. Geballe, S. Geller, E. Corenzwit: Phys. Rev. **95**, 1435 (1954).
6. W. Weber: Physica B **126**, 217 (1984).
7. L. R. Testardi: Rev. Mod. Phys. **47**, 637 (1975).
8. M. Weger and I. B. Goldberg: In *Solid State Physics*, ed. by H. Ehrenreich, F. Seitz, D. Turnbull (Academic, New York, 1973), Vol. 28, p. 1.
9. J. Labbé and J. Friedel: J. de Physique **27**, 153 and 303 (1966).
10. R. Flükiger, H. Küpfer, J. L. Jorda, J. Muller: IEEE Trans. Magn. **23**, 980 (1987).
11. H. Devantay, J. L. Jorda, M. Decroux, J. Muller, R. Flükiger: J. Mater. Sci. **16**, 2145 (1981).
12. M. Suenaga: In *Superconductor Materials Science: Metallurgy, Fabrication and Applications*, ed. by S. Foner, B. B. Schwartz (Plenum, New York, 1981).
13. J. P. Charlesworth, I. MacPhail, P. E. Madsen: J. Mater. Sci. **5**, 580 (1970).
14. Gmelin: Handbook of Inorganic Chemistry (Springer, Berlin, Heidelberg, 1987), system No 49, part B2: "Nb".
15. L. H. Allen, M. R. Beasley, R. H. Hammond, J. P. Turneaure: IEEE Trans. Magn. **23**, 1405 (1987).
16. D. A. Rudman, F. Hellman, R. H. Hammond, M. R. Beasley: J. Appl. Phys. **55**, 3544 (1984).
17. W. Wiedemann: Z. Physik **151**, 307 (1958).
18. A. R. Kaufman and J. J. Pickett: Bull. Am. Phys. Soc. **15**, 833 (1970).

19. C. F. Old and I. MacPhail: J. Mater. Sci. **4**, 202 (1969).
20. a. L. Rinderer, E. Saur, J. Wurm: Z. Physik **174**, 405 (1963).
    b. E. J. Saur and J. P. Wurm: Naturwissenschaften **49**, 127 (1962).
21. P. Kneisel, H. Küpfer, W.Schwarz, O. Stolz, J. Halbritter: IEEE Trans. Magn. **13**, 496 (1977).
22. G. Arnolds, H. Heinrichs, R. Mayer, N. Minatti, H. Piel, W. Weingarten: IEEE Trans. Nucl. Sci. **26**, 3775 (1979).
23. G. Müller, P. Kneisel, D. Mansen, H. Piel, J. Pouryamout, R. W. Röth: In Proc. 5th European Particle Acclerator Conference, Sitges, Barcelona, ed. by S. Myers et al. (IOP, Bristol, Philadelphia, 1996), Vol. 3, p. 2085.
24. a. D. Dasbach, G. Müller, M. Peiniger, H. Piel, R. W. Röth: IEEE Trans. Magn. **25**, 1862 (1989).
    b. M. Peiniger, M. Hein, N. Klein, G. Müller, H. Piel, P. Thüns: In Proc. 3rd Workshop on RF Superconductivity, ed. by K. W. Shepard, Argonne National Laboratory, Report ANL-PHY-88-1 (1988), p. 503.
25. M. Perpeet, M. A. Hein, G. Müller, H. Piel, J. Pouryamout, W. Diete: J. Appl. Phys. **82**, 5021 (1997).
26. C. A. Neugebauer: J. Appl. Phys. **35**, 3599 (1964).
27. a. L.H. Allen, M. R. Beasley, R. H. Hammond, J. P. Turneaure: IEEE Trans. Magn. **19**, 1003 (1983).
    b. L. H. Allen, W. J. Anklam, M. R. Beasley, R. H. Hammond, J. P. Turneaure: IEEE Trans. Magn. **21**, 525 (1985).
28. J. R. Gavaler, M. A. Janocko, J. K. Hulm, C. K. Jones: J. Vac. Sci. Technol. **8**, 180 (1971).
29. L. R. Testardi, J. H. Wernick, W. A. Royer, D. D. Bacon, A. R. Storm: J. Appl. Phys. **45**, 446 (1974).
30. J. Steuer: Phys. Stat. Sol. A **114**, 665 (1989).
31. a. C. T. Wu, R. T. Kampwirth, J. W. Hafstrom: J. Vac. Sci. Technol. **14**, 134 (1977).
    b. R. T. Kampwirth, J. W. Hafstrom, C. T. Wu: IEEE Trans. Magn. **13**, 315 (1977).
32. A. Andreone, A. Cassinese, A. Di Chiara, M. Ivarone, F. Palomba, A. Ruosi, R. Vaglio: J. Appl. Phys. **82**, 1736 (1997); IEEE Trans. Appl. Supercond. **7**, 1772 (1997).
33. J. J. Hanak: In *Metallurgy of Advanced Electronic Materials*, ed. by G. E. Brock (Wiley, New York, 1963), p. 161.
34. M. Peiniger: Dissertation, University of Wuppertal, Report WUB-DIS 89-1 (1989).
35. a. J. A. Strozier, D. L. Miller, O. F. Kammerer, M. Strongin: J. Appl. Phys. **47**, 1611 (1976).
    b. J. M. Dickey, M. Strongin, O. F. Kammerer: J. Appl. Phys. **42**, 5808 (1971).
36. a. B. Hillenbrand, Y. Uzel, K. Schnitzke: Appl. Phys. **23**, 237 (1980).
    b. B. Hillenbrand, H. Martens, H. Pfister, K. Schnitzke, Y. Uzel: IEEE Trans. Magn. **13**, 491 (1977).
37. D. Mansen: University of Wuppertal, External Report WUD 95-33 (1995).
38. H. H. Farrell, G. H. Gilmer, M. Suenaga: Thin Solid Films **25**, 253 (1975); J. Appl. Phys. **45**, 4025 (1974).
39. P. Kneisel, H. Küpfer, O. Stolz, J. Halbritter: Adv. Cryog. Eng. **24**, 442 (1978).
40. W. D. Kingery, H. K. Bowen, D. R. Uhlmann: *Introduction to Ceramics* (Wiley, New York, 1976).
41. W. DeSorbo: Phys. Rev. **132**, 107 (1963).

42. A. Andreone, A. Cassinese, M. Iavarone, R. Vaglio, I. I. Kulik, V. Palmiere: Phys. Rev. B **52**, 4473 (1995).
43. W. L. Holstein: J. Phys. Chem. **97**, 4224 (1993).
44. Z.-Y. Shen: *High-Temperature Superconducting Microwave Circuits* (Artech House, Boston, London, 1994).
45. a. W. L. Holstein, L. A. Parisi, D. J. Kountz, C. Wilker, A. L. Matthews, P. N. Arendt, R. C. Taber: IEEE Trans. Magn. **27**, 1568 (1991).
    b. D. W. Face, F. M. Pellicone, R. J. Small, L. Bao, M. S. Warrington, C. Wilker: IEEE Trans. Appl. Supercond. **9**, June 1999.
46. R. B. Hammond, G. V. Negrete, L. C. Bourne, D. D. Strother, A. H. Cardona, M. M. Eddy: Appl. Phys. Lett. **57**, 825 (1990).
47. R. Beyers and T. M. Shaw: Solid State Physics **42**, 135 (1989).
48. *Physical Properties of High Temperature Superconductors II*, ed. by D. M. Ginsberg (World Scientific, Singapore, 1990).
49. A. M. Portis: *Electrodynamics of High-Temperature Superconductors*, Lecture Notes in Physics (World Scientific, Singapore, 1992), Vol. 48.
50. *Advances in Superconductivity*, Vol. VI (Springer, Tokyo, 1994).
51. C. N. R. Rao and J. Gopalakrishnan: *New Directions in Solid State Chemistry* (Cambridge University Press, Cambridge, 1986).
52. C. Thomsen: In *Light Scattering in Solids VI. Topics in Applied Physics 68*, ed. by M. Cardona, G. Güntherodt (Springer, Heidelberg, 1991).
53. J. D. Jorgensen: Physics Today **44**, 34 (1991).
54. B. Batlogg: Physics Today **44**, 44 (1991).
55. E. D. Specht, C. J. Sparks, A. G. Dhere, J. Brynestad, O. B. Cavin, D.M. Kroeger, H. A. Oye: Phys. Rev. B **37**, 7426 (1988).
56. a. B. W. Veal, A. P. Paulikas, H. You, H. Shi, Y. Fang, J. W. Downey: Phys. Rev. B **42**, 6305 (1990).
    b. J. D. Jorgensen, B. W. Veal, A. P. Paulikas, L. J. Nowicki, G. W. Crabtree, H. Claus, W. K. Kwok: Phys. Rev. B **41**, 1863 (1990).
    c. J. D. Jorgensen, M. A. Beno, D. G. Hinks, L. Soderholm, K.J. Volin, T. L. Hitterman, J. D. Grace, I. K. Schuller, C. U. Segre, K. Zhang, M. S. Kleefisch: Phys. Rev. B **36**, 3608 (1987).
57. V. Z. Kresin and S. A. Wolf: Phys. Rev. B **51**, 1229 (1995).
58. N. Phillips, J. P. Emerson, R. A. Fisher, J. E. Gordon, B. F. Woodfield, D. A. Wright: J. Supercond. **7**, 251 (1994).
59. V. Z. Kresin, A. Bill, S. A. Wolf, Yu. N. Ovchinnikov: In *The Gap Symmetry and Fluctuations in High-T_c Superconductors*, NATO-ASI Series B: Physics Vol. 371, ed. by J. Bok, G. Deutscher, D. Pavuna, S. A. Wolf, (Plenum Press, New York, 1998), p. 55.
60. K. C. Hass: Solid State Physics **42**, 213 (1989).
61. H. Claus, M.Braun, A. Erb, K.Röhberg, B. Runtsch, H. Wühl, G. Bräuchle, P. Schweib, G. Müller-Vogt, H. v. Löhneysen: Physica C **198**, 42 (1992).
62. R. J. Cava, B. Batlogg, C. H. Clem, E. A. Rietman, S. M. Zahurak, D. Werder: Nature **329**, 423 (1987).
63. C. C. Tsuei: Physica A **168**, 238 (1990).
64. D. M. Newns, H. R. Krishnamurty, P. C. Pattnaik, C. C. Tsuei, C. C. Chi, C. L. Kane: Physica B **186–188**, 801 (1993).
65. W. E. Pickett: J. Supercond. **4**, 397 (1991).
66. J. Halbritter: J. Appl. Phys. **68**, 6315 (1990).
67. B. Avenhaus: PhD thesis, University of Birmingham (1996).

68. a. S. Hensen: Dissertation, University of Wuppertal (1999).
    b. S. Orbach-Werbig: Dissertation, University of Wuppertal, Report WUB-DIS 94-9 (1994).
    c. S. Hensen, S. Orbach-Werbig, G. Müller, H. Piel, N. G. Chew, J. A. Edwards, R. G. Humphreys: In *Applied Superconductivity*, Inst. Phys. Conf. Ser. No 148, ed. by H. C. Freyhardt (IOP, Bristol, 1993), p. 1053.
69. N. G. Chew, J. A. Edwards, R. G. Humphreys, J. S. Satchell, S. W. Goodyear, B. Dew, N. J. Exon, S. Hensen, M. Lenkens, G. Müller, S. Orbach-Werbig: IEEE Trans. Appl. Supercond. **5**, 1167 (1995).
70. J. Geerk, G. Linker, O. Meyer: Materials Science Reports **4**, 193 (1989).
71. R. W. McCallum: J. Metals, January 1989, p. 50.
72. J. S. Wallace and B. A. Bender: J. Supercond. **1**, 153 (1988).
73. P. Chaudari, J. Mannhart, D. Dimos, C. C. Tsuei, C. C. Chi, M. M. Oprysko, M. Scheuermann: Phys. Rev. B **41**, 4038 (1990); Phys. Rev. Lett. **60**, 1653 (1988).
74. D. E. Farrell, B. S. Chandrasekhar, M. R. DeGuire, M. M. Fang, V. G. Kogan, J. R. Clem, D. K. Finnemore: Phys. Rev. B **36**, 4025 (1987).
75. O. Ishii, T. Konaka, M. Sato, Y. Koshimoto: Jap. J. Appl. Phys. **29**, L2177 (1990).
76. N. McN.Alford, T. W. Button, M. J. Adams, S. Hedges, B. Nicholson, W. A. Phillips: Nature **349**, 680 (1991).
77. a. M. Hein, F. Hill, G. Müller, H. Piel, H. P. Schneider, M. Strupp: IEEE Trans. Appl. Supercond. **3**, 1745 (1993).
    b. M. Hein, S. Kraut, E. Mahner, G. Müller, D. Opie, H. Piel, L. Ponto, D. Wehler, M. Becks, U. Klein, M. Peiniger: J. Supercond. **3**, 323 (1990).
    c. M. Hein, G. Müller, H. Piel, L. Ponto, M. Becks, U. Klein, M. Peiniger: J. Appl. Phys. **66**, 5940 (1989).
78. K. Oka, K. Nakane, M. Ito, M. Saito, H. Unoki: Jap. J. Appl. Phys. **27**, L1065 (1988).
79. T. Aselage and K. Keefer: J. Mater. Res. **3**, 1279 (1988).
80. F. Licci, P. Tissot, H. J. Scheel: J. Less-Common Met. **150**, 201 (1989).
81. M. Maeda, M. Kadoi, T. Ikeda: Jap. J. Appl. Phys. **28**, 1417 (1989).
82. H. Murakami, T. Suga, T. Noda, Y. Shiohara, S. Tanaka: Jap. J. Appl. Phys. **29**, 2720 (1990).
83. F. Greuter, P. Kluge-Weiss, H. Zimmermann, C. Schüler: Physica C **153–155**, 361 (1988).
84. S. J. Rothman, J. L. Routbort, U. Welp, J. E. Baker: Phys. Rev. B **44**, 2326 (1991).
85. a. R. H. Hammond and R. Bormann: Physica C **162–164**, 703 (1989).
    b. R. Bormann and J. Nölting: Appl. Phys. Lett. **54**, 2148 (1989).
86. a. T. B. Lindemer, F. A. Washburn, C. S. MacDougall: Physica C **196**, 390 (1992).
    b. R. Feenstra, T. B. Lindemer, J. D. Budai, M. D. Galloway: J. Appl. Phys. **69**, 6569 (1991).
87. R. G. Humphreys, J. S. Satchell, N. G. Chew, J. A. Edwards, S. W. Goodyear, S. E. Blenkinsop, O. D. Dosser, A. G. Cullis: Supercond. Sci. Technol. **3**, 38 (1990).
88. W. Vook: Int. Metals Rev. **27**, 209 (1982).
89. G. Zinsmeister: In *Basic Problems in Thin Film Physics*, ed. by R. Niedermeyer and H. Mayer (Vandenhook and Ruprecht, Göttingen, 1966).
90. D. Walton: J. Chem. Phys. **37**, 2182 (1962).
91. W. K. Burton, N. Cabrera, F. C. Frank: Philos. Trans. Roy. Soc. A **243**, 299 (1951).

92. a. E. Bauer and H. Poppa: Thin Solid Films **12**, 167 (1972).
    b. E. Bauer: Z. Kristallographie **110**, 372 (1958).
93. J. A. Thornton: J. Vac. Sci. Technol. **12**, 830 (1975).
94. F. C. Frank: Adv. Phys. **1**, 91 (1952); Discussions Faraday Soc. **5**, 48 (1949).
95. P. Bennema and G. H. Gilmer: In *Crystal Growth: An Introduction*, ed. by P. Hartman (North-Holland, Amsterdam, 1973).
96. R. McCormack, D. deFontaine, G. Ceder: Phys. Rev. B **45**, 12976 (1992).
97. A. Oral, S. J. Bending, R. G. Humphreys, M. Henini: Supercond. Sci. Technol. **10**, 17 (1997).
98. a. V. Matijasevic, P. Rosenthal, K. Shonohara, A. F. Marshall, R. H. Hammond, M. R. Beasley: J. Mater. Res. **6**, 682 (1991).
    b. C. B. Eom, J. Z. Sun, B. M. Lairson, S. K. Streiffer, A. F. Marshall, K. Yamamoto, S. M. Anlage, J. C. Bravmana T. H. Geballe: Physica C **171**, 354 (1990).
99. E. Osquiguil, M. Maenhoudt, B. Wuyts, Y. Bruynseraede: Appl. Phys. Lett. **6**, 1627 (1992).
100. *Deposition Technologies for Films, Coatings – Developments and Applications*, ed. by R. F. Bunshah (Noyes, USA, 1982).
101. M. Leskala, J. K. Truman, C. H. Mueller, P. H. Holloway: J. Vac. Sci. Technol. A **7**, 3147 (1989).
102. C. H. Stoessel, R. F. Bunshah, S. Prakash, H. R. Fetterman: J. Supercond. **6**, 1 (1993).
103. S. Miyazawa, Y. Tazoh, H. Asano, Y. Nagai, O. Michikami, M. Suzuki: Adv. Mater. **5**, 179 (1993).
104. P. R. Broussard and S. A. Wolf: J. Cryst. Growth **91**, 340 (1988).
105. D. Bäuerle: *Laser Processing and Chemistry* (Springer, Berlin, Heidelberg, New York, 1996).
106. D. Dijikkamp, T. Venkatesan, X. D. Wu, S. A. Shaheen, N. Jisrawi, Y. H. Min-Lee, W. L. McLean, M. Croft: Appl. Phys. Lett. **51**, 619 (1987).
107. a. B. Roas, L. Schultz, G. Endres: Appl. Phys. Lett. **53**, 1557 (1988).
     b. B. Roas: Dissertation, University Erlangen-Nürnberg (1990).
108. N. Klein, G. Müller, H. Piel, B. Roas, L. Schultz, U. Klein, M. Peiniger: Appl. Phys. Lett. **54**, 757 (1989).
109. a. F. G. Hill, M.Getta, M. A. Hein, G. Müller, H. Piel, B. Dam, S. Beuven, J. Schubert, W. Zander: Appl. Supercond. **5**, 249, (1998).
     b. F. G. Hill: Dissertation, University of Wuppertal (in preparation).
110. B. Dam, N. J. Koeman, J. H. Rector, B. Stäuble-Pümpin, U. Poppe, R. Griessen: Physica C **261**, 1 (1996).
111. L. Lynds, B. R. Weinberger, D. M. Potrepka, G. G. Peterson, M. P. Lindsay: Physica C **159**, 61 (1989).
112. a. J. Schubert and W. Zander: Forschungszentrum Jülich, unpublished (1997).
     b. B. Stritzker, J. Schubert, U. Popper, W. Zander, U. Krüger, A. Lubig, C. Buchal: J. Less-Common. Met. **165**, 279 (1990).
113. M. Lorenz, H. Hochmuth, D. Natusch, H. Börner, G. Lippold, K. Kreher, W. Schmitz: Appl. Phys. Lett. **68**, 3332 (1996).
114. a. R. G. Humphreys, J. S. Satchell, N. G. Chew, J. A. Edwards, S. W. Goodyear, M. N. Keene, S. J. Hedges: Materials Science, Engineering B **10**, 293 (1991).
     b. N. G. Chew, S. W. Goodyear, J. A. Edwards, J. S. Satchell, R. G. Humphreys: Appl. Phys. Lett. **57**, 2016 (1990).
115. a. P. Berberich, J. Tate, W. Dietsche, H. Kinder: Appl. Phys. Lett. **53**, 925 (1988).
     b. W. Prusseit: Dissertation, Technichal University of Munich (1994).

116. M. Hein, S. Hensen, G. Müller, S. Orbach, H. Piel, M. Strupp, N. G: Chew, J.A. Edwards, S. W. Goodyear, J. S. Satchell, R. G. Humphreys: In *High T_c Superconductor Thin Films*, ed. by L. Correra (Elsevier, North-Holland, Amsterdam, 1992), p. 95.
117. a. H. Kinder, P. Berberich, B. Utz, W. Prusseit: IEEE Trans. Appl. Supercond. **5**, 1575 (1995).
   b. R. Semerad, B. Utz, W. Prusseit, H. Kinder: In *Applied Superconductivity*, Inst. Phys. Conf. Ser. No 148, ed. by D. Dew-Hughes (IOP, Bristol, 1995), p. 847.
118. B. Chapman: *Glow Discharge Processes – Sputtering, Plasma Etching* (Wiley, New York, 1980).
119. A. C. Anderson and R. L. Slattery: Applied Supercond. Conf. Boston (1994), unpublished.
120. U. Poppe, J. Schubert, R. R. Arons, W. Evers, C. H. Freiburg, W. Reichert, K. Schmidt, W. Sybertz, K. Urban: Solid State Commun. **66**, 661 (1988).
121. a. G. Müller, B. Aschermann, H. Chaloupka, W. Diete, M. Getta, M. Hein, S. Hensen, F. Hill, M. Lenkens, S. Orbach-Werbig, T. Patzelt, H. Piel, J. Rembesa, H. Schlick, T. Unshelm, R. Wagner: IEEE Trans. Appl. Supercond. **5**, 1729 (1995).
   b. M. Lenkens, B. Aschermann, S. Hensen, M. Jeck, S. Orbach, H. Schlick, H. Chaloupka, G. Müller, H. Piel: J. Alloys & Compounds **195**, 559 (1993).
122. G. B. Stringfellow: *Organometallic Vapor-Phase Epitaxy – Theory, Practice* (Academic, Boston, 1989).
123. H. Yamane, H. Masumoto, T. Hirai, H. Iwasaki, K. Watanabe, N. Kobayashi, Y. Muto, H. Kurosawa: Appl. Phys. Lett. **53**, 1548 (1988).
124. F. Schmaderer and G. Wahl: Proc. 7th ECVD Conference, June 1989, Perpignan, France (1989).
125. a. O. Thomas, A. Pisch, E. Mossang, F.Weiss, R. Madar, J. P. Senateur: J. Less-Common Met. **164 & 165**, 444 (1990).
   b. C. Dubordieu, G. Delabouglise, O. Thomas, J. P. Senateur, D. Chateigner, P. Germi, M. Pernet, S. Hensen, S. Orbach, G. Müller: In *Applied Superconductivity*, ed. by H. C. Freyhardt (DGM, Oberursel, 1993), p. 1081.
126. J. Geerk, U. Kaufmann, W. Banert, H. Rietschel: Phys. Rev. B **33**, 1621 (1986).
127. K. E. Kihlstrom, D. Mael, T. H. Geballe: Phys. Rev. B **29**, 150 (1984).
128. A. Barone and G. Paterno: *Physics and Applications of the Josephson Effect* (Wiley, New York, 1982).
129. A. I. Braginski: Bull. Pol. Acad. Sci. **45**, 57 (1997).
130. K. K. Likharev: *Dynamics of Josephson Junctions and Circuits* (Gordon and Breach, Philadelphia, 1991).
131. R. Gross, L. Alff, A. Beck, O. M. Froehlich, D. Koelle, A. Marx: IEEE Trans. Appl. Supercond. **7**, 2929 (1997).
132. a. J. Gao, Y. Boguslavskii, B. B. Klopman, D. Terpstra, G. J. Gerritsma, H. Rogalla: IEEE Trans. Appl. Supercond. **3**, 2034 (1993).
   b. J. Gao, W. A. M. Aarnink, G. J. Gerritsma, H. Rogalla: Physica C **171**, 126 (1990).
133. R. Simon, J. B. Bulman, J. F. Burch, S. B. Coons, K. P. Daly, W. D: Dozier, R. Hu, A. E. Lee, J. A. Luine, C. E. Platt, M. J. Zani: Appl. Phys. Lett. **58**, 543 (1991).
134. K. Herrmann, Y. Zhang, H. M. Mück, J. Schubert, W. Zander, A. I. Braginski: J. Appl. Phys. **78**, 1131 (1995).
135. C. L. Jia, B. Kabius, K. Urban, K. Herrmann, J. Schubert, W. Zander, A. I. Braginski: Physica C **176**, 211 (1992).

136. J. Z. Sun, W. J. Gallagher, A. C. Callegari, V. Foglietti, R. H. Koch: Appl. Phys. Lett. **63**, 1561 (1993).
137. a. H. R. Yi, D. Winkler, Z. G. Ivanov, T. Claeson: IEEE Trans. Appl. Supercond. **5**, 2778 (1995).
b. H. R. Yi, Z. G. Ivanov, D. Winkler, Y. M. Zhang, H. Olin, P. Larsson, T. Claeson: Appl. Phys. Lett. **65**, 1177 (1994).
138. B. Aschermann: Dissertation, University of Wuppertal (1997).
139. B. A. Aminov, B. Aschermann, M. A. Hein, F. Hill, M. Lorenz, G. Müller, H. Piel: Phys. Rev. B **52**, 13631 (1995).
140. T. Frey, C.C. Chi, C. C. Tsuei, T. Shaw, F. Boszo: Phys. Rev. B **49**, 3483 (1994).
141. N. Newman, W. G. Lyons: J. Supercond. **6**, 119 (1993).
142. a. M. A. Hein: In *Studies of High-Temperature Superconductors*, ed. by A. Narlikar (Nova Sciences, New York, 1995), Vol. 18, p. 141.
b. H. J. Chaloupka, M. A. Hein, G. Müller: In *High-T_c Microwave Superconductors and Applications*, ed. by R. B. Hammond and R. S. Withers: Proc. SPIE **2156**, 36 (1994).
143. M. J. Lancaster: *Passive Microwave Device Applications of High-Temperature Superconductors* (Cambridge University Press, Cambridge, 1997).
144. E. A. Fitzgerald: Mater. Sci. Rep. **7**, 87 (1991).
145. a. G. Kästner, C. Schäfer, S. Senz, D. Hesse, M. Lorenz, H. Hochmuth, M. Getta, M. A. Hein, T. Kaiser, G. Müller: IEEE Trans. Appl. Supercond. **9**, June 1999.
b. G. Kästner, D. Hesse, M. Lorenz, R. Scholz, N. D. Zakharov, P. Kopperschmidt: Phys. Stat. Sol. A **150**, 381 (1995).
146. a. A. G. Zaitsev, G. Ockenfuss, R. Wördenweber: In *Applied Superconductivity 1997*, Inst. Phys. Conf. Ser. No 158, ed. by H. Rogalla and D. H. A. Blank (IOP, Bristol, 1997), Vol. 1, p. 25.
b. R. Wördenweber, J. Einfeld, R. Kutzner, A. G. Zaitsev, M. A. Hein, T. Kaiser, G. Müller: IEEE Trans. Appl. Supercond. **9**, June 1999.
147. R. W. Simon, C. E. Platt, A. E. Lee, G. S. Lee, K. P. Daly, M.S. Wire, J.A. Luine, M. Urbanik: Appl. Phys. Lett. **53**, 2677 (1988).
148. Y. S. Touloukian, R. H. Kirby, R. E. Taylor, T. Y. R. Lee: In *Thermal Expansion – Non-Metallic Solids* (Plenum, New York, 1977), Vol. 13.
149. J. Krupka, R. G. Geyer, M. Kuhn, J. H. Hinken: IEEE Trans. Microwave Theory Techn. **42**, 1886 (1994).
150. T. Konaka, M. Sato, H. Asano, S. Kubo: J. Supercond. **4**, 283 (1991).
151. a. T. E. Harrington, J. Wosik, S. A. Long: IEEE Trans. Appl. Supercond. **7**, 1861 (1997).
b. T. P. Hogan, J. Wosik, L. M. Xie, J. H. Miller, T. Harrington, A. G. Zaitsev, R. Wördenweber: IEEE Trans. Appl. Supercond. **9**, June 1999.
152. H. E. Weaver: J. Phys. Chem. Solids **11**, 274 (1959).
153. R. Brown, V. Pendrick, D. Kalokitis, B. H. T. Chai: Appl. Phys. Lett. **57**, 1351 (1990).
154. G. Rupprecht and R. O. Bell: Phys. Rev. **125**, 1915 (1962).
155. Y. Kobayashi, J. Dato, K.Yajima: Trans. IEICE **E72**, 290 (1989).
156. C. Zuccaro, M. Winter, N. Klein, K. Urban: J. Appl. Phys. **82**, 5695 (1997).
157. R. Schwab, R. Spörl, P. Severloh, R. Heidinger, J. Halbritter: In *Applied Superconductivity 1997*, Inst. Phys. Conf. Ser. No 158, ed. by H. Rogalla and D. H. A. Blank (IOP, Bristol, 1997), Vol. 1, p. 61.
158. P. A. Smith and L. E. Davis: Electron. Lett. **28**, 424 (1992).
159. J. Talvacchio, G.R. Wagner, S. H. Talisa: Microwave Journal, July (1991), p. 105.

160. B. P. Gorshunov, G.V. Kozlov, S. I. Krasnosvobodstev, E. V. Pechen, A. M. Prokhorov, A. S. Prokhorov, O. I. Syrotnsky, A. A. Volkov: Physica C **153–155**, 667 (1988).
161. V. B. Braginsky, V. S. Ilchenko, Kh. S. Bagdassarov: Phys. Lett. A **120**, 300 (1987).
162. R. Heidinger: J. Nucl. Mater. **212–215**, 1101 (1994); **173**, 243 (1990).
163. P. Merchant, R. D. Jacowitz, K. Tibbs, R. C. Taber, S. S. Laderman: Appl. Phys. Lett. **60**, 763 (1992).
164. a. X. D. Wu, S. R. Foltyn, R. E. Muenchhausen, D. W. Cooke, A. Pique, D. Kalokitis, V. Pendrick, E. Belohoubek: J. Supercond. **5**, 353 (1992).
    b. X. D. Wu, R. C. Dye, R. E. Muenchhausen, S. R. Foltyn, M. Maley, A. D. Rollett, A. R. Garcia, N. S. Nogar: Appl. Phys. Lett. **58**, 2165 (1991).
165. A. G. Zaitsev, R. Kutzner, R. Wördenweber: Appl. Phys. Lett. **67**, 2723 (1995).
166. B. F. Cole, G. C. Liang, N. Newman, K. Char, G. Zacharchuk, J. Martens: Appl. Phys. Lett. **61**, 1727 (1992).
167. D. K. Fork, F. A. Ponce, J. C. Tramontana, N. Newman, J. M. Phillips, T. H. Geballe: Appl. Phys. Lett. **58**, 2432 (1991).
168. A. H. Cottrell: In *Physics of Metals*, ed. by P. B. Hirsch (Cambridge University Press, Cambridge, 1975), Vol. 2.
169. a. M. Lorenz, H. Hochmuth, D. Natusch, G. Lippold, V. L. Svetchnikov, T. Kaiser, M. A. Hein, R. Schwab, R. Heidinger: IEEE Trans. Appl. Supercond. **9**, June 1999.
    b. M. Lorenz, H. Hochmuth, J.Frey, H. Börner, J. Lenzner, G. Lippold, T. Kaiser, M. A. Hein, G. Müller: In *Applied Superconductivity 1997*, Inst. Phys. Conf. Ser. No 158, ed. by H. Rogalla and D. H. A. Blank (IOP, Bristol, 1997), Vol. 1, p. 283.
170. a. A. G. Zaitsev, G. Ockenfuss, D. Guggi, R. Wördenweber, U. Krüger: J. Appl. Phys. **81**, 3069 (1997).
    b. A. G. Zaitsev, R. Kutzner, R. Wördenweber, T. Kaiser, M. A. Hein, G. Müller: J. Supercond. **11**, 361 (1998).
171. R. Semerad, K. Irgmaier, H. Kinder, M. Hein, T. Kaiser: unpublished (1997).
172. J. Geerk, G. Linker, M. Hein, T. Kaiser: unpublished (1997).
173. W. Prusseit, S. Corsepius, F. Baudenbacher, K. Hirata, P. Berberich, H. Kinder: Appl. Phys. Lett. **61**, 1841 (1992).
174. G. A. Samara: J. Appl. Phys. **68**, 4214 (1990); Phys. Rev. B **13**, 4529 (1976); Phys. Rev. **165**, 959 (1968).
175. C. Zuccaro, I. Ghosh, K. Urban, N. Klein, S. Penn, N. Alford: IEEE Trans. Appl. Supercond. **7**, 3715 (1997).
176. a. A. Baumfalk, H. J. Chaloupka, S. Kolesov, M. Klauda, C. Neumann: IEEE Trans. Appl. Supercond. **9**, June 1999.
    b. M. A. Hein, B. A. Aminov, A. Baumfalk, H. J. Chaloupka, F. Hill, T. Kaiser, S. Kolesov, G. Müller, H. Piel: In *Applied Superconductivity 1997*, Inst. Phys. Conf. Ser. No 158, ed. by H. Rogalla and D. H. A. Blank (IOP, Bristol, 1997), Vol. 1, p. 319.
177. M. Sparks, D. F. King, D. L. Mills: Phys. Rev. B **26**, 6987 (1982).
178. I. Bunget and M. Popescu: *Physics of Solid Dielectrics*, Material Science Monograph 19 (Elsevier, Amsterdam, 1984).
179. C. Zuccaro, M. Winter, N. Klein, S. Penn, N. McN.Alford, P. Filhol, G. Forterre: In *Applied Superconductivity 1997*, Inst. Phys. Conf. Ser. No 158, ed. by H. Rogalla and D. H. A. Blank (IOP, Bristol, 1997), Vol. 1, p. 69.
180. D. Nghiem, J. T. Williams, D. R. Jackson: IEEE Trans. Microwave Theory Tech. **39**, 1553 (1991).

181. J. Kessler, R. Dill, P. Russer: IEEE Trans. Microwave Theory Tech. **39**, 1566 (1991).
182. O. Vendik and E. Kollberg: Microwave & RF **7**, 118 (1993).
183. D. Kaparkov, V. Sherman, I. Vendik, M. Gaidukov, V. Osadchiy, P. Petroc, S. Razumov: In *Applied Superconductivity 1995*, Inst. Phys. Conf. Ser. No 148, ed. by D. Dew-Hughes (IOP, Bristol, 1995), p. 1139.
184. Proc. ESA/ESTEC Workshop on Space Applications of High-Temperature Superconductors, 27–28 April 1993, ESTEC, Nordwijk, ed. by M. Guglielmi and E. Armandillo. ESA WPP-052 (1993).
185. A. Porch, B. Avenhaus, F. Wellhöfer, P. Woodall: In *Applied Superconductivity*, Inst. Phys. Conf. Ser. No 148, ed. by D. Dew-Hughes (IOP, Bristol, 1995), p. 1039.
186. a. J. C. Booth, J. A. Beall, L. R. Vale, D. A. Rudman, R. H. Ono: IEEE Trans. Appl. Supercond. **9**, June 1999.
 b. J. C. Booth, J. A. Beall, R. H. Ono, F. J. B. Stork, D. A. Rudman, L. R. Vale: Extended Abstracts, Internat. Conf. Supercond. Electronics, ed. by H. Koch, Berlin, Germany (1997).
187. G. W. Mitschang: IEEE Trans. Appl. Supercond. **5**, 69 (1995).
188. F. Reif: *Fundamentals of Statistical and Thermal Physics* (Mc-Graw-Hill, New York, 1995).
189. R. F. Barron: *Cryogenic Systems* (Oxford University Press, London, 1985).
190. a. R. Radebaugh: In *Application of Cryogenic Technology 10*, ed. by J. P. Kelley (Plenum, New York, 1991), p. 1.
 b. For a recent review see R. Radebaugh: IEEE Trans. Appl. Supercond. **9**, June 1999.
191. P. J. Kerney and M. Nisenoff: Cryogenics **35**, 405 (1995).
192. M. Nisenoff: In *Cryocoolers 8*, ed. by R. G. Ross, Jr. (Plenum, New York, 1995), p. 913.
193. R. A. Ackermann: Superconductor Industry **6**, 15 (1993).
194. M. Nisenoff, F. Patten, S. A. Wolf: In *Cryocoolers 9*, ed. by R. G. Ross, Jr. (California Institute of Technology, Pasadena, 1996).
195. G. Kaiser, M. Thürk, P. Seidel: Adv. Cryo. Eng. **40** (1995); **39**, 1281 (1993).
196. G. Thummes, R. Landgraf, F. Giebeler, M. Mück, C. Heiden: Proc. Cryogenic Eng. Conf., CEC/ICMC, Columbus, Ohio, July 17–21, 1995.
197. For the relevant trends compare Contributions to *Cryocoolers*, ed. by R. G. Ross, Jr. (Plenum, New York), Volumes 8 (1995) and 9 (1996), and, e. g., Proceedings of the 15th Internat. Cryogenic Eng. Conf., Cryogenics **34** (1994).
198. R. A. Byrns: Proc. 4th Workshop on RF Superconductivity, ed. by Y. Kojima, KEK Report 89-21, KEK, Tsukuba, Japan, 329 (1990).
199. D. Proch and C. H. Rode: Proc. 5th Workshop on RF Superconductivity, ed. by D. Proch, DESY M-92-01, Deutsches Elektronen Synchrotron, Hamburg, Germany, 1049 (1992).
200. A. Matlashov, Y. Zhuravlev, V. Slobodchikov, N. Bondarenko, A. Bakharev, D. Rassi: Proc. 9th Intern. Conf. Biomagn., August 1993, Vienna, Austria.
201. A. Cochran, G. B. Donaldson, L. N. C. Morgan, R. M. Bowman, K. J. Kirk: Brit. J. NDT **35**, 173 (1993).
202. R. Radebaugh: Jap. J. Appl. Phys. **26**, Supplement 26-3, 2076 (1987) and Adv. Cryo. Eng. **35**, 1191 (1990).
203. M. Wilson: Superconductor Industry **6**, 30 (1993).
204. W. E. Gifford and R. Longsworth: Adv. Cryo. Eng. B **10**, 69 (1965).
205. G. Thummes, F. Giebeler, C. Heiden: In *Cryocoolers 8*, ed. by R. G. Ross, Jr. (Plenum, New York, 1995), p. 383.

206. E. I. Mikulin, A. A. Tasarov, M. P. Shkrebyonock: Adv. Cryo. Eng. **29**, 629 (1984).
207. S. Zhu and P. Wu: Cryogenics **30**, 514 (1990).
208. Special Issue: Adv. Cryo. Eng. **43** (1998).
209. a. G. Thummes, C. Wang, C. Heiden: Cryogenics **37**, 159 (1997).
    b. C. Wang, G. Thummes, C. Heiden: Cryogenics **37**, 159 (1997).
210. M. Nisenoff, S. A. Wolf, J. C. Ritter, G. Price: Physica C **209**, 263 (1993).
211. T. Kawecki, D. R. Mahony, S. S. Chappie: Proc. Space Cryogenics Workshop (1995).

Chapter 6

1. M. A. Hein: Supercond. Sci. Technol. **10**, 867 (1997) and Particle Accelerators **53**, 135 (1996).
2. H. J. Chaloupka: In *Applications of Superconductivity*, NATO-ASI Series, ed. by H. Weinstock (Kluwer, Dordrecht, 1999).
3. K. K. Likharev: *Dynamics of Josephson Junctions and Circuits* (Gordon and Breach, Philadelphia, 1991).
4. G. L. Matthaei, L. Young, E. M. T. Jones: *Microwave Filters, Impedance-Matching Networks, and Coupling Structures* (Artech House, Dedham, Massachusetts, 1980), Chap. 2.
5. Chapter 8 of Ref. 4.
6. S. B. Cohn and J. Shimizu: Quarterly Progress Report 2, SRI Project 1114, Contract DA36-039 SC-63232, Stanford Research Institute, Menlo Park, California (May 1955).
7. Chapter 11 of Ref. 4.
8. Chapter 4 of Ref. 4.
9. R. Saal: *Handbook of Filter Design* (Dr. Alfred Hüthig, Heidelberg, 1988).
10. P. L. Tschebyscheff: Mémoires de l'Académie Impériale des Sciences de St. Pétersbourg. Sixième Série. Sciences mathématiques et physiques. Tome VII, 199 (1859).
11. V. Belevich: Wireless Engineer **29**, 106 (1952).
12. H.J. Orchard: Wireless Engineer **30**, 3 (1953).
13. S. B. Cohn: Proc. IRE **47**, 1342 (1959).
14. H. W. Bode: *Network Analysis and Feedback Amplifier Design* (D. Van Nostrand, New York, 1945), p. 216.
15. S. Butterworth: Wireless Engineer, (1936) p. 536.
16. W. Cauer: Math. Zeitschrift **38**, 1 (1933).
17. R. Levy: IEEE Trans. Microwave Theory Tech. **24**, 172 (1976).
18. R. R. Mansour, B. Jolley, S. Ye, F. S. Thomson, V. Dokas: IEEE Trans. Microwave Theory Tech. **44**, 1322 (1996).
19. M. A. Hein, C. Bauer, W. Diete, S. Hensen, T. Kaiser, G. Müller, H. Piel: J. Supercond. **10**, 109 (1997).
20. S. Kolesov, H. Chaloupka, A. Baumfalk, T. Kaiser: J. Supercond. **10**, 179 (1997).
21. S. Kolesov: University of Wuppertal, unpublished (1997).
22. M. Reppel, H. Chaloupka, S. Kolesov: In *Applied Superconductivity 1997*, Inst. Phys. Conf. Ser. No 158, ed. by H. Rogalla and D. H. A. Blank (IOP, Bristol, 1997), Vol. 1, p. 323.
23. a. B. A. Aminov, H. Piel, M. A. Hein, T. Kaiser, G. Müller, A. Baumfalk, H. J. Chaloupka, S. Kolesov, H. Medelius, E. Wikborg: IEEE Trans. Appl. Supercond. **9**, June (1999).

b. M. A. Hein, B. A. Aminov, A. Baumfalk, H. J. Chaloupka, F. Hill, T. Kaiser, S. Kolesov, G. Müller, H. Piel: In *Applied Superconductivity 1997*, Inst. Phys. Conf. Ser. No 158, ed. by H. Rogalla and D. H. A. Blank (IOP, Bristol, 1997), Vol. 1, p. 319.
24. G.-C. Liang, D. Zhang, C.-F. Shih, M. E. Johannson, R. S. Withers, D. E. Oates, A. C. Anderson, P. Polakos, P. Mankievich, E. deObaldia, R. E. Miller: IEEE Trans. Microwave Theory Tech. **43**, 3020 (1995).
25. Chapter 15 of Ref. 4.
26. N. Newman and W. G. Lyons: J. Supercond. **6**, 119 (1993).
27. Z.-Y. Shen: *High-Temperature Superconducting Microwave Circuits* (Artech House, Boston, London, 1994).
28. M. J. Lancaster: *Passive Microwave Device Applications of High-Temperature Superconductors* (Cambridge University Press, Cambridge, 1997).
29. R. Withers: In *Superconducting Devices*, ed. by S. T. Ruggiero and D. A. Rudman (Academic, Boston, 1990), p. 227.
30. T. VanDuzer: In *Superconducting Electronics*, NATO ASF Ser. F/59, ed. by H. Weinstock and M. Nisenoff (Springer, Berlin, Heidelberg, 1989), p. 285.
31. a. A. W. Kleinsasser: In *The New Superconducting Electronics*, ed. by H. Weinstock and R. W. Ralston (Kluwer, Dordrecht, 1993), p. 249.
b. A. W. Kleinsasser and W. J. Gallagher: In Ref. 29, p. 325.
32. R. Gross, B. Gerdemann, L. Alff, T. Bauch, A. Beck, O. M. Froehlich, D. Koelle, A. Marx: Appl. Supercond. **3**, 443 (1995).
33. H. Chaloupka and H. Piel: Proc. 3rd Workshop on RF Superconductivity, ed. by K. W. Shepard, Argonne National Laboratory, Report ANL-PHY-88-1 (1988), p. 273.
34. H. Chaloupka: J. Supercond. **5**, 403 (1992).
35. H. J. Chaloupka, M. A. Hein, G. Müller: In *High-T_c Microwave Superconductors and Applications*, ed. by R. B. Hammond and R. S. Withers, Proc. SPIE **2156**, 36 (1994).
36. F. W. Patten and S. A. Wolf: IEEE Trans. Appl. Supercond. **5**, 3203 (1995).
37. Superconductivity – A Global Perspective, Status report of the 4th ISIS meeting, July 1995, Washington D.C., USA; Contacts: *Consortium of European Companies Determined to Use Superconductivity*, CONECTUS, Eynsham, Withye, Oxon OX8 1TL, United Kingdom; or *Council on Superconductivity for American Competitiveness*, CSAC, 1250 24th Street N.W., Suite 300, Washington DC 20037, USA; or *International Superconductivity Technology Center*, ISTEC, Eishin Kaihatsu Bldg, 34-3 Shimbashi 5-Chome, Minato-Ku, Tokyo, 105 Japan.
38. M. Paetsch: *Mobile Communications in the US and Europe: Regulation, Technology and Market* (Artech House, Boston, 1993).
39. W. C. Y. Lee: *Mobile Cellular Telecommunication Systems* (McGraw-Hill, New York, 1989).
40. N. J. Boucher: *The Cellular Radio Handbook*, 3rd Ed (Quantum, Mill Valley CA, 1995).
41. *Microwave and millimeter wave applications of high-temperature superconductivity*, Special Issue, IEEE Trans. Microwave Theory Tech. **44**, 1193–1392 (1996).
42. Z. X. Ma, H. Wu, P. Polakos, P. Mankievich, D. Zhang, G. Liang, A. Anderson, P. Kerney, B. Andeen, R. Ono: Extended Abstracts, Internat. Supercond. Electronics Conf., ed. by H. Koch (Berlin, Germany, 1997), p. 128.
43. D. G. Smith, P. S. Gerner, C. P. Zuhoski: IEEE Trans. Appl. Supercond. **7**, 3710 (1997).

References 383

44. S. J. Fiedziuszko, J. A. Curtis, S. C. Holme, R. S. Kwok: IEEE Microwave Theory Tech. **44**, 1248 (1996).
45. a. T. Patzelt, B. Aschermann, H. Chaloupka, U. Jagodzinski, B. Roas: Electron. Lett. **29**, 1578 (1993).
    b. T. Patzelt, B. Aschermann, H. Chaloupka, U. Jagodzinski, G. Gieres, B. Roas: In *Applied Superconductivity*, ed. by H. C. Freyhardt (DGM, Oberursel, 1993), p. 991.
46. D. Zhang, G. C. Liang, C. F. Shih, M. J. Johansson, R. S. Withers: IEEE Trans. Microw. Theory Tech. **43**, 3030 (1995).
47. M. J. Scharen, D. R. Chase, A. M. Ho, A. O'Braid, K. R. Raihn, R. J. Forse: IEEE Trans. Appl. Supercond. **7**, 3744 (1997).
48. R. J. Dinger: J. Supercond. **3**, 287 (1990).
49. R. C. Hansen: IEEE Trans. Aerosp. Electron. Syst. **26**, 345 (1990).
50. a. I. B. Vendik, V. O. Sherman, A. A. Svishchev: IEEE Trans. Appl. Supercond. **9**, June 1999.
    b. O. G. Vendik, A. B. Kozyrev, I. B. Vendik: Extended Abstracts, Internat. Supercond. Electronics Conf., ed. by. H. Koch (Berlin, Germany, 1997), p. 80.
    c. O. G. Vendik and S. G. Kolesov: J. Phys. III France **3**, 1659 (1993).
51. A. I. Braginski: Bulletin Polish Acad. Sciences, Technical Sciences **45**, 57 (1997).
52. M. Klauda, C. Neumann, T. Kässer: Bosch Telecom and Robert Bosch GmbH, private communication (1997).
53. C. Kittel: *Introduction to Solid State Physics* (Wiley, New York, 1968); (German edition: R. Oldenbourg, München, 1988).
54. R. Simon: Superconductor Industry **8** (2), 18 (1995).
55. W. A. Edelstein, G. H. Glover, C. J. Hardy, R. W. Redington: Magn. Res. Med. **3**, 604 (1986).
56. D. I. Hoult and P. C. Lauterbur: J. Magn. Reson. **34**, 425 (1979).
57. D. I. Hoult and R. E. Richards: J. Magn. Reson. **24**, 71 (1976).
58. Z. H. Cho, C. B. Ahn, S. C. Juh, H. K. Lee, R. E. Jacobs, S. Lee, J. H. Yi, J. M. Jo: Med. Phys. **15**, 815 (1988)
59. R. S. Withers, B. F. Cole, M. E. Johannson, G. C. Liang, G. Zaharchuk: In *High-T$_c$ Microwave Superconductors and Applications*, ed. by R. B. Hammond and R. S. Withers, Proc. SPIE **2156**, 27 (1994).
60. a. R. S. Withers, G. C. Liang, B. F. Cole, M. Johansson: IEEE Trans. Appl. Supercond. **3**, 2450 (1993).
    b. W. H. Wong, R. S. Withers, R. Nast, V. Kotsubo, M. E. Johansson, H. D. W. Hill, L. F. Fuks, K. A. Delin, B. Cole, A. L. Brooke, W. W. Brey, A. Barfknecht, W. A. Anderson: Proc. Cryo. Eng. Conf. (Columbus, OH, July 1995).
61. J. G. vanHeteren, T. W. James, L. C. Bourne: Magn. Res. Med. **32**, 396 (1994).
62. H. Okada, T. Hasegawa, J. G. van Heteren, L. Kaufman: J. Magn. Reson. **107**, 158 (1995).
63. R. D. Black, T. A. Early, P. B. Roemer, O. M. Mueller, A. Mogro-Campero, L. G. Turner, G. A. Johnson: Science **259**, 793 (1993).
64. M. Vester, F. Steinmeyer, B. Roas, G. Thummes, K. Klundt: Proc. Int. Soc. for Mag. Res. in Medicine, SMRM, Vancouver, B.C., April 12–18 (1997), p. 1528.
65. B. Avenhaus, A. Porch, F. Huang, M. J. Lancaster, P. Woodall, F. Wellhöfer: Electron. Lett. **31**, 985 (1995).
66. H. Chaloupka, M. Jeck, S. Kolesov, O. Vendik: Proc. 22nd Europ. Microwave Conf., Helsinki (1992), p. 189.

384   References

67. B. Karasik, I. I. Milostnaya, M. A. Zorin, A. I. Elantev, G. N. Gol'tsman, E. M. Gershenzon: IEEE Trans. Appl. Supercond. **5**, 3042 (1995).
68. M. A. Heusinger, A. D. Semenov, R. S. Nebosis, Y. P. Gousev, K. F. Renk: IEEE Trans. Appl. Supercond. **5**, 2595 (1995).
69. F. Hegmann, D. Jacobs-Perkins, S. H. Moffat, C. C. Wang, R. A. Hughes, M. Currie, P. M. Fauchet, T. Y. Hsiang, J. S. Preston, R. Sobolewski: Appl. Phys. Lett. **67**, 285 (1995).
70. Y. Nagai, N. Suzuki, O. Michikami: IEICE Trans. Electron. **E77C**, 1229 (1994).
71. a. K. F. Raihn, N. O. Fenzi, G. L. Hey-Shipton, E. R. Saito, P. V. Loung, D. L. Aidnik: IEEE Trans. Microwave Theory Tech. **44**, 1374 (1996).
    b. E. R. Soares, K. F. Raihn, N. O. Fenzi, G. L. Matthaei: IEEE Trans. Appl. Supercond. **5**, 2276 (1995).
72. a. D. Galt, C. Price, J. A. Beall, R. H. Ono: Appl. Phys. Lett. **63**, 3078 (1993).
    b. J. A. Beall, R. H. Ono, D. Galt: J. C. Price: IEEE MTT-S Digest, 1421 (1993).
73. B. Karasik, I. I. Milostnaya, M. A. Zorin, A. I. Elantec, G. N. Gol'tsman, E. M. Gershenzon: IEEE Trans. Appl. Supercond. **5**, 3042 (1995).
74. H. Piel: In *Applied Superconductivity 1995*, Inst. Phys. Conf. Ser. No 148, ed. by D. Dew-Hughes (IOP, Bristol, 1995), Vol. 1, p. 1.
75. O. G. Vendik, E. Kollberg, S. S. Gevorgian, A. B. Kozyrev, O. I. Soldatenkov: Electron. Lett. **31**, 654 (1995).
76. D. C. DeGroot, J. A. Beall, R. B. Marks, D. A. Rudman: IEEE Trans. Appl. Supercond. **5**, 2272 (1995); D. Galt, J. C. Price, J.A. Beall, T. E. Harvey: ibid., p. 2575.
77. B. Lax and K. J. Button: *Microwave Ferrites and Ferrimagnetics* (McGraw-Hill, New York, 1962).
78. a. D. E. Oates and G. F. Dionne: IEEE Trans. Appl. Supercond. **9**, June 1999.
    b. G. F. Dionne, D. E. Oates, D. H. Temme, J. A. Weiss: IEEE Microw. Theory Tech. **44**, 1361 (1996).
79. S. Schornstein, I. S. Ghosh, N. Klein: In *Applied Superconductivity 1997*, Inst. Phys. Conf. Ser. No 158, ed. by H. Rogalla and D. H. Blank (IOP, Bristol, 1997), Vol. 1, p. 267.
80. S. Kolesov, H. Chaloupka, A. Baumfalk, F. J. Goertz, M. Klauda: Extended Abstracts, Internat. Supercond. Electronics Conf., ed. by H. Koch (Berlin, Germany, 1997), p. 272.
81. I. Ghosh, D. Schemion, N. Klein: Proc. IEEE Int. Frequency Control Symp., Orlando, May 28–30 (1997).
82. a. D. Jedamzik, R. Menolascino, M. Pizarroso, B. Salas: IEEE Trans. Appl. Supercond. **9**, June 1999.
    b. European Research Project AC115: *Superconducting Systems for Communications*, SUCOMS, (1997). Contact: D. Jedamzik, GEC-Marconi Materials Technology Ltd.
83. a. J. S. Hong, M. J. Lancaster, R. B. Greed, D. Voyce, D. Jedamzik, J. A. Holland, H. J. Chaloupka, J. C. Mage: IEEE Trans. Appl. Supercond. **9**, June 1999.
    b. J. S. Hong and M. J. Lancaster: IEEE Trans. Microwave Theory Techn. **44**, 2099 (1996).
    c. M. J. Lancaster and J. S. Hong: IEE Colloquium on Microwave Components for Communication Systems, 23rd April 1997, University of Bradford.
84. S. W. Goodyear, R. G. Humphreys, J. S. Satchell, N. Parker, M. J. Wooliscroft: IEEE Trans. Appl. Supercond. **9**, June 1999.

85. M. Reppel, H. Chaloupka, J. S. Hong, D. Jedamzik, M. Lancaster, J.-C. Mage, B. Marcilhac: Proc. ACTS Mobile Commun. Summit, (1998), Vol. 1, p. 1.
86. a. C. Wilker, Z.-Y. Shen, V. X. Nguyen, M. S. Brenner: IEEE Trans. Appl. Supercond. **3**, 1457 (1993).
    b. Z.-Y. Shen, C. H. Wilker, P. Pang, W. L. Holstein, D. Face, D. J. Kountz: IEEE Microw. Theory Tech. **40**, 2424 (1992).
87. a. N. Klein, C. Zuccaro, U. Dähne, H. Schulz, N. Tellmann, R. Kutzner, A. G. Zaitsev, R. Wördenweber: J. Appl. Phys. **78**, 6683 (1995).
    b. N. Tellmann, N. Klein, U. Dähne, A. Scholen, H. Schulz, H. Chaloupka: IEEE Trans. Appl. Supercond. **4**, 143 (1994).
88. a. N. Klein, S. Schornstein, I. S. Ghosh, D. Schemion, M. Winter, C. Zuccaro: IEEE Trans. Appl. Supercond. **9**, June 1999.
    b. S. Schornstein, I. S. Ghosh, N. Klein: IEEE MTT-S Internat. Symp. Digest, 1319 (1998).
89. G.-C. Liang, D. Zhang, C. F. Shih, M. E. Johannson, R. S. Withers, A. C. Anderson, D. E. Oates: IEEE Trans. Appl. Supercond. **5**, 2652 (1995).
90. H. Chaloupka, M. Jeck, B. Gurzinski, S. Kolesov: Electron. Lett. **32**, 1735 (1996).
91. A. P. Jenkins, A. P. Bramley, D. J. Edwards, D. Dew-Hughes, C. R. M. Grovenor: J. Supercond. **11**, 5 (1998).
92. B. A. Aminov, A. Baumfalk, H. J. Chaloupka, M. A. Hein, F. Hill, T. Kaiser, S. Kolesov, G. Müller, H. Piel, H. Medelius, E. Wikborg: IEEE MTT-S Internat. Symp. Digest, Vol. 1, p. 363 (1998).
93. T. Kaiser, B. A. Aminov, A. Baumfalk, H. J. Chaloupka, A. Cassinese, M. A. Hein, S. Kolesov, H. Medelius, G. Müller, M. Perpeet, H. Piel, E. Wikborg: J. Supercond. **12**, 343 (1999).
94. A. C. Anderson, H. Wu, Z. X. Ma, P. A. Polakos, P. M. Mankievich, A. Barfknecht, T. Kaplan: Contribut. Appl. Supercond. Conf., ASC'98, Palm Desert, California (1998).
95. A. Baumfalk, H. J. Chaloupka, S. Kolesov, M. Klauda, C. Neumann: IEEE Trans. Appl. Supercond. **9**, June 1999.
96. N. Klein, N. Tellmann, C. Zuccaro, P. Swiatek, H. Schulz: In *Applied Superconductivity 1995*, Inst. Phys. Conf. Ser. No 148, ed. by D. Dew-Hughes (IOP, Bristol, 1995), p. 743.
97. L. Hao, J. C. Gallop, F. Abbas: In *Applied Superconductivity 1997*, Inst. Phys. Conf. Ser. No 158, ed. by H. Rogalla and D. H. A. Blank (IOP, Bristol, 1997), Vol. 1, p. 275.
98. H. Weinstock and M. Nisenoff (Eds.): *Superconducting Electronics* (Springer, Berlin, 1989).
99. H. Weinstock and R. W. Ralston (Eds.): *The New Superconducting Electronics* (Kluwer, Dordrecht, 1993).
100. B. D. Josephson: Phys. Lett. **1**, 251 (1962).
101. D. E. McCumber: J. Appl. Phys. **39**, 2503, 3113 (1968).
102. W. C. Stewart: Appl. Phys. Lett. **12**, 277 (1968).
103. A. M. Portis: *Electrodynamics of High-Temperature Superconductors*, Lecture Notes in Physics, Vol. 48 (World Scientific, Singapore, 1993).
104. A. Barone and G. Paterno: *Physics and Applications of the Josephson Effect* (Wiley, New York, 1982).
105. S. S. Tinchev: Physica C **222**, 173 (1994).
106. V. Polushkin, S. Uchaikin, S. Knappe, H. Koch: IEEE Trans. Appl. Supercond. **5**, 2790 (1995).

107. a. E. Il'ichev, V. Zakosarenko, R. P. J. Ijsselstein, V. Schultze, H.-G. Meyer, H. E. Hoenig, H. Hilgenkamp, J. Mannhart: Phys. Rev. Lett. **81**, 894 (1998).
    b. E. Il'ichev, V. Zakosarenko, R. P. J. Ijsselstein, V. Schultze, H.-G. Meyer: In *Applied Superconductivity 1997*, Inst. Phys. Conf. Ser. No 158, ed. by H. Rogalla and D. H. A. Blank (IOP, Bristol, 1997), Vol. 1, p. 567.
108. S. W. Goodyear, N. G. Chew, R. G. Humphreys, J. S. Satchell, K. Lander: IEEE Trans. Appl. Supercond. **5**, 3143 (1995).
109. V. Ambegaokar and A. Baratoff: Phys. Rev. Lett. **10**, 486 (1963).
110. T. A. Fulton and D. E. McCumber: Phys. Rev. **175**, 585 (1968).
111. a. D. M. Ginsberg, R. E. Harris, R. C. Dynes: Phys. Rev. B **14**, 990 (1976).
    b. R. E. Harris, R. C. Dynes, D. M. Ginsberg: Phys. Rev. B **14**, 993 (1976).
112. D. G. McDonald, R. L. Peterson, C. A. Hamilton, R. E. Harris, R. L. Kautz: IEEE Trans. Electron Devices **27**, 1945 (1980).
113. F. Auracher and T. VanDuzer: J. Appl. Phys. **44**, 848 (1973).
114. S. Shapiro, Phys. Rev. Lett. **11**, 80 (1963).
115. A. J. Dahm, A. Denenstein, T. F. Finnegan, D. N. Langenberg, D. J. Scalapino: Phys. Rev. Lett. **20**, 859 (1968).
116. T. Nagatsuma, K. Enpuku, F. Irie, K. Yoshida: J. Appl. Phys. **54**, 3302 (1983).
117. J. C. Swihart: J. Appl. Phys. **32**, 461 (1961).
118. D. Winkler, Y. M. Zhang, P. A. Nilsson, E. A. Stepantsov, T. Claeson: Phys. Rev. Lett. **72**, 1260 (1994).
119. Ya. G. Ponomarev, B. A. Aminov, M. A. Hein, H. Heinrichs, V. Z. Kresin, G. Müller, H. Piel, K. Rosner, S. V. Tchesnokov, E.G. Tsokur, D. Wehler, K. Winzer, A. V. Yarygin, K. T. Yusupov: Physica C **243**, 167 (1995).
120. J. C. Gallop: *SQUIDs, the Josephson Effect and Superconducting Electronics* (Adam Hilger, Bristol, 1991).
121. J. Clarke: In *SQUID Sensors: Fundamentals, Fabrication and Applications*, ed. by H. Weinstock (Kluwer, Dordrecht, 1996).
122. R. L. Peterson: J. Appl. Phys. **50**, 4231 (1979).
123. J. M. Rowell, M. Gurvitch, J. Geerk: Phys. Rev. B **24**, 2278 (1981).
124. A. I. Braginski and J. Talvacchio: In *Superconducting Devices*, ed. by S. T. Ruggiero and D. A. Rudman (Academic, San Diego, California, 1990), Chap. 8, p. 273.
125. R. Gross: IEEE Trans. Appl. Supercond. **7**, 2929 (1997).
126. R. Gross and B. Mayer: Physica C **180**, 235 (1991).
127. P. W. Anderson and J. M. Rowell: Phys. Rev. Lett. **10**, 230 (1963).
128. G. B. Donaldson, A. Cochran, McA. McKirdy: ibid. Ref. 121, p. 599.
129. J. P. Wikswo, Jr.: ibid. Ref. 121, p. 629.
130. Y. Zhang, Y. Tavrin, H.-J. Krause, H. Bousack, A. I. Braginski, U. Kalberkamp, U. Matzander, M. Burghoff, L. Trahms: J. Appl. Supercond. **3**, 367 (1995).
131. See references in Ref. 121.
132. J. Clarke: Phil. Mag. **13**, 115 (1966).
133. M. B. Ketchen, T. Kopley, H. Ling: Appl. Phys. Lett. **44**, 1008 (1984).
134. L. N. Vu and D. J. VanHarlingen: IEEE Trans. Appl. Supercond. **3**, 1918 (1993).
135. A. Mathai, D. Song, Y. Gim, F. C. Wellstood: IEEE Trans. Appl. Supercond. **3**, 2609 (1993).
136. J. Kirtley, M. B. Ketchen, K. G. Stawaisz, J. Z. Sun, W. J. Gallagher, S. H. Blanton, S. J. Wind: Appl. Phys. Lett. **66**, 1138 (1995).
137. P. Müller: In *Advances in Solid State Physics*, Vol. 34, ed. by R. Helbig (Vieweg, Braunschweig, Wiesbaden, 1994), p. 1.
138. P. Müller: Physica B **222**, 385 (1996).

139. G. J. Chen and M. R. Beasley: IEEE Trans. Appl. Supercond. **1**, 140 (1991).
140. K. K. Likharev and V. K. Semenov: IEEE Trans. Appl. Supercond. **1**, 3 (1991).
141. K. K. Likharev: Czech. J. Phys. **46**, suppl. S6, 3331 (1996).
142. a. K. H. Gundlach: In *Applied Superconductivity*, Inst. Phys. Conf. Ser. No 148, ed. by D. Dew-Hughes (IOP, Bristol, 1995), p. 757.
    b. K. H. Gundlach: ibid. Ref. 121, p. 259.
143. a. Y. Y. Divin, U. Poppe, K. Urban, O. Y. Volkov, V. V. Shirotov, V. V. Pavlovskii, P. Schmueser, K. Hanke, M. Geitz, M. Tometti: IEEE Trans. Appl. Supercond. **9**, June 1999.
    b. Y. Y. Divin, V. V. Pavlovskii, O. Y. Volkov, H. Schulz, U. Poppe, N. Klein, K. Urban: IEEE Trans. Appl. Supercond. **7**, 3426 (1997).
    c. Y. Y. Divin, O. Y. Polyansky, A. Y. Schulman: IEEE Trans. Magn. **19**, 613 (1983).
144. M. A. Tarasov, A. Ya. Shul'man, G. V. Prokopenko, V. P. Koshelets, O. Y. Polyanski, I. L. Lapitskaya, A. N. Vystavkin, E. L. Kosarev: IEEE Trans. Appl. Supercond. **5**, 2686 (1995).
145. Y. Y. Divin, S. Y. Larkin, S. E. Anischenko, P. V. Khabaev, S. V. Korsunsky: Int. J. IR&MM Waves **14**, 1367 (1993).
146. V. P. Koshelets, S. V. Shitov, L. V. Filippenko, A. M. Baryshev, H. Golstein, T. DeGrauw, W. Luinge, H. Shaeffer, H. VanDe Stadt: Appl. Phys. Lett. **68**, 1273 (1996).
147. Y. Taur, J. H. Classen, P. L. Richards: IEEE Trans. Microwave Theory Tech., **22**, 1005 (1974).
148. K. Suzuki, K. Hayashi, M. Fujimoto, K. Yamaguchi, S. Yoshikawa, Y. Enomoto: In *High-T_c Microwave Superconductors and Applications*, ed. by R. B. Hammond and R. S. Withers, Proc. SPIE **2156**, 69 (1994).
149. Y. Fukumoto, H. Kajikawa, R. Ogawa, Y. Kawate: Jap. J. Appl. Phys. **31**, L1239 (1992).
150. M. A. Hein: In *Studies of High-Temperature Superconductors*, Vol. 18, ed. by A. Narlikar (Nova Sciences, New York, 1995), Chap. 5, p. 141.
151. a. O. Harnack, S. Beuven, M. Darula, H. Kohlstedt, M. Tarasov, E. Stephansov, Z. Ivanov: IEEE Trans. Appl. Supercond. **9**, June 1999.
    b. G. Kunkel: Dissertation, Research Center Jülich and University of Wuppertal (1995).
    c. S. Beuven: Dissertation, Research Center Jülich and University of Wuppertal (1996).
152. J. Konopka, I. Wolff, S. Beuven, M. Siegel: IEEE Trans. Appl. Supercond. **5**, 2443 (1995).
153. D. J. Durand, J. Carpenter, E. Ladizinsky, L. Lee, C. M. Jackson, A. Silver, A. D. Smith: IEEE Trans. Appl. Supercond. **2**, 33 (1992).
154. J. H. Takemoto-Kobayashi, C. M. Jackson, C. L. Pettiette-Hall, J. F. Burch: IEEE Trans. Appl. Supercond. **2**, 39 (1992).
155. C. M. Jackson, J. H. Kobayashi, E. B. Guillory, C. Pettiette-Hall, J. F. Burch: J. Supercond. **5**, 417 (1992).
156. T. Patzelt, M. A. Hein, H. Chaloupka, U. Jagodzinski, S. Schmöe: University of Wuppertal, unpublished (1994).
157. S. J. Berkowitz, C. F. Shih, W. H. Mallison, D. Zhang, A. S. Hirahara: IEEE Trans. Appl. Supercond. **7**, 3056 (1997).
158. D. Jäger and F. J. Tegude: Appl. Phys. **15**, 393 (1978).
159. P. K. Tien and H. Suhl: Proc. IRE **46**, 700 (1958).
160. A. L. Cullen: Nature **181**, 332 (1958).
161. A. Davidson: IEEE Trans. Magn. **17**, 103 (1981).

162. J. S. Martens, J. R. Wendt, V. M. Hietala, D.S.Ginley, C. I. H. Ashby, T. A. Plut, M. P. Siegal, S. Y. Hou, J. M. Phillips, G. K. G. Hohenwarter: J. Appl. Phys. **72**, 5970 (1992).
163. R. L. Kautz: In *Metrology at the Frontiers of Physics and Technology*, ed. by L. Crovini and T. J. Quinn (North-Holland, Amsterdam, 1992), p. 259.
164. H. G. Meyer, H. J. Köhler, G. Wende, A. Chwala: IEEE Trans. Appl. Supercond. **5**, 2907 (1995).
165. F. X. Hebrank, P. Gutmann, F. Müller, E. Vollmer, J. Niemeyer: IEEE Trans. Appl. Supercond. **5**, 2911 (1995).
166. S. P. Benz and C. A. Hamilton: Appl. Phys. Lett. **68**, 3171 (1996).
167. J. Lukens: In *Superconducting Devices*, ed. by S. T. Ruggiero and D. A. Rudman (Academic, New York, 1990), p. 135.
168. S. Han, B. Bi, W. Zhang, J. E. Lukens: Appl. Phys. Lett. **64**, 1424 (1994).
169. P. A. A. Booi and S. P. Benz: Appl. Phys. Lett. **68**, 3799 (1996).
170. M. Siegel: In *Advances in Superconductivity VII*, ed. by K. Yamafuji, and T. Morishita (Springer, Tokyo, 1995), p. 1097.
171. R. L. Kautz: IEEE Trans. Appl. Supercond. **5**, 2702 (1995).
172. S. Beuven, M. Darula, J. Schubert, W. Zander, M. Siegel, P. Seidel: IEEE Trans. Appl. Supercond. **5**, 3288 (1995).
173. J. Edstam, G. Brorsson, E. A. Stepantsov, H. K. Olsson: IEEE Trans. Appl. Supercond. **5**, 3276 (1995).
174. a. M. Darula, T. Doderer, S. Beuven: Supercond. Sci. Technol. **12**, R1 (1999).
b. A. M. Klushin, W. Prusseit, E. Sodtke, S. I. Borovitskii, L. E. Amatuni, H. Kohlstedt: Appl. Phys. Lett. **69**, 1634 (1996).
175. R. Gerdemann, K. D. Husemann, R. Gross, L. Alff, A. Beck, B. Elia, W. Reuter, M. Siegel: J. Appl. Phys. **76**, 8005 (1994).
176. A. H. Silver and J. E. Zimmerman: Phys. Rev. **157**, 317 (1967).
177. T. Ryhänen, H. Seppä, R. Ilmoniemi, J. Knuutila: J. Low-Temp. Phys. **76**, 287 (1989).
178. H. Koch: In *Sensors – A Comprehensive Book Series in 8 Volumes*, ed. by W. Göbel, J. Hesse, J. N. Zemel (VCH, Weinheim, 1989).
179. J. P. Wikswo: IEEE Trans. Appl. Supercond. **5**, 74 (1995).
180. I. M. Dmitrenko, G. M. Tsoi, V. I. Shnyrkov, V. V. Kartsovnik: J. Low-Temp. Phys. **49**, 417 (1982).
181. B. Chesca: J. Low-Temp. Phys. **110** (1998).
182. a. M. Mück: In *Superconductive Devices and Circuits*, ed. by R. A. Buhrman, J. Clarke, K. Daly, R. H. Koch, J. A. Luine, R. W. Simon, Proc. SPIE **2160**, 86 (1994).
b. M. Mück, J. Clarke, C. Heiden: J. Appl. Phys. **75**, 4588 (1994).
183. G. J. Ehnholm: J. Low-Temp. Phys. **29**, 1 (1977).
184. a. M. A. Hein, M. Gottschlich, S. Schmöe, H. Piel, E. Sodtke, Y. Zhang: In *Applied Superconductivity 1995*, Inst. Phys. Conf. Ser. No 148, ed. by D. Dew-Hughes (IOP, Bristol, 1995), p. 1533.
b. M. A. Hein, B. A. Aminov, H. Piel, M. Strupp, Y. Zhang, A. I. Braginski: IEEE Trans. Appl. Supercond. **5**, 2501 (1995).
c. M. A. Hein, M. Strupp, G. Müller, H. Piel, Y. Zhang, A. I. Braginski: In *Superconductive Devices and Circuits*, ed. by R. A. Buhrman, J. Clarke, K. Daly, R. H. Koch, J. A. Luine, R. W. Simon, Proc. SPIE **2160**, 108 (1994).
185. Y. Zhang, W. Zander, J. Schubert, F. Rüders, H. Soltner, M. Banzet, N. Wolters, X. H. Zheng, A. I. Braginski: Appl. Phys. Lett. **71**, 704 (1997).
186. M. A. Hein and S. Schmöe: University of Wuppertal, unpublished (1995).

187. V. K. Kornev, K. K. Likharev, O. V. Snigirev, Ye. S. Soldatov, V. V. Khanin: Radio Engineering and Electronic Physics **25**, 122 (Silver Spring, Maryland, 1980).
188. Y. Zhang, M. Gottschlich, H. Soltner, E. Sodtke, J. Schubert, W. Zander, A. I.Braginski: Appl. Phys. Lett. **67**, 3183 (1995).
189. L. D. Jackel and R. A. Buhrman: J. Low-Temp. Phys. **19**, 201 (1975).
190. Y. Zhang, N. Wolters, X. H. Zeng, J. Schubert, W. Zander, H. Soltner, H. R. Yi, M. Banzet, F. Rüders, A. I. Braginski: Applied Superconducitivity (1999).
191. S. Erné, H. D. Hahlbohm, H. Lübbig: J. Appl. Phys. **47**, 5440 (1976).
192. M. A. Hein, H. J. Chaloupka, J. Mitterer, P. Seidel: Electron. Lett. **32**, 940 (1996).
193. P. Seidel, M. A. Hein, H. J. Chaloupka: University of Wuppertal, unpublished (1996).
194. H. Taub and D. L. Schilling: *Principles of Communication Systems* (McGraw-Hill, New-York, 1986).

Index

AC Josephson effect 321, 331
anomalous field effect 159, 222
- Josephson junctions 229
- magnetic scattering 164
- nonequilibrium 164, 229
antennas 65, 67, 304
- magnetic resonance systems 304

BCS theory 19
- energy gap 21
- Hamilton operator 20
- Nb_3Sn 74
- scaled 34, 77
- YBCO 90
boundary resistance *see* thermal resistance 133

Carnot efficiency 274, 310
- coefficient of performance 276
- technical efficiency 275
chemical vapor deposition 256, 258
coherence length 4, 82
- Ginzburg–Landau 5, 111
- Nb_3Sn 70
- Pippard 5
coherence peak 25, 27, 75, 85
conductivity 23, 76
- BCS theory 24
- complex 2, 3, 10, 12, 16, 215
- Drude model 3
- effective-medium 160
- equivalent circuit 11
- flux-flow 215
- heat *see* thermal 133, 135
- Josephson junction 14
- lower limit 40, 98
- normal 2
- thermal 196
Cooper pairs 3, 27
- lifetime 28
- nonequilibrium 108
coupled-grain models 159, 179

critical field 104, 218, 310
- anisotropic 128
- d-wave symmetry 129
- effective 216
- empirical criterion 179
- global quench 136
- local quench 136, 142, 144
- lower 106, 117, 209
- Nb_3Sn 166, 169, 196
- superheated 104, 115, 166
- thermodynamic 106, 111
- thin film 127
- upper 112
- YBCO 166, 169, 183
critical state 179
critical-state model 124, 207, 211, 217
critical thickness 266
cryocoolers 273
- appropriate for HTS 275
- Gifford–McMahon 279
- pulse tube 280
- recuperative 275
- regenerative 275
- reliability 277, 279
- Stirling 279
current–field relation 2, 3, 11, 112, 122, 163
- BCS theory 23
- Josephson junction 13, 14, 15
- local 4, 6, 12, 71, 111
- nonlocal 4, 6
current–phase relation 319

defects 132, 136, 185, 203, 204, 232, 311
demagnetization 129, 200, 214
density of states 19, 24, 27, 29, 33, 240
- BCS superconductor 22
- d-wave symmetry 38
- two-band model 36
depinning frequency 122, 215, 217
- Nb_3Sn 210

392 Index

- YBCO   210
duty cycle   61, 147, 220

Eliashberg equations   32, 35
energy gap   22, 26, 28, 33, 36, 249
- anisotropic   30, 42, 82
- HTS   41
- Nb$_3$Sn films   80
- time scale   108
- YBCO films   86
epitaxial films   255, 264

filters   283, 303, 312
- bandwidth   285
- Butterworth   294
- Cauer   295
- characteristic function   289, 295
- coupling   286, 293
- design   288
- insertion loss   285
- low-pass prototype   291
- lumped-element   313
- narrow-bandwidth   307
- power handling   298, 314
- quasi-elliptic   296
- reflection coefficient   289
- ripple   287
- skirt   285
- superconducting   297
- transducer loss ratio   285
- transmission coefficient   284
- Tschebyscheff   290
flux penetration   204, 209, 214
- discontinuity lines   204, 207
- inhomogeneous   207
- magneto-optic imaging   204
frequency intermodulation   150
- Josephson junction   155
- SQUIDs   339
- third-order   152
- YBCO films   173

gapless superconductivity   28, 35, 74, 249
Ginzburg–Landau
- parameter   5, 105, 112
- phase diagram   117
- relaxation time   113
- surface impedance   113
- theory   104, 110
grain boundaries   14, 83, 160, 162, 245
- artificial   260
- step-edge   261
grain size   83, 160, 242, 245

- Nb$_3$Sn films   78
granularity   74, 82, 85, 89, 93, 222
- transmission line model   15, 160
group delay   50, 289, 292, 297, 298

Hilbert transform spectroscopy   328

inductivity   11, 14, 16
- Josephson   13

Josephson devices   283, 318, 326
- microwave detection   327
- mixers   328
- phase shifters   329
- SQUID   322
- voltage controlled oscillator   331
Josephson junctions   12, 81, 83, 125, 318
- characteristic voltage   12, 320
- critical field   106
- engineered   161, 319
- inductance   125, 319
- plasma frequency   321
- preparation   260, 324
- Shapiro steps   321, 327
- surface impedance   125, 320
- switching time   108
Josephson transmission line   15, 329
Josephson voltage standards   331

laser scanning   190
liquefied gases   273, 278
- cooling capacity   278
- cryostats   278
- latent heat   275
London equation   3, 119
loss tangent
- dielectric   9, 49, 265, 268, 311
- differential   46, 109, 113, 123, 124, 142, 200
lumped-element circuits   303

magnetic field   103, 176
magnetic impurities   28, 35
magnetic pairbreaking   30
magnetic scattering see magnetic impurities   164
magnetic vector potential   1
Maxwell's equations   1
mean free path   3, 4, 12, 79, 82
microwave heating   103, 131, 197, 231, 311, 316
- dynamic   138

microwave SQUIDs 332
- demodulation 342
- dispersive 323
- equivalent circuit 322
- hysteretic 334
- inductance 335
- magnetometers 322, 337, 341
- quantization 342
- resistance 335
- scattering parameters 333
- transmission coefficient 336
miniaturization 268, 272, 300, 303, 314
multimode analysis 186

nonlinear response 179
- adiabatic 104
- dynamic 104, 146, 176
- in parallel fields 216
- in perpendicular fields 199, 202
- magnetic 105, 224
- Nb$_3$Sn films 194
- parametrization 177
- thermal 105, 225
- thermal-magnetic feedback 219, 232
- Tl-2212 180
- Tl-2223 180
- Y-123 179
normal electrons 3
   see also quasi-particles

order parameter 21, 28, 110
- anisotropic 30
- complex-valued 28
- d-wave symmetry 23, 38, 250
- two-band model 35
overcriticality 109, 147, 149
oxygen stoichiometry 94, 207, 311

packaging 272, 311
pair breaking 163, 179, 229
   see also magnetic pair breaking
passive devices 299
- low-phase-noise oscillators 317
- magnetic resonance 302
- mobile communication 302
- novel circuits 312
- satellite-based communication 302
- superconductivity industry 302
penetration depth 2, 4, 19, 24, 46, 74, 85, 88, 105, 166, 301
- Campbell 122
- d-wave symmetry 39, 130
- Ginzburg–Landau 107, 111
- grain boundaries 14

- HTS 11, 82, 90, 93
- Josephson 13, 14
- Nb$_3$Sn 11, 71, 80
permeability 46
permittivity 2, 16, 46, 59, 265, 268, 300
- complex 267
phase diagram
- Nb$_3$Sn 242
- oxygen stoichiometry
- Y-123 249, 252, 253
physical vapor deposition 256, 257, 259
pinning 124, 202, 209, 214, 217
power
- circulating 45, 63, 103, 284, 298
- dissipated 7, 44, 51, 63, 103, 107, 131, 157, 185, 190, 298
- incident 51, 298
- radiated 52, 296
- reflected 52
- transmitted 51, 298
pulsed laser deposition 256, 257

quality factor 44, 283, 300, 310
- loaded 49, 284
- R-L-C circuit 47
- transmission line 48
- unloaded 49, 56, 284
quasi-particles 20, 24, 85
- scattering 12, 27, 36, 85, 97, 108
- spin-flip scattering 165

resistively shunted junction 13
- RSJ model 126, 161, 319
resistivity 11, 14, 16
- flux flow 122
- Nb$_3$Sn 80, 241
resonant frequency 44, 53, 57, 191, 285
- Slater's theorem 46, 193
resonators 43
- cavity 55, 57
- characteristic impedance 45
- coaxial 65
- coupling coefficients 51
- CW operation 53, 235
- dielectric 59, 66
- disk 62, 186, 191, 316
- Fabry–Perot 66, 187
- nonlinear response 174
- parallel-plate 62
- quality factor 44, 296
- scattering parameters 50
- transient-time analysis 55, 147

Index

skin depth   2, 4, 85
skin effect
- anomalous   4
- normal   4, 18
sputtering   256, 258
SQUID (superconducting quantum interference device)   322
   see also microwave SQUIDs
- microscope   66
stored energy   44, 190, 284
substrates   265
superconductors
- A15   240
- d-wave symmetry   90, 97
- granular   14, 84
- high-temperature   34
- Nb_3Sn 68, 240
- perovskite   247
- s-wave symmetry   97, 102
- strong-coupling   32, 77, 86
- Tl   83
- two-band model   34
- type-I   118
- type-II   118
- Y-123   83
superheating   114
- dynamic   148
superimposed DC fields   219
supersaturation   253, 259
surface impedance   1, 5, 23
- active imaging   66
- anisotropic superconductor   30
- BCS theory   26
- critical state   124
- d-wave symmetry   40, 97
- dielectric   9, 267
- finite film thickness   8
- flux-flow   120, 210, 215, 222
- frequency dependence   18, 34, 92
- granular films   16
- hysteresis in magnetic field   199
- Josephson junction   14, 126, 160
- Josephson stripline   15
- magnetic impurities   29
- measurement   55, 63
- metallic cap layers   10
- Nb_3Sn 68
- nonlinear   103
- normal conductor   6
- passive imaging   64
- penetration of flux   206
- perturbation scanning   67
- strong-coupling   33
- superconductor   6
- two-band model   36
- YBCO   84, 97, 226
surface pinning   115
surface reactance   6, 7, 46, 286
   see also penetration depth
surface resistance   6, 7, 45, 74, 285, 300, 311
- dynamic   157
- effective   185
- frequency dependence   18, 76
- global heating   138
- local defects   141
- Nb_3Sn 73
- nonlinear   171, 174
- residual   16, 40, 74, 86
- Y-123 films   85, 172
switched devices   306

thermal imaging see thermometry   189
thermal resistance   107, 135, 136, 189
thermometry   65, 198
tin-vapor diffusion   69, 243
transmission lines   48, 271
tuning elements   307
- dielectric   310
- digital   308
- ferroelectric   308
- magnetic   309
- superconducting   310
twinning   255
two-band model   34, 90, 94, 97, 249
- d-wave symmetry   37, 101
two-fluid model   3, 6, 11, 15, 25, 90, 229

Springer Tracts in Modern Physics

140 **Exclusive Production of Neutral Vector Mesons at the Electron-Proton Collider HERA**
By J. A. Crittenden 1997. 34 figs. VIII, 108 pages

141 **Disordered Alloys**
Diffusive Scattering and Monte Carlo Simulations
By W. Schweika 1998. 48 figs. X, 126 pages

142 **Phonon Raman Scattering in Semiconductors, Quantum Wells and Superlattices**
Basic Results and Applications
By T. Ruf 1998. 143 figs. VIII, 252 pages

143 **Femtosecond Real-Time Spectroscopy of Small Molecules and Clusters**
By E. Schreiber 1998. 131 figs. XII, 212 pages

144 **New Aspects of Electromagnetic and Acoustic Wave Diffusion**
By POAN Research Group 1998. 31 figs. IX, 117 pages

145 **Handbook of Feynman Path Integrals**
By C. Grosche and F. Steiner 1998. X, 449 pages

146 **Low-Energy Ion Irradiation of Solid Surfaces**
By H. Gnaser 1999. 93 figs. VIII, 293 pages

147 **Dispersion, Complex Analysis and Optical Spectroscopy**
By K.-E. Peiponen, E.M. Vartiainen, and T. Asakura 1999. 46 figs. VIII, 130 pages

148 **X-Ray Scattering from Soft-Matter Thin Films**
Materials Science and Basic Research
By M. Tolan 1999. 98 figs. IX, 197 pages

149 **High-Resolution X-Ray Scattering from Thin Films and Multilayers**
By V. Holý, U. Pietsch, and T. Baumbach 1999. 148 figs. XI, 256 pages

150 **QCD at HERA**
The Hadronic Final State in Deep Inelastic Scattering
By M. Kuhlen 1999. 99 figs. X, 172 pages

151 **Atomic Simulation of Electrooptic and Magnetooptic Oxide Materials**
By H. Donnerberg 1999. 45 figs. VIII, 205 pages

152 **Thermocapillary Convection in Models of Crystal Growth**
By H. Kuhlmann 1999. 101 figs. XVIII, 224 pages

153 **Neutral Kaons**
By R. Beluševič 1999. 67 figs. XII, 183 pages

154 **Applied RHEED**
Reflection High-Energy Electron Diffraction During Crystal Growth
By W. Braun 1999. 150 figs. IX, 222 pages

155 **High-Temperature-Superconductor Thin Films at Microwave Frequencies**
By M. Hein 1999. 134 figs. XIV, 395 pages

156 **Growth Processes and Surface Phase Equilibria in Molecular Beam Epitaxy**
By N.N. Ledentsov 1999. 17 figs. VIII, 84 pages

157 **Deposition of Diamond-Like Superhard Materials**
By W. Kulisch 1999. 60 figs. X, 191 pages

Springer and the environment

At Springer we firmly believe that an international science publisher has a special obligation to the environment, and our corporate policies consistently reflect this conviction.

We also expect our business partners – paper mills, printers, packaging manufacturers, etc. – to commit themselves to using materials and production processes that do not harm the environment. The paper in this book is made from low- or no-chlorine pulp and is acid free, in conformance with international standards for paper permanency.